THE ORGANIZATION OF LOGISTIC SUPPORT SYSTEMS

The Organization of Logistic Support Systems

The proceedings of a conference held in Luxembourg
under the aegis of the NATO Scientific Affairs Committee

Edited by: R. BUREAU
General Manager, NATO Maintenance and Supply Agency

AMERICAN ELSEVIER PUBLISHING COMPANY, INC.
New York, 1972

Published in the United States by
AMERICAN ELSEVIER PUBLISHING COMPANY, INC.
52 Vanderbilt Avenue, New York, N.Y. 10017

First printed 1972

Library of Congress Catalog Card Number: 70-141206
International Standard Book Number: 0-444-19615-3

I.B.M. Computer Typesetting by Print Origination, Liverpool, England.

Printed and bound in Great Britain

Foreword

This book constitutes the Proceedings of a Conference sponsored and organised by NATO and held in Luxembourg from 19 to 22 May 1970.

It is a kind of sequel to a similar conference held in Athens from 4 to 8 September 1967 on "The Applications of Operational Research to Large Scale Provisioning Systems". More specifically, it is the extension or the development of a particular aspect of the subject which attracted attention in the discussions held during the Athens Conference[1]

Is it fair to claim that the theme–undeniably an ambitious one–has been fully examined? At any rate, a beginning has certainly been made, and the papers read, together with the discussions they prompted, make a particularly useful contribution to this very wide field, while providing at the same time a review of the work carried out so far.

The large number of participants at the conference is in itself evidence of the interest aroused by this subject among operational research students and logisticians in the civil and military fields.

R. BUREAU
Conference Director

[1] J.J. FERRIER (Ed.), *Large Scale Provisioning Systems,* The English Universities Press Ltd, London.

Avant-Propos

Cet ouvrage constitue le Compte-Rendu d'une Conférence organisée sous les auspices de l'OTAN, à Luxembourg du 19 au 22 mai 1970.

Elle est en quelque sorte le prolongement d'une Conférence similaire, organisée à Athènes du 4 au 8 septembre 1967, avec pour thème: "Les applications de la recherche opérationnelle aux grands systèmes d'approvisionnement". Plus précisément, prolongement ou développement d'un aspect particulier qui fut évoqué au cours des échanges de vues que permit la Conférence d'Athènes[1].

Le sujet, en fait très ambitieux, a-t-il été traité? On peut penser qu'il a en tout cas été abordé et que les communications présentées, ainsi que les discussions qui ont suivi, constituent une contribution à ce vaste problème, et une mise au point des travaux dans ce domaine.

L'assistance particulièrement nombreuse à cette Conférence porte en tout cas témoignage de l'intérêt suscité par la question dans le monde des chercheurs opérationnels et des logisticiens du secteur et militaire.

R. BUREAU
Directeur de la Conférence

1 *J.J. FERRIER,* Large Scale Provisioning Systems, *The English Universities Press Ltd, London.*

Table of Contents

Discours d'Accueil
Speech of Welcome

Monsieur le Ministre Eugene Schauss Ministre de l'Interieur du Gouvernement Luxembourgeois

Messieurs les Généraux, Mesdames, Messieurs

Le Grand Duché de Luxembourg est très heureux d'accueillir une Conférence OTAN qui siégera sur son territoire durant une semaine.

Situé au centre d'une Europe naguère déchirée par de nombreuses guerres, le Luxembourg est foncièrement pacifiste et particulariste. C'est dire· qu'il a été et demeure par conviction autant que par nécessité, au centre de la construction européenne. C'est dire aussi qu'il a adhéré à l'OTAN dès sa formation, non seulement avec sa raison mais avec son coeur.

Bien entendu, nous sommes parfaitement conscients de la relativité des choses et du très faible poids de nos efforts dans le domaine militaire. Il était normal que nous tachions de compenser cet état de choses par une participation plus active dans d'autres secteurs.

C'est ainsi que nous avons hébergé avec plaisir une Agence de l'OTAN, la NAMSA, que nous abritons depuis début 1968. C'est ainsi que nous avons vu avec faveur que Luxembourg était adopté par le Comité Scientifique de l'OTAN comme siège de votre Conférence.

Vous n'attendrez certainement pas de mois que je vous parle de l'objet de vos travaux, les problèmes posés par l'Organisation et la mise en place des grands systèmes de soutien logistique. Mais permettez-moi cependant, Mesdames et Messieurs, quelques réflexions d'ordre général.

Je suis frappé du bond prodigieux accompli ces dernières années dans le domaine qui est le vôtre. Quand on songe aux campagnes de Napoléon, qui ne sont cependant pas si lointaines, on se croirait encore dans la préhistoire: le guerrier est à pied ou à cheval et emporte avec lui son bagage. Si la guerre franco-allemande de 1870 a vu la consécration du chemin de fer, la seconde guerre mondiale l'utilisation intensive du char, du camion automobile et de l'aviation, pour ne pas trop insister sur la bombe atomique, nous assistons aujourd'hui au triomphe de l'électronique et des ordinateurs. Les développements de la technique ont donc donné une dimension nouvelle aux problèmes logistiques.

Je constate aussi que dans votre domaine l'influence de la technique différencie de moins en moins ce qui est proprement militaire de ce qui est proprement civil. L'évolution est parallèle dans les deux secteurs, la logistique militaire utilise les mêmes moyens, les mêmes concepts que la logistique civile, et on peut même dire que les nécessités militaires stimulent, en logistique également, les réalisations civiles.

Mais il ne faut pas cacher que les techniques deviennent de plus en plus compliquées, les horizons de plus en plus vastes. On peut dès lors très sérieusement s'arrêter à une question d'ordre philosophique: la complication ne s'oppose-t-elle pas jusqu'à un certain point à la souplesse d'adaptation. L'heureux dénouement d'Appollo XIII ne nous incite-t-il pas à insister davantage sur le facteur souplesse et faculté d'adaptation des systèmes?

Je dois bien me contenter de poser le problème et vous laisser le soin de le résoudre. Quoi qu'il en soit, au nom du Gouvernement Luxembourgeois, je redis combien nous sommes heureux que cette Conférence se déroule à Luxembourg et je vous souhaite beaucoup de succès dans vos travaux. Nul doute que vos échanges de vues seront fructueux.

Allocution d'Ouverture: Des Systemes et des Hommes

Opening Address: Of M.I.S. and Men

le Commissaire Général de la Marine, J. FERRIER

Le Général Bureau, en me faisant l'honneur de me réserver le coup d'envoi de cette conférence, m'a expressément demandé de la situer dans le prolongement de celles qui l'ont précédée. C'était m'inviter, en quelque sorte, à vous raconter mes campagnes. Les vieux marins ne résistent jamais à une telle prière. Coleridge l'a abondamment démontré.

J'appartiens en effet, comme plusieurs d'entre vous, à l'héroique phalange des vétérans d'Athènes. C'est là, au pied de l'Acropole, dans la gloire de Minerve, déesse de la sagesse casquée, que furent livrés, il y a six ans, puis il y a trois ans, sous la double bannière de la Recherche Opérationnelle et de l'O.T.A.N., de fiers combats pour l'optimisation du soutien logistique.

La conférence initiale eut lieu en Septembre 1964. Elle avait pour titre: *Les applications de la recherche opérationnelle aux problèmes militaires d'approvisionnement et de fourniture de services** L'exposé introductif en avait été confié à Mr Murray A. Geisler, de la RAND. Cette estimable corporation n'avait guère plus de douze ans à l'époque. Aujourd'hui, bien qu'elle compte parmi les plus vénérables dames d'honneur de la Recherche Opérationnelle, elle est à peine sortie du "teen age", de ces heureuses années d'apprentissage pendant lesquelles, si l'on n'atteint pas encore au sommet de la sagesse, on fait du moins provision d'expérience, en apprenant par exemple, comme l'enseigne Mr Durbin, à se méfier du "bottom up". Mr Geisler nous fit quant à lui une belle démonstration dans le style "top down". Du haut du Sinaï Californien où la RAND s'était déjà hissée en escaladant de 1957 à 1963 ses célèbres "laboratory problems" L P $_{I}$ à L P $_{IV}$, il sut donner au peuple élu de la logistique que nous voulions être un aperçu de la Terre Promise et l'espérance d'y pénétrer bientôt, à la remorque de ce que les Français appellent l'Informatique et qui, si l'on en croit la traduction récemment proposée par Mr Kennedy au IVème congrès de l'IFORS à Venise**, n'est rien d'autre que l'application de la méthode scientifique à l'élaboration des systèmes d'information du management, désormais mondialement connus sous le sigle M.I.S.

Parlant aussitôt après Mr Geisler—vous voyez qu'en six ans j'ai avancé d'une place—j'eus la chance de pouvoir présenter, d'entrée de jeu, à l'admiration du peuple, un M.I.S. ou pour reprendre formule de notre co-vétéran, le Major Høgberg, un L.M.I.C.S. qui

* Application of operational research methods to military servicing and supply problems.

** Mr D.M. Kennedy, président de la session "informatics", crut bon de définir dans son discours d'ouverture ce mot nouveau-né, traduit du Français "informatique" et récemment introduit dans la langue du management, version américaine. Il proposa: "informatics is the application of scientific methodology to the design of information systems for improved decision making". La dernière précision paraît inutile, ou insuffisante, car le traitement de l'information débouche non pas seulement sur la décision, mais aussi bien sur la connaissance, le contrôle et le jugement.

venait précisément d'être mis au point et en place par la Marine française pour son matériel d'emploi courant. Le seul problème restant en suspens était celui de la fixation d'un estimateur de l'efficacité logistique globale pour le matériel réputé essentiel, en donnant à ce qualificatif une valeur absolue et non pas relative[1]. J'espérais que tout le monde brûlerait d'en débattre. Il fallut attendre une communication de Mr Rosenman, faite le dernier jour, pour que la discussion s'anime sur ce point capital. Six ans apres, il vous apparaîtra, en écoutant MM. Baglow, Combes, Rawson, Miller, Geraerds, Beare et Curtis, Simon et Boll, Buck, Rougeris, Fent, Ayliffe, et Green que les débats portant sur les méthodes d'évaluation et les lois de composition des performances ou contre-performances du soutien logistique sont et demeurent enfin largement ouverts. C'est une question qui fera couler encore beaucoup d'encre. Elle est au coeur des modèles quantitatifs, des méthodes de classement purement analytique des politiques ou des structures, comme celle présentée par Mr Combes et, surtout, des grandes simulations qui seules pourront venir à bout des tenaces illusions dans le domaine que nous avons choisi d'explorer.

Malgré cette première ouverture sur le monde des systèmes réellement intégrés, la conférence de 1964 se pencha surtout sur des problèmes, je ne dirais pas mineurs, mais sectoriels. Il est normal que les petits garçons commencent par grimper aux arbres avant de dominer les forêts. Leurs naïves escalades peuvent d'ailleurs les amener assez haut et à voir assez loin. Ainsi je me rappelle avec enchantement l'intervention d'un gentleman très britannique qui, fort de 30 années d'expérience vécue au contact d'avions cloués au sol par toutes les avaries possibles–Dieu sait si elles peuvent être nombreuses et variées–eut le talent de nous élever au-dessus des nuages tout en conservant les pieds sur la terre. Sur un thème voisin de celui qui a inspiré Mme Giovanna Piu[2] il nous fit sentir combien il est hasardeux d'obtenir, même en s'aidant de questionnaires savamment établis et imprimés, des réponses précises, c'est-à-dire des informations de base exactes, complètes, non subjectives. Or de telles informations sont la raison d'être des M.I.S. et le pain quotidien de tous nos systèmes logistiques. Il dépend des moissonneurs que les systèmes les plus subtils et les mieux agencés échappent ou non au désastreux bégaiement qui, en redoublant le M.I.S., transforme un *management information system* en ce que Mr Russel Ackoff appelle un *management misinformation system*[3], avec toutes les suites hautement catastrophiques qui peuvent résulter d'un si déplorable avatar.

Trois ans plus tard, Mr Sumner pinçait la même corde et lançait la conférence de 1967 dans le vent de l'Histoire en y donnant le signal de la première contestation.

Cette conférence avait pour titre et pour thème "les grands systèmes d'approvisionnement". Il y avait progrès puisque l'on passait des problèmes aux systèmes et même aux grands systèmes, en somme de l'étude des mouvements à celle des machines. Progrès d'autant plus remarquable que le mot système, s'il n'était pas un nouveau

1 "Réalisation d'une gestion centralisée et partiellement automatisée de stocks de matériel militaire de nature commerciale; transmission et exploitation des données; recherche d'estimateurs d'efficacité logistique."(Actes de la Conférence de 1964.)

2 *How to succeed in improving reliability by actually trying,* John D. North (Boulton Paul Aircraft Ltd).

3 R.L. Ackoff (1967): Management Misinformation systems (*Mgmt. Sci.* 14 (4)).

venu, ne s'était pas encore enrichi de tout ce que lui ont apporté les conquêtes récentes de la téléinformatique, les percées conceptuelles de la biocybernétique et les travaux de ceux que l'on commence à appeler, du moins en France, les "systématiciens".

Son contenu n'en était pas moins immense, déjà, dans le domaine qui nous intéresse. Ce fut l'occasion de l'inventorier. Nous eûmes droit à la description, agréablement commentée, d'une bonne vingtaine de systèmes nés ou à naître, de tous types, et couvrant tout les horizons de la logistique depuis la prévision des besoins en pièces de rechange jusqu'à la cannibalisation des pièces en service et à l'évaluation des résultats finaux, avec en vedette Américaine, présentée par M. Craig C. Sherbrooke, le système METRIC dont ce fut la grande première[1]

Dans son discours d'ouverture, Robert G. Brown avait choisi de nous parler de l'architecture des systèmes et des tensions internes qui y prennent naissance du fait d'inévitables oppositions intellectuelles entre des administrateurs plus ou moins ineptes et des savants plus ou moins irresponsables. Il montra que pour surmonter ces oppositions, pour franchir ces interfaces, il n'est d'autre recours qu'en un dialogue véritablement ouvert et que les courbes d'échange sont le meilleur sinon le seul instrument d'un tel dialogue.

J'aurais pu vous resservir ce discours, dont j'ai quelque peu plagié le titre[2], car il a trouvé en vous tous de profondes résonances. Vous êtes nombreux à souhaiter avec Messieurs Williams, Page, Rawson, le renforcement et même l'élargissement d'un dialogue qui, du sommet à la base, devrait associer dans une compréhension commune de nos systèmes logistiques non seulement les hommes de science et les administrateurs mais aussi les constructeurs, les réparateurs et les distributeurs, les "mission planners", les dirigeants et les exécutants, les officiers d'état-major et les combattants de première ligne voire, comme cela vient d'être splendidement démontré à Houston, les directeurs de vol et les naufragés de l'espace. En outre chacun d'entre vous attaque et s'efforce de surpasser, par une savante dialectique, l'une ou l'autre des oppositions majeures qui tourmentent ce monde de contradiction où "les deux" n'est jamais une réponse acceptable, de la part d'une grande personne, quand il faut choisir "lequel". Non seulement l'opposition traditionnelle entre le coût et l'efficacité si du moins on parvient à mesurer celle-ci, mais aussi l'opposition entre l'urgence et la sécurité, entre les deux échelons de Mr Alberts et les trois de MM. Conrath et Burke, entre le régime de paix et le régime de conflit avec Mr Rougerie, entre les moyens de transport classiques et les camions géants de Mr Bresson, entre les investissements initiaux soit en appareils neufs soit en moyens de réparation avec Mr Sutton, entre les réparations coup par coup et les révisions périodiques avec MM. Hough et O'Connor, l'entretien préventif et l'entretien curatif avec Mr Geraerds, le supply et la maintenance avec MM. Hurst, Landau et Baglow, la maintenance et le remplacement (Geraerds), les bancs

[1] Voir *Large Scale Provisioning Systems*: J. Ferrier éditeur, E.U.P. London; et Dunod Paris: *Les grand systèmes d'approvisionnement.*

M.E.T.R.I.C.	:	Multi Echelon Technique for Recoverable Item Control
A.F.R.A.M.S.	:	Air Force Recoverable Assembly Management System
C.A.R.E.S.	:	Computation and Research Evaluation System
U.A.D.P.S.	:	Uniform Automatic Data Processing System
SIMSYS	:	Système de Simulation de Systèmes. etc. . .

[2] *Managers, scientists and systems,* R.G. Brown.

d'essais manuels et les bancs d'essais automatiques avec MM. Baker, Morris et Rawson, les structures horizontales et les structures verticales, centralisées ou décentralisées, avec Mr Høgberg, le "top down" et le "bottom up" (Durbin, le transport routier et "l'air lift" (Conrath et Burke), les réparations faites à bord ou à terre (Rawson, Williams), "on site" ou "off site" (Buck), l'entreprise publique et les contrats privés, la spécialisation technique et la valeur combattante des équipages, l'eau douce et l'eau salée pour circuits d'incendie, les bras et les robots, les constructions individualisées et les juxtapositions de modules géants, avec Mr Williams, les procédures américaines et les habitudes militaires des autres nations avec Mr Hormann, et même, avec Mr Miller, l'opposition délicieusement mathématique entre l'approche primale et l'approche duale du problème N.O.R.S.

Par dela toutes ces oppositions d'ordre pratique ou théorique, technique ou budgétaire, logistique ou opérationnel, le Commandant Sallantin a su découvrir d'autres oppositions tout à fait fondamentales, d'ordre naturel, qui semblent imposer des règles universelles au jeu de la transmission et du déchiffrement des messages physiques, biologiques, linguistiques, philosophiques et, par conséquent, à l'échange non seulement des signaux mais des idées. Des circonstances indépendantes de sa volonté empèchent malheureusement le Capitaine de Frégate Sallantin de présenter la communication qui a été annoncée sur *L'empirisme et la Méthode en matière de soutien logistique.* Mais son oeuvre demeure et ceux d'entre vous qui pourront se procurer ses petits livres bleus auront la joie d'aller plus loin sur la piste, sinon de Mao, du moins de Lao Tseu, Leibniz, Raymond Lulle et Pythagore.

Mais revenons en 1967. Notre deuxième conférence, placée sous le patronage de Minerve, fut gratifiée par Jupiter de deux coups de tonnerre.

Le premier fut frappé par Mr Sumner. Il dénonça, avec toute l'autorité reconnue au représentant d'U.S. Eucom, la fragilité des systèmes monumentaux édifiés sur l'argile de la nature humaine et la vanité de vouloir faire toujours mieux en traitant toujours plus savamment des informations toujours plus raffinées, alors que les hommes chargés de récolter à la base des informations exactes et pertinentes, ce bien désormais souverain, risquent de comprendre de moins en moins ce qu'on attend d'eux et de s'en désintéresser de plus en plus. Cela jeta un froid. La foudre vint, et la grêle, avec le second coup de tonnerre, frappé celui-ci par quelqu'un que vous connaissez tous.

Le Général Bureau, en effet, fit observer que l'on avait fort bien tourné autour du buisson mais sans aller au fond des choses et que des problèmes fondamentaux n'avaient pas même été effleurés. Ces problèmes concernent l'organisation, le déploiement et le démarrage des grands systèmes avec, en première ligne, la question toujours controversée du nombre optimal des échelons et celle, souvent escamotée, du niveau optimal des stocks initiaux.

Et voilà pourquoi nous sommes ici!

Car le Général Bureau fut pris au mot et au lasso par le Comité Scientifique de l'O.T.A.N. qui le pria de vouloir bien organiser une Conférence afin de combler les graves lacunes qu'il avait si pertinemment dénoncées. Méfiez-vous, Messieurs, Mr Knight vous regarde. Il vous surveille, tout prêt à épingler au calendrier de 1973 le premier d'entre vous qui s'abandonnera à des réflexions aussi téméraires.

Le Général Bureau accepta cette croix et il a fait appel à nous tous pour que nous l'aidions à la porter. Vous avez fraternellement répondu, à raison de 25% d'anciens compagnons d'Athènes, ce qui est un gage de continuité et de 75% de nouveaux enrôlés, ce qui est une promesse d'ouverture.

Parviendrez-vous à aller plus haut, à creuser plus profond, à voir plus loin, comme on vous invite à le faire, à remodeler pour le meilleur et non point pour le pire les M.I.S. ou les L.M.I.C.S.? L'avenir nous le dira par la bouche de Mr Tyler Port qui aura le redoutable honneur de tirer les leçons de ce festin de l'esprit. Au moment de passer à table, un coup d'oeil sur le menu risque de donner l'impression que l'on sert toujours les mêmes plats. Comment pourrait-il en être autrement? Les problèmes classiques de fiabilité (Piu), de files d'attente et de "pipe-line factor" (Sutton), d'optimisation des stocks sous contraintes financières ou spatiales (Barnard et Seger), de planification des transports (Blumberg), d'évaluation du coût et de l'efficacité, de budgétisation (Mr Thu), d'estimation des besoins futurs (Peguy et Sahnoun), et même de cannibalisation avec Mr Hirsch; qui est un dangereux récidiviste de cette cuisine particulière, occupent plus de la moitié de la carte. C'est à juste titre car, s'il est vrai que la modernisation est un processus continu, pour citer une phrase prononcée il y a six ans par Mr Geisler, il n'est pas exagéré de faire tous les trois ans le point de l'avance réalisée dans l'étude et l'application des mécanismes et modèles fondamentaux d'où découle tout le reste, à commencer par les grandes simulations qui, dans ce domaine des grands systèmes, nous ouvrent le chemin le plus sûr sinon le plus court vers la Vérité.

Cependant, la première partie du programme est plus spécifiquement consacrée aux thèmes directeurs qui font l'originalité de la présente conférence. A côté de choses qui ont déjà été dites, mais qu'il est toujours bon de répéter, nous y avons rencontré des points de vue orginaux, des idées neuves, des témoignages intéressants et des prises de position vigoureuses, comme celles tendant à élargir le dialogue et à resserrer l'interdépendance entre tous les partenaires du grand jeu de la logistique. On y trouve, bien entendu, quelques nouveaux sigles pour ceux qui en font collection. Nous avons déjà salué le L.M.I.C.S. de Mr Høgberg, mais le S.P.L.I.C.E. du Colonel Page mérite à coup sûr la place d'honneur qu'un marin ne saurait lui contester, sur la grand'vergue. On y trouve surtout un réel effort de surpassement afin de s'élever des modèles aux métamodèles, de l'analyse à la synthèse des systèmes, de la pratique à la théorie de la maintenance, des solutions de principe aux principes mêmes des solutions, des idées reçues à leur mise en cause, de la confrontation des points de vue à l'organisation du dialogue, du conformisme à la contestation puis à la participation. Il nous appartient à tous, compagnons d'une même cordée, de réussir cette belle escalade vers les temples sereins, chantés il y a 2000 ans par Lucrèce, d'où les sages peuvent contempler avec complaisance le tumulte parfaitement organisé des opérations savamment soutenues, qu'elles soient civiles, militaires ou, pour reprendre l'image du poète, maritimes: *Sauve mari magno, turbantibus aequora ventis.*

Craignons pourtant de nous élever si haut et si droit, dans les sphères silencieuses des systèmes, que nous perdions de vue les pauvres exécutants, demeurés prisonniers de l'espace tangentiel du Commandant Sallantin, du plan trivial de Mr Koestler*, de la terre des hommes de Saint Exupéry, de ce monde rampant, plein de bruit et de fureur,

* Voir *Le Yogi et le Commissaire:* les mémoires d'Hillary.

ô Shakespeare, mais le seul en vérité où il se passe des choses réelles. Craignons de ne plus entendre le cri d'alarme lancé il y a trois ans à Athènes et, à force de marcher le nez tendu vers les étoiles de l'optimisation, craignons de nous retrouver, comme l'astronome de la fable, emprisonnés au fond d'un puits, même si c'est un puits de science.

Permettez-moi de conclure sur un souvenir personnel datant de ma toute première campagne. Cela se passait à l'Ecole Polytechnique, à l'époque déjà bien lointaine où les "jeunes savants", comme on nous appelait alors avec une affectueuse ironie, étaient unanimement fiers d'être des militaires, de porter un uniforme, d'obéir à une discipline et d'avoir comme devise "pour la patrie, les sciences et la gloire". On nous faisait payer ces divers avantages en nous apprenant le maniement des armes, en particulier celui de l'épée. Notre instructeur, l'adjudant-chef Eckmann, avait fait ses classes et même la grande guerre (celle de 1914-18) du côté Allemand. Cette singularité qui lui valait toute notre respectueuse considération se traduisait par un extrême souci de la perfection. J'éprouvais, comme beaucoup de jeunes mathématiciens, une certaine difficulté à tenir mon épée bien droite au moment du "Présentez Armes! " et j'entends encore la voix câline de l'adjudant Eckmann m'exhortant à la redresser: "La lame verticale, Mr Ferrier, bien verticale, encore un peu plus verticale! Mais non! N'exagérez pas! "

Messieurs! Elevons nos lames, nos esprits, nos observatoires et les piliers de nos temples à la verticale. Mais sans exagération! Sachons nous ménager une perspective un peu rasante, pour ne point perdre le sens du relief, donc de la réalité des choses, de leur éloignement et de leur pesanteur. C'est indispensable, même quand on va visiter la Lune. Lorsque nous allâmes pour la première fois visiter Minerve, à Athènes, en 1964, le grand péché contre l'esprit de la recherche opérationnelle, celui contre lequel fulminaient toutes les têtes bien remplies d'algorithmes, était de *sous-optimiser.* La crainte d'y succomber nous a conduits vers les systèmes très fortement intégrés qui eurent les honneurs de la deuxième conférence. Au moment d'aller encore plus haut et plus loin dans la voie de la perfection, prenons conscience qu'un autre danger nous menace, qu'un autre péché nous guette: celui de *suroptimiser.* Nous nous y exposons chaque fois que nous prenons le risque de fonder notre conception des M.I.S. sur une fausse conception, une "misconception" des hommes. Afin de mieux y échapper, ne manquons jamais d'introduire dans nos cocktails cybernétiques à base d'esprit de géométrie et d'esprit de système, un soupçon d'esprit de charité et de finesse, l'angustura du management.

The Systems Approach to Integrated Logistics Systems Study

Henry J. Williams*

1. INTRODUCTION–What is "Integrated Logistics Support"?

The subareas that are generally considered to make up the elements of ILS are:

FUNDING	TECHNICAL DATA
PROCUREMENT	PERSONNEL REQUIREMENTS
STANDARDIZATION	TRAINING
RELIABILITY	MODIFICATION
MAINTAINABILITY	SUPPORT SYSTEM OPERATION
TRANSPORTATION	SUPPORT EQUIPMENT
STORAGE	FACILITIES
MAINTENANCE	DISPOSAL
SUPPLY	

Since this list looks just about like the whole world, I should probably point out what is *not* included in ILS. In a major weapon system procurement, the two main non-ILS elements are: (1) the basic hardware design and (2) the initial hardware procurement. However, since the operation of the system and all subsequent hardware modifications are ILS functions, ILS may account for about 80% of the life cycle costs of the system. This leads me into the subject of the historical place of logistics in the scheme of things.

2. THE HISTORICAL APPROACH TO LOGISTICS

Since logistics has been concerned primarily with the development of plans for the support, and the support itself, of hardware and personnel systems, it has been an "after the fact" discipline. By that I mean that the logistician never got into the act until the system was fully defined or designed.

In particular, a designer would develop the hardware system with little or no consideration of the complexities or cost of supporting it. His only concern was the operational performance of the system assuming, of course, that it would always be in good working order. When the design and development was complete, the logistician was brought in to determine what maintenance, spares, documentation, test equipment, facilities, etc. were needed to support the equipment. This, of course, brought about some unfortunate situations. One of these situations is the need to maintain and supply a large number of different equipments all performing the same function. Because of the severe affect that inadequate logistics can have on the status of equipment, standardization has become an important goal in designing weapon systems. This is only a partial solution to the problem, however, since most new

* This paper was originally prepared by Dr. Donald Stewart while he was with Planning Research Corporation.

weapon systems must operate in an environment with old systems developed under different situations. For example, the U.S. Navy has a serious problem inherited from its past procurement practices; namely, 30% of all the equipment in the fleet is one of a kind–also referred to as "onesies." Also, on a typical navy ship one might find myriads of different kinds of fractional horsepower electric motors, pumps, and other such components. Whereas only two or three sizes might satisfy all the function on board a ship, one frequently finds 20 or 30 types which vary in size, manufacture, or configuration.

Perhaps you can imagine the support problems associated with this situation. First, there is the problem of stocking and supplying repair parts for all these different equipments. Next, there is the problem of having the maintenance skills available to maintain these different items. This implies the need for myriads of different technical manuals, training schools and curricula, technical data to document all these configurations, special test and support equipment, etc. One can readily see that this "after the fact" approach to logistic system design can have severe effects on the logistic life cycle costs. The major cause of this situation was that in earlier procurements, the goal was "least cost procurement," and this implied the minimization of only the design and initial hardware investment. Thus, by shielding logisticians from the design process 20% of the life cycle costs were tightly controlled while the remaining 80% went uncontrolled.

This tendency to shield logistics people from the details of the design process extended even further. For instance, logisticians were usually shielded from the details of the mission and told only what they had to know to do their job, and frequently, not even that. This led to the situation where the supply specialist who was developing a plan for, say, a ship, did not know whether the ship would be operating with other ships or alone. This fact alone would have a great impact on the supply plan such as providing duplicate supplies on each ship. Since the supply specialist was not given mission details, a determination of possible impact could not be made. Another result of this narrow view was that the logistician never looked at the impact of the system under examination on the overall goals of the organization. Therefore, all systems had to be assumed to be of equal importance. Assuming equal importance for all systems is very wasteful because it does not lend itself to making an economical allocation of resources.

Still discussing the history of logistics, we find that most logistic plans were designed for static systems. Specifically, little effort was made to include in the logistic system design a mechanism for changing the plan to reflect changing requirements. And yet, obviously, supply plans and maintenance plans would be quite different for an extended war as opposed to a quick-reaction conflict. In a quick-reaction war one might not even need to provide support to the equipment but would just throw it away. In a war of attrition the logistic systems probably would mean the difference between victory and defeat. The importance of the uncontrolled 80% of life cycle costs has finally become accepted.

3. THE NEW APPROACH TO ILS

What are we doing about these problems in our new approach to logistics system design? The most important change being brought about is the changing of logistics from an "after the fact" discipline to a "before the fact" discipline. This is done in three ways:

(i) Providing design guidelines concerning logistic factors, e.g., habitability, safety, availability requirements, stowage, facilities, manning, standardization.

(ii) Timely evaluation of the interactions and input of logistic elements to other considerations.

(iii) Integration of logistic factors into total system design through tradeoffs.

This is not meant to imply that all the problems in this area have been solved. Actually, there are still obvious difficulties to be overcome. A designer cannot effectively incorporate logistic considerations into this design unless he has a good appreciation of ILS. Hence, it is essential that he become better educated. In the short run, it is the responsibility of the logistician to indoctrinate him; in the long run, it must be the responsibility of engineering schools to impart the knowledge and appreciation of logistics through his training. In addition, it is important that future logisticians get a basic engineering education to improve their ability to communicate with design engineers.

Unfortunately, there are no simple solutions to the incorporation of logistic factors into design. Ultimately, the design must come out of a designer's head. Therefore, the logistician's first job is to indoctrinate and train the designers with logistic criteria. In order to perform this training function and the new design review function, it is necessary to develop guidelines with which to evaluate "good practice" in design. This will come mainly with experience.

Another important change to logistics has been brought about by the development and application of mathematical and statistical methods. Through the use of mathematical models great strides have been made in the development of more effective and lower cost logistic systems.

The next important change that has happened to logistics is the recognition that the various subareas within logistics have important interactions with one another and should be treated as a unified discipline. This discipline has been given the name "ILS" for Integrated Logistics Systems and merely implies that the need for a "total system approach" in the development of logistic plans is recognized.

Thus, all of the separate logistic disciplines that I mentioned earlier have been organized into a single discipline largely because of the interactions among them. This brings me to the most important function of the systems analyst in the design of an integrated logistic system–analysis of logistic tradeoffs.

4. TRADEOFF CONSIDERATIONS

The analysis of logistic tradeoffs provides an abundance of opportunities for the operations researcher to apply mathematical and statistical methods. By this I mean quantifying and comparing the various course of action which present themselves in the design and operation of each of the subsystems in a major weapon system. To illustrate, I will again use the example of a naval ship design, each of these tradeoff considerations could be applied equally well to any other major system with some minor modifications.

Some of the areas in which tradeoffs are made in ship design include:

(i) Standardization–There are three types of standardization we are concerned with: (1) intra-ship, (2) intra-class, and (3) intra-navy. These each have different tradeoff considerations. For instance, increasing intra-ship standardization has a direct affect on the onboard stocking costs and the availability of repair parts. This results in improved ship effectiveness through increased ship availability resulting from shortened supply and maintenance times. However, if standardization is increased to the point where innovation is seriously affected, then ship effectiveness is decreased. Improving intra-class standardization has a smaller affect on onboard stocking levels but decreases supply time when other ships of the class are stationed nearby. Intra-navy standardization has the least affect on a particular ship's effectiveness, the main result being increased supply system costs.

(ii) Reliability and Maintainability *vs* Maintenance and Personnel–The interaction here can be seen in that equipment which has high R or M would have low maintenance personnel requirements; hence, one can trade off people for redundancy.

(iii) Maintenance Plan and Concept *vs* Training Requirements–Maintenance plans which incorporate a throw-away maintenance concept have the smallest training requirements, whereas those plans requiring bit and piece replacement have the highest training requirements. Thus we can trade off training for design sophistication or supply cost.

(iv) Reliability and Maintainability *vs* Support Equipment, Supplies, etc.–In this case, one can exchange support equipment for redundant equipments or more maintainable equipments. This also can bring about savings in crew since they are required to operate the support equipment.

(v) Maintenance Concept *vs* Ship Effectiveness–There is an obvious tradeoff in maintenance personnel assigned onboard vs. the tender vs. the shipyard. However, since the manning level is the most important single factor affecting ship costs, the least-cost ship would have all maintenance, except for catastrophic failures, performed ashore. This concept would have a significant impact on reducing the size and cost of the hull. Unfortunately, it provides an overall ship availability and effectiveness which is inadequate for a warship. One answer is to increase the onboard maintenance manning, however, there are other factors which can be traded off against increased manning. For instance, instead of piece part replacement the equipment can be modularized, and whole modules can be replaced thus reducing the need for maintenance skills and manning. This approach has some unfortunate results, however. The level of standardization of spares is sharply reduced causing a large increase in spares also have a high rate of obsolescence. Some of these costs, however, can in turn be traded off with the decreased training costs and the reduction of special tools, test, and support equipment.

(vi) Automation–In its simplest form automation reduces the need for operational manpower onboard a ship. However, since the increased complexity of equipment implies higher failure rates and maintenance requirements, this implies a higher level of maintenance manning, parts, technical manuals, and test equipment. Therefore, automation features usually have to provide a higher level of ship capability to trade off against the reduction of ship availability. There are other interactions we should consider also such as increasing redundancy to improve the ship's availability. Hence, in the end we are trading off people—their weight, volume, and cost against redundant automation equipment and its weight, volume, and cost.

(vii) Team Training—In order to maintain a high level of combat proficiency in a ship's crew, the ship must engage in frequent team training exercises. These exercises involve the use of other ships, submarines, and airplanes as targets. The crew then performs all its combat functions as a team. It acquires, tracks, identifies, calculates firing solutions, and simulates the firing of weapons. The crew is then graded according to its timeliness and accuracy. The costs of these training services (planes, ships, etc.) are high, but there are various tradeoffs that can be made. The effectiveness of the training can be reduced by pretending there is a target. Also, radio controlled planes and helicopters can be utilized to simulate aircraft, and torpedoes can simulate submarines. Another approach is to incorporate additional hardware in the design that can inject signals into the sensors which will give identical radar and sonar displays as the real targets. Thus, we can trade off the cost of operating planes and ships against the investment in onboard simulation devices.

(viii) Automatic Testing—Here the savings in maintenance personnel can be traded off against the investment cost of an automatic test system.

(ix) Stocking *vs* Transportation—If a ship stocked every part onboard to protect itself against all failures, the stocking costs would be prohibitive. Therefore, based on failure rates, essentially, and other factors one can trade off stocking costs against the cost of high-speed transportation in order to set inventory levels.

(x) Fresh Water Fire System—Ships have traditionally utilized salt water in their firemains. However, microscopic animal life is ingested into the system, and shellfish grow in all the valves. Whenever the valves are opened and closed, these shellfish ruin the valve seats thereby requiring extensive replacement at ship overhaul. One can, therefore, trade off the cost of circulating fresh water in the system against the continual cost of replacing valves.

(xi) Exotic Paints and Coatings—There are exotic coatings available which are much more costly than ordinary paints. However, thay can prevent the rusting of the hull and the growth of barnacles over extended periods of time. Therefore, one can trade off this additional cost against the reduced costs of scraping and painting the hull.

(xii) Podule Concept—Because of our rapidly changing technology, the electronic and weapon systems onboard a ship obsolesce much more rapidly than the rest of the ship. This means that a warship will undergo several major modifications during its lifetime. These modifications can sometimes be almost as costly as the original ship. However, one could conceivably design a combat information center or other major part of the ship as a removable podule. Therefore, to modernize the ship one just replaces this podule with a more up-to-date one. This podule design increases acquisition cost significantly, but this can be traded off against the savings in modernization costs.

5. INTEGRATION

I have mentioned a few of the areas of logistics where significant tradeoffs can be made. There are obviously many more. An appropriate question is, "How do we integrate and control all these tradeoffs in a meaningful way?" Basically, there are three ways this can be done:

(i) *Selection*

Choosing those areas which are reasonably independent enables suboptimizations to be performed without causing serious interactions with other areas. If one were to try to enumerate and evaluate all the combinations of maintenance plans, manning, training, supply, etc., the task gets completely out of hand. Even if there were only ten alternative plans for each logistic area, the number of combinations to be evaluated would be in the millions. One of the important problems that faces us, then, is how to reduce the number of alternatives to a manageable number. There are some convenient methods available to help do this. One of these is the use of physical constraints or limits; for instance, we can eliminate all manning alternatives which exceed the capability of the plane/ship/tank, etc. Another method is to relate several parameters; for instance, it probably is not meaningful to consider simultaneously: (1) bit and piece replacement and a crew with low skill level or low requirements and (2) large maintenance crew and the throw-away maintenance concept. Although these methods are an aid, it is still necessary to use judgment in selecting logistic alternatives which cover all reasonable choices and yet are manageable in number.

(ii) *Model Building*

Performance and availability models provide a powerful tool for integrating areas which have significant interactions and tradeoffs.

(iii) *Iteration*

The performance and iteration of analyses at ever increasing levels of detail is an important aid to damping out the variations caused by interactions and provide for a converging solution.

In summary, I have discussed how logistics has been performed historically in piecemeal fashion thereby causing potential tradeoffs to remain unrecognized or at least difficult to perform. The advent of Integrated Logistics as a discipline has speeded up the recognition and performance of tradeoffs bringing about improved weapon system performance at lower life cycle cost.

DISCUSSION

Buck

Is there not another level of standardization than all-Navy, namely, all Defense or all Government and is this not a significant level in your experience?

Williams

Yes, there are levels of standardization such as all Defence and all Government. There can be great payoffs from standardization at these levels. Obvious are such items as less unit cost, decreased stock level requirements, ease of interchange of components and/or maintenance support. However there can also be great dangers such as over-emphasis on standardization with resulting rigidity in design with an accompanying lack of flexibility in operations. Also when you reach more encompassing levels of standardization, you in turn reach more encompassing masses of red tape and officialdom. Inertia resists attempts to standardize and if once standards are agreed upon, the same inertia tends to resist modernization or any other change.

There can be benefits from high order standardization but also great danger. Therefore a logistician must approach the idea with intuition and caution.

Guney

Influence of competitive industry on standardization?

Williams

Very definitely industry has an influence on the decisions made as to what direction a specific procurement will go. When the economic stakes are high enough, great political pressure can also be brought to bear. Some of the larger procurements will run into millions of dollars.

If the standard item is to be designated from existing sources then there will be great competition to have the specifications written so strictly that they will either exclude all but a single source or so broadly that they will include all sources. If a product is to be developed then the competition is just as keen to land the contract, but here the logistician may seek low cost and supplier competence.

If the procurement is for a multi-national group such as NATO, then pressures can come from Governments as well as companies.

Alberts

To what extend do you believe the complex of special logistics models can be integrated to describe a "Total Logistics System" model?

Williams

I am afraid that my experience has been that special models tend to remain just that, special. Unless models are designed to be integrated into a larger system they are seldom compatible. The best you may be able to do is to modify existing models so that at the least they may "talk to one another".

Logistic Planning and Management in the Electronic Age

Lt Col M.B. Page

1. INTRODUCTION

In the ten years since the decision to end National Service, the British Army has been in a state of transition from a manpower intensive Army of the 1939–45 kind to an equipment intensive force. The extent of its adjustment has been greater than in the Royal Navy or Royal Air Force in the same period, as they have always been more equipment orientated than the Army.

This rapid expansion of equipment has occurred mainly in the field of electronics, due to the increased use of guided weapons for air defence and in the anti-armour role, the many new applications for radar and television and the increased scales of radio.

Such equipments are generally much more complex than their predecessors in their design, construction, maintenance and repair. Because the aim is to exploit new technologies to the full, the research and design staff are operating on the fringes of knowledge and technique. This militates against accurate budgetting and programming, but it is safe to say that such equipments are invariably expensive. It also means that experience in the maintenance and repair of the current generation equipment is of little relevance for the future when a major change in technology is involved. Another major consideration is that electronic equipment is prone to randon failures that cannot be forestalled by means of cyclic overhauls. The most important characteristic is that diagnosis of faults involves the use of complex test equipment, and that subsequent repair involves the use of highly trained manpower. As there is increased reliance on communications and surveillance equipment, a high rate of availability and a high rate of assurance are necessary. On the credit side to the logistician is that the equipment lends itself to the extensive use of modular repair and as many of the components are lightweight, they can be flown to centralised repair facilities.

2. AIM

The aim of this talk is to describe the system of management being introduced by the British Army to plan and develop the logistic support for complex equipment.

3. ORGANISATION NOTE

Within the British Army, the Master General of the Ordnance (MGO) is responsible for research, development and production of equipment, except that in the electronic field, our Ministry of Technology invariably acts as his agent. The Quarter Master General (QMG) is responsible for equipment currently in service, so there is a transfer of responsibility when the equipment is deployed. Since the planning of logistic support must begin while the equipment is being designed, the QMG must be aware of what is going on within MGO'S department and must ensure that the activities of both staffs, plus those of the Ministry of Technology are closely coordinated, in order to be able to guarantee that the equipment will be deployed to time and that its logistic backing will be complete in all respects. This is what Project SPLICE is designed to do.

4. LOGISTIC SUPPORT

Before describing the system in detail, I wish to review briefly the main elements of any logistic plan. These are:-

(i) *Equipment Management Policy*. This defines the nature, depth and estimated frequency of repair to which the equipment is to be subject in service; it is based on the reports of the Maintenance Advisory Groups who work in conjunction with the Research and Development authorities. It is developed almost concurrently with the design of the equipment. As it is evolved, action can be initiated to prepare other aspects of the logistic plan in detail.

(ii) *Quantitative Requirements.* These will be funded as percentages of Unit Entitlement cost in the first instance, but later will be expressed in terms of equipments and spares by quantities and types. The list includes:-

(a) Additional equipments for:-
Initial Maintenance
Repair Pools
War Maintenance Reserves.

(b) Initial Spares Requirements for all stocking points.

(c) Repair Pools of repairable major assemblies and ancillaries.

(d) Scales of tools, spare parts, assemblies and ancillaries to be scaled and held:-
With each equipment
Per platoon, company or battalion.

(e) Specialist tools, gauges and test equipment. This last category can be expensive, costing up to 10% of the total inventory investment.

(f) War Maintenace Reserves of spares.

(g) Specialist buildings for storage and for workshops.

(iii) *Technicians.* Must be selected and trained.

(iv) *Publications.* These are sponsored by a variety of authorities and include:-
Complete Equipment Schedules
User Handbooks
Electrical and Mechanical Engineering Regulations
Technical Ammunition Bulletins
Illustrated Spare Parts Catalogue.

In something as complex as a guided weapon systems the requirements in these categories multiply up. We have to consider:-

(i) The basic weapon system—launchers, trackers, etc.

(ii) The missiles.

(iii) The associated vehicles.

(iv) Scales of radio, including vehicle harnesses and vehicle installation kits.

(v) Mobile workshops in support of the units concerned—and their specialised requirements.

It all adds up to an immense task of planning and coordination.

5. PROJECT SPLICE

5.1 General

A systems approach to the task is clearly necessary. Project SPLICE, which is designed to ensure such an approach, stands for the Systematic Planning of Logistics for the Introduction of Complete Equipment.

Its basic principle is one of team work, in that all of the agencies involved provide representatives who combine as a SPLICE team under the chairmanship of a Q(G4) staff officer. Their mission is to plan and coordinate the logistic support for the system and to monitor its development. Representatives of the equipment sponsor branch, and the appropriate finance branch, belong to the team. If the equipment is to be used by more than one Service and is to be Single Service Managed by the Army, then the other user Services will be invited to join the SPLICE team. For instance, the Air Force Department is represented on the SPLICE pilot project.

The main management technique used is network analysis. They are three types of charts:-

(a) The master logic chart.

(b) A milestone chart.

(c) A Project Activities Relationship Diagram (PARD).

I shall describe these in some detail.

5.2 Master Logic Chart

Preparation of the master logic chart can begin at the time of the drafting of the General Staff Requirement, when decisions are taken on the budgetary arrangements necessary to provide the logistic support.

All the procedural steps are plotted out on a network. Figure 1 gives a simplified impression of the start of a network, dealing with one major component of a weapon system. It could be the tracker, launcher, a radar, the test and repair vehicle or any other element. Each poses its own set of logistic problems which are identified on the network.

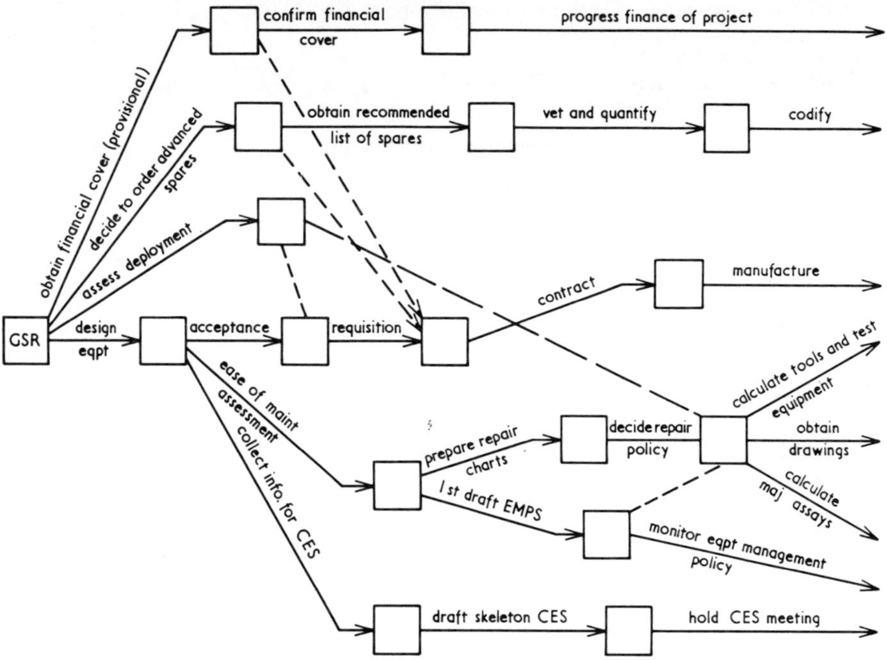

Figure 1 Master logic chart–the start of a simplified SPLICE network

The first step is to get the logic of the system right; to take a simple example, you cannot requisition spares until:-

(a) The repair concept has been established.

(b) Repairable assemblies have been identified.

(c) Spares have been codified.

(d) Spares scales have been assessed.

(e) Taking the probable deployment into account, a detailed calculation of two years requirements, to include initial stocking elements for units, depots and workshops, has been made.

When the logic is right, the next step is to assess timings for each activity, so that the total time through the net work is known. Moving on to the end of the network, the purpose of our time schedule is apparent. You will see from Figure 2 that, after the final event on each line, the network draws together at the Q Readiness Date (QRD) or logistic readiness date if you prefer. In the example shown it falls in its correct place, before the Confidential Policy Date for the equipment; this is the date by which it is planned to introduce the new equipment.

If it did not do so, then the next step would be to re-examine the longest line, in terms of time, through the network (The Critical Path) to see where time can be saved. This is an examination of timings in the first instance; but it can include a re-assessment of

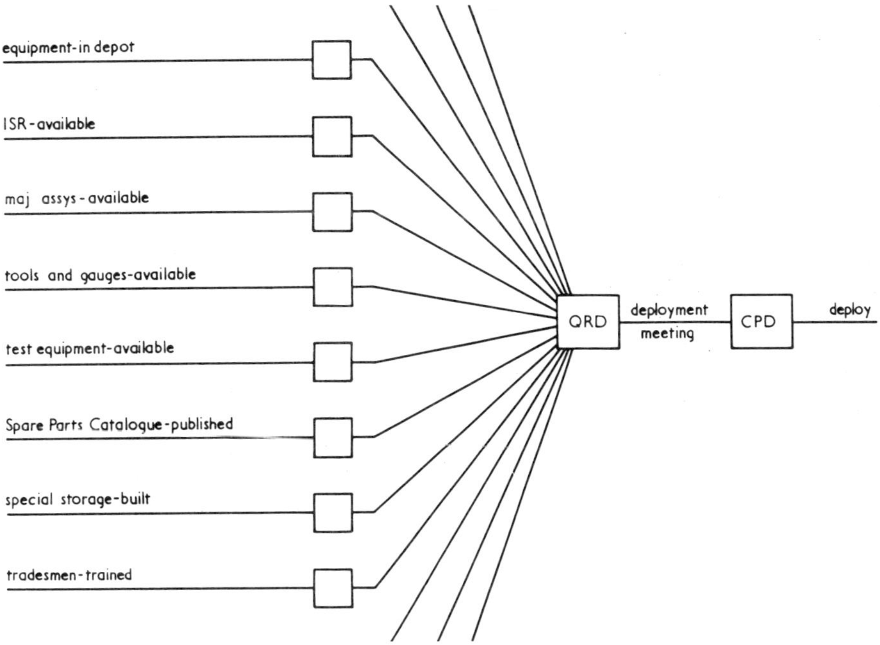

Figure 2 Master logic chart—the end of a simplified SPLICE network

method. For example, the advanced spares ordering technique could be used, so that a small quota of spares is made available in advance of the main order, based on forecasts of usage data provided by the manufacturer. The network is up-dated by computer when changes become necessary. The computer identifies the new Critical Path again at each re-run.

If the QRD cannot be manipulated into its proper place, then either:-

(i) Deployment of the equipment must be delayed with consequences for:
Operational readiness
Training
The extended logistic support of the current generation of equipment.

(ii) Deployment may go ahead, perhaps on a restricted basis, in the knowledge of the precise degree of logistic risk being run.

Figures 1 and 2 are simple examples to illustrate the principles. A complete network is much more extensive. To give you an idea of the scale of the activity on our pilot project:-

	As at 1 Jan 70	*Now*
The master Logic chart has	1200 events	1,900
Of which coordination is necessary for	900 events	1,300
The various sub-networks have	6000 events	9,500

5.3 Milestone Chart

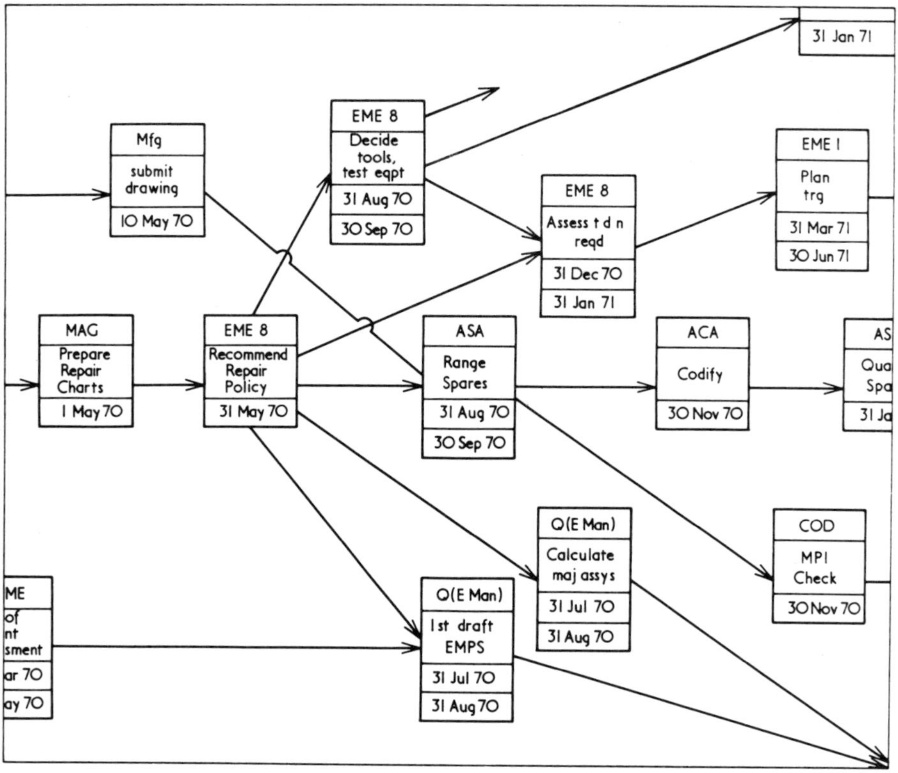

Figure 3 Extract from a simplified SPLICE milestone chart

The master logic chart gives us our appreciation and outline plan. To translate it into a coordinated plan the milestone chart is used. An example is shown in Figure 3. Each major event is shown, together with:-

(a) The branch or agency responsible

(b) The latest date for completion. If two dates are shown, this indicates "float-time".

These dates are arrived at using a "management by objectives" approach to the individual members of the team, within the constraints and strategy of the network.

The milestone chart serves as a progress chart, so that the team can monitor the development of logistic support in furtherance of their plan.

5.4 Projective Activities Relationship Diagram (PARD)

This is the executive portion of the teams instruction to its individual members. An example is shown in Figure 4. It is the milestone chart re-drawn so that each branch or

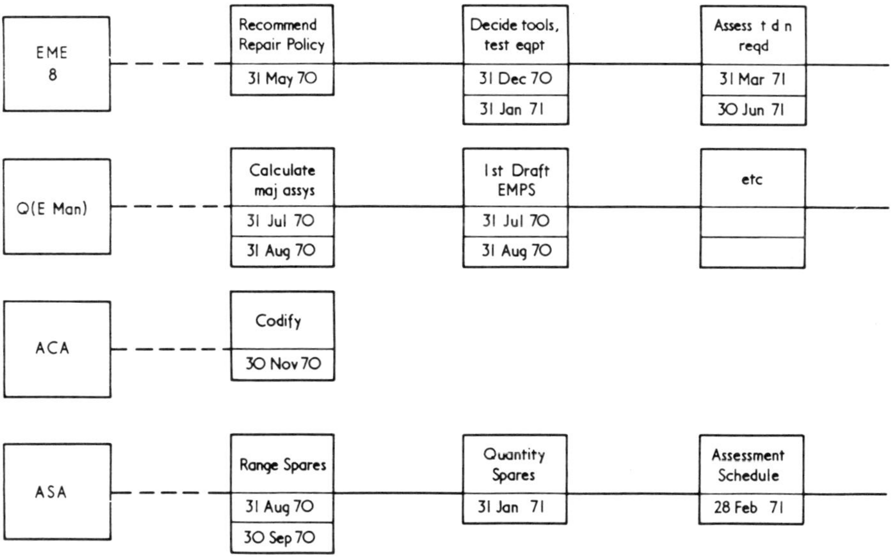

Figure 4   Extract from a simplified SPLICE PARD chart

agency has a clear statement of its responsibilities in sequence. The PARD enables the branches and agencies to develop their own sub networks and plan their own work within the time frames they have agreed. As none of them is engaged on one project alone, they can plan their work according to priorities by means of the PARDs.

5.5 Deployment Meeting

With the aid of the forcasts provided by SPLICE it is possible for the logisticians to state their position clearly at the Deployment Meeting. This is a meeting conducted by the MGOs staff at two-star level for major equipment. The equipment sponsor, representatives of the User Arm and the Operations staffs and the principal logistic officers attend. If a decision is taken to commence deployment on an agreed date, the responsibility for in-Service Maintenance is assumed by the Q Staff and Services from that date.

5.6 Relationship to Other Projects

In parallel with Project SPLICE we are developing Project VESPER, which is a computerised system for the management of in-Service technical equipment. We foresee an interface between SPLICE, and VESPER, so that any difficulties or delays with the development and production of any new equipment will generate a response from VESPER in terms of the logistic implications for its equivalent in the current generation. At the least this response could be a print-out of a spares requisition to cover the extended in-Service life of the present weapon or communication system.

6. SUMMARY

The complexity of modern electronic equipment necessitates a new approach to the planning of its logistic management. This planning must begin as early as possible and all of the branches and agencies must be coordinated so that they work to a common goal—the QRD, which should precede the planned deployment date. The method used we call SPLICE. It involves:-

(a) A systems approach to the problem

(b) Teamwork

(c) The use of network analysis, with the network up-dated by computer

(d) The determination of branch and agency objectives

(e) Monitoring and feed-back

(f) Coordination of the progress of the new equipment with the logistic state of the current generation equipment, thereby achieving an economic transition from one generation to the next.

The results obtained from the pilot project to date are sufficiently encouraging to have led to the application of SPLICE to a variety of other important equipments, even before the techniques are fully refined, and to the formal adoption of some of its provisions in advance of the remainder of SPLICE.

DISCUSSION

Volten

In view of the emphasis on random failures and highly trained manpower for the maintenance of complex equipment, would you agree that there are different outcomes to the "Tradeoffs" in Maintainability versus personnel lost between different armies, dependant upon whether you have a regular army, or national service and a conscripted army.

Do you think it possible for this difference to influence decisively the logistic structure and even the choice of weapon systems.

How does this factor affect standardization in NATO.

Page:

This is a very important question that gets straight to the heart of the problem. I am expressing only a personal view when I say that I believe that the impending advent of complex and costly equipment was a major factor in leading the British Government to decide to discontinue National Service.

All soldiers need simple and reliable equipment. No equipment sponsor sets out consciously to develop a complex equipment. I take the view that we are in a transitional stage when dealing with the current generation of electronic equipment. It may well be that succeeding generations of equipment will be more robust and reliable, reflecting the basic military requirement and hence may be more suited to being used by National Servicemen.

The important thing is to ensure that the equipment design selected is within the capacity of the force concerned to use it and to maintain it efficiently. To take an extreme case, well outside NATO, there are the problems the Russians are experiencing in equipping the Egyptian Army. The equipment that an army can use efficiently is related directly to the Technological base of the community that produces the Army. When an advanced technology does not exist, the cost of the training programme needed to adapt an army to accepting sophisticated equipment may well be prohibitive. This is the extreme situation that has no direct application within NATO.

Due to their varying policies towards National Service, NATO nations are beginning to vary in the ability of their armies to accept highly complex equipment. There is little advantage in giving a soldier an intensive training for the majority of his National Service since he will be operationally effective for a short period only and his training standards will deteriorate when he goes in the reserve.

However, it may still be possible, within NATO, to standardise on minor systems and on parts of major systems. It may well be that NATO can assist by developing agencies that will be able to supplement national efforts, both in training and in logistics. The advanced training of technicians and the flexibility and economy to be obtained through carrying out supply and major repairs on an international scale merit further study.

The short answer to your question, therefore, is that this might well be only an interim problem, until more reliable electronic equipment becomes available. I would not like to guess how long this "interim" will be. The basic military requirement remains that we want simple, reliable equipment. I am hopeful that NATO agencies will be able to supplement national efforts, particularly during this interim period, so that the impetus towards standardisation is not lost.

Rosenman

How was this process managed prior to SPLICE?
How have the managers adapted to the new system?
Can you tell as yet whether the new system is beneficial?

Page:

The functions that I have outlined all existed prior to SPLICE. The Supply and Repair organizations developed their own systems, with a loose coordination by the Q staff. The Q staff tended to direct its attention to the Initial Spares Requirement, or as the United States Forces term it more correctly, Concurrent Spares. While this did not matter so long as the equipment was relatively simple difficulties arose when complex equipment came along.

We then lacked both a methodology and a computerized system for handling the data. These are now available, so SPLICE became possible.

Not all managements have reacted in the same way to SPLICE. Some were quick to see the advantages; others tended to regard it as an encroachment on their own liberty and freedom of action. They also saw the danger of their own staffs being involved in work outside their own sphere. These obstacles will disappear as confidence in the system develops. There is no doubt that the system is already proving beneficial. As logisticians we already know much more about the equipment that forms our pilot project than we do (or have done) about any other equipment at a somewhat similar stage of development. We have already been able to influence the development of the weapon system concerned to a limited extent.

Young:

Is there any intention for SPLICE procedures to systematically influence *design* to produce lower support costs?

Page:

Not yet, but I believe that this is the next natural area into which to develop the system. It is important that we consolidate what we have done so far and establish SPLICE as it is already conceived, before we undertake exploitation tasks.

Having sounded that cautionary note, there will be, I believe, great scope for the logistician here. I have in mind two equipments that were developed for the British Army, but were later sold to other Armies. The redesign work that went into producing the sale-pattern equipment has produced, in both cases, a more simple, reliable and more readily maintainable equipment than the in-service model used by the British Army. I cite these two as examples of what can be achieved to meet specific logistics situations.

It is important to be able to quantify the logistic penalties so that the cost-effectiveness of offsetting them can be expressed in similar terms to those used for sales purposes.

O'Connor:

The operation of a system as described implies the need for the head of the project to be thoroughly committed to the methods, and to be able to enforce compliance from and if necessary indoctrinate, other agencies. How has the Army Department overcome these problems in the functionally structured organization which exists in the U.K.?

Page:

The audience may well be wondering why MOD London has to come to Luxembourg to discuss its internal problems!

The leader of the SPLICE team must have complete confidence in the method he is using. He must inculcate the team spirit into his team, motivating them to the common goal, which is the QRD.

He has many aids, particularly the networks. They impose a discipline on the system, on the personnel participating and on the use of time. This discipline will grow, as each member of the team will be fully aware of the consequences of delay on the part of the agency whose spokesman he is. In the final analysis I do not think he would be able to face his team-mates knowing that he had let them down. If there is slippage he will always feel obliged to show how lost time can be made up again.

In any vertically organised structure, horizontal links are necessary, so that day-to-day business can be conducted. This is recognised in SPLICE which formalises, to a limited extent, the inter-disciplinary links that we all know exist.

Ferrier:

1) Que signifient VESPER et FORWARD

2) Avant d'apprendre a réparer le matériel, il faut apprendre au personnel à s'en servir. Cela est souvent le plus long à démarrer même dans une armée de métier. Quelle est l'expérience de Monsieur Page sur ce point, en ce qui concerne le Q.R.D.

Page:

VESPER is the computerised system for the management of the in-service inventory of equipment. It is based on the Central Inventory Control Point that is being established. It will deal with:

(a) Depot stock holdings of equipment and assemblies, both serviceable and repairable.

(b) A continuous census of all vehicles and equipments deployed to units.

(c) By means of continuous asset/liability calculations, ensuring that there is a correct logistic balance between the deployed fleet and the resources to support it, both in the short and the long terms.

(d) In conjunction with FORWARD, the planning of our base overhaul programmes. The supply organisation will formulate the repair requirement and the repair organisation will identify its capacity to fulfil these taks. The contribution of the Q Staff will be to provide:

(1) Information relating to the availability of replacement equipments and when.

(2) Priorities for repair.

We should receive from the computers, on a Material Budgetary Analysis Document, the answer to the question whether to repair or to scrap and replace.

VESTER stands for:

*V*EHICLES and
*E*QUIPMENTS
*S*PARE
*P*ARTS
*E*CONOMICS and
*R*EPAIR

FORWARD is operated by our repair organisation. It provides information on equipment reliability and spares usage. It also provides the repair organisation with a wide variety of management data. Its contribution, in conjunction with VESPER, in the base repair planning field I have already mentioned.

FORWARD stands for:

*F*EEDBACK
*O*F
*R*ELIABILITY
*W*ORKSHOP
*A*ND
*R*EPAIR
*D*ATA

To the second part of the question, the training of technicians can be one of the longest logistic tasks and could well influence the Q Readiness Date. This is because the repair techniques must be developed first, then a system of training. Apart from selecting the technicians, which can be a major problem of its own, there is the important task of providing them with equipment on which to do their training.

It is always possible for them to start on a pre-production model while the Service-pattern production models are being manufactured. But, before their training courses are complete, they will need a Service-pattern equipment to work on for a while.

Baglow:

Have you been able to get reliable forecasts of the duration of the events in your network, or have you had some nasty surprises?

Page:

The establishment of realistic timings is one of the biggest and most important tasks. It is difficult because everyone, particularly the research and development staffs, are working on the fringes of technology. When they design something their first aim is to produce a workable solution; then its reliability and maintainability must be assessed and improved upon. This all takes time. Inevitably their assessments of the time required are optimistic.

When such slippage does occur, of the kind I have described, this will produce a reaction on the main equipment itself and on its planned deployment date. They, in turn, will affect the Q Readiness Date. What is more important to the logistician in this context, are the implications of this slippage on your current pattern of equipment. You will have been running down your stocks of spares and reducing your repair resources, only to be faced suddenly with a need to maintain the equipment in service for a period of six months, a year, or more, longer than you had planned, at a time when it is at its least reliable.

To follow the main line of your question, and to refer back to Admiral Ferrier's question for a moment, you may have a case of an equipment being deployed and it proving to be less reliable than had been expected. In such a case you may be faced with a heavy base repair load before your own Service base repair facilities have been developed. You may then have to go back to your contractor for repair, assuming that he is willing to undertake the work. If he has further orders coming along, he may not be willing to get involved.

It is also worth considering the advantages likely to be obtained by the Contractor carrying out some of your technician training for you. But both training and repair are contractual problems. It is always worth considering whether there should be clauses to these effects in your contracts.

Høgberg:

From a Logistics Systems Development point of view, how do you ensure integration between SPLICE, VESPER, FORWARD?

Page:

VESPER and FORWARD are concerned with in-service equipment. The coordination of the Supply Organisation's computers on which VESPER is to be based, with those of the Repair Organisation's FORWARD is primarily a matter for the Q (i.e. G4) staff. But both organisations have many years of experience in coordinating their activities in Automatic Data Processing and have established their own liaison committee.

SPLICE must have links with VESPER so that the progress in development of the new generation equipment can be related to the scale of logistic support still needed for the current equipment. Members of the SPLICE team will usually have a responsibility, either direct or indirect, for the in-service equipment, so that cross links are established.

Ultimate responsibility for coordination rests with the Equipment Management Policy Committee, which draws together under Q (i.e. G4) staff chairmanship representatives of supply, repair, finance, costing and user interests at a sufficiently senior level within MOD to ensure close coordination and the resolution of difficulties that cannot be cleared by normal staff processes.

Stone:

Do you see any major obstacles to the adaptation of the SPLICE approach to multinational equipment projects, apart from the fundamental one of political will to cooperate? Would there be a rôle for NAMSA here?

Page:

You mention the limitation of political will. It may well be that there is a security factor here, in that a nation may be unwilling to disclose certain design information until work has proceeded beyond a specific stage or until a firm sales deal has been made.

Most nations, and particularly the smaller ones, are concerned with the sales prospects for nationally designed equipment, so I foresee these problems being overcome in time.

I am sure that NAMSA has a big part to play in providing an integrated logistic philosophy that will be of assistance to designers. There appears to be considerable scope for the application of SPLICE techniques to developing and progressing international logistic systems as well as national ones, provided the common logistic philosophy is first agreed.

Logistique des Armées en Compagne—Evolution, Developpements futurs

Maurice Bresson

"La tactique militaire est semblable à l'eau; car l'eau obéissant à son cours naturel, descend des lieux élevés et se précipite vers le bas. De même, dans la guerre, le procédé permettant d'éviter ce qui est fort et d'attaquer ce qui est faible.

L'eau modèle son cours d'après le terrain sur lequel elle coule; le soldat construit sa victoire en fonction de l'ennemi qui se trouve devant lui."

SUN TSE, *L'ART DE LA GUERRE*

1. INTRODUCTION

L'exposé se propose d'analyser les problèmes posés par le soutien logistique des armées en campagne pour différents types de conflits.

En 1967, au séminaire OTAN de Münich sur les développements récents des modèles de combat, un modèle analytique a été présenté concernant les effets des contraintes logistiques sur le mouvement, la vitesse et la puissance des armées en campagne dans une situation de conflit classique.

La confrontation du modèle avec les données numériques extraites de la Seconde Guerre Mondiale, montre un excellent accord entre les faits historiques et l'outil théorique, en dépit de la simplicité de ce dernier.

Toutefois, l'objectif poursuivi par le modèle présenté a Münich, était seulement de répondre à la question:

"A l'échelon stratégique, le rythme des opérations que l'on peut observer pour la période 1940-45, sur différents théâtres de guerre, peut-il s'expliquer en grande partie par les contraintes logistiques? "

Pour répondre à cette question, c'est-à-dire pour vérifier une hypothèse, il convenait donc d'introduire dans le modèle, des effectifs, des besoins logistiques, des caractéristiques et des performances pour les véhicules de transports conformes à la réalité d'alors. Sur le plan méthodologique, il suffisait que le modèle soit capable de tenir compte des principaux groupes de facteurs quantifiables disponibles dans les documents historiques.

En satisfaisant à ces conditions, on pouvait montrer, chiffres à l'appui, qu'effectivement les contraintes logistiques ont conditionné le rythme des opérations, en fixant des bornes supérieures précises dans l'espace et le temps aux concepts et aux ambitions tactiques.

Ce type de résultats est particulièrement net pour les armées aux effectifs nombreux et à la logistique très importante des armées anglo-saxonnes d'une part, et des armées soviétiques d'autre part, pour la période 1943-1945.

A titre indicatif, au 30 avril 1945, le Corps Expéditionnaire américain en Europe (non inclus les autres armées alliées), comprenait 3 065 505 hommes, 710 650 véhicules, 80 000 véhicules blindés (chars, chasseurs de chars, automitrailleuses, transports blindés et véhicules semi-chenillés), 50 000 pièces d'artillerie, 3 750 000 armes de petit calibre (armes individuelles et armes automatiques).

A la même époque, du côté soviétique, les trois Fronts chargés de l'attaque sur Berlin, (non compris les cinq Fronts agissant dans les autres secteurs du Théâtre), comprenaient: 1 593 800 combattants, 6 160 canons automoteurs, 23 585 canons et mortiers tractés, 95 383 camions de transport, 6 696 avions pour les grandes unités participant directement à l'assaut. Si l'on compte l'ensemble des unités du même secteur déployées sur les arrières des armées et des trois fronts, les chiffres précédents deviennent: 2 500 000 hommes, 6 250 chars, 41 600 canons, 7 500 avions.

Ces chiffres font très clairement percevoir l'ampleur des besoins logistiques des grandes armées de cette période.

Le modèle du symposium de Münich rendait notamment compte de façon nette, du caractère *discontinu* des offensives, dont le rythme se trouvait réglé par la constitution de dépôts logistiques successifs, fonctionnant à la manière des réservoirs de barrages hydrauliques, exigeant des délais importants avant d'être remplis. L'offensive déclenchée ("réservoir plein") s'engouffrait le plus souvent de façon torrentielle durant sa période initiale (une fois gagnée la phase de rupture) pour se ralentir peu à peu, et finir par épuiser sa puissance de choc lorsque le centre de gravité des éléments de tête se trouvait à quelques 300-500 km des bases de départ.

Il est évidemment d'un grand intérêt pratique de comprendre dans quelle mesure le rythme des offensives et l'amplitude de leurs élongations maximales pour les conditions d'aujourd'hui et de demain, seraient encore réglés par le jeu rigoureux des contraintes logistiques. Le modèle présenté ici, qui est une extension du modèle de Münich, se propose de répondre à cette interrogation.

De plus, l'effort qui a été entrepris donne un aperçu du bénéfice que l'on pourrait escompter de modèles plus généraux, élaborés suivant des concepts analogues, mais construits pour tirer pleinement parti des ressources offertes aujourd'hui par l'ordinateur.

On pourrait alors, dès le temps de paix, effectuer des calculs englobant dans une formulation unique, la multitude de paramètres caractérisant par exemple la logistique des adversaires aux prises sur un même Théâtre de guerre.

Enfin, la maîtrise d'un tel outil, mis au point pendant les périodes de paix, devrait se révéler extrêmement utile en situation de conflit, pour suivre en temps réel le déroulement des opérations et permettre au Commandement intéressé de prendre à tout instant les meilleurs décisions, concernant notamment la répartition et l'économie judicieuse de ses forces, en dépit du rythme d'évolution de la bataille qui peut se révéler brutalement sans commune mesure avec celui des conflits précédents.

2. DESCRIPTION DU MODELE

2.1 Généralités

Le modèle doit pouvoir s'adapter à différents types de guerre, actuels ou futurs, tels qu'on peut les imaginer aujourd'hui, compte tenu du rapport des forces en présence, des doctrines en vigueur et de leur évolution possible, des progrès technologiques concernant notamment les moyens d'acheminer la logistique des armées en campagne. Les véhicules de transport classiques (du type camion) ont fait l'objet de progrès assez lents, mais continus, depuis la seconde guerre mondiale. Leur perfectionnement est un des soucis constants des armées les plus importantes. De nombreuses variantes du véhicule logistique à roue ont été étudiées, en vue notamment de l'affranchir de ses servitudes d'itinéraires qui limitent sa mobilité aux routes ou, à la rigueur, aux chemins, ainsi que des difficultés de franchissement des coupures. De plus, les logisticiens songent évidemment à mettre en oeuvre d'autres ressources que les véhicules à roues ou à chenilles, tels que: avions gros porteurs, hélicoptères lourds, et, ultérieurement, véhicules à coussin d'air, sans omettre les possibilités offertes par la pose d'oléoducs permettant sur les arrières des grandes unités, une progression continue des sources de carburant, dont le flux indispensable constitue sans doute le chapitre le plus important pour la logistique des unités blindées et mécanisées.

L'ensemble des paramètres introduits dans le modèle répond à ces multiples préoccupations.

2.2 Présentation du problème

A l'instant $t = 0$, (Cf. fig. 1) Rouge prend l'offensive contre Bleu sur l'axe $0x$. Sa base de ravitaillement principal est en R_0, que nous prenons comme origine des coordonnées. Ses éléments de tête franchissent la ligne de contact L_1 à l'instant initial $t = 0$. La base de Bleu est en B_0 et l'on pose $R_0B_0 = D$, distance exprimée en kilometre. Nous

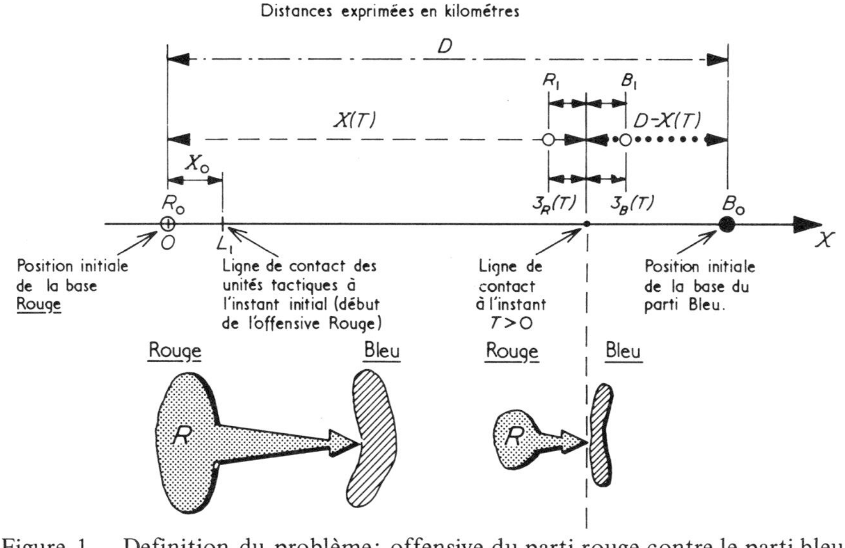

Figure 1 Definition du problème: offensive du parti rouge contre le parti bleu

considérons en fonction de la variable courante t la position $x(t)$ des éléments avancés de Rouge au cours de leur avance. Nous supposons qu'à chaque instant $T > 0$ le ravitaillement s'effectue en R_1 et B_1 pour les unités tactiques Rouges et Bleues respectivement, c'est-à-dire à une certaine distance de la ligne de contact commune des éléments avancés des deux camps.

2.3 Définition des principaux paramètres (à chaque instant fixé $T > 0$)

–Forces Rouges

$mox\ (T)$ = Nombre d'unités tactiques rouges qui peuvent être normalement ravitaillées au point R_1 de coordonnée: $X(T) - z_R(T)$

Ce chiffre est donc le nombre maximum d'unités Rouges en état de combattre au point x, compte tenu de la seule contrainte logistique. Bien entendu *d'autres contraintes,* stratégiques, tactiques, ou même politiques *peuvent diminuer ce nombre, mais elles ne peuvent l'augmenter. "On peut faire moins bien, mais on ne peut faire mieux".*

$M_0(T)$ = Nombre total d'unités logistiques en route au moment où les unités tactiques de pointe arrivent en $x(T)$. *Nous avons supposé un système "banalisé" dont les véhicules, suivant le principe d'une noria continue, se déplacent individuellement ou par petites rames.* (Par unités logistiques on entend donc un élément appartenant à un ensemble de moyens discontinus, par exemple des camions, des hélicoptères, des avions, etc.). A la limite, une unité logistique peut être réduite à un véhicule isolé.

p = Poids total d'un véhicule logistique à pleine charge.

On a:

$$p = p_0 + C_M$$

avec:

p_0 = poids du véhicule vide, avec l'équipage, l'eau et le lubrifiant du moteur, l'outillage de bord et les pièces de rechange.

C_M = Charge utile maximale définie comme suit:

C_M = Charge utile (telle qu'elle est définie de façon classique) + *le plein des réservoirs de carburant* et éventuellement celui de réservoirs supplémentaires ou de nourrices de réserve.

Posons:

$$\varphi = \frac{C_M}{p} = \frac{\text{Charge utile maximale}}{\text{Poids total maximal à pleine charge}}$$

(Remarquons que φ est généralement une fonction croissante de p).

P_t = Puissance du véhicule exprimée en chevaux à l'instant t, avec $P_t \leqslant P_\eta$ où P_η désigne la puissance maximale sur l'arbre.

On a:

$$P_t = f\,[p_t, V(t)]$$

formule dans laquelle:

p_t = Poids total du véhicule à l'instant t et

$V(t)$ = Vitesse du véhicule au même instant t.

Pour des véhicules logistiques à vitesse modérée, on pourra souvent négliger la résistance de l'air, en air calme, ainsi que celle du vent, de la résistance aux viragees, etc., de sorte que l'on a alors l'expression simple:

$$P_t = \left(\frac{n + \rho}{100\,\mu}\right) \frac{p_t \cdot V(t)}{75}$$

formule dans laquelle:

η = Coefficient qui dépend du profil de la route, et peut être considéré comme proportionnel à la pente exprimée en pourcentage lorsque le véhicule gravit des côtes.

ρ = Coefficient dépendent de la résistance de roulement. Sa valeur varie avec l'état de la chaussée, la qualité des bandages, la vitesse du véhicule. Toutefois ρ varie peu autour d'une valeur moyenne et l'on peut admettre en première approximation: $\rho = 3 \pm 0{,}5$ en observant que ρ peut parfois s'abaisser jusque'à 2,2.

μ = Rendement mécanique de la transmission. (On prend souvent $\mu = \frac{2}{3}$ à 0.7).

Avec cette dernière valeur on peut donc adopter la formule suivante:

$$P_t = \left(\frac{\rho + n}{18\,900}\right) p_t \cdot V(t)$$

dans laquelle P_t est exprimée en chevaux, p_t en kilogrammes, $V(t)$ en kilomètres par heure.

Si l'on prenait $\mu = \frac{2}{3}$, la formule deviendrait:

$$P_t = \left(\frac{\rho + n}{18\,000}\right) p_t \cdot V(t)$$

On pourrait donc adopter de façon générale:

$$\boxed{P_t = \left[\frac{\rho + n}{(185 \pm 5) \times 10^2}\right] \cdot p_t \cdot V(t)}$$

Soit maintenant:

C_s = Consommation spécifique du moteur du véhicule logistique, exprimée en kilogramme/ch/heure.

Le taux de consommation à l'instant t d'un véhicule de poids total p et roulant à la vitesse $V(t)$ est:

$$P_t \cdot C_s = C_s f\,[p_t, V(t)]$$

Pendant un intervalle de temps T fixé, la consommation est alors:

$$C_s \int_0^T f\,[p_t, V(t)]\,dt$$

Si l'on adopte pour P_t la formule approchée du paragraphe précédent, on a pour une vitesse constante V soutenue pendant l'intervalle $[0, T]$:

$$\left(\frac{\rho + n}{18\,500}\right) VC_s \int_0^T p_t\,dt$$

En supposant que pendant cet intervalle $[0, T]$, la variation du poids total du véhicule due à la seule consommation de carburant est négligeable, on a évidemment:

$$C(T) = (\text{Consommation pendant le temps } T) = \frac{\rho + n}{18\,500} VC_s p_{to} T,$$

avec:

p_{to} = Poids total du véhicule à l'instant initial.

Désignons par:

C_{ms} = Consommation moyenne par véhicule et par heure de route.

C_{ms_1} = Consommation moyenne par véhicule et par heure de route, *lorsque le véhicule est chargé.*

C_{ms_2} = Consommation moyenne par véhicule et par heure de route, *lorsque le véhicule n'est pas chargé.*

Sur un trajet aller et retour, dans lequel l'aller s'effectue chargé et le retour à vide (sauf le carburant nécessaire pour la route), et en supposant les itinéraires aller et retour sensiblement identiques, on peut prendre:

$$C_{ms} = \frac{C_{ms_1} + C_{ms_2}}{2} = \left(\frac{\rho + n}{18\,500}\right) C_s \left(\frac{V_{m_1}\, p + V_{m_2}\, p_o}{2}\right)$$

avec:

V_{m_1} = Vitesse moyenne au cours du trajet aller et

V_{m_2} = Vitesse moyenne au cours du trajet retour.

Posons:

T_m = Total des temps morts nécessaires pour les opérations de chargement et de déchargement.

$\mathcal{H}_1$ = Fraction du temps d'un trajet aller (à l'exclusion des temps morts) pendant lequel chaque unité logistique Rouge (ou véhicule) est effectivement en mouvement. Ce coefficient est introduit pour tenir compte des contraintes sur la circulation, telles l'interdiction de rouler de jour si le parti Rouge ne possède pas une maïtrise aérienne suffisante . . .

$\mathcal{H}_2$ = Fraction du temps d'un trajet retour, définie de façon identique à $\mathcal{H}_1$

$\tau_m(T)$ = Besoin logistique de chacune des unités tactiques Rouges parvenues en x. Ce besoin s'exprime en poids par unité de temps; par exemple en kilogramme par heure.

Soit $\gamma(t)$ la vitesse d'avance des points de ravitaillement des unités tactiques (points R_1 et B_1 de la figure 1 pour Rouges et Bleus respectivement). Cette vitesse sera supposée en général peu différente dans chaque zone de combat de la vitesse des unités tactiques au contact.

En définissant:

$X(t)$ = Distance des éléments au contact à l'instant $T > 0$, en prenant pour origine commune des coordonnées, la base logistique du parti Rouge, l'instant initial $t = 0$, (celui du début de l'offensive Rouge), on a:

$$X(T) = X_o + \int_0^T \gamma \quad (t)\; dt\,, \qquad (T > 0)$$ ou:

X_0 = Distance séparant la ligne de contact des unités tactiques adverses, du dépôt des unités logistiques du parti Rouge, à l'instant initial de l'offensive de ce dernier. Les élongations logistiques, ou distances séparant les dépôts des points de livraison ont alors pour valeur:

$$X(T) - z_R(T) = \int_0^T \gamma \quad (t)\; dt - \left[z_R(T) - X_o \right]$$ (pour le parti Rouge)

et:

$$D - X(T) = (D - X_0) + \left[z_R(T) - z_B(T) \right] - \int_0^T \gamma \quad (k)\, dt$$ (pour le parti Bleu)

expressions dans lesquelles:

$z_R(T)$ et $z_B(T)$ sont les distances séparant la ligne de contact des points de ravitaillement R_1 et B_1 pour Rouge et Bleu respectivement.

Toutes les distances $X(T)$, X_0, $z_R(T)$, $z_B(T)$, D sont exprimées en kilomètres.

2.4 Principales formules

En partant de la notion d'équilibre logistique, c'est-à-dire en supposant qu'à chaque instant T les unités tactiques reçoivent la quantité de produits logistiques correspondant à leurs besoins, on peut alors établir des relations montrant explicitement comment sont liés:

- les effectifs des unités tactiques,
- le nombre et les performances des véhicules logistiques,
- les élongations et la qualité des itinéraires, etc.

Nous nous placons donc en régime "stationnaire". Il est évident que ce régime "stationnaire" ne s'instaure qu'au bout d'un certain nombre d'heures à partir du début de l'offensive, lorsque les Grandes Unités Tactiques auront épuisé leurs réserves de carburant et consommé leurs dotations initiales en munitions.

Suivant ces idées de base, on a établi les formules suivantes:

Nombre maximal d'unités tactiques qui peuvent être soutenues au point de vue logistique.

$$m_{ox}(T) = \frac{A_1 \int_0^T \gamma \quad (t)\, dt + B_1}{\int_0^T \gamma \quad (t)\, dt + C_1}$$

avec:

$$\begin{cases} A_1 = -\dfrac{M_0(T)}{C_m(T)} \left(\dfrac{\mathcal{H}_1 \mathcal{H}_2 V_{m_1} V_{m_2}}{\mathcal{H}_1 V_{m_1} + \mathcal{H}_2 V_{m_2}} \right) \left[C_s \left(\dfrac{\rho + n}{18\,500} \right) (2 - \varphi)\, p \right] \\ B_1 = \dfrac{M_0(T)}{C\, m(T)} \left(\dfrac{\mathcal{H}_1 \mathcal{H}_2 V_{m_1} V_{m_2}}{\mathcal{H}_1 V_{m_4} + \mathcal{H}_2 V_{m_2}} \right) \left[\left(z_R - X_0 \right) C_s \left(\dfrac{\rho + n}{18\,500} \right) (2 - \varphi) + \varphi \right] p \\ C_1 = T_m \left(\dfrac{\mathcal{H}_1 \mathcal{H}_2 V_{m_1} V_{m_2}}{\mathcal{H}_1 V_{m_1} + \mathcal{H}_2 V_{m_2}} \right) - \left(z_R - X_0 \right) \end{cases} \quad (1)$$

Cas particulier de la formule (1)

Si l'on a: $z_R = X_0$ (c'est-à-dire que la distance séparant les points de ravitaillement des unités tactiques de la ligne de contact sur laquelle elles sont parvenues, reste sensiblement constante depuis le début de l'offensive).

et:

$$\begin{cases} \mathcal{H}_1 = \mathcal{H}_2 = \mathcal{H}_m \\ \text{avec:} \\ V_{m_1} = V_{m_2} = V_{m_1} \end{cases}$$

(les fractions du temps où les véhicules sont effectivement en mouvement, sont pratiquement les mêmes sur les itinéraires aller et retour, de même que les vitesses aller et retour);

et si l'on pose de plus:

$$C_{ms} = C_s \left(\frac{\rho + n}{18\,500} \right) \left(\frac{V_{m_1} p + V_{m_2} p_0}{2} \right)$$

= consommation horaire moyenne par véhicule, on obtient alors:

$$A_1 = -\frac{M_0(T)}{C_m(T)} \left(\mathcal{H}_m C_{ms} \right)$$

$$B_1 = \frac{M_0(T)}{C_m(T)} \left(\frac{\mathcal{H}_m V_m C_m}{2} \right)$$

$$C_1 = \frac{\mathcal{H}_m V_m T_m}{2}$$

de sorte que la formule (1) devient alors une formule (1 bis) de même forme, mais dans laquelle les valeurs des coefficients sont calculées au moyen des expressions A_1, B_1, et C_1 données ci-dessus.

$$\text{(1 bis)} \qquad m_{ox}(T) = \frac{A_1 \int_0^T \gamma\ (t)\,dt + B_1}{\int_0^T \gamma\ (t)\,dt + C_1}$$

La formule 1 bis
se réduit alors la celle présentée au symposium de Munich, lorsqu'on y a remplacé $X(T)$ par: $\int_0^T \gamma\ (t)\,dt$ et en supposant $M_0\ (T)$ et $C_m\ (T)$ constants pendant l'intervalle des temps $[O, T]$.

L'intérêt de â formule (1) qui comporte au total 15 paramètres, est de relier des groupes de facteurs relevant de domaines très divers:

Besoins des unités tactiques	$m_{ox}(T)$	est la borne supérieure des unités tactiques qui peuvent être effectivement soutenues au point de vue carburant, munitions, vivres, etc.
	$C_m(T)$	exprime le besoin logistique global moyen de chaque combattant de l'effectif m_{ox}.
Qualités intrinsèques du parc logistique	M_o(T)	est la richesse du parc logistique (exprimée en nombre de véhicules à l'instant T).
	c_s, ρ, φ, P	Caractérisent un ensemble de qualités intrinsèques qui sont parmi les plus importantes pour chaque véhicule logistique.
Facteurs résultant eux-mêmes de processus complexes	π	dépend du profil et de la qualité des itinéraires.
	$\mathcal{H}_1, \mathcal{H}_2$	dépendent de la qualité des itinéraires, de l'organisation de la circulation, des actions possibles de l'ennemi sur ces itinéraires.
	V_{m_1} et V_{m_2}	dépendent à la fois des qualités intrinsèques des véhicules, de l'organisation de la circulation routière, de la discipline de route des vehicules, de la qualité des itinéraires, de l'action possible de l'ennemi sur les itinéraires et sur les véhicules eux-mêmes.
	T_m	dépend de l'organisation des voies d'accès aux dépôts et aux points de livraisons, de la circulation routière, des moyens de manutention, de l'action possible de l'ennemi sur les dépôts et les bases.
	$z_r(T)$ et X_0	dépendent des conditions tactiques et des possibilités adverses sur les dépôts pour la phase initiale et sur les points de livraisons pour les phases ultérieures.

Distance de consomption logistique X_C:

$$(2) \qquad X_c = \frac{18\,500\,\varphi}{c_s\,(\rho + \mathrm{n})\,(2 - \varphi)} + \left[z_R\,(T) - X_0 \right]$$

formule dans laquelle X_C est la distance maximale à laquelle des unités tactiques peuvent être soutenues par des véhicules de caractéristiques intrinsèques: cs, φ, ρ, sur un itinéraire caractérisé par n, avec une situation tactique à laquelle est associée la distance $[\, z_R - x_0 \,]$, *sans création de dépôts intermédiaires.*

En d'autres termes, chaque véhicule consomme sur l'itinéraire X_C (aller + retour) l'équivalent de sa charge utile maximale en carburant.

Densité linéaire maximale D (T) des unités tactiques Rouges qui peuvent être effectivement soutenues au point de vue logistique, (densité "frontale").

Soit $y(T)$, la largeur de la zone d'action de l'ensemble des grandes unités tactiques Rouges considérées à l'instant $T > 0$, (Cf. fig. 2).

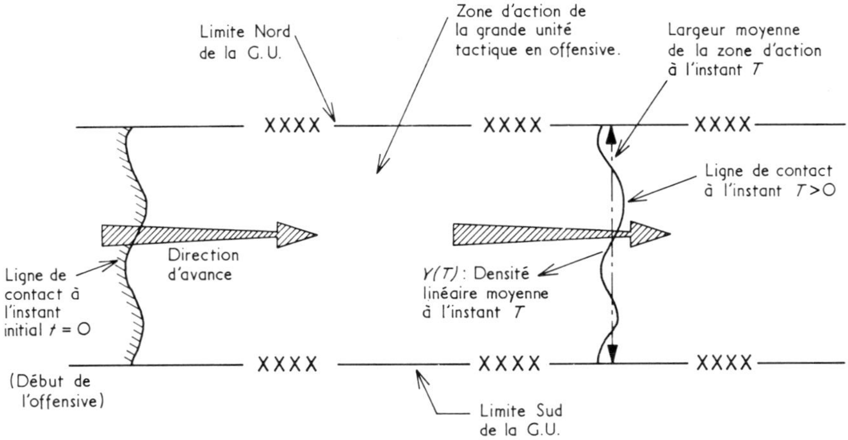

Figure 2 Densité linéaire moyenne (frontale)

On a alors:

$$(3) \qquad D\,(T) = \frac{A_1 \int_0^T \gamma\,(t)\,dt + B_1}{y\,(T) \left[\int_0^T \gamma\,(t)\,dt + C_1 \right]}$$

Formule dans laquelle $D(T)$ est la densité linéaire moyenne maximale réalisable à l'instant T, si l'on tient compte des seules contraintes logistiques, tandis que les coefficients A_1, B_1, C_1 ont les mêmes valeurs que dans les formules (1) ou (1 bis). Les valeurs $D(T)$ s'expriment en effectifs par kilomètre.

Nombre moyen $\Delta(T)$ de véhicules logistiques en mouvement à un instant T fixé

Supposons que dans la zone des grandes unités tactiques Rouges considérées, il existe h circuits indépendants (par circuit on entend un itinériare aller et un itinéraire retour différents), on à défaut, h circuits organisés de telle sorte que leurs trafics respectifs n'interfèrent pas sensiblement les uns avec les autres, à l'exception de quelques portions limitées de trajets.

Désignons par $\Delta(T)$ le nombre moyen de véhicules logistiques par kilomètre d'itinéraire "montant" ou "descendant" à l'instant T.

On a: (4)

$$\Delta(T) = \frac{M_0(T)}{2h} \cdot \frac{V_{m_4} + V_{m_2}}{V_{m_1} V_{m_2} T_m + \left(\frac{\mathcal{H}_1 V_{m_1} + \mathcal{H}_2 V_{m_2}}{\mathcal{H}_1 \mathcal{H}_2}\right)\left\{\int_0^T \gamma(t)\,dt - \left[z_R(r) - X_0\right]\right\}}$$

ou $\Delta(T)$ est la densité moyenne du flux de véhicules logistiques, exprimée en nombre moyen de véhicules par kilomètre.

Nombre $M_O(T)$ de véhicules logistiques nécessaires pour soutenir un effectif tactique $m_{ox}(T)$, à l'instant T

(5)

$$M_0(T) = m_{ox}(T) \quad \frac{\int_0^T \gamma(t)\,dt + C_1'}{A_4' \int_0^T \gamma(t)\,dt + B_1'}$$

avec:

$$\begin{cases} A_1' = -\left(\frac{1}{C_m}\right)\left(\frac{\mathcal{H}_1 \mathcal{H}_2 V_{m_4} V_{m_2}}{\mathcal{H}_1 V_{m_1} + \mathcal{H}_2 V_{m_2}}\right)\left[C_s \left(\frac{\rho + n}{18\,500}\right)\left(2 - \varphi\right) p\right] \\ B_1' = \left(\frac{1}{C_m}\right)\left(\frac{\mathcal{H}_2 \mathcal{H}_2 V_{m_1} V_{m_2}}{\mathcal{H}_1 V_{m_1} + \mathcal{H}_2 V_{m_2}}\right)\left[\left(z_R - X_0\right) C_s \left(\frac{\rho + n}{18\,500}\right)\left(2 - \varphi\right) + \varphi\right] p \\ C_1' = T_m \left(\frac{\mathcal{H}_2 \mathcal{H}_2 V_{m_1} V_{m_2}}{\mathcal{H}_1 V_{m_1} + \mathcal{H}_2 V_{m^2}}\right) - \left(z_R - X_0\right) \end{cases}$$

3. APPLICATIONS DES FORMULES

3.1 Valeurs des paramètres

Il a fallu assigner un certain nombre de valeurs plausibles aux paramètres figurant dans les formules en fonction de différentes données stratégiques, géographiques, logistiques, technologiques.

Les besoins logistiques par homme et par heure sont approximativement de 3 kg, dans lesquels les carburants entrent dans une proportion variant de 1/2 à 3/4 dans les phases d'exploitation.

Le volume des effectifs terrestres varie de 750 000 à 2 500 000 hommes. Les élongations logistiques extrêmes pour le théâtre européen occidental sont, à partir de la ligne de l'Oder, de 1000 km pour les principaux centres industriels, de 2000 km pour l'ensemble de l'Europe moins la péninsule ibérique et de 3000 km avec ladite péninsule.

Sur le théâtre chinois à partir de l'Amour on a des élongations variant de 2000 à 4500 km.

Les grandes unités blindées soviétiques sont prévues pour abattre des étapes quotidiennes de 100 km avec, pour le théâtre occidental, des objectifs à 10 jours, 3 semaines et 1 mois pour les 3 zones définies plus haut.

Ceci implique un système logistique souple, s'appuyant sur des bases situées en territoire "ami", puissamment protégées, partiellement enterrées et soigneusement camouflées, alimentant un flux continu de petites rames et même de véhicules circulant individuellement.

Bien entendu, un réseau d'oléoducs poussés au plus près des unités en marche contribuerait encore à soulager les véhicules logistiques et à maintenir le rythme de l'avance.

Quant aux performances des véhicules logistiques, nous avons considéré deux types de transporteurs:

. un camion classique de 12 T à pleine charge dont 3,6 T de charge utile, équipé d'un moteur de 100 à 150 CV, consommant 160 à 165 g de fuel par cheval et par heure, avec un rendement de transmission voisin de 0,7 et un coefficient de roulement de l'ordre de 2,5. Ce véhicule a deux vitesses moyennes effectives de 50 et de 25 km/h;

. un camion de 100 T à pleine charge dont 60 T de charge utile, équipé d'un diesel de qualités spécifiques analogue au précédent, soutenant des vitesses moyennes de 25 km/heure pendant de longues périodes (équipage double), appartenant à la classe des Dumper utilisés sur les grands chantiers. Ce véhicule serait doté d'un $\varphi = \frac{\text{charge utile}}{\text{poids total à pleine charge}}$ très élevé.

3.2 Résultats numériques: réseaux de courbes

Six réseaux de courbes ont été tracés. Les quatre premiers, calculés au moyen de la formule (1) donnent les valeurs de m_{OX} (effectifs tactiques qui peuvent être soutenus convenablement au point de vue logistique) en fonction du temps T. Chacun de ces quatre réseaux comporte dix courbes, correspondant aux valeurs suivantes du parc de véhicules logistiques:

$$M_O(T) = 10^4, 2 \times 10^4, 3 \times 10^4, 4 \times 10^4, 5 \times 10^4, 6 \times 10^4, 7 \times 10^4, 8 \times 10^4, 9 \times 10^4 \text{ et } 10^5.$$

Pour chacune des courbes, la vitesse d'avance moyenne des unités tactiques (ligne de contact) a été prise égale à 4 km/heure, soit 96 km/jour ≈ 100 km/jour.

– Le premier réseau a été calculé avec les valeurs suivantes:

(i) besoins logistiques spécifiques des unités tactiques:
$Cm(T)$ = constante = 3 kg/homme/heure (72 kg/homme/jour);

(ii) données sur le véhicule logistique:
ρ = 2,4 (bon roulement),
n = 0,6 (profil peu accidenté de l'itinéraire),
C_s = 0,162 kg/ch/heure (consommation excellente pour un moteur diesel d'aujourd'hui).[1]
φ = 0,3,
p = 12 000 kg (véhicule de la classe 10-12 tonnes à pleine charge),
C_M = φp = 3 600 kg (charge utile normale–bien qu'un peu faible–pour un véhicule de cette catégorie),
Vm_1 = 40 km/heure (vitesse moyenne soutenue sur le trajet aller, c'est-à-dire véhicule pratiquement à pleine charge),
Vm_2 = 60 km/heure, (vitesse moyenne soutenue sur le trajet de retour, c'est-à-dire véhicule vide).

(iii) données sur l'organisation et les possibilités du trafic logistique:
Tm = 4 h (total des temps morts de chargement, déchargement, ainsi que les délais d'attente au voisinage des dépôts et des points de livraisons),

$\mathcal{H}_1$ = $\mathcal{H}_2$ = 3/4 ce qui signifie qu'en dehors des temps morts, chaque véhicule roule les 3/4 du temps: par exemple, sur un trajet demandant au total 24 heures, on peut avoir 4 heures de délais de chargement, déchargement et d'attente, 15 heures de route effective, 5 heures pour les repos des conducteurs et l'entretien minimum des véhicules, (ce dernier chiffre nous a semblé proche des possibilités limites des équipages, surtout lorsque l'effort se prolonge plusieurs jours consécutifs).

(iv) données sur l'organisation du dépôt initial et des points de ravitaillement:
$[z_R(T) - X_0] = 0$, c'est-à-dire que la distance séparant les dépôts et les points de livraison de la ligne de contact, reste sensiblement constante tout au long de l'offensive, pour des raisons de sécurité par exemple.

–Le second réseau a été calculé avec les mêmes valeurs des paramètres que le précédent, sauf pour les paramètres suivants:

n = 3,6 (profil d'itinéraire accidenté, équivalent en moyenne au franchissement de rampes à 3 ou 4%),
$V_{m_1} = V_{m_2}$ = 25 km/h (en plus du profil accidenté des itinéraires, mauvais état de

[1] Consommation possible pour un diesel normalement évolué dans un avenir assez proche.

la chaussée, attente aux abords de coupures plus ou moins mal rétablies etc.),

$\mathcal{H}_1 = \mathcal{H}_2$ = 1/4, (en raison du climat d'insécurité sur les lignes logistiques, seules certaines fractions du temps sont effectivement utilisées pour rouler). Ce climat d'insécurité peut résulter par exemple d'attaques aériennes, de celles de commandos aéroportés ou d'éléments de gnérillas, des bombardements, sur les carrefours ou les points de passage obligés au moyen de missiles sol-sol à têtes classiques, chimiques ou nucléaires.

–Le réseau n° 3 reprend les mêmes valeurs que pour le second réseau, sauf que le véhicle logistique est un très gros porteur avec: $\varphi = 0{,}6$, $p = 10^5$ kg, de sorte que sa charge utile maximum $C_M = 60$ tonnes.

–Le réseau n° 4 reprend les mêmes valeurs que pour le réseau 3, sauf que les véhicules logistiques peuvent rouler en permanence, c'est-à-dire que:

$$\mathcal{H}_1 = \mathcal{H}_2 = 1$$

Ces conditions peuvent être réalisées, par exemple avec un double équipage à bord de chaque gros porteur, dans une ambiance de faible insécurité sur les lignes logistiques ainsi qu'avec une excellente organisation de la circulation.

Avec ces quatre familles de cas très différents, notamment pour ce qui concerne le type de véhicule logistique, on peut espérer "encadrer" certains aspects caractéristiques du soutien logistique de grandes armées blindées et mécanisées, lancées à fond de train pour exploiter les effets encore inconnus des bordées nucléaires tirées au profit des forces terrestres.

–Les réseaux n° 5 et n° 6 calculés avec la formule (4) donnent, en fonction du temps, la densité linéaire $\Delta(T)$ de véhicules logistiques en mouvement. Chacun de ces réseaux comporte 9 courbes correspondant respectivement aux valeurs suivantes des paramètres:

$$\begin{cases} \text{3 valeurs pour } \dfrac{M_0}{h}\text{, soit } 5 \times 10^3,\ 10^3,\ \dfrac{10^4}{15}\text{, multipliées par:} \\ \text{3 valeurs pour } \mathcal{H}_1 = \mathcal{H}_2 = \mathcal{H}_m = 1, \dfrac{3}{4}, \dfrac{1}{4}. \end{cases}$$

Le réseau n° 5 correspond à:

$$V_{m_1} = V_{m_2} = 50 \text{ km/heure.}$$

Le réseau n^0 6 correspond à:

$$V_{m_1} = V_{m_2} = 25 \text{ km/heure.}$$

Les valeurs des paramètres qui ont été prises identiques pour les deux réseaux sont:

$$\begin{cases} T_m = 4 \text{ heures} \\ \gamma \quad (t) = \text{constante} = 4 \text{ km/heure (ou 96 km/jour} \approx 100\text{km/jour)} \\ [\ z_R(T) - X_0] = 0 \end{cases}$$

L'intérêt des réseaux 5 et 6, qui mettent en évidence la densité du flux logistique, est de donner un premier élément quantitatif, pour apprécier la vulnérabilité d'un courant d'importance vitale sur les arrières des armées.

En effet, l'attaque du courant lui-même ou de certains points clés des itinéraires permettant de bloquer ces derniers, au moins temporairement, sera d'autant plus efficace que la densité du flux sera plus élevée.

3.3 Utilisation des réseaux de courbes et interprétation des résultats

La définition d'une opération stratégique pourrait se résumer très schématiquement, de la façon suivante:

3.3.1 En situation offensive:

(i) *sur un objectif stratégique donné*, par exemple un objectif géographique tel qu'un fleuve, une portion de rivage, un complexe industriel ou démographique, une ligne de montagnes . . . . .);

(ii) *dans un délai donné*, à compter du déclenchement de l'offensive;

(iii) *amener à pied d'oeuvre les forces suffisantes,* permettant à ces dernières d'accomplir à partir de l'instant de leur mainmise sur l'objectif, la séquence des missions qui leur ont été fixées dans le cadre général du plan d'opération (par exemple achever la destruction des forces ennemies en retraite ou encerclées, être en mesure de stopper les contre-offensives adverses, exploiter les ressources militaires ou autres offertes par l'objectif qui a été occupé).

En fait, un plan de campagne est, dans sa phase initiale, l'assemblage et la coordination dans l'espace et le temps, d'opérations stratégiques, telles que nous les avons définies schématiquement.

A l'échelon d'une opération, si l'on passe de l'analyse qualitative à des expressions quantitatives, on obtient deux familles de grandeurs:

L'objectif et les délais impartis pour la conquête de celui-ci, fixent, le rythme de l'opération, c'est-à-dire la fonction γ (t), suivant notre terminologie, ou tout au moins, un ordre de grandeur de la valeur moyenne de cette fonction.

L'expression quantifiée des forces "suffisantes" est plus délicate. Elle présuppose normalement une hypothèse sur le niveau des forces résiduelles adverses, une fois l'objectif conquis. Cette évaluation est le coeur même de la tâche délicate des Etats-Majors. Dans le cadre de notre analyse nous supposons, pour simplifier, que les forces seront considérées comme "suffisantes", si le rapport global des forces amies et des forces ennemies, exprimées toutes deux en termes d'effectifs, dépasse une certaine valeur critique R_c.

Cette simplication serait valable si tous les autres facteurs étaient comparables: les dotations, la valeur du commandement et la qualité de l'organisation, les matériels, les liaisons, l'instruction, le moral des troupes, etc.

Cependant, l'expérience historique montre que si les campagnes se prolongent suffisamment, en dépit de la compexité du problème, les Etats-Majors parviennent à définir des barèmes semi-empiriques très valables. Ces barèmes indiquent notamment des rapports des forces nécessaires, permettant d'escompter des chances raisonnables de succès, tel que le rapport des effectifs en hommes, celui des chars, de l'artillerie, des avions, etc.

Notre démarche qui, dans le cadre de l'exposé, a essentiellement un but méthodologique, s'inspire donc, malgré son caractère très simplifié, d'un processus réaliste, dont le schéma est donné par la figure 9.

Pour le commandement logistique de l'opération, le problème est alors complètement défini: comment alimenter un effectif $m_{Ox}(T)$ parvenu à l'instant T, en un point distant des bases logistiques d'une longueur $x(T)$, sachant que chaque unité (hommes) de l'effectif consomme en moyenne $C_m(T)$ par unité de temps.

Nous n'avons pas abordé le problème des sources ou de la composition optimale des dépôts. Les formules et les réseaux de courbes correspondants, qui sont donnés à titre illustratif concernent seulement la phase transport. Par exemple le réseau des courbes de la figure 3, permet par lecture directe une estimation du nombre de véhicules logistiques ayant des caractéristiques données dans le premier reseau $(C_s, \rho, \varphi, p, V_m)$ qui seraient nécessaires pour soutenir, à partir d'une base située sur l'ODER une offensive vers l'Ouest mettant en action des armées blindées ou complètement mécanisées, avançant à un rythme moyen de 90 à 100 km par jour.

Sur une ligne de contact passant par Luxembourg, un effectif global de 1 million d'hommes affectés au théâtre de Centre Europe, pourrait être soutenue par environ 45 000 véhicules logistiques du type considéré et dans les conditions indiquées (Point A de la figure).

Sur la frontière Pyrénéenne, un effectif de 1/2 million d'hommes pourrait être soutenu par environ 65 000 véhicules logistiques, toujours dans les conditions précédentes (Point B de la figure), Enfin, à hauteur de Gibraltar ou de Lisbonne, un effectif de 200 000 hommes exigerait un parc de véhicules de l'order de 45 000, comme pour le premier exemple.

Ces chiffres peuvent paraître relativement modestes pour ce qui concerne le nombre de véhicules nécessaires à mettre en oeuvre.

Mais nous avons relevé à dessein l'exemple numérique précédent, afin de mettre en évidence la sensibilité des résultats aux variations des conditions de circulation et d'écoulement du flux logistique et, de souligner par conséquent, l'intérêt que la défense pourrait avoir à perturber ces conditions.

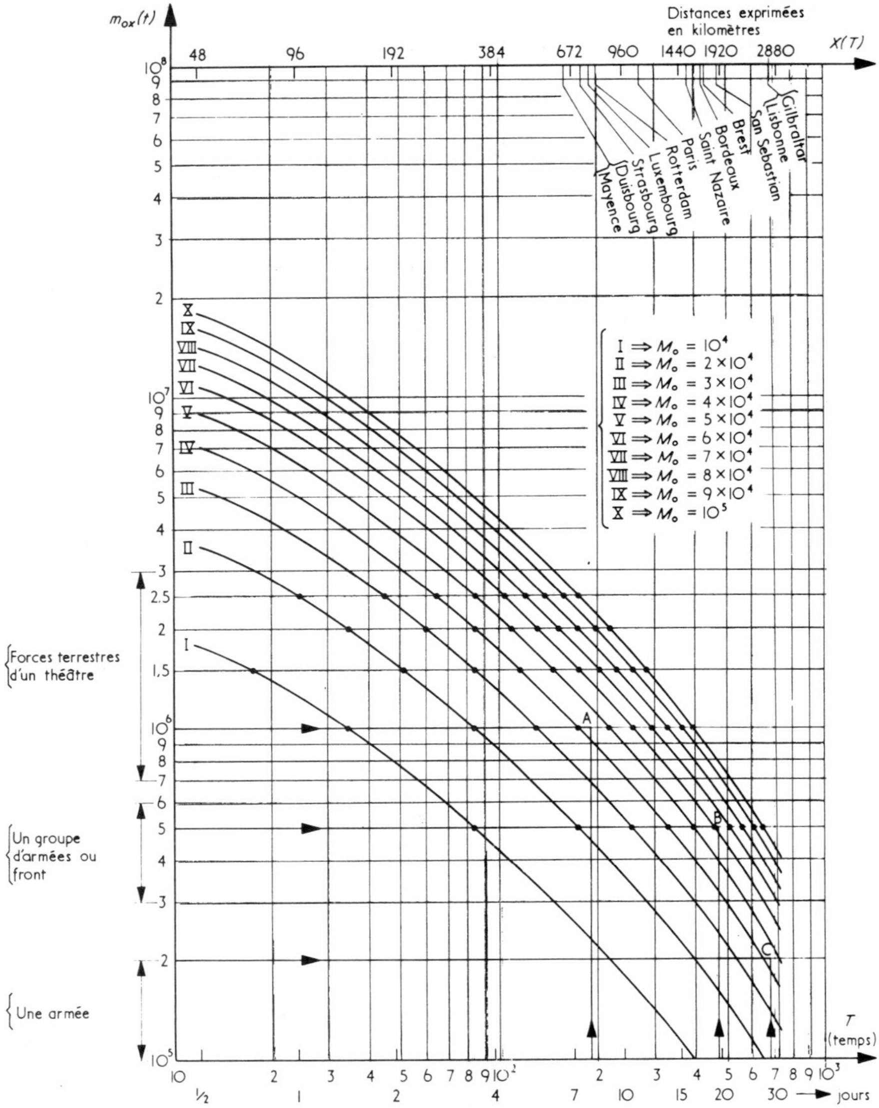

Figure 3 Possibilités de soutien logistique
$V_{m_1} = 40$ km/h, $V_{m_2} = 60$ km/h, $\mathcal{H}_m = ¾$

Le réseau de courbe suivant (fig. 4) correspond au même véhicule logistique, mais avec un profil de route plus difficile (n = 3,6) et surtout à des conditions de circulation sévères ($\mathcal{H}_1 = \mathcal{H}_2 = ¼$ et $V_{m_1} = V_{m_2} = 25$ km/h), qui pourraient résulter à la fois d'un réseau routier endommagé et de fréquentes interventions adverses contre le flux lgoistique (attaques aériennes, attaque de commandos ou de guerilla, etc.) ne permettant aux véhicules que de circuler 1/4 du temps, au lieu des 3/4 comme précédemment, lorsque les contraintes étaient seulement d'ordre interne (fatigue des équipages, entretien des véhicules).

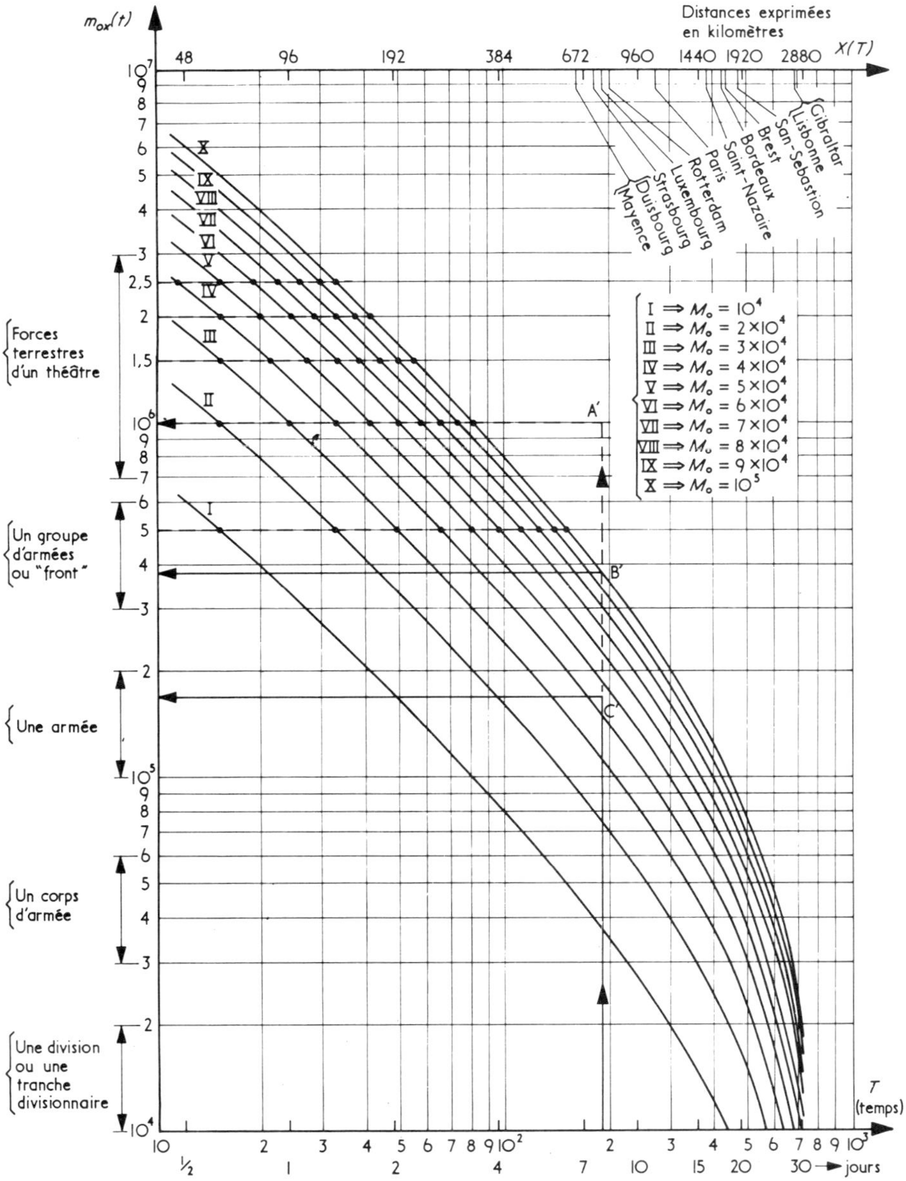

Figure 4 Possibilités de soutien logistique
$V_m = 25$ km/h, $\mathcal{H}_1 = ¼ = \mathcal{H}_2$

A l'examen de ce réseau on constate, toujours en prenant la ligne de contact passant par Luxembourg comme référence, qu'un effectif de 1 million d'hommes ne pourrait être soutenu même avec 100 000 véhicules, loin de là, alors que 45 000 du même type suffisaient dans l'exemple précédent.

Le calcul exact montre que 266 244 véhicules logistiques seraient nécessaires pour soutenir 1 million d'hommes (point A' de la figure 4). Avec 100 000 véhicules, moins de 400 000 hommes pourraient être soutenus (point B'). Enfin, avec 45 000 véhicules, suffisant dans l'exemple précédent, 170 000 hommes seulement pourraient être soutenus au point de vue logistique, à hauteur d'une ligne de contact passant par Luxembourg en partant des dépôts de l'ODER, (point C' de la figure 4). A hauteur des Pyrénées les 65 000 véhicules capables dans l'exemple précédent de soutenir un Groupe d'Armées de 500 000 hommes, ne pourraient plus soutenir qu'un Corps d'Armée de 60 000 hommes. Enfin, à hauteur de Gibraltar ou de Lisbonne, les 45 000 véhicules suffisant dans l'exemple précédent pour soutenir une armée de 200 000 hommes, ne seraient plus en mesure que de soutenir une division d'environ 11 000 hommes.

Les variations importantes que l'on peut observer d'un exemple à l'autre, en les traduisant en termes militaires concrets: Groupe d'Armées, Armée, Corps d'Armée, Division, soulignent l'intérêt qu'un défenseur, dans une optique d'économie de ses forces, aurait à utiliser judicieusement une fraction de ces dernières pour amenuiser le flux logistique considérable que réclament les unités blindées ou complètement mécanisées dont la capacité de choc et la vitesse d'exploitation rendent le freinage direct si difficile, même en supposant que la défense puisse mettre en oeuvre de puissants feux nucléaires tactiques.

Tous les résultats que nous venons de mentionner ont été obtenus avec un même type de véhicule logistique militaire courant à l'heure actuelle, à savoir le camion de 10 à 12 tonnes à pleine charge, et de 3,5 à 4 tonnes de charge utile. Il a paru intéressant de calculer aussi les possibilités que donnerait un véhicule logistique très différent, existant lui aussi à l'heure actuelle, mais se situant dans la gamme des gros porteurs à vocation civile comme nous l'avons déjà indiqué. Dans un but d'exploration, nous avons fait volontairement abstraction des difficultés d'emploi opérationnel possible pour une telle classe de véhicules logistiques.

Avec ce type de véhicule, sur un Théâtre européen, tous les problèmes logistiques que nous avons évoqués (soutien d'un million d'hommes sur la "ligne de contact Luxembourg", soutien d'un demi million sur les Pyrénées et de 200 000 à Lisbonne ou à Gibraltar) pourraient être résolus avec moins de 20 000 véhicules (Points A", B", C" de la figure 5, soit de l'ordre de 17 à 19 000 véhicules) même en ne circulant qu'un quart du temps à la vitesse moyenne modeste de 25 km/heure, c'est-à-dire dans les mêmes conditions difficiles que pour le second des exemples précédents.

Il n'est pas sans intérêt d'observer qu'un théâtre comme celui d'Europe Occidentale compris entre la ligne de l'ODER et l'ATLANTIQUE, est, au point de vue du soutien logistique des armées blindées et mécanisées, à la taille d'un parc d'une vingtaine de milliers de véhicules gros porteurs (dans la classe 60 tonnes de charge utile) à condition que chacun d'eux soit capable de rouler au moins 1/4 du temps à une vitesse moyenne soutenue d'au moins 25 km/heure.

Bien entendu, les possibilités d'un éventuel envahisseur seraient encore augmentées s'il disposait en plus d'un tel parc de véhicules, d'oléoducs capables de prendre en charge les besoins majeurs en carburants ainsi que d'avions gros porteurs relayés par hélicoptères lui permettant de soutenir au moins les pointes blindées avancées, les têtes

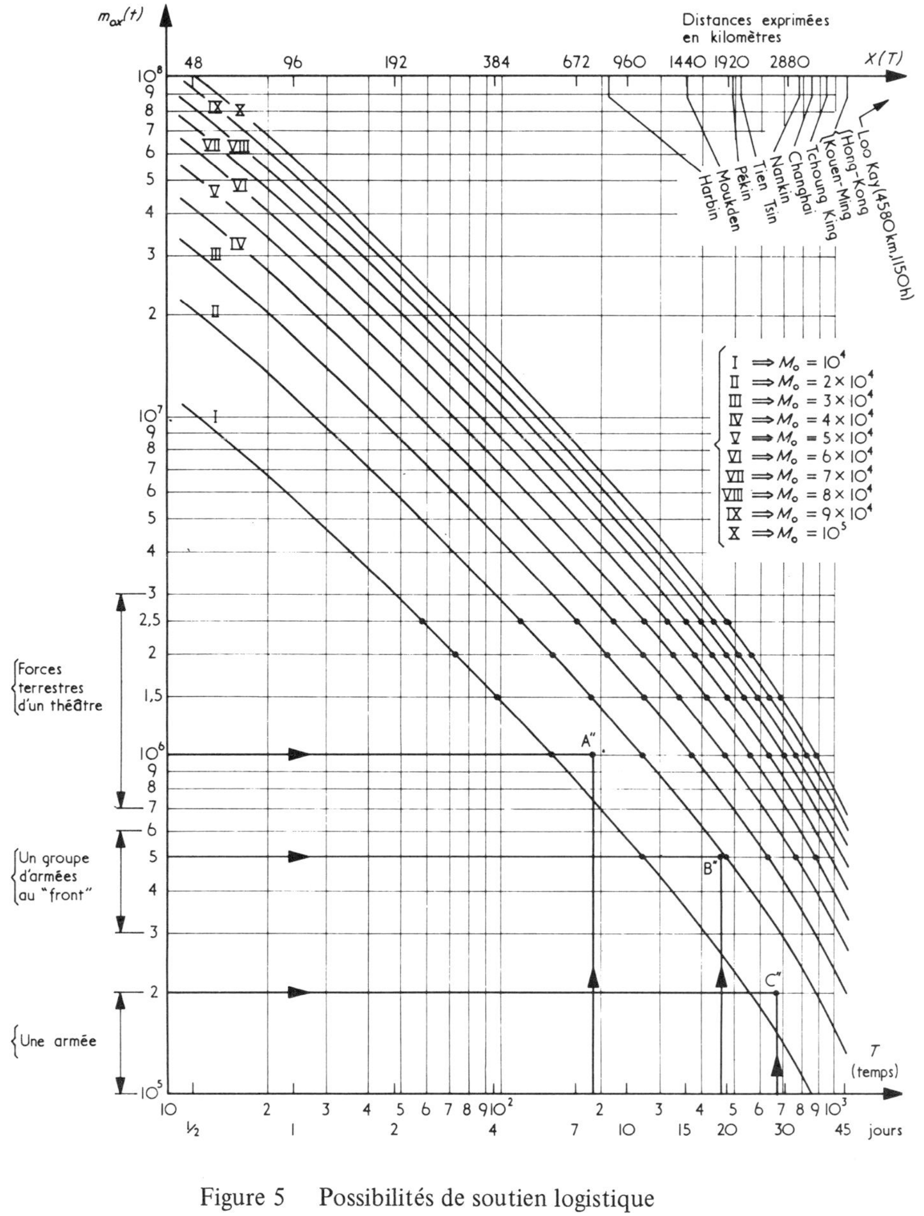

Figure 5 Possibilités de soutien logistique

$$\varphi = 0.6, \mathcal{H}_1 = 1/4 = \mathcal{H}_2$$

de point sur les fleuves, les divisions aéroportées, les unités isolées ou momentanément encerclées, les unités chargées de s'emparer des PC importants et des dispositifs de télécommunications ainsi que des ports ayant une importance logistique, de certaines villes et noeuds de communications, des aérodromes, des bases de missiles et des dépôts ...

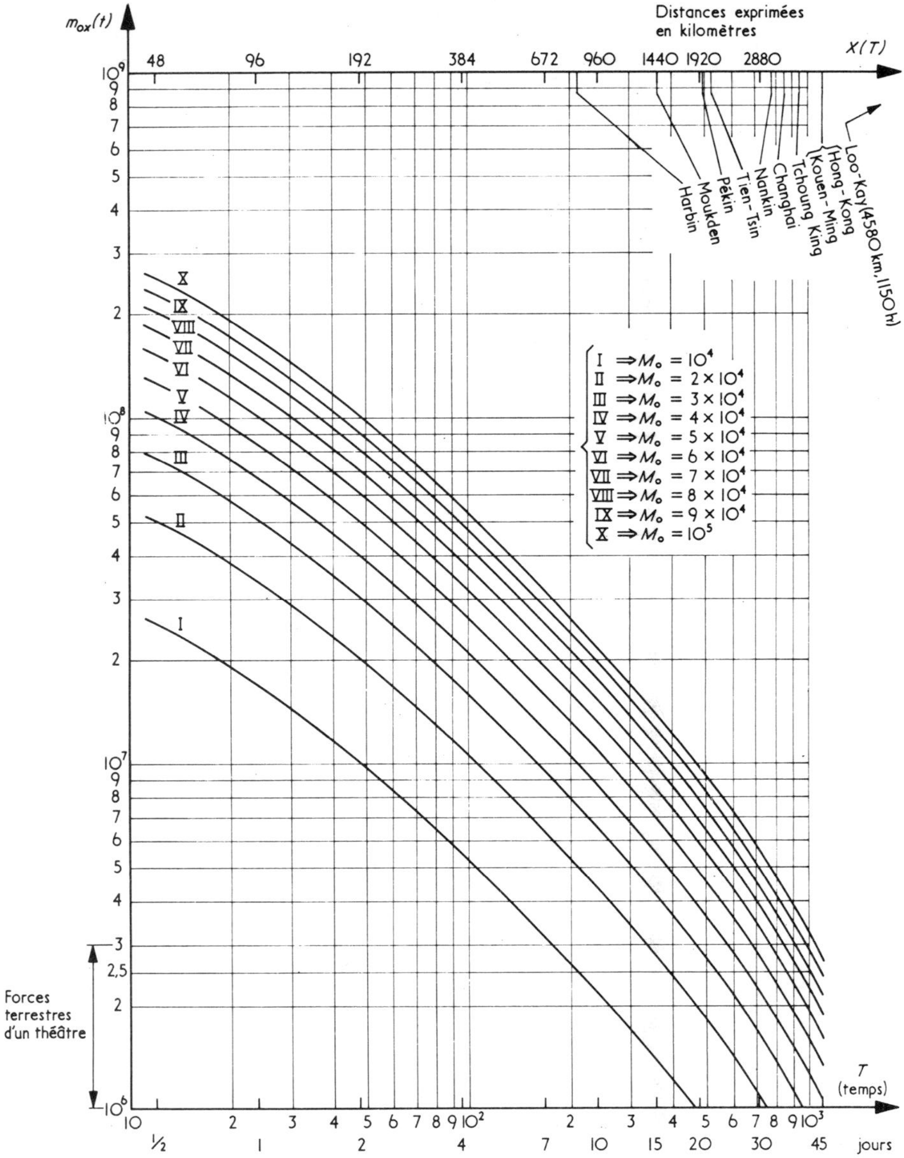

Figure 6 Possibilités de soutien logistique
$\varphi = 0.6$, $\mathcal{H}_m = 1$

Si l'on considère maintenant un théâtre de guerre beaucoup plus vaste que celui d'Europe Occidentale, comme le Théâtre Chinois, compté à partir des bases russes de Sibérie en bordure du fleuve Amour, on constate en examinant les figures 3 et 4, que le soutien logistique d'armées importantes, est pratiquement impossible si l'envahisseur éventuel souhaitait atteindre toute la profondeur du théâtre, en ne disposant que du *véhicule logistique courant, considéré dans les deux premiers exemples*, dès que les conditions de circulation deviendraient un peu difficiles (n = 3,6), même avec un parc de véhicules numériquement énorme, puisque les distances de consomption logistique sont respectivement de l'ordre de 3 360 km à 3 810 km, pour des véhicules de la classe 12 à 13 tonnes, avec 3,6 et 4,2 tonnes de charge utile.

Il semble donc que la conquête d'un vaste Théâtre d'opération à l'échelle de l'Asie et non plus seulement à celle d'un "petit cap du continent asiatique"[1] comme l'Europe, exigerait, entre autres moyens, le recours à des familles nouvelles de transports terrestres du moins si le conquérant souhaitait exploiter à fond la supériorité que lui procure un parc d'environ 100 000 véhicules de combat blindés, avec les possibilités correspondantes de marches et de manoeuvres foudroyantes, comparées aux possibilités des armées à prédominance d'infanterie dont la Chine dispose encore presque exclusivement à l'heure actuelle.

Le nombre de véhicules gros porteurs (60 tonnes de charge utile) nécessaire à l'échelon du Théâtre Chinois, dépend évidemment de l'état des routes, des conditions de circulation et de sécurité des itinéraires.

L'examen du réseau de courbes de la figure 5 montre, par exemple, qu'avec un parc de 50 000 véhicules, même en se plaçant dans des conditions difficiles, l'envahisseur serait en mesure de soutenir 600 000 hommes sur une ligne de contact passant au Sud de Shangaï vers 3 200 km des dépôts et des bases du fleuve Amour.

Dans de bonnes conditions de circulation (25 km/heure, maintenus jour et nuit dans interruption), ce même parc de 50 000 véhicules lourds permettrait de soutenir, sur la même ligne au-delà de Shangaï, près de 2,5 millions d'hommes. Enfin, pour ravitailler, vers la frontière du Vietnam Nord (c'est-à-dire à 4 600 km de ses bases) une armée de 100 000 hommes, on trouve qu'il suffirait d'un parc variant de 4 230 véhicules lourds pour les meilleures conditions de circulation à 16 920 véhicules du même type dans des conditions de circulation difficiles (circulation 1/4 du temps).

Pour résumer notre analyse des résultats, en peut avancer qu'un parc composé de véhicules logistiques militaires courants à l'heure actuelle, numériquement comparable au parc des grandes armées de l'époque 1943–1945, permettrait encore de soutenir, à l'échelle du Théâtre Européen, les puissantes formations blindées ou mécanisées dont peut disposer l'URSS et ses satellites, dans la mesure où l'écoulement du flux logistique indispensable aux grandes unités de l'avant ne serait pas bouleversé par l'action des défenseurs.

Cet effort sur la logistique de l'assaillant nous apparaît comme une des clés possibles d'une défense efficace.

Mais la tâche du défenseur dans son action visant à paralyser le flux logistique adverse serait d'autant plus ardue que la densité de véhicules adverses serait plus faible (c'est-à-dire que le nombre de véhicules, en route à un instant donné, serait plus petit), et que ceux-ci seraient plus aptes à circuler en tous terrains même lorsque le terrain aura été bouleversé. On reconnaît dans cette perspective l'intérêt du véhicule gros porteur pour l'assaillant.

[1] Expression de Paul Valéry dans *La crise de l'Esprit*, publié en 1919 dans la revue l'Athénaeum à Londres.

Enfin, à l'échelle de l'Asie, et plus particulièrement de la Chine, il semble bien que seuls les systèmes logistiques à base de véhicule gros ou très gros porteurs, soient en mesure de soutenir les armées à la fois puissantes et nombreuses qui seraient indispensables au conquérant, s'il souhaitait assurer sa mainmise sur les masses démographiques de ce continent, à moins d'avoir réduit ces dernières à peu de chose au moyen d'armes de destructions massives.

Dans tous les cas, il est d'un intérêt certain d'ausculter et de suivre les tendances soviétiques en matière logistique tout autant qu'en matière de doctrine stratégique et d'évolution des matériels tactiques.

3.3.2 En situation défensive:

un plan stratégique peut se définir, toujours en restant très schématique, de façon exactement symétrique à celui de l'offensive:

(i) *Sur un objectif d'importance stratégique donné* (défini par la géographie ou de toute autre manière),

(ii) *Dans un délai donné*,

(iii) *Etre en mesure de disposer au voisinage de cet objectif de forces suffisantes* pour s'opposer à l'exécution de la mission des forces que l'adversaire aura pu amener à pied d'oeuvre sur l'objectif dans le délai imparti.

Sur l'objectif, les forces réelles dont peut disposer le commandement du parti en offensive peuvent s'évaluer numériquement, ainsi:

Inf. [Nbre de G.U. Tactiques à pied d'oeuvre à l'instant T; $m_{OX}(T)$]; en rappelant que:

$m_{OX}(T)$ = nombre maximal d'unités qui peuvent être soutenues au point de vue logistique à l'instant T.

Cette expression signifie que l'on ne compte comme unités tactiques réellement efficaces, que celles qui peuvent disposer d'une logistique suffisante.

Dans un plan stratégique parfaitement organisé et se déroulant sans à coup, les deux nombres de l'expression précédente entre crochets restent évidemment sensiblement égaux.

Si l'on adopte cette définition, on constate que l'objectif du défenseur consiste à minimiser la valeur de:

$$[\text{G.U. Tactiques}; M_{OX}(T)]$$

de façon à obtenir l'inégalité:

$$(\text{forces du défenseur}) > \frac{Inf.\ [\ \text{G.U. Tactiques}; M_{ox}(T)\ ]}{\mathcal{R}_c}$$

expression dans laquelle les termes Grandes Unités Tactiques, ainsi que $m_{ox}(T)$ concernent le camp se trouvant en situation offensive et où le nombre $\mathcal{R}_c$ est le rapport de force critique précédemment défini, comme un indice permettant au parti en offensive de réussir sa mission avec un niveau de probabilité qu'il estime suffisant.

Dans le cadre de notre exposé nous laissons de côté l'aspect tactique, c'est-à-dire l'action du défenseur contre les Grandes Unités Tactiques du parti en situation offensive.
Il est clair que l'un des modes d'action possibles du défenseur consiste à essayer "d'étouffer" la logistique adverse, ce qui revient en suivant notre schématisation numérique, à essayer de faire tendre $m_{ox}(T)$ vers zéro.

L'étude de cette action est évidemment un vaste sujet, du reste passionnant, mais qui sort lui aussi du cadre du présent exposé.

Nous nous sommes bornés à indiquer les premiers jalons d'une telle investigation.

L'attaque d'un flux de communications (système de transport) dépend de la richesse en itinéraires du réseau considéré et de la densité de véhicules que l'on peut trouver à chaque instant sur chaque itinéraire du réseau. La vulnérabilité du système sera évidemment croissante avec la densité globale $(\frac{M_o}{h})$ par itinéraire.

Dans cette expression, M_o représente le nombre total de véhicules logistiques affectés à un secteur donné disposant au plus de h circuits indépendants entre les dépôts et les points de livraison aux unités tactiques de l'avant.

Les figures n° 7 et n° 8 sont constituées de deux réseaux de courbes, correspondant chacun à deux gammes de vitesses (50 km/heure et 25 km/heure) pour 9 valeurs distinctes des couples $\left\{ (\frac{M_o}{h}), \mathcal{H}_m \right\}$

Pour chaque courbe, la densité linéaire moyenne Δ(T) (nombre moyen de véhicules logistiques par kilomètre d'itinéraire) est donnée en fonction du temps T, pour des forces ayant pris l'offensive à l'instant $t = 0$.

L'examen des deux réseaux, (dont le but est de fournir des éléments quantifiés permettant au parti sur la défensive de calculer les moyens d'attaque qui lui seraient nécessaires, pour paralyser le flux logistique adverse), permet de dégager les points suivants:

–Pour $(\frac{M_o}{h})$ fixé, la densité $\Delta(T)$ qui conditionne les opportunités d'attaquer des véhicules en mouvement change peu avec la vitesse moyenne V_m des véhicules logistiques, même lorsque celle-ci varie d'un facteur de 2.

–Avec des véhicules circulant indépendamment–ou par petities rames–ce qui est une de nos hypothèses de base; la *dilution* du flux logistique est rapidement assez importante. Par exemple avec 10 000 vehicules logistiques dans un secteur d'une centaine de kilomètre de large ou davantage, disposant de 10 à 15 circuits routiers, au bout de 1 à 2 jours d'offensive la densité linéaire $\Delta(T)$ tombe à moins de 1 véhicule par kilomètre ($\mathcal{H}_m$ = ¼, circulation ¼ du temps)

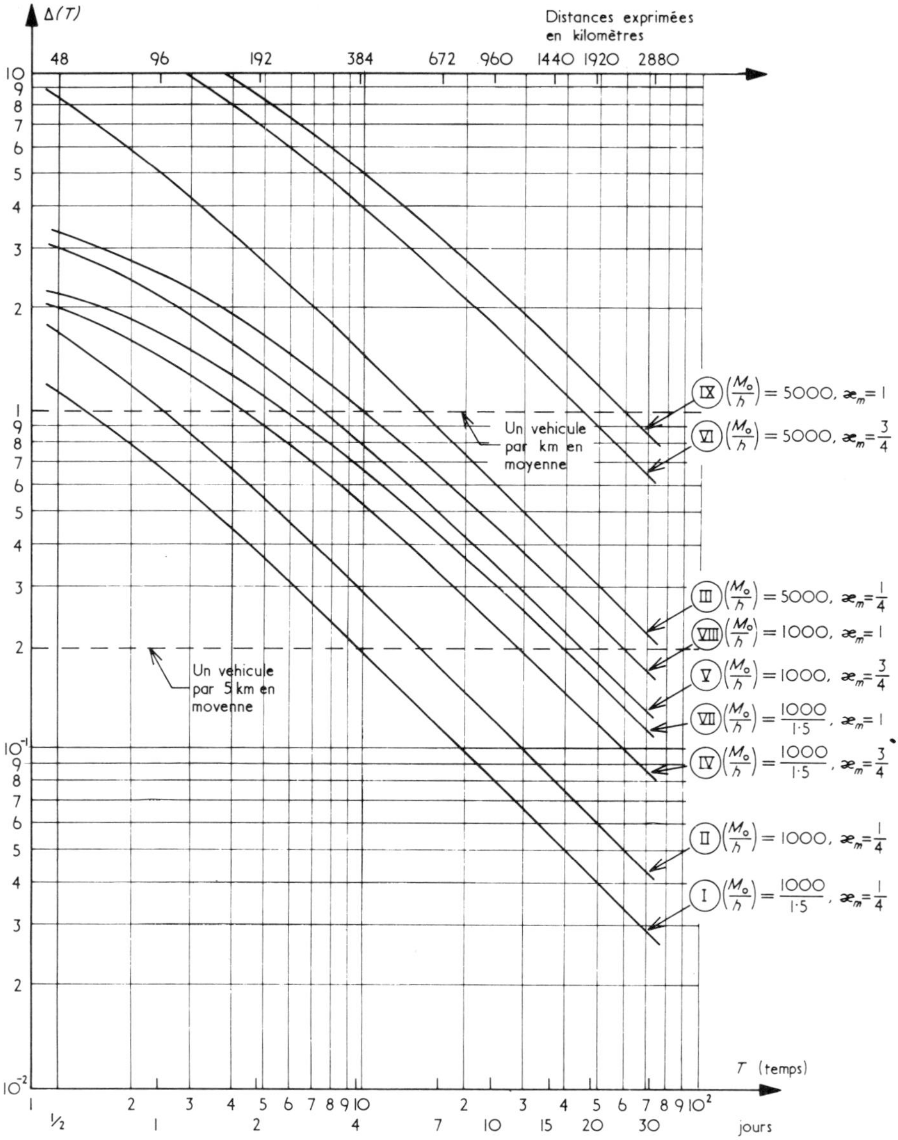

Figure 7 Densité linéaire moyenne des véhicules en mouvement sur les itineraires
(V_m = 50 km/h)

Pour $\mathcal{H}$ = ¾, une dilution de moins de 1 véhicule au kilomètre est atteinte au bout de 2 à 4 jours. Enfin, toutes choses égales par ailleurs, la dilution de 1 véhicule par 5 km est atteinte au bout de 4 à 7 jours lorsque $\mathcal{H}$ = ¼ et entre 12 à 19 jours lorsque $\mathcal{H}$ = ¾. L'examen des réseaux de courbes montre qu'avec $\mathcal{H}$ = ¼ èt V_m = 25 km/h, un parc de 10 000 véhicules logistiques courants peut assurer au bout de 1 à 2 jours le soutien de forces comprises entre 180 000 hommes et 330 000 hommes (soit des effectifs

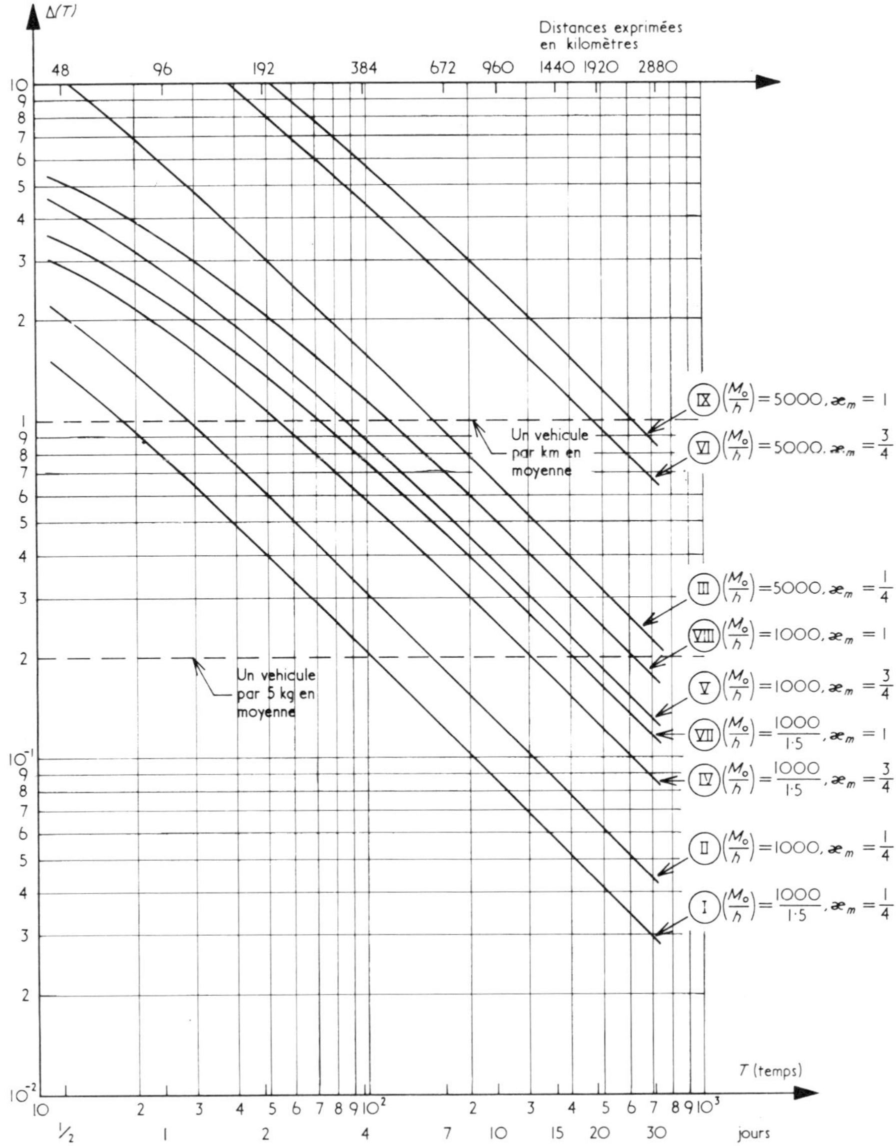

Figure 8 Densité linéaire moyenne des véhicules en mouvement sur les itineraires (V_m = 25 km/h)

correspondant à une armée ou à un front respectivement). Dans le cas de $\mathcal{H}$ = ¾ et V_m = 40 à 60 km/h, ces mêmes 10 000 véhicules logistiques pourraient soutenir au bout de 4 jours environ 450 000 hommes, c'est-à-dire un effectif de groupe d'armées ou de front. Il s'agit donc de chiffres ayant des ordres de grandeur plausibles quant aux effectifs, au nombre de véhicules logistiques utilisés, à la largeur des secteurs, ainsi qu'à la densité d'un réseau routier du type Europe Occidentale.

Toujours en se plaçant du point de vue du défenseur, s'efforçant d'annihiler la puissance de choc des Grandes Unités Blindées ou Mécanisées d'un adversaire lancé dans une offensive à fond de train, par l'asphyxie de la logistique de ce dernier, les remarques précédentes permettent d'avancer quelques suggestions pratiques.

–La dilution du flux de véhicules montre que, même en utilisant des armes à grand rayon de destruction instantanée (de l'ordre du kilomètre ou davantage), le nombre de véhicules logistiques détruits à chaque coup pourra être très faible, c'est-à-dire de l'ordre de quelques véhicules par projectile. *L'attaque systématique directe* des véhicules logistiques proprement dit au moyen d'armes à grand rayon létal instantané apparaît donc comme une politique de tir peu rentable à priori.

–Par contre, les attaques précises renouvelées par séquences, contre tous les véhicules repérables sur des tronçons d'itinéraires de 50 à 100 km seraient déjà beaucoup plus rentables. Il s'agit d'attaques qui pourraient être effectuées par des avions d'appui au sol tirant au canon automatique ou par salves de roquettes contre tout ce qu'ils peuvent détecter sur un itinéraire, de jour comme de nuit.

–Bien entendu, les attaques du type précédent seraient encore beaucoup plus payantes, si le flux de transport adverse comportait des goulots d'étranglement provoquant la création de files d'attente, par exemple au voisinage des coupures, des fleuves, de carrefours interdits, de zones minées ou contaminées, des traversées de rues encombrées par les éboulis provenant d'agglomérations partiellement détruites.

–Le mode d'attaque le plus rentable semble bien être celui de points clés du réseau de communications utilisé par l'adversaire, bien que le choix des objectifs à détruire puisse poser un problème cruel lorsqu'il s'agit d'un réseau de communications très dense, situé en territoire national ou en territoire ami.

Quantitativement, nous avons vu que la densité $\Delta(T)$ du flux logistique adverse varie, comme l'inverse de h. Chaque fois que l'un des h circuits existants primitivement dans un secteur sera coupé, la densité moyenne du flux dans les $(h - 1)$ circuits restants augmentera, c'est-à-dire que pour chaque mission d'attaque effectuée sur une fraction d'itinéraire de longueur donnée, le nombre moyen d'objectifs offerts augmentera. On peut évaluer numériquement cet effet sur les réseaux de courbes des figures 7 et 8 puisque si M_0 reste fixé la diminution de h, entraîne l'accroissement de $(\frac{M_0}{h})$.

–En réalité, l'attaque judicieuse de certains points du réseau de communications emprunté par l'ennemi peut avoir des effets plus importants que ceux qui peuvent être escomptés d'après le paragraphe précédent, car ces effets peuvent être cumulatifs.

L'augmentation de la densité linéaire de trafic par diminution de h, favorise au voisinage des fleuves, des coupures, des carrefours, la création de files d'attentes de véhicules et d'embouteillages, ce qui devrait avoir pour effet d'augmenter les chances de détection et d'attaque de ces rassemblements, qui offrent à leur tour la possibilité supplémentaire de créer par des amoncellements de véhicules détruits de nouvelles coupures sur le réseau.

Il convient d'ailleurs de remarquer que si des files d'attentes importantes et nombreuses pouvaient être provoquées, l'emploi d'armes à grand rayon létal instantané (armes de destruction massive), reprendrait alors tout son intérêt.

Dans cette perspective, les zones probables d'embouteillage devraient faire l'objet d'un système complet de prévisions, d'observations et de tirs préparés, dont le rendement pourrait être très élevé, surtout si l'on parvenait à déclencher le processus multiplicatif d'engorgement auquel on a fait allusion, jusqu'à ce que les effets de paralysie d'abord localisés, se rejoignent pour aboutir finalement à l'asphyxie complète du réseau de communications.

4. QUESTIONS A APPROFONDIR

L'ampleur du problème n'a permis que de jeter les premiers jalons d'une analyse quantitative approfondie de la logistique des forces terrestres en opérations.

En fonction des résultats obtenus jusqu'à présent, énumérons quelques-uns des sujets d'étude qui paraissent les plus fondamentaux:

–*L'évolution quantitative des besoins logistiques des armées en campagne* et plus particulièrement lorsqu'on s'éloigne des conditions de conflit classique pour lesquelles on possède des précédents historiques. Les estimations les plus importantes par leurs conséquences, semblent concerner les besoins des grandes unités blindées ou mécanisées en ambiance nucléaire d'une part et les besoins des unités engagées en ambiance de guerre subversive d'autre part. Les annexes du présent mémoire n'ont pu qu'avancer quelques valeurs numériques plausibles extrapolées à partir des données historiques et en se basant sur des réflexions d'ordre assez qualitatif sur le développement des armements et des Armées.

–*L'évolution technologique des moyens de transport logistique.* Nous avons vu que, même en restant dans le domaine des véhicules terrestres à roues, les problèmes logistiques peuvent recevoir dès à présent, des solutions très différentes suivant les types de véhicules adoptés. Bien entendu, la gamme des véhicules terrestres utilisables à des fins logistiques ne peut que s'accroître. De nombreux projets ont d'ailleurs été entrepris à ce sujet: véhicules articulés, véhicules à chenilles, véhicules à trains de roulement spéciaux, véhicules flottables, véhicules amphibies . . . etc.

On peut également montrer les possibilités, toujours en expansion du transport aérien pour soutenir notamment les pointes avancées des armées terrestres. Par exemple, 12 avions Antonov–AM-22 "COCK" peuvent à eux seuls, soutenir 50 000 hommes à 800 km des bases de départ, chaque avion étant relayé par 8 hélicoptères type MIL–MI-6 "HOOK".

Enfin des solutions comme les oléoducs de campagne ne sont déjà plus des nouveautés, puisqu'elles ont contribué de façon importante au soutien en carburant des unités motorisées de la dernière période de la Seconde Guerre Mondiale.

Le véhicule composite, utilisant à la fois le coussin d'air (pour les parties plates des itinéraires, y compris les fleuves, les lacs, les zones inondées ou marécageuses . . .) et un train de roulement du type chenillé par exemple pour les fractions de parcours accidentés, offre, à prix élevé pour le moment, des perspectives séduisantes permettant d'entrevoir un véhicule logistique, pratiquement indépendant du réseau routier et des moyens de franchissement toujours si vulnérables.

–La vulnérabilité des flux logistiques des adversaires en présence. En se plaçant au point de vue du défenseur, nous avons mis l'accent sur l'intérêt pour ce dernier d'étudier à fond la vulnérabilité du système logistique adverse. Compte tenu, par exemple, des caractéristiques des armées soviétiques, avec leur taux élevé de consommation logistique, il nous semble qu'une conception stratégique recherchant le rendement global maximal pour un ensemble de forces aéro-terrestres placées au moins initialement en situation défensive, pourrait analyser les méthodes d'asphyxie de la logistique adverse avec autant de fruit que l'engagement proprement dit des Grandes Unités tactiques.

–Cette manière d'envisager le problème se situe dans une orientation de pensée préconisée depuis longtemps par LIDDELL HART sous le terme général de "Stratégie de l'Approche Indirecte".

Il reste évidemment à démontrer qu'une telle stratégie peut être envisagée sur un Théâtre d'opérations aussi exigü que celui d'Europe Occidentale.

Au-delà des principes stratégiques, si satisfaisants qu'ils soient sur le plan intellectuel, le choix des meilleures orientations ne peut s'effectuer longtemps en avance, dès le temps de paix, qu'au moyen de calculs aussi rigoureux que possible.

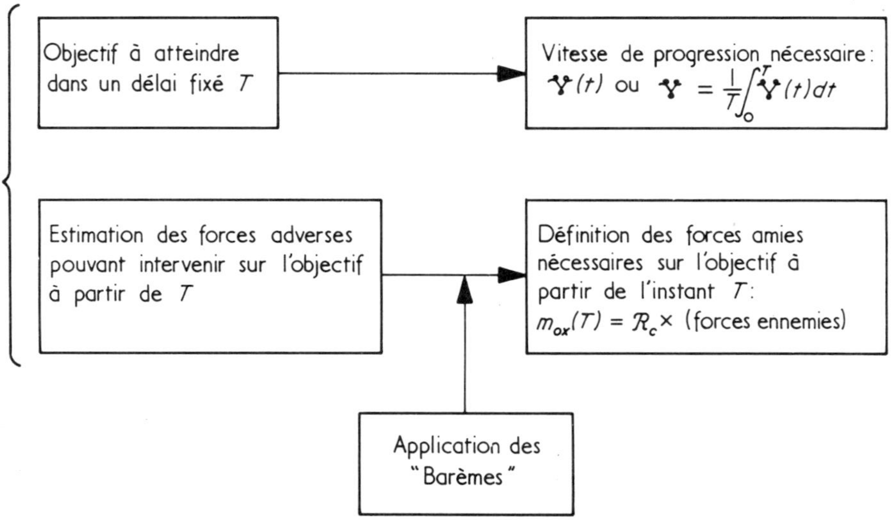

Figure 9 Schéma quantitatif d'un plan stratégique (offensive)

Suivant cette optique, la Société AERO a d'ailleurs eu l'occasion d'aborder l'élaboration de Modèles Mathématiques, sur la vulnérabilité des réseaux de communications, sur les moyens d'attaques nécessaires pour déconnecter certaines fractions d'un réseau, sur le flux logistique maximal survivant dans un réseau en fonction du nombre d'attaques auxquelles il a été soumis, de leur puissance, du choix de leurs points d'application et de la rémanence plus ou moins durable de leurs effets. Bien entendu, la résolution complète de ces problèmes exige que l'on soit en mesure d'obtenir ou de calculer des valeurs significatives sur le taux de détection des moyens d'attaque en fonction de la densité, de la nature, de la vitesse, de la discipline de route des véhicules adverses pris pour objectif, des probabilités conditionnelles d'attaque et de destruction des véhicules lorsque ces derniers ont été détectés, d'évaluer les conséquences des destructions obtenues pour la création d'embouteillages aux abords des fractions d'itinéraires qui ont été coupées ou obstruées . . . etc.

Toujours en se plaçant au point de vue du défenseur, le maintien de son propre réseau logistique apparaît aussi comme un sujet vaste et préoccupant, avec ses longues lignes de communications maritimes ou aériennes, ses points de rupture de charge à la jonction de deux types de moyens de transport, sur lesquels l'adversaire ne manquerait pas de concentrer son attention lors de la phase d'élaboration de ses plans stratégiques et ultérieurement une partie de son potentiel de destruction, dès le début des opérations actives.

Dans ce domaine également, la seconde partie de notre conférence de Münich, jetait les bases d'un modèle mathématique s'efforçant d'intégrer les facteurs maritimes et terrestres et de mettre en relief leurs interactions.

Il serait évidemment fécond de pouvoir aboutir en fin de compte à un modèle global permettant de représenter de façon quantitative l'évolution et les capacités résiduelles des systèmes logistiques des deux adversaires au cours du temps, sur un Théâtre de guerre important, pour différentes hypothèses sur le type de conflit.

La poursuite d'un tel objectif peut paraître ambitieuse. Mais, compte tenu des résultats déjà obtenus avec un arsenal mathématique assez modeste, la réussite d'une telle entreprise paraît à la portée d'une bonne équipe de chercheurs opérationnels pouvant faire largement appel aux ressources offertes par l'informatique et les ordinateurs disponibles dès à présent.

5. CONCLUSION

" . . . Pour déterminer le résultat des combats avant même que ceux-ci aient commencé, il est nécessaire d'atteindre les trois buts suivants:

a) couper les lignes de communications de l'ennemi et paralyser ainsi l'organisation de sa logistique;

b) le couper de ses lignes de retraite, et saper ainsi sa volonté en détruisant son moral;

c) frapper ses postes de commandement, ses centres administratifs et rompre son système de télécommunications, détruisant ainsi les liaisons entre son cerveau et ses membres.

L'analyse de ces trois buts confirme le bien fondé de l'axiome de Napoléon: Tout le secret de l'art de la guerre réside dans l'aptitude à se rendre maître des communications".*

* B.H. Liddell Hart *Histoire Mondiale de la Stratégie* Annexe II du Général Yigael Yadin, Chef d'Etat-Major des Armées Israéliennes en 1949.

L'étude dont nous vous avons présenté les premiers éléments s'est essentiellement attachée au volet logistique du triptyque d'inspiration napoléonienne posé par le Général YADIN comme axiomatique de la stratégie.

L'apport de notre modèle peut se scinder schématiquement en deux rubriques: l'aspect Méthodologie et la mise en évidence de quelques faits saillants.

5.1 Sous l'angle méthodologique:

–Il est possible d'élaborer un modèle simple, qui permette de rendre compte de façon quantitative du rythme des opérations sur différents théâtres de guerre au cours du deuxième conflit mondial, en mettant essentiellement l'accent sur les contraintes d'ordre logistique.

Le modèle primitif, en quelque sorte validé par sa confrontation avec les données historiques peut être développé, de façon à permettre des prévisions sur le rythme possible des opérations futures, sur différents Théâtres et pour des hypothèses variées sur le type de conflit considéré. L'extension du modèle permet d'intégrer des facteurs tels que:

. doctrines stratégiques et vitesses des Grandes Unités tactiques des adversaires possibles;

. contraintes d'ordre géographique et longueur des lignes de communication;

. effectifs mis en oeuvre;

. besoins logistiques pour différents types de conflits et différents types d'unités tactiques;

. conditions de survie des réseaux de communications aux efforts de destruction des antagonistes;

. performances et développements futurs des véhicules logistiques terrestres;

. effets résultant de l'introduction de nouveaux moyens de transport au profit de la logistique des armées en opération: nouveaux véhicules terrestres, transports continus (type oléoducs), transports aériens.

–Le modèle analytique, même dans sa forme développée, n'est évidemment qu'une première étape. Toutefois, dans son état actuel, ce travail permet de mettre en évidence les aspects les plus importants qu'il convient d'approfondir pour tirer pleinement profit de ce type d'investigation.

Un des points les plus complexes du problème d'ensemble, est l'analyse de cette partie de la bataille qui a pour objet la destruction des courants logistiques des adversaires en présence. En particulier, sur un théâtre de la taille de l'Europe Occidentale, est-il possible d'asphyxier le flux logistique ennemi suffisamment vite, afin que le puisssance réelle de ses Grandes Unités tactiques, réduite par l'effort adverse sur leur soutien, soit à la mesure des forces tactiques du défenseur encore en état de combattre?

Pour conclure sur l'aspect méthodologique, il nous semble clair qu'après avoir analysé et compris les mécanismes de base, qui gouvernent les problèmes de soutien logistique des armées en campagne, l'étape suivante consiste dans l'élaboration d'un modèle global capable de rendre compte à la fois des courants logistiques des deux adversaires et de leurs efforts réciproques pour les paralyser.

Bien entendu, le nombre de paramètres, l'ampleur des calculs à effectuer (même si chacun d'eux est assez simple), exigent que le modèle soit conçu dès le départ pour être programmé sur ordinateur.

L'emploi de ce dernier devrait permettre d'explorer toutes les parties du domaine des solutions envisageables qui apparaissent au planificateur comme les plus fécondes ou les plus redoutables, de tester la sensibilité de ces solutions aux variations de certains paramètres ou de certaines hypothèses, bref de se livrer à loisir, dès le temps de paix à l'investigation approfondie d'un riche sujet dans lequel stratégie et technique interfèrent étroitement et qui serait de nature à décourager toute tentative de connaissance approfondie ne faisant pas appel à des méthodes très puissantes.

5.2 Sous l'angle des faits intéressants, on peut retenir les points suivants:

–Sur le Théâtre d'Europe Occidentale, *le nombre de véhicules logistiques de type classique, nécessaires pour soutenir une offensive blindée*, mettant en jeu des effectifs considérables tout en soutenant un rythme d'avance de l'ordre de 100 km par jour, *paraît parfaitement à la mesure des forces terrestres soviétiques*, même si les difficultés de circulation de ces véhicules se révélaient assez grandes.

–Mais il convient de souligner que l'observation précédente, n'est valable qu'à la condition "sine qua non" que le défenseur soit incapable d'infliger au flux logistique adverse un taux suffisant de destruction et de freinage, pouvant d'ailleurs aboutir comme cas limite à sa paralysie presque complète.

–Cette dernière perspective se rapproche des vues exprimées par le Maréchal SOKOLOWSKI, dans l'ouvrage de stratégie publié sous sa direction à Moscou en 1962 et dans lequel il affirme que des opérations d'envergure sur le Théâtre Européen ne sont concevables qu'en ambiance nucléaire.

Néanmois, il tient pour assuré que, même dans l'hypothèse de tirs nucléaires massifs, le "Blitz-Krieg" blindé reste non seulement possible mais encore indispensable du point de vue de l'offensive. Bien entendu, pour assurer le soutien logistique des grandes unités blindées ou mécanisées lancées à fond de train sans marquer d'arrêt avant d'avoir atteint leurs objectifs stratégiques, l'ouvrage du Maréchal SOKOLOWSKI propose des solutions nouvelles, notamment la pose rapide d'oléoducs de campagne à un rythme de progression voisin de celui des unités tactiques ainsi que l'emploi massif du transport aérien.

Même en restant dans le cadre des transports terrestres, il semble bien que la vulnérabilité d'un flux logistique puisse être grandement diminuée par l'emploi de véhicules véritablement appropriés à leur mission. Les performances requises pour de tels véhicules sont assez *aisées à définir, sinon à réaliser*: capacités tous terrains, aptitude (si possible) à franchir sans aide fleuves et coupures, possibilité de traverser sans dommages graves les zones contaminées, très grande robustesse, confort poussé pour des équipages appelés à effectuer de longues étapes, entretien mécanique minimum, basse consommation spécifique, bonne vitesse moyenne soutenue sur les terrains favorables, très bon rapport (charge utile/charge totale) et finalement une certaine capacité d'auto défense. Il convient de souligner aussi, qu'un fort rapport (charge utile/charge totale) rend minimum le nombre de véhicules permettant d'écouler un courant logistique donné, ce qui a pour effet de diminuer à chaque instant la densité de véhicules effectivement en route (nombre moyen de véhicules par kilomètre d'itinéraire) et par conséquent de faire décroître le taux de détection et de destruction par les moyens adverses du flux logistique considéré dans son ensemble. La vulnérabilité globale d'un flux peut donc être très nettement diminuée par l'augmentation des performances des véhicules, même si la probabilité de destruction de chaque véhicule (lorsque ce dernier a été détecté) est du même ordre que la probabilité correspondante pour un véhicule moins performant.

–L'ouvrage précédemment cité du Maréchal SOKOLOWSKI n'évoque pas les possibilités offertes par l'évolution des véhicules logistiques terrestres, dérivés des véhicules courants à l'heure actuelle en fonction des progrès technologiques qui peuvent pourtant être fort importants, notamment dans la catégorie gros ou très gros porteur.

Toutefois, les perspectives offertes par de tels véhicules nous ont paru mériter un examen attentif et c'est pourquoi nous leur avons accordé implicitement ou explicitement une assez large place dans notre travail.

Sur un Théâtre de guerre, comme le serait la Chine, le nombre de véhicules logistiques classiques nécessaire pour soutenir jusqu'au bout de puissantes offensives blindées ou mécanisées, serait énorme, voire *infini, lorsqu'il s'agit des confins du territoire, puisque ces derniers sont situés à des distances des bases russes pour lesquelles joue la consomption logistique* (la distance de consomption étant celle pour laquelle la consommation en carburant de chaque véhicule sur un trajet aller et retour représente un poids égal à celui de la charge utile maximale).

Bien entendu, on peut imaginer la création de dépôts intermédiaires, mais la création de ces derniers exige du temps et oblige à diviser une campagne en phases successives avec des pauses, ce qui exclut donc la perspective d'une offensive foudroyante terrassant l'adversaire d'un seul élan, telle que l'imagine la doctrine soviétique, du moins à l'échelon du Théâtre Européen.

Ces observations, prenant pour point de départ de simples calculs logistiques, ne sont pas dénuées d'intérêt pour comprendre par exemple l'attitude des Russes vis à vis de la Chine.

Le Théâtre Européen, même en le considérant dans sa totalité reste, malheureusement pour lui, parfaitement à la taille d'une entreprise soviétique. Le Théâtre Chinois ne l'est pas, à moins de repenser la doctrine de guerre, de disposer de moyens d'extermination massifs, d'imaginer de nouveaux instruments de mainmise sur les objectifs stratégiques, d'élaborer de puissants outils permettant le soutien lointain des armées, notamment pour résoudre leurs problèmes logistiques.

. Sur un Théâtre Asiatique, les contraintes géographiques résultant des distances et de la faiblesse des réseaux de communications, combinées avec l'obligation de contrôler et de dominer de grandes masses humaines, exigeraient le soutien d'effectifs importants dans des conditions logistiques extrêment difficiles. Les calculs effectués au cours de l'étude montrent cependant qu'une flotte de véhicules terrestres ayant les caractéristiques indiquées plus haut, elle-même complétée par des transports aériens lourds, serait numériquement concevable à l'échelle d'une très grande puissance comme l'URSS qui s'est montrée, à l'heure actuelle, capable de se constituer un parc d'environ 100 000 véhicules de combat.

–Nous ignorons évidemment quelle est la politique russe en matière de conception et de développement pour son parc de véhicules logistiques, tant pour les exigences d'un Théâtre Européen que pour celles d'un théâtre Asiatique! Pour ce qui concerne la position chinoise nous relevons dans la revue de Défense Nationale de Mars 1970: "La revalorisation des forces classiques (de la Chine populaire) est sans doute une conséquence directe des affrontements de 1969 . . . ., il a été décidé d'accroître la puissance de feu de l'Armée de Libération populaire en construisant des armements lourds et *d'améliorer sa mobilité par des achats massifs de camions* . . . . ."

Comme le montre cette dernière remarque, en dépit de la disparité des régimes politiques et des doctrines militaires, le fait logistique finit par s'imposer.

Bien que parfois sous-estimée, la logistique des Armées a toujours constitué une condition "sine qua non" de leur puissance. Cela reste particulièrement vrai pour les grandes armées entièrement blindées ou mécanisées de l'adversaire éventuel sur le Théâtre Européen. Sa logistique est à la fois un chaînon essentiel de sa puissance combattive et un point de faiblesse possible.

Cet aspect mérite d'être analysé avec soin, chiffres à l'appui, surtout pour celui des deux adversaires qui se trouverait réduit par les circonstances politiques, la géographie ou le rapport des forces en présence, à adopter une attitude défensive au moins dans la phase initiale d'un conflit.

Il est généralement admis que la forme défensive de la lutte est la plus difficile sur le plan conceptuel. Cette difficulté même ne peut que nous inciter à déployer toutes les ressources de "l'intelligence en guerre", surtout à une époque de l'histoire où des instruments nouveaux, tels que l'ordinateur, offrent à l'esprit humain un pouvoir multiplicateur dont on commence seulement à entrevoir les effets.

REFERENCES

1. American Enterprise in Europe. The Role of the SOS in the Defeat of Germany. Printed Under the Direction of Chief Information and Education Division, USFET, 1945.

2. ALLARD, Colonel. Le Ravitaillement en munitions. Campagne 1943–1945. Conférence faite à BERNE le 27 Avril 1947. Ministère de la Guerre, E.M.A. 4ème bureau.

3. LALLEMAND . Les rampes limites[1] *Bulletin d'Information Technique et Scientifique, Section Technique de l'Armée,* Octobre 1951, pp. 549–551.

   (1) La Vie automobile, n° 1 443, Septembre 1951, pp. 242–248.

4. Headquarters, Department of the Army. Nuclear Weapons Employment *Department of the Army Pamphlet n⁰ 39–1,* Mai 1959.

5. KIRIN, I., Colonel. Speed in the Attack, *Military Review*, Août 1961, pp. 76–81.

6. LIDDELL HART, B.H. *Histoire mondiale de la Stratégie,* Editions PLON, 1962.

7. SOKOLOVSKII, V.D. *Soviet Military Strategy.* The Rand Corporation Translation. Prentice-Hall, Inc. 1963.

8. X.X.X. L'adaptation des forces terrestres soviétiques à une guerre nucléaire, *Revue de Défense Nationale,* Février 1966, pp. 212–223.

9. Exposition de matériel d'armement terrestre, Tome 2: Mobilité. Direction Technique des Armements Terrestres, Matériels présentés à l'exposition de Satory, 1967.

10. BAUMEISTER, T., MARKS, L.S., (directeurs de l'édition). *Standard Handbook for Mechanical Engineers.* McGraw Hill Book Co. 1967, édition révisée.

11. TCHOUIKOV, V.I., Marechal, *Feu sur Berlin.* Presses de la Cité, Traduction de R. JOUAN, 1968.

12. WOOD, D., *Conflict in the Twentieth Century.* The Institute for Strategic Studies, ADELPHI papers, n° 48, Juin 1968.

13. Les unités mécanisées dans les Armées étrangères URSS *L'Armée*, N° 91, Octobre 1969, pp. 92–99, n° Spécial.

14. *The Military Balance 1969–1970.* The Institute for Strategic Studies, Londres, 1969.

15. République populaire de Chine: Campagne de préparation à la guerre. *Revue de Défense Nationale,* Mars 1970, p. 502.

16. TAYLOR, J.W.R., Missiles 1969, *Flight International,* 14 Novembre 1968.

DISCUSSION

Buck: What do the letters "A.E.R.O" stand for?

Bresson: Le sigle de la Société A.E.R.O à laquelle j'appartiens signifie:

> *A*utomation,
> *E*lectronique,
> *R*echerche,
> *O*perationnelle.

Ferrier: Les avions et les camions très gros porteurs seront vraisemblablement escortés. Faut-il les considérer comme des objectifs "mous" ou des objectifs "durs"?

Bresson: Les avions gros porteurs (ainsi que les Hélicopteres de "livraison avancée") sont intrinsequement des objectifs "mous". Evidemment, le commandement qui les utilise aura certainement conscience de leur vulnérabilité. Il s'efforcera en conséquence, dans toute la mesure de ses moyens disponibles, de les escorter et de les protéger, d'autant plus que les chargements logistiques transportés seront plus précieux pour la conduite de la bataille. Les avions gros porteurs et les hélicopteres de transport logistique (et d'évacuation) auront ainsi tendance à devenir des objectifs moins vulnérables, sinon durs. Il n'en demeure pas moins que ces objectifs très payants devraient faire l'objet d'une attention et d'un effort particulier de la part d'un commandement soucieux de diminuer la force vive de son adversaire par la paralysie de la logistique de ce dernier.

Par exemple, les 110 tonnes de charge utile d'un avion du type ANTONOV–22, transportées, en un seul voyage, peuvent représenter les ravitaillements essentiels pour un régiment de 100 chars moyens, pour une journée de combat. (Par char: 800 litres de carburant, soit 650 kg plus 200 kg d'emballage pour le carburant, 250 kg de munitions, soit environ 10 coups complets, ce qui représente un total d'environ 1100 kg par char).

Il est cependant bien clair qu'un tel avion, une fois déjouée l'action de son écran de protection, est à la merci d'un rafale de canon de 30 mm, aussi bien que d'un missile Air-Air, voire d'une charge nucléaire tactique n'éclatant pas trop loin, tandis que la mise hors de combat d'un régiment de T-62, même avec des moyens nucléaires considérables est un "travail tactique" très rude. La "coriacité" comparée des deux types d'objectif doit donc apparaître assez clairement à la lumière de cet exemple schematique. Il n'en reste pas moins, qu'un véritable modèle serait seul en mesure de préciser quantitivement la vulnérabilité réelle des systèmes logistiques en fonction de leur protection et du type d'attaque auxquels ils pourraient être soumis.

Cette dernière remarque est également valable pour les "flottes" de camions gros porteurs en observant que leur nombre éventuel et leur dispersion sur les itinéraires, empêchera probablement de leur affecter une protection comparable à celle qui devrait être consentie pour un objectif aussi précieux, et relativement rare, qu'un avion de 100 à 150 T de charge utile.

Arentsen: Did it make any difference if you had automatic supply or supply on demand?

Bresson: Dans le modèle nous n'avons pas établi de distinction explicite entre logistique à la demande et logistique automatique, puisque notre premier but était de calculer les moyens nécessaires pour acheminer un certain flux logistique (d'ailleurs fonction du temps), sans nous préoccuper de savoir si ces besoins s'exprimaient en raison d'un niveau moyen prévu "a priori" ou au contraire tenaient compte de la consommation observée et des prévisions adaptées aux circonstances. On peut toutefois mentionner que l'utilisation du modèle est particulièrement simple, lorsque l'on peut faire une évaluation a priori du taux moyen de consommation par unité tactique (ou par homme) et par unité de temps. Cette manière de procéder, qui ne saurait prétendre à la détermination des meilleures politiques logistiques, permet de définir dans ses grandes lignes, les besoins logistiques globaux, les moyens nécessaires pour les acheminer, le rythme auquel les opérations tactiques doivent être menées sans inconvénients graves pour l'acheminement du flux logistique, et enfin fournir une base numérique pour l'évaluation de la vulnérabilité du réseau de soutien logistique et les moyens à prévoir pour assurer sa survie.

Solstrand: It was mentioned that the *need* for certain types of logistics was an input parameter for the model. How have these *needs* been established, and with what size of the corresponding *uncertainties*? Or to put it another way: The performance of the army is a function of the supply rate of either type of logistics–ranging from 100% effectiveness when the supply is unlimited to 0% when there is no supply at all. At what level of performance were the needs established?

Bresson: Nous n'avons pas considéré explicitement dans le Modèle la dégradation des performances tactiques résultant d'un approvisionnement logistique insuffisant. Nous avons simplement introduit, comme une fonction du temps, un certain taux de consommation unitaire, estimé comme satisfaisant en moyenne, puis calculé en fonction de ce paramètre les moyens nécessaires pour acheminer en temps utile le flux logistique correspondant.

Votre question, d'ailleurs fort pertinente, ouvre un vaste champ d'investigation qui dépasse le cadre de notre modèle actuel. En effet, si l'on affecte (du reste avec un certain degré d'arbitraire) l'indice 100% à un taux de consommation estimé comme pleinement satisfaisant, la diminution de ce taux entraîne simultanément, non seulement une décroissance de la valeur tactique des grandes unités, mais encore une diminution des exigences concernant les moyens logistiques à mettre en oeuvre. Dans quelle mesure un commandement stratégique peut-il accepter cette diminution de la probabilité de remplir sa mission tactique au prix d'un allégement (ou d'une insuffisance insurmontable) de son réseau logistique?

La réponse à ce genre de question présuppose évidemment l'établissement de modèles non plus simplement logistiques mais encore tactiques, permettant en particulier de déterminer les fonctions de décroissance d'efficacité tactique (expression qu'il n'est pas simple de préciser quantitativement) en raison du niveau logistique disponible variant par exemple de l'indice 100 à l'indice zero.

On peut d'ailleurs se poser le problème inverse du précédent: Un commandement au niveau stratégique, ne disposant à un moment donné que d'un flux logistique insuffisant pour l'ensemble des grandes unités tactiques disponibles, doit-il ravitailler à 100% un nombre limité d'unités chargées par conséquent de toutes les missions essentielles ou bien doit-il au contraire ravitailler à n% ($n < 100$) la totalité de ses unités, sachant pertinemment que pour chacun d'elle la capacité tactique sera inférieure à celle que l'on aurait pu escomptée avec des troupes pleinement pourvues?

Historiquement parlant, c'est à ce genre de décision et de choix d'une politique optimale que devait faire face le Commandement Supprême Interallié sur le théâtre européen à la fin de l'été 1944.

La résolution rigoureuse de tels problèmes, exige de façon sine qua non, l'établissement de ces fameux diagrammes montrant pour chaque unité la dégradation des performances tactiques escomptées en fonction de la décroissance du flux logistique.

C'est ce type de problème (et de difficulté) que vous avez sans doute abordé? Il se situe exactement à la jonction des problèmes logistiques et tactiques. C'est à dire que son importance stratégique est extrême. Malheureusement son ampleur est telle, qu'il ne pouvait guère être question de l'aborder avec quelque utilité au cours d'un exposé se situant dans le cadre d'une Conférence consacrée essentiellement à l'aspect logistique.

Combining Analytic and Simulation Models to Relate Logistic Operations to Weapon System Performance

R. M. Simon
C. H. Ball

1. INTRODUCTION

Because of the complexity of modern weapon systems and their logistic support operations, it has become increasingly more difficult to evaluate the effect of system configuration and changes on weapon system effectiveness. In this paper the structure of a computer simulation developed to study these relationships is described. This simulation can be used to produce cost-effectiveness curves relating weapon system performance to spare part replenishment budgets, and hence can be of use in the preparation or evaluation of budgetary requests. Operational measures of weapon system performance are used (e.g., readiness, rather than fill-rates or backorders), and both the transient and steady-state effects of spare-part replenishment budgets are obtained. The measurement system may also be utilized to evaluate the effect of alternative methods of initial provisioning, alternative stockage policies, expansion of repair facilities, alternative transportation schedules, changes in part reliabilities, altered levels of weapon system activity, and a myriad of other alterations to the weapon system or support system.

The development of this measurement system was co-sponsored by the U. S. Navy Strategic Systems Projects Office and U. S. Navy Supply Systems Command under contract with G.E./TEMPO. The weapon system of immediate interest was the fleet of ballistic missile nuclear submarines of the U. S. Navy.

In Section 2, a model of the fleet ballistic missile (FBM) weapon system and its support system is described, and in Section 3 the measure of weapon system effectiveness is presented. In Section 4 it is shown how judicious simplifying assumptions are used to decompose the problem in a natural manner into tractable subproblems. Instead of one intractable simulation, an analytical "readiness evaluator" and a sequence of efficient single-part-type simulations are obtained.

2. THE OPERATION/SUPPORT SYSTEM

The model which we shall consider is depicted in Figure 1. The system is composed of three echelons. The first echelon consists of 36 ballistic missile nuclear submarines (SSBNs).[1] The second echelon consists of four submarine tenders, each serving nine submarines. The third echelon consists of the CONUS supply complex of stock points, inventory control points, and repair facilities. The model described here is for "push" items, where third echelon procurements are made on a system-wide basis, and there is a great deal of redistribution among the third echelon stock points. For "pull" items there is no redistribution and procurements are made by the stockpoints themselves.

1 Any number of submarines is permitted.

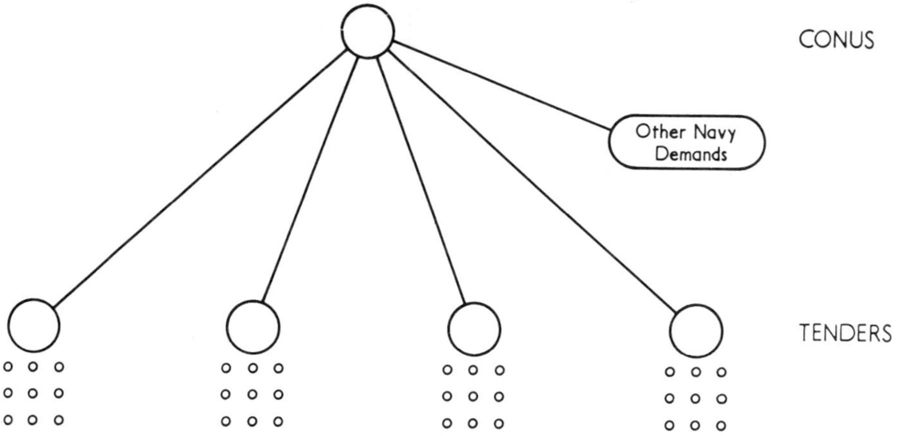

Figure 1

The budgetary process is also different for pull items, and the model for pull items will not be described here.

The operation of the system, briefly, is as follows. Each submarine goes through a cycle of 60 days on patrol and 30 days on refit. The cycles are staggered so that one submarine arrives and one departs from any given tender every ten days. A submarine carries 16 missiles, and the complex equipments needed to operate a submarine and launch the missiles contain many different types of parts. A part type (FSN)[2] may have several applications in a submarine and each submarine carries a certain number of spares (possibly zero) of each FSN it uses. During patrol, parts fail and are replaced by spares. If the number of failures of an FSN during patrol exceeds the number of spares on board the submarine at the start of patrol, then some of the missiles may become non-launchable. On the other hand, because of built-in redundancies, certain shortages may exist without causing any missiles to become inoperable.

The engineering structure of a submarine can be represented by a series-parallel diagram such as in Figure 2 (a real engineering diagram is of course much more complex). Each vertex represents a location of a part, and the number in the vertex represents the FSN to be found in that location. An arrow represents the nose of a missile, and a missile is launchable if there exists a path from the root (o) of the graph to the corresponding arrow such that each location on that path contains an operable part of the specified type. The ship system of a submarine will be represented in such a diagram in the branch affecting all 16 missiles.

When a submarine returns from patrol to its tender, it requisitions parts to replace those that failed. A submarine manages each FSN i with a (s_i, S_i) inventory policy peculiar to that FSN; that is, when the number of spares are installed applications falls to s_i or below, requisitions are placed to bring the level up to S_i. There is continual review of the inventory position as long as the submarine is in refit. Most FSNs have an $(S_i\text{-}1,\ S_i)$ inventory policy. Whether or not $s_i = S_i\text{-}1$, we shall refer to S_i as the Coordinated Shipboard Allowance List (COSAL) quantity.

A tender stocks spares of each type that might be requested by the submarines it

[2] Federal stock numbered item.

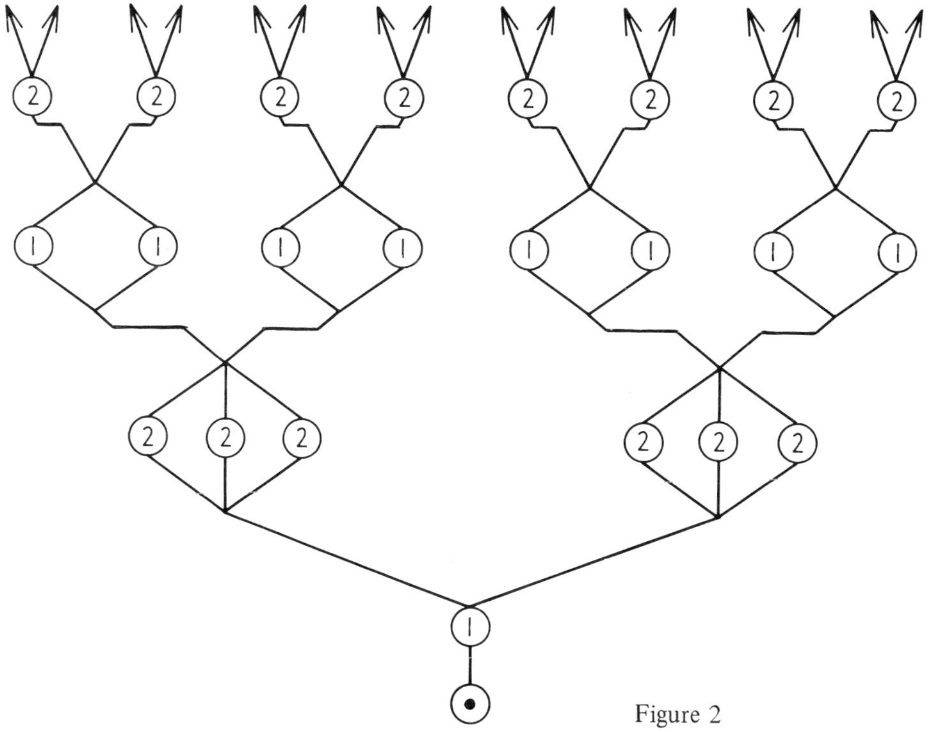

Figure 2

serves. It also stocks spares for its self-support. The tender stock is replenished from CONUS by ship approximately every 30 days. The replenishment policies used by the tender are of the continuous review (s,Q) type (often $Q = 1$).[3] If a submarine requisition cannot be completely filled from tender stocks, and if the need for the requisition is great enough, the tender will pass to CONUS a priority request for the unfilled portion. Parts requested on a priority basis arrives before the end of refit, it will be placed on the submarine. Otherwise, it will be held until that submarine returns for its next refit. If the need for the unfilled portion is not great enough to justify a priority requisition, then the demand will be backlogged at the tender. When the tender receives its next normal resupply from CONUS, the requested units will be held for the submarine.

CONUS stock is used to resupply the tenders, to meet priority requests, to supply other activities (such as training), and to supply all other Navy weapon systems. The FBM program represents only a small part of total naval operations and most FSNs are used by several weapon systems. For budgetary purposes, the entire set of FNSs is partitioned into cogs (cognizant groups). Since the cogs are funded independently and transference of funds across cogs is prohibited, we need only concern ourselves with cogs containing at least on FSN of importance to the operation of the submarines.

[3] Using a continuous review (s, Q) replenishment policy, whenever the asset position falls to s or below, an order for a multiple of Q units is placed. The multiple is the largest multiple which does not bring the asset position above $s + Q$. The asset position is defined to be the number of units on hand, plus on order, plus in repair on the tender, minus backlogged demand from the submarines.

Such FSNs will be called missile related (mr) FSNs. In a cog containing missile related FSNs, the non-missile related (nmr) FSNs in the cog cannot be completely ignored because of the common budgeting within the cog. That is, at the start of a fiscal year, a budget B is allocated for CONUS replenishments of FSNs in the cog from exogenous suppliers. Based on this allocated budget and current third echelon assets,[4] the third echelon (s, Q) replenishment policies are calculated for all FSNs in the cog. Actually, the Q's are set independently of the budget or the asset levels, by[5]

$$Q_i = \sqrt{8A_iD_i/I_ic_i} \tag{1}$$

where

A_i = procurement set up cost

D_i = expected quarterly demands[6]

I_i = holding cost (interest plus obsolescence plus storage)

c_i = unit price

i = subscript denoting FSN i.

The reorder points (s_i) do depend on the budget and the asset positions. The method of setting the (s_i) is called the inventory review. The objective is to set the reorder points so that the expected total expenditure for the year for CONUS replenishments of cog items does not exceed the allocated cog budget. The inventory review is accomplished as follows.

For a trial value of the shortage cost λ, the reorder point for FSN i is determined as the smallest non-negative integer s_i such that

$$\sum_{k=s_i+1}^{\infty} P_i(k) \leq I_iQ_ic_i/(4E_iD_i\lambda), \tag{2}$$

where E_i denotes the relative essentiality of FSN i, and $p_i(k)$ is the probability mass function for the number of demands for FSN i on CONUS during a CONUS procurement lead time. Expression (2) is obtained by balancing the holding and shortage costs

4 The asset position of an FSN at CONUS is the total number of units on hand at CONUS stock points, plus the number of units on order from suppliers (manufacturers, DSA, GSA), plus the total number of repairable carcases in CONUS repair facilities, minus the total stockpoint due-outs.

5 There are over-rides to this economic lot size formula for low demand items.

6 For repairable items, D_i is replaced by $D_i(1-r_i)$, $(1-r_i)$ being the attrition rate of FSN i.

during a procurement lead time. Having determined the (s_i), the expected third echelon replenishment cost through the year is calculated. For FSNi this is approximated by

$$E [\$_i \mid a_i] = [\frac{4D_i - (a_i - s_i)}{Q_i}] \, Q_i c_i \qquad (3)$$

where a_i denotes the third echelon asset position at the time of the inventory review. The budget B is then compared with

$$\sum_i E [\$_i \mid a_i] , \qquad (4)$$

where the summation is taken over all of the FSNs in the cog If expression (4) exceeds B by a pre-specified amount ϵ, λ is decreased and the procedure is repeated. A value of λ is finally selected such that

$$\sum_i E [\$_i \mid a_i] \approx B \ . \qquad (5)$$

This λ determines the reorder points for FSNs in the cog through inequality (2).

Although the cog budget is allocated on an annual basis, the inventory reviews are performed quarterly. At the start of each quarter, a new λ is selected so as to make the expected expenditure through the remainder of the fiscal year equal to the remaining budget (that part of the initial budget that has not already been spent for CONUS replenishment). Thus, at the start of the second quarter, "3" is used in expression (3) rather than "4", and a_i is the current asset position. In expression (5), B represents the remaining budget.

Because of the random nature of demands on the third echelon, the cog budget may be depleted before the end of the year. If this happens, then all CONUS replenishment buys from this point through the end of the year are postponed till the start of the next fiscal year. At the start of the next fiscal year, all postponed buys are processed according to the old replenishment policies. The new budget is decremented by this amount, and the first quarter inventory review is performed.[7]

3. A PERFORMANCE MEASURE

Let $M_s(t)$ denote the random number of missiles operable[8] at time t on submarine s. Then a measure of strategic readiness $R(t)$ at time t is

$$R(t) = \sum_s M_s(t) \qquad (6)$$

[7] An alternative budgeting procedure which can be treated by the simulation is the following. The allocated annual budget is divided into four quarterly budgets. The inventory review is performed at the start of each quarter, but if the quarterly budget is exceeded, then replenishments cease till the start of the next quarter. Excess quarterly budgets are carried over to the next quarter, but not to the next fiscal year.

where the summation is over those submarines which are "on station," thus excluding those submarines which are either in refit or are traveling between their tenders and their station-keeping positions. A missile is "launchable" if it is operable and its submarine is on station.

Even though the FBM program has a single purpose, i.e., a retalitory threat, it does not necessarily have a unique performance measure. Rather than define strategic readiness as above, the number of launchable missiles could have been normalized by dividing by the number of missiles on station, thus obtaining values of strategic readiness lying between 0 and 1.

Another measure of strategic readiness might be the pair

$$\left[\sum_{s} M_s(t), \ \sum_{s} 1 \right] \tag{7}$$

where again the summation is over the submarines on station. If the submarine patrol schedules are somewhat irregular, the number of missiles launchable, as well as the number of missiles on station, are both of interest. Still another measure of strategic readiness might be a time average of $R(t)$. These examples show that equation (6) is fundamental to many measures of readiness. Therefore, in this paper we adopt equation (6) as *the* measure of strategic readiness.

Since $R(t)$ is a random variable, we are interested in its expected value

$$\begin{aligned} E\,[R(t)] &= E\left[\sum_{s} M_s(t) \right] \\ &= \sum_{s} E[M_s(t)] \,. \end{aligned} \tag{8}$$

For fixed t, the $M_S(t)$ are dependent random variables because the cog budget constraints affect the spares on board at the start of each patrol. Even though the $M_S(t)$ are dependent, the expectation of a sum of random variables is the sum of the expectations. Thus equation (8) shows that the expected number of missiles launchable $E[M_S(t)]$ at time t by SSBN s is the basic building block in studying expected strategic readiness.

In the next section we shall describe a technique for estimating $E[M_S(t)]$ for each SSBN and for selected times t_j.

[8] "Operable" will mean not inoperable due to lack of spares; that is, do not wish to count missile down-time due to planned maintenance, or the time necessary to locate a failure and install a spare. We also do not count down-time because of applications which are not replaceable during patrol. These types of down-time are not chargeable to the logistics support system.

4. THE STRUCTURE OF THE MEASUREMENT SCHEME

4.1 The Support/Operational Decomposition

Using conditional expectation, expression (8) may be written

$$E\left[M_s(t)\right] = E_{N_{s,k(t)}}\left[E_{D_{s,t}}\left[M_s(t) \mid N_{s,k(t)}\right]\right]. \tag{9}$$

$N_{s,k(t)}$ denotes the random vector of numbers of spares of each missile related FSN on board submarine s at the start of its patrol in process at time t (denoted the $k(t)$th patrol). $D_{s,t}$ denotes the random vector of failures of each missile related RSN occuring during the $k(t)$th patrol of submarine s up to time t. Expression (9) suggests that the problem can be decomposed into a conditional expected readiness calculation and a supply system simulation. Finding the conditional expectation of $M_s(t)$ for a fixed kit of spares $N_{s\,k(t)}(t)$ is a reliability problem. The support system effects $M_s(t)$ only through $N_{sk(t)}$. Analytical models are available [1], [2], [3] to calculate

$$E_{D_{s,t}}\left[M_s(t) \mid N_{s,k(t)}\right]. \tag{10}$$

Having available methods for calculating or estimating (10), we need only be concerned with methods for producing realizations of the quantities

$$N_{s,k(t)} \tag{11}$$

from their proper distributions. Suppose that for fixed s and $k(t)$, say s and k, realizations $N_{s,k}^{(1)}, N_{s,k}^{(2)}, \ldots, N_{s,k}^{(m)}$ have been samples from the distribution of $N_{s,k}$. Then an unbiased estimate of $E\left[M_s(t)\right]$ is obtained from

$$E\left[M_s(t)\right] = \frac{1}{m} \sum_{l=1}^{m} E_{D_{s,k}}\left[M_s(t) \mid N_{s,k}^{(l)}\right]. \tag{12}$$

4.2 The Support System Simulator

4.2 1. Basic Structure

Let us now focus attention on methods of sampling realizations of the quantities (11) from their proper distributions. The distributions of the quantities (11) are quite complicated in the FBM support system. We have tried to make simplying assumptions which will render the system more amenable to analysis, but which will not significantly alter the important characteristics of the system. The following two assumptions are made in this spirit

> For any two distinct FSNs i and j, demands on the third echelon for FSNs i and j form independent stochastic processes. This implies that failures of FSNs i and j on submarine s are statistically independent. (13)

> The number of failures of each FSN during a patrol is assumed to have the distribution of the number of incidents of a renewal process (i.e., if a failure occurs, there is a spare available so that the process can continue) but truncated at $A_{s,i} + N_{s,k,i}$, where $A_{s,i}$ denotes the number of applications of FSN i on submarine s, and $N_{s,k,i}$ denotes the number of spares of FSN i on board submarine s, at the start of its k'th patrol. (14)

Assumption (14) permits us to generate "virtual failures" for all missile related FSNs for the entire simulated period before any simulation is done. The virtual failure generator produces failures as for renewal processes. The truncation of these virtual failures occurs during the simulation.

The motivation for assumption (13) is that it permits the support system to be decomposed by cog; that is, since the cogs are funded independently, assumption (13) implies that the random vectors N_{s_1,k_1,C_1} and N_{s_2,k_2,C_2} are independent for $C_1 \neq C_2$ and for any k_1, k_2, s_1, s_2 (where $N_{s\,k,C}$ denotes the vector of number of spares of missile related FSNs in cog C on board submarines s at the start of its kth patrol). This result reduces very significantly the storage requirements for the supply system simulator. The supply system simulator need only treat one cog at a time. The simulator samples realizations of

$$N_{s,k,C} \tag{15}$$

For fixed cog C, we run such a simulation to sample (15) for all s and for those k of interest (that is, for our time horizon of interest).

The decompositions thus far discussed are represented graphically in Figure 3. The figure indicates that the measurement scheme consists of a supply system simulator and a conditional expected readiness calculator. The supply system simulator consists of a sequence of single cog simulations, each simulation for the entire time horizon of interest. In each single cog simulation the virtual demand generator first produces realizations of the number of virtual failures for each missile related FSN in the cog for each patrol of each submarine. These are all generated before any simulation of the cog takes place. The logistics simulation of the cog consists of feeding all demands for missile related FSNs into the upper two echelons of the supply system. The supply system is simulated in detail. Inventory reviews are performed at the start of each quarter, and CONUS replenishments may have to be postponed at various times because of lack of funds. Demands for non-missile related FSNs also enter conus and utilize the major of the cog budget. During the simulation, the quantities (15) are saved. In determining the demands placed by a submarine on its tender during refit, the virtual demands may have to be truncated.

4.2 2. Aggregation Within a Cog

Because of the large number of FSNs in some cogs, a simulation of the supply system even one cog at a time is computationally infeasible. But the supply system has an interesting structural aspect which can be utilized. In most cogs, the missile related FSNs constitute only a very small percentage of the total cog population. The large majority of the FSNs, the non-missile related FSNs, cannot be ignored because of the common funding within the cog. Since the non-missile related FSNs are of interest

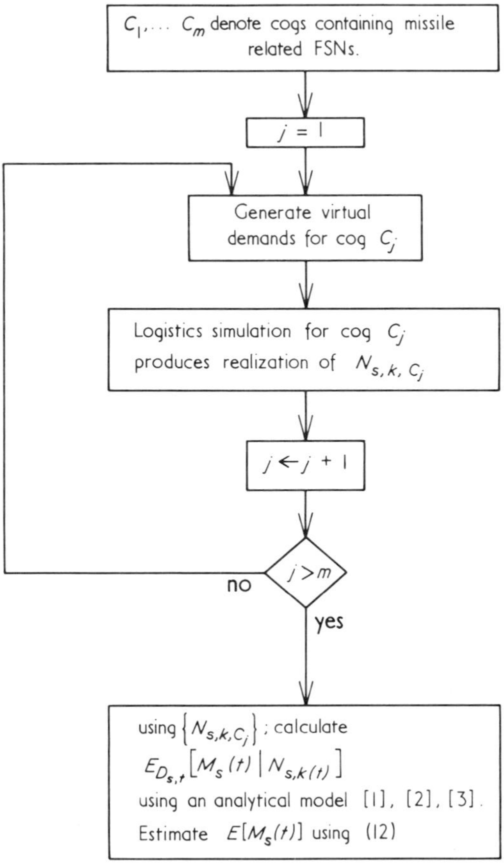

Figure 3

only in terms of their total dollar pull on the third echelon as a function of time, a technique has been developed to treat these FSNs in aggregate. In this way, the logistics simulation itself need only concern itself with the flow of the missile related FSNs. Consider the stochastic process

$$\left\{ \$_{nmr,p}(t)\ ,\ 0 \leq t \leq 1 \right\} \ , \tag{16}$$

where $\$_{nmr,p}(t)$ represents the dollar value of demand for third echelon purchases of non-missile related FSNs in the cog from the start of quarter p of the simulation up to time t, where t=1 denotes the end of quarter p. The phrase "dollar value of demand . . ." indicates the amount that would be spent by the third echelon for replenishment if orders were never postponed due to lack of funds. The probability laws governing (16) depend upon the initial asset positions at CONUS at the start of quarter p for non-missile relates FSNs in the cog, and upon the third echelon replenishment policies during the quarter. The non-missile related FSNs in a cog affect the missile related FSNs in the cog only through the stochastic process (16). We are not concerned with the flow of individual non-missile related FSNs through the supply system.

Hence, if we can obtain the probability laws governing (16) without simulating non-missile related FSNs, then it is very desirable to treat them in aggregate.

In the logistic simulation, information about the stochastic processes (16) is required for two purposes. First, at the start of each quarter an inventory review is performed. The re-order points of all FSNs in the cog are set by means of the technique described in Section II to make the expected CONUS cog replenishment expenditure for the remainder of the year fall within a tolerance of the remaining cog budget. This presents us with the problem of estimating the expected replenishment expenditure for the remainder of the year for the aggregate of non-missile related FSNs in the cog as a function of the shortage cost of the inventory review technique. This estimate must be made without knowing the asset levels of the non-missile related FSNs at the time of the review, if they are not to be simulated.

Secondly, the stochastic processes (16) are used in the logistic simulation as follows. For each quarter we need a realization of (16) for that quarter to reflect the utilization of the cog budget by the non-missile related FSNs. As time proceeds in the simulation, the cog budget must be continuously decremented by the amount of this realization.

Let us now present our method of handling these problems, and of thereby treating the non-missile related FSNs of the cog in aggregate. We first note that for any FSN i, specifying a shortage cost λ, determines the third echelon replenishment policy $(s_i(\lambda), Q_i)$ that will be used. For a third echelon asset position of "a," the value of $E\,[\$_i(a,\lambda,q)$ can be estimated from (3) with "q" replacing "4", or the following exact expression can be used

$$E\left[\$_i(a,\lambda,q)\right] = c_i Q_i \sum_{n=0}^{\infty} (n+1)\, Pr\left[nQ_i + a_i - s_i(\lambda) \leq D_i(q) < (n+1)Q_i + a_i - s_i(\lambda)\right], \tag{17}$$

where $D_i(q)$ denotes the randon demand on CONUS over the next q quarters for units of FSN i. The exact expression for $Var\ \$_i(a,\lambda,q)$ is

$$Var\left[\$_i(a,\lambda,q)\right] = (c_i Q_i)^2 \sum_{n=0}^{\infty} (n+1)^2\, Pr\left[nQ_i + a - s_i(\lambda) \leq D_i(q) < (n+1)Q_i + a - s_i(\lambda)\right] - E^2\left[\$_i(a,\lambda,q)\right]. \tag{18}$$

For the geometric distribution of demand, a closed form expression can be obtained for the mean and variance of $\$_i(a,\lambda,q)$ [4]. For high demand items, expression (18) can be approximated by

$$Var\left[\$_i(a, \lambda, q)\right] \approx c_i^2 \quad Var\left[D_i(q)\right] \tag{19}$$

Let us first consider the inventory review problem. Let λ_2 denote the trial shortage cost, and suppose that there are q quarters remaining in the year. Let $\lambda_1, \lambda_0, \lambda_{-1}, \ldots$ denote the shortage costs used for the cog the previous and penultimate quarters, respectively. If non-missile related FSN i is a high demand item, then its current third echelon asset position will be approximately uniformly distributed on the set $\{ s_i(\lambda_1) + 1, \ldots, s_i(\lambda_1) + Q_i \}$ [5]. Thus, we estimate the expected third echelon expenditure for this item over the next q quarters using a policy $(s_i(\lambda_2)\, Q_i)$ to be

$$E[\$_i(\lambda_1, \lambda_2, q)] \approx \frac{1}{Q_i} \sum_{j=1}^{Q_i} E[\$_i(s_i(\lambda_1) + j, \lambda_2, q)] \tag{20}$$

If FSN *is* is a slow moving item, we do not wish to make this uniformity assumption. In such a case, $\lambda_1, \lambda_0, \lambda_{-1}, \ldots$ are used to estimate the distribution of the current asset position. The quantity $E[\$_i(a, \lambda_2 q)]$ can thereby be estimated with "a" unknown.[9] We shall not discuss these estimators here, but shall merely denote the estimator used by [6]

$$E[\$_i(\bar{\lambda}_1, \lambda_2, q)]. \tag{21}$$

In the same way that (21) is obtained

$$Var[\$_i(\bar{\lambda}, \lambda_2, 1)] \tag{22}$$

is obtained as an estimator of $Var[\$_i(a, \lambda_2, 1)]$ with the current asset position "a" unknown except that $\lambda = (\lambda_1, \lambda_0, \lambda_{-1}, \ldots)$ were used in setting the third echelon replenishment policies for the previous periods. For a high demand FSN, we have analogous to (20) that $Var[\$_i(\lambda, \lambda_2, 1)]$

$$= \frac{1}{Q_i} \sum_{j=1}^{Q_i} Var[\$_i(s_i(\lambda_1) + j, \lambda_2, 1)] + \frac{1}{Q_i} \sum_{j=1}^{Q_i} \left\{ E[\$_i(s_i(\lambda_1) + j, \lambda_2, 1)] - E[\$_i(\lambda_1, \lambda_2, 1)] \right\}^2$$

$$= \frac{1}{Q_i} \sum_{J=1}^{Q_i} \left\{ Var[\$_i(s(\lambda_1) + j, \lambda_2, 1)] + E^2[\$_i(s(\lambda_1) + j, \lambda_2, 1) \right\}$$

$$-E^2[\$_i(\lambda_1, \lambda_2, 1)]. \tag{23}$$

9 For very slow moving items, the re-order point will be independent of the imputed shortage cost because of an over-ride on expression (2). Such items are ordered on a 1-1 basis and cause no difficulty since their asset positions are constant.

Prior to any simulation or generation of virtual demands, the quantities

$$\mu_{nmr}(\bar{\lambda}, \lambda_2, q) = \sum_{i \epsilon nmr} E\,[\,\$_i(\bar{\lambda}, \lambda_2, q)\,]\,, \tag{24}$$

$$\sigma^2_{nmr}(\bar{\lambda}, \lambda_2, 1) = \sum_{i \epsilon nmr} Var\,[\,\$_i(\bar{\lambda}, \lambda_2, 1)\,] \tag{25}$$

are tabulated as functions of λ, λ_2, and q (for variance we need only $q = 1$). This calculation can be performed from the data base and does not depend upon simulation results. At the start of each quarter during the simulation, an inventory review is performed as previously described except that for a trial shortage cost λ_2, the aggregate dollar pull through the remainder of the year for all of the non-missile related FSNs is estimated by $\mu_{nmr}(\lambda, \lambda_2, q)$ where q denotes the number of quarters remaining in the year.

Because of our independence assumption (13), and because of the very large number of non-missile related FSNs in a cog, it follows from the Central Limit Theorem that $\$_{nmr,p}(1)$ is approximately normally distributed with mean

$\mu_{nmr}(\bar{\lambda}, \lambda_2, 1)$ and variance

$\sigma^2_{nmr}(\bar{\lambda}, \lambda_2, 1)$;[10] that is,

$$\$_{nmr,p}(1) \sim N(\mu_{nmr}(\bar{\lambda}, \lambda_2, 1), \sigma^2_{nmr}(\bar{\lambda}, \lambda_2, 1)). \tag{26}$$

During the simulation, after performing the inventory review at the start of a quarter to determine an acceptable λ_2, a realization of (16) is generated for the quarter by sampling $\$_{nmr,p}(1)$ from (26) and then forming

$$\hat{\$}_{nmr,p}(t) = t\$_{nmr,p}(1)\quad,\quad 0 \leq t \leq 1. \tag{27}$$

This assumes a linear rate of CONUS spending throughout the quarter for non-missile related FSNs.[11] Methods have also been devised to introduce more variability into this dollar pull, e.g., a normal process with independent increments.

Thus, the logistics simulator need only simulate the missile related FSNs in the cog. At the start of each quarter an inventory review is performed. Since the missile related FSNs are simulated, their asset levels are known and expression (3) can be used

[10] Data analysis has shown that in most cogs there are a few non-missile related FSNs whose variance dominates the summation (25). These FSNs are taken out and simulated along with the missile related FSNs in the cog in order to improve the accuracy of the normal approximation.

[11] If this is the first quarter of a year, then before the inventory review is performed or the non-missile related dollar pull is generated, the third echelon postponed buys for the previous year (if there are any) are placed according to the old replenishment policies.

(replacing "4" by the appropriate "*q*"). For the non-missile related FSNs, expression (24) is used. Iteration on λ is performed until the expected CONUS expenditure through the remainder of the year falls within a tolerance of the remaining cog budget. Having thereby obtained an acceptable λ for the quarter, call it λ_2, a dollar pull on CONUS for non-missile related FSNs is generated for the quarter. This is done by sampling $\$_{nmr,p}(1)$ from (26) and then utilizing (29). The demands of the missile related FSNs are then simulated for the quarter. The cog budget is continuously decremented[12] as dictated by (27). If the budget is depleted during the quarter, then CONUS replenishment buys are postponed till the start of the next fiscal year. In any event, at the start of the next quarter the procedure is repeated.

4.2.3 Decomposition Within a Cog

In the previous sections we have shown how the logistics simulation can be performed one cog at a time with the set of non-missile related FSNs in the cog are treated in aggregate. Although the missile related FSNs in a cog generally constitute only a few percent of the total cog population, in some cogs there may be as many as several thousand missile related FSNs. An FSN has many attributes (unit price, demand characteristics, SSBNs it is used on, procurement lead time, etc.), hence simulating the missile related FSNs in one cog involves the storage and utilization of a considerable amount of information. Unless an exceptional amount of internal storage is available, it is necessary to store this information on an external device such as disk or drum. This is very undesirable, because the constant transference of information between external and internal storage will be very time consuming.

In this section we indicate an approximate technique for decoupling the missile related FSNs in a cog so that the logistic simulation can be performed one missile related FSN at a time; each single FSN simulation being for the entire time horizon of interest (several years). In each single FSN simulation there is no need to have communication with external storage (except to take as input the virtual demands generated beforehand). This technique utilizes a preprocessor which determines a shortage cost λ for the cog for each quarter of the entire simulated period. These shortage costs determine the CONUS replenishment policies for all FSNs in the cog for each quarter of the simulated period (by(1) and (2)). The preprocessor also computes the time during each year at which the cog budget is exhausted (for those years in which it is exhausted). The output of the preprocessor permits the logistics simulations to be performed one missile related FSN at a time, for the entire length of the simulated period. During this simulation inventory reviews are not performed nor are budgets decremented, because reorder points and times of budget depletion are already furnished.

In order for the preprocessor to accomplish the above two tasks without approximation, it is necessary that the exact dollar pulls on the third echelon be known for each quarter of the simulated period. This can be accomplished for the non-missile related FSNs, because an aggregate dollar pull can be generated directly on the third echelon. But for the simulated FSNs we have only a tape of virtual demands for the

[12] If the budget is depleted during the third quarter, then no inventory review is necessary at the start of the fourth quarter.

simulation run. This tape of demands can be transformed exactly into dollar pull only by running the supply system simulation, because many of these demands occur at the first and second echelons and do not directly generate demands for third echelon stock. In the preprocessor the following observation has been utilized.

> For most cogs, the total amount of money spent per year for third echelon replenishments of missile related FSNs is quite small compared to the total dollar pull for all FSNs in the cog.

This observation implies that the time of budget depletion depends almost entirely on the dollar pull of the non-missile related FSNs in the cog. Also, in performing inventory reviews, the dominant component of the expected expenditure comes from the non-missile related FSNs. If the above observation holds in the extreme, then in the preprocessor the missile related FSNs can be neglected. We shall not assume such an extreme case, but shall approximate the dollar pull from missile related FSNs in the aggregate manner described for the non-missile related FSNs.

Before any demand generation or simulation is performed, tabulate from the data base the quantities

$$\mu(\bar{\lambda}, \lambda_2, q) = \sum_{i \epsilon c} E[\$_i(\bar{\lambda}, \lambda_2, q)] \quad , \tag{28}$$

$$\mu_A(\bar{\lambda}, \lambda_2, q) = \sum_{i \epsilon A} E[\$_i(\bar{\lambda}, \lambda_2, q)] \quad , \tag{29}$$

$$\delta_{A^2}(\bar{\lambda}, \lambda_2, 1) = \sum_{i \epsilon A} Var[\$_i(\bar{\lambda}, \lambda_2, 1)] \quad , \tag{30}$$

where the first summation is over all of the FSNs in the cog and the other two summations are over the set A of aggregated FSNs. The set A consists of all FSNs in the cog except those few with "large relative variance." An FSN j is said to have a large relative variance if $c_j^2 \; Var\,[Dj(1)]$ is a considerable percentage ($\geqslant$ 5%) of $\sum_{i \epsilon C} c_i^2 \; Var\,[D_i(1)]$ where c_i and $D_i(1)$ are the unit price and random quarterly demand on CONUS respectively.

After the functions (28), (29), and (30) are tabulated, virtual demands are generated for the missile related FSNs in the cog, and for those non-missile related FSNs with large relative variance. The preprocessor is then run. The operation of the preprocessor is as follows. We are initially furnished with the vector λ of shortage costs used for the cog in the several quarters preceding the start of the simulated period, and with the effective budget B for the first simulated year (the allocated budget minus the value of postponed buys for the previous year). An inventory review is performed to obtain an acceptable shortage cost λ_2 for the first quarter of the first simulated year. λ_2 is selected so that

$$\mu(\bar{\lambda}, \lambda_2, q) \approx B \tag{31}$$

where q = 4. An aggregate dollar pull is now generated for the first quarter. This is a total for all FSNs in the cog, and is denoted by

$$\left\{ \$_p(t), \qquad 0 \leq t \leq 1 \right. \tag{32}$$

The subscript p indexes the quarters, and $t = 1$ corresponds to the end of the quarter. A realization of (32) is generated as follows

(a) Sample x from $N(\mu_A(\bar{\lambda}, \lambda_2, 1), \quad \sigma_A^2(\bar{\lambda}, \lambda_2, 1))$.

(b) Process the tape of virtual demands for the FSNs in the cog but not in the set A to approximate their total dollar pull for the quarter.[13] Let y denote this amount.

(c) Let $\hat{\$}_p(t) = t(x+y), \quad 0 \leq t \leq 1,$

be the realization of (32) for the quarter.

Having generated the aggregate dollar pull for the first quarter, we check to make sure that the budget was not exhausted. Assuming that it was not, the effective budget for the remainder of the year is $B\text{-}\$_1(1)$. This is the new B. λ_2 is included in the new λ. We return to expression (31) and perform an inventory review to obtain λ_2 for the second quarter. In using expression (31), $q = 3$ is now used since there are three quarters remaining in the year. An acceptable λ_2 is obtained and then an aggregate dollar pull for the second quarter is generated using the procedure above.[14] We proceed to the third quarter in the same way as we proceeded from the first quarter to the second quarter. After processing the third quarter, we continue on to the fourth quarter.

Let us consider what happens if the budget is depleted during the year. Suppose, for example, that after processing the third quarter, we find that $\$_3(1) > B$, where B denotes the remaining budget at the start of the third quarter. The $t_{\hat{e}} = B/\hat{\$}_3(1)$

13 The dollar pull for the non-missile related FSNs not in A can be obtained exactly. This is because the demands for these items as given on the virtual demand tape feed directly into CONUS and we can keep track of their CONUS asset positions (there are but a few such items). Thus, at the start of any quarter of the preprocessor operation we know the CONUS asset position for such an FSN, the inventory review gives the CONUS replenishment policy for the quarter, and the virtual demand tape gives the number of demands on CONUS during the quarter for the FSN. The number of CONUS purchases during the quarter can thus be calculated. For the missile related FSNs we cannot obtain the exact dollar pull, because the demands on the virtual demand tape may not all be placed with CONUS during the quarter. This is because they are "virtual demands" which may have to be truncated during the simulation, and also because a demand placed with the tender will not be demanded of CONUS till the tender asset position falls to the tender reorder point, perhaps during a later quarter. If the tender uses an $(S\text{-}1, S)$ replenishment policy, then this latter effect does not exist. In processing the virtual demand tape for a missile related FSN, we assume that all demands for the FSN during the quarter are placed with CONUS, and we thus treat them in the same manner as the non-missile related FSNs.

14 The set A does not change.

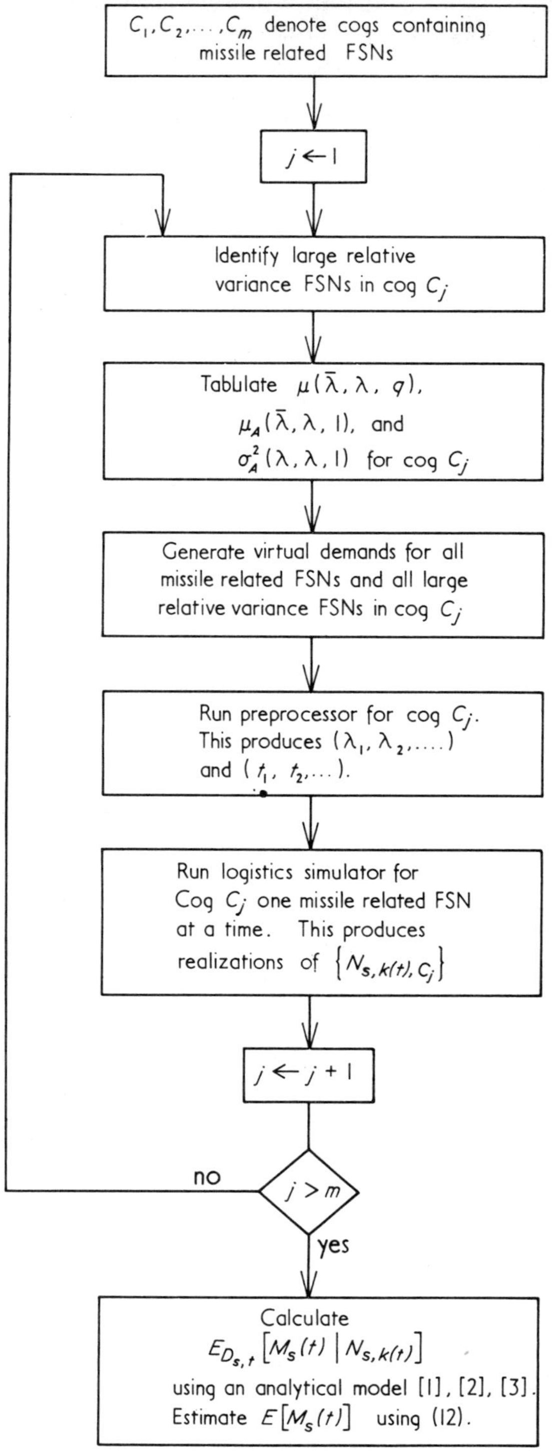
$C_1, C_2, \ldots, C_m$ denote cogs containing missile related FSNs
$j \leftarrow 1$
Identify large relative variance FSNs in cog C_j
Tabulate $\mu(\bar{\lambda}, \lambda, q)$, $\mu_A(\bar{\lambda}, \lambda, 1)$, and $\sigma_A^2(\lambda, \lambda, 1)$ for cog C_j
Generate virtual demands for all missile related FSNs and all large relative variance FSNs in cog C_j
Run preprocessor for cog C_j. This produces $(\lambda_1, \lambda_2, \ldots)$ and $(t_1, t_2, \ldots)$.
Run logistics simulator for Cog C_j one missile related FSN at a time. This produces realizations of $\{N_{s,k(t),C_j}\}$
$j \leftarrow j+1$
no
$j > m$
yes
Calculate $E_{D_{s,t}}\left[M_s(t) \mid N_{s,k(t)}\right]$ using an analytical model [1], [2], [3]. Estimate $E[M_s(t)]$ using (12).

Figure 4

denotes the time during the third quarter at which the budget was exhausted. We record $t_1 = 6 + 3t_e$ as the time (in months) during the first year at which the cog budget was exhausted. We also record $P = (1\text{-}t_e)\hat{\$}_3\ (1)$ as the dollar value of postponed buys in the third quarter. An inventory review is not performed at the start of the fourth quarter, but rather, λ_2 is set equal to the λ_2 for the third quarter. λ is updated and a realization of (32) is generated for the fourth quarter in the usual way. Since there is no money available, this aggregate dollar pull only increases the dollar value of postponed buys, $P \leftarrow P + \hat{\$}_4(1)$.

The next year in the preprocessor is treated in the same way as the first year, etc.

The output of the preprocessor is a sequence of shortage costs for the cog, one for each quarter of the simulated period, and a time during each year at which third echelon buys are postponed till the start of the next year. The logistics simulation may then be run one missile related FSN at a time, each single FSN simulation being for the entire length of the simulated period. Those non-missile related FSNs which were not included in the class A of aggregated FSNs are really not simulated. They were simulated only in the sense that realizations of their demands were produced by the virtual demand generator for use in the preprocessor.

REFERENCES

1. Hirsch, W.M., M. Meisner, and C. Boll, "Cannibalization in Multicomponent Systems and the Theory of Reliability," *Naval Research Logistics Quarterly,* Vol. 15, No. 3, pp. 331-359.

2. Simon, R.M., "The Reliability of Multi-State Systems Subject to Cannibalization," Report AM-69-1, School of Engineering and Applied Science, Washington University, St. Louis, Missouri.

3. Howard, W.J., "A Multi-Echelon Markov Model for Relating Supply System Performance to Fleet Readiness," TEMPO-General Electric Company, 67TMP-123, Santa Barbara, California, December 1967.

4. Johnson, C., and H. Stanfield, Unpublished Notes, TEMPO-General Electric Company, Santa Barbara, California.

5. Simon, R.M., "The Uniform Distribution of Inventory Position for Continuous Review (*s, Q*) Policies," The RAND Corporation, P3938, September 1968.

6. Final Technical Manual, Project MARIS, TEMPO-General Electric Company, Santa Barbara, California, August 1969.

DISCUSSION

Buck: Is there any way of taking $ from one "COG" to another when effectiveness may be improved?

Truelove:*We considered only current Navy policy, for which each COG budget is fixed. However, obviously, this could be done.

Olivier: The shortage cost is computed from the budget and the civilian industries appreciate the chance of the Navy. However, if the method allows for good sharing of the allocated budget, which procedure is used to determine the m.r. items?

Truelove: N.m.r. items are determined by function–however, there is provision for "Essentiality" of each item (see the paper).

* The paper by Messrs. Simon and Ball was presented by Dr. Truelove

A Model for the Evolution of the Royal Navy's Upkeep System

K.J. Rawson

Busy people are not quickly persuaded that they might be able to do their jobs better. There is a deep reluctance to sweep away existing organizations which have been through the settling phase during which they will have experienced the painful process of accommodating themselves to the people available to run them. Scepticism of good ideas before they have become practical proposals is understandable. Rarely is there an opportunity to begin system design with a blank sheet of paper.

On the other hand, it is often unwise to improve patches of an organization without a general look at the whole. It is folly, for example, to try to introduce a management information system without re-examining, fundamentally, the form of management that it is to serve.

What seems to be needed is a simple, preferably pictorial, plan of evolution recognizing the present and the future. It needs to be comprehensible to all levels of managers, to specialists and to administrators. Changes need to be seen in the context of the overall plan with its complex interfaces and all levels of managers must be able to interest themselves in the development.

1. TRADITIONAL MATERIAL MANAGEMENT

Over the years, the management of material for the Royal Navy has evolved upon lines traditional to many large owners. Management has been effected by large organizations converging at the very top. These organizations have dealt successfully in-house with concept, design, procurement, operation, logistic support maintenance, renovation, finance, manpower and disposal. Design and procurement of ships, weapons and submarines have been largely contained within their functional boundaries thus encouraging much professional expertise to flourish. The maintenance and the logistic support organizations have likewise developed their particular skills. Pride and loyalty have created fine esprits de corps which has, however, tended to encourage isolation of each organization.

The material objective implicit in situations of this type is basically technological excellence. Prime emphasis is placed upon levels of capability which are achievable, within the bounds of scientific knowledge, by ships and weapons. There is an insistence on high quality for its own sake. Constraints imposed by the other organizations are resisted. Pursuit of such an objective does assume

(a) that the upkeep system can ensure that the performance is always available for use

(b) that high quality is always worthwhile.

Times have passed when we can afford to make such assumptions. Few would now venture into major design exercises without assurance that their products will be adequately available for use, nor would they specify high quality without assurance that it is directed towards a necessary standard of reliability.

Yet isolation of the different parts of a material organization encourages these questionable presumptions. It is tempting to those charged with performance design to consider the upkeep problems remote enough (unless they are specifically charged with them) to be largely ignored; engineers who have no direct responsibility for finance will inevitably satisfy their professional pride by specifying only prime quality, scientists asked only for high levels of capability will approach the barriers of science without constraint; logisitics organizations will stockpile heavily so that they will never be caught out of stock; maintenance facilities will be managed on the basis of ample spare capacity to meet any eventuality. Financial constraint imposed upon such situations creates in-fighting. The top board of management which is the only place where the organizations converge, are overwehlmed with detail.

In an attempt to overcome these difficulties of isolation, procedures, techniques and tools (often mistaken for panaceas), are applied to the organizations to create bridges amongst them. Thus value engineering seeks to bridge design, procurement and cost; quality assurance seeks to bridge procurement and operation; reliability engineering seeks to bridge design, maintenance and logistics; configuration control seeks to bridge procurement, logistics and renovation; design disclosure devices seek to bridge design and maintenance. There are many more such important devices. The Committee and the Working Party are also used as bridges, often extensively and on issues fundamental to overall policy. Unfortunately such tools often blur responsibilities and create conflicts of loyalties.

Traditional managers of design, becoming aware of the loss of availability due to the unreliability associated with increasing complexity, are often tempted to impose a general policy of repair-by-replacement. Combination of the policies of technological excellence and repair-by-replacement can, of course, be prohibitively expensive.

It takes courage for the top management to question a basic philosophy of many years standing which they have themselves served during their careers and have now inherited to administer. This is particularly so when that traditional philosophy has been the emotional one of 'nothing but the best for the fighting man'. Yet financial and political pressures force them to turn their thoughts to ensuring an adequate system effectiveness for the money and men available. It is fortunate that predictive techniques and data retrieval systems have been developed in time to permit such a change. The change is, in fact, more fundamental than at first it appears.

A first definition of the upkeep system for the Royal Navy, as it happens to exist today is given in Fig.1. It is defined by the eleven resources which contribute to upkeep and each resource is, of course, itself a system. The capital investment in this system is measured in thousands of millions of pounds and tens of thousands of men. The table indicates how each of the resources would be measured in terms of the two elements of its effectiveness, capability and availability. The final column indicates the variables common to the resources, which provide the means of optimization.

Upkeep System Resources	Measurement of Resource in terms of its: Levels of Performance	Availability For use	Costs
1. Self Maintenance (1st Line)	Facilities, skills, standards of output, efficiency (items 1–5)	Place, numbers, periodicity, queueing, capital utilization (items 1–4)	Men, Money, Time
2. Afloat support (2nd Line)			,,
3. Base support (3rd Line)			,,
4. Dockyard and private yards (4th Line)			,,
5. Survey/Material Assessment			,,
6. Original design	Behaviour	Reliability Maintainability	,,
7. Original production	Quality	Extent of application of quality assurance	,,
8. Logistics	Module size	Risk of stock out, echelons, numbers	,,
9. Personnel	Skill levels, numbers	Leave, drafting, shift working	,,
10. Training	Standard of instruction	Throughput	,,
11. Information			
a. Management Data	Depth, type	Periodicity distribution	,,
b. Reference	,, ,,	,, ,,	

Figure 1 The Royal Navy's upkeep system in terms of its resources

Recognition of the system in this form is useful, in that, it does begin to indicate problem areas. To those with a knowledge of current responsibilities, for example, it demonstrates that the resources are not all responsible to the same top manager; nor are they required to optimize outside of their own boundaries; they are required to spend annually money and men which has been allocated to them on the basis of their own history of spending. However, such recognition is not in itself sufficiently persuasive to encourage us to embark upon a fundamentally new philosophy and organization. The doubts expressed in the USA [1] on the wisdom of such fundamental sudden changes, and the massive effort apparently required to pursue them, are sufficient to deter even the greatest enthusiasts in countries with more limited effort available.

However, the philosophy of cost effectiveness, be it expressed in the form of the WSEIAC approach or in the housewife's consumer association magazine, is fundamentally appealing. Indeed, even though it may be alleviated by the need for flexibility in military posture, its application is inevitable.

System engineering, like operational research is, too often, regarded as a collection of useful analytical tools [e.g. 2, 3]. Such tools cannot be applied without perception of the entire system, its inertia, and an objective for it. What is needed is a model which will permit a disciplined, evolutionary change in organization and outlook. This paper offers a possible model.

2. DECISION AREAS

The table in Fig.1, suitably developed, is helpful as the basis for management of a large support system once it has been created. Within a fixed budgetary ceiling, trade off amongst the resources can be effected to improve its output. Its output is, of course, the material state of the Navy. Thus expenditure on one resource must be counterbalanced by savings on another, the exchange resulting in a higher total output. The table reminds us how unwise it is, if our interest is in the whole system, to attempt to optimize one resource–such as the logistics, however large–without taking cognizance of its interfaces with the other resources.

In order to examine these interfaces, it is necessary to look at those processes which create the system output in the first place and which affect this material state of our ships throughout their lives. Fig.2 identifies those activities and their interrelationships in the form of a network describing the whole life of a ship. Each activity identifies some process which is undertaken to change the material state of the ship, during its whole life. A more detailed diagram of Fig. 2 is shown opposite.

The conventions used are as follows. Small circles enclose individual activities. Boxes enclose constraints or policies when they are inputs and enclose statements when they are outputs. Lines represent communication between activities. Each activity is therefore wholly prescribed by its inputs and its outputs, the standards of each needing definition. Large circles group together similar activities for convenient reference but have no great significance. The main chronological divisions are shown by the top lines, giving the extent of the phases concerned with concept, design, productions, deployment and refit.

The network as reproduced here, is not complete but it is in sufficient detail to demonstrate its uses:-

(a) it enables a long and complex material life to be perceived,

(b) it permits job definition and forms the basis for accountable management in the material field,

(c) it forms the basis for programming and control,

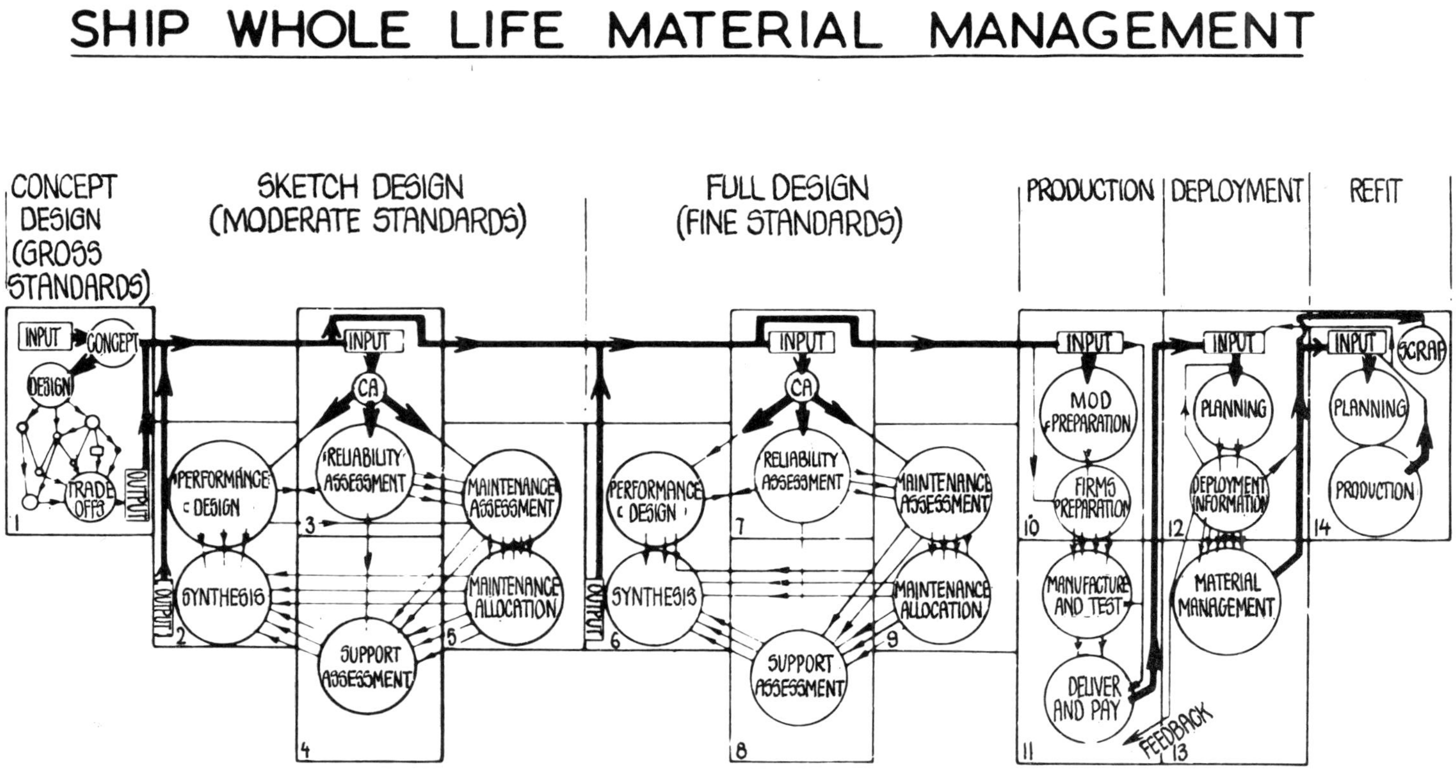

Figure 2 Ship whole life material management decision network

(d) it demonstrates how to formalise communications and to use them as a basic for control and documentation.

By showing the sequence of activities, the dependence of one manager upon another is clear. In the sketch design phase, for example, the broad cycle of events is

the creation of block diagrams
design for performance
reliability assessment
maintenance assessment
maintenance allocation
support assessment
optimization

Capability design by traditional methods [4], is seen to have important interfaces with other activities and cannot therefore be performed in isolation, if total system effectiveness is the objective. The processes are plainly iterative, carried out in successively greater depth. Logistics is observed to be well advanced in this sequence and dependent vitally upon the reliability and maintenance assessments. The logistics organization may therefore be expected to be based on critical interfaces with the organizations responsible for predicting reliability, maintenance and repair and also usage and consumption. It will need to be dependent upon successive updating of those predictions until actual experience is fed back during the deployment phase through the information system.

The network of Fig. 2 has been compiled for a total ship. A large part of the ship is, of course, composed of equipment and it may be presumed that many of the activities described will substantially comprise summations of the characteristics of equipment which are determined outside of specific ship applications. It is probably the activities at equipment level therefore which have the major effect upon the upkeep resources.

An activity network at equipment level is given in Fig.3. The same conventions apply as were used in Fig.2.

Fig.3 represents the ideal sequence of activities which should be put in motion at the design stage in order to decide in what replaceable elements an equipment should be designed. It is also applicable if these and certain of the other variables become invariable for reasons outside the designer's control. In the most general case there are four variables

(a) the size of replaceable element, (including the whole equipment), for repair purposes,

(b) the method of testing for failure, (ranging from visual to automatic test equipment),

(c) the characteristics of the repair resources (e.g. servicing lines, capacity),

(d) the spares stock out risk.

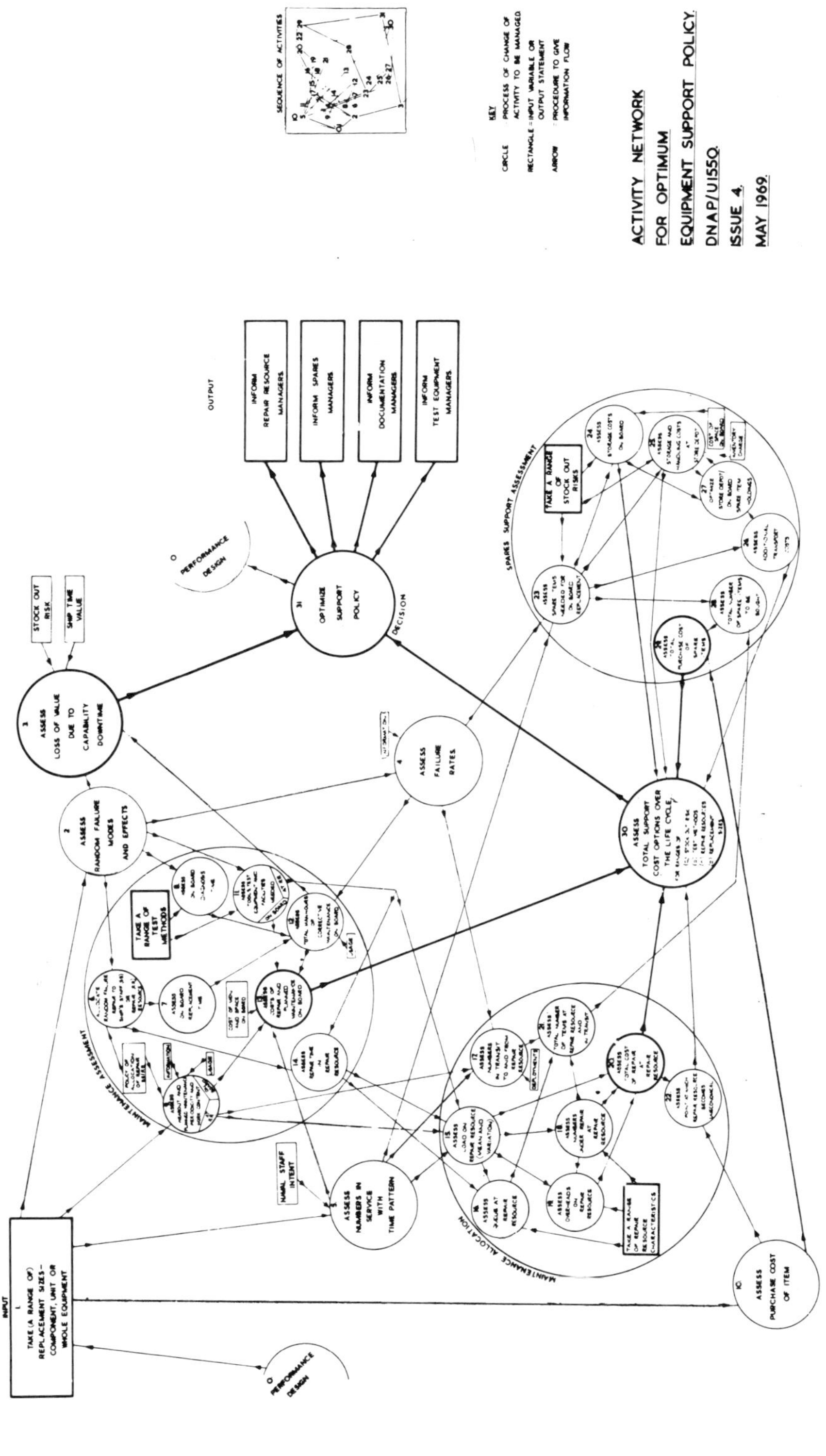

Figure 3 Activity network for optimum equipment support policy

These variables are not, in fact, independent and common sense will indicate that many of the combinations are impossible or inappropriate. In practice, therefore, the number of sensible combinations will be relatively small.

It has been presumed that each element of the equipment will exhibit a predictable failure density which is the superposition of an exponential distribution (called 'random failure') and Gaussian type distributions (called 'wearout'). If the two are taken together using, perhaps, a Weibull distribution, the first sequence of activities will require modification.

The first task is the prediction of the random failure modes, their effects and their rates of occurrence. This makes possible a group of activities, labelled 'maintenance assessment'–for example, diagnosis time, repair time, wearout period, corrective maintenance effort, tools and facilities needed. These facilities culminate in the assessment of the costs of planned and corrective maintenance onboard ship and the load which the equipment will place upon the upkeep system.

A sequence of activities follows for the repair resource. Assessments are made of the load falling upon it, the queue, the numbers under repair, its economic breakpoint, the numbers in transit etc. leading to the total cost at the repair resource for different repair resource characteristics.

Spares support assessment is triggered by the failure rate assessment and the numbers expected in service. The scaling of spares at the various echelons of supply are familiar problems to the operational researcher and these activities culminate in the assessment of the total purchase price of spares. Stock out risk is the variable introduced here.

The three costs–of repair on board, of repair at the resource, and of spares stocks–can now be added to give total support costs.

The network demonstrates the marriage of total support cost with value of the capability downtime. This results in a decision based upon optimization or trend or even judgement and this decision, perhaps following iteration of this whole process, is conveyed to the line managers responsible for documentation, repair resources, spares provisioning and test equipment. The designer, constrained by this decision can proceed to achieve the required capability standards and failure rates.

3. USES OF THE NETWORKS

The networks basically describe those activities which are necessary to a full prediction of the whole life of an article so that the whole life may be committed during design, with deliberation and on the basis of economy. They therefore represent a means of managing their prediction, according to the best information then available, and managing the different phases of their material lives, as the information improves.

Identified decision areas can now be allocated to managers who can be presented with precise terms of reference. In the case of the ship network, a project manager responsible for the entire design, for example, would expect to issue terms of reference–or, indeed, job specifications to give full accountability–to sub managers responsible for performance design, reliability assessment, maintenance assessment etc.

They in their turn would appoint lesser managers for each smaller activity. Sometimes one manager would embrace more than one activity if he would otherwise be under loaded or if communications were difficult between the activities.

Communications between managers can be formalized–and minimized, once they have been identified. They can be scheduled so that they may form the basis of a control system [e.g. 5, 6]. Each activity can be scrutinized to find out whether it costs more in time and money than it is worth. The decisions taken during the process can be formally recorded to form part of the documentation and configuration control.

While a project organization is probably very attractive for a whole ship design process, it may not be thought so for an equipment. The network assists the determination of a suitable organization by grouping together activities which require great expertise, high professionalism, large capital investment in machines, or the use of existing organization. Thus, it is an evolutionary tool in that, if the cohesion of existing organizations is more important than the project, they may exist together, at least for the present. The identification of the work to be performed does permit the appointment of a manager to do it and an examination of the compatibility of this work with any other loyalties that he possesses.

4. OPTIMIZATION

The equipment network attempts to isolate the four most influential variables in the problem. There are many more that could have been introduced. Young [7] tackles the problem of defining the best type of electronics printed circuit board for use in the Royal Navy by considering over 40 variables associated with equipment design, operation and support. This requires a computer program of some complexity.

The model described in this paper, while not obviating such an approach for special application, is intended to encourage an evolution of perception, organization, responsibility and procedure from those pertinent to technological excellence towards those pertinent to whole life economy. Initial application of such new philosophy would probably be crude. But with the adoption of the philosophy and the means to implement it through the model, there could, without regular disruptive re-organization, be a progression from the crude, manual optimization to computor operation of many variables as the former are found inadequate by the personal development of those people actually responsible for the implementation.

There are, of course, dangers in making a piecemeal approach. It does not forbid a resource sub optimization which could be damaging to overall optimization. However, these networks do permit the total system manager to observe the interfaces and to make a judgement on whether a sub-optimization is likely to be advantageous.

For example, sub-optimizations dealing with the best arrangement of spares echelons [e.g. 8] would probably be entirely acceptable, while the determination of the levels of stock would, in isolation from the repair resources, be unacceptable. Imperfections may well be introduced in this way. It is important, however, to accept that existing systems have considerable inertia. The sudden imposition of a novel, generalized process causes resentment to the people who have to implement it. People change slowly; training, while necessary, must not be presumed to transliterate their thoughts overnight. Furthermore, the more complex the problem, the more information is

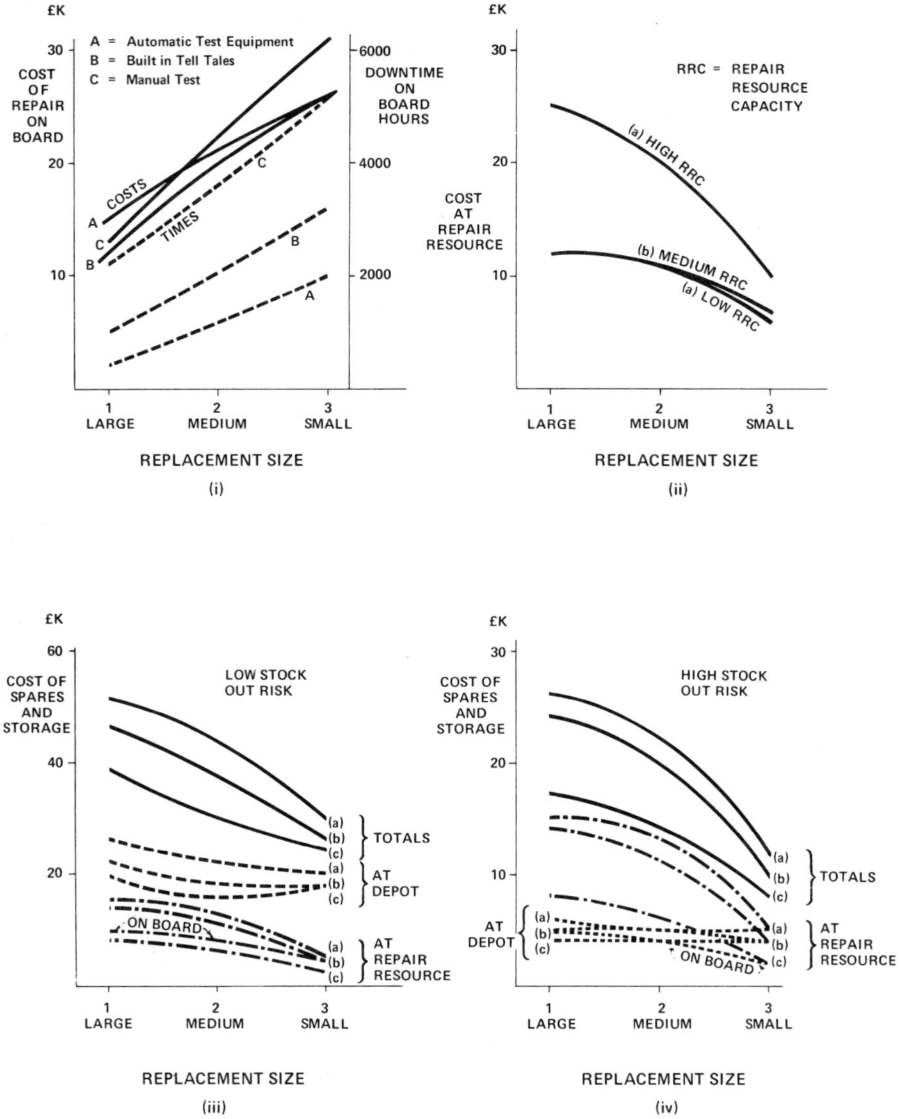

Figure 4

needed to feed it and the more operators are needed to operate it; these are expensive investments which must be judged worthwhile in advance of their commitment.

It has not been possible yet to obtain information to apply the equipment network in its entirety. To demonstrate its application, however, fictitious information has been used to produce the four parts of Fig.4, (which are consistent). The exercise presumes that the designer has the choice of three possible sizes of replaceable element for his equipment; size 1 represents the whole equipment; size 2 represents large module or sub-assembly replacement; size 3 represents small element or component replacement.

There is also a choice of test equipment to go on board, affecting diagnosis time. *A* represents automatic test equipment and therefore rapid diagnosis; *B* represents built-in tell-tales; *C* represents manual test and prolonged diagnosis time.

The choice of repair resource may arise from the number of service channels, number of employees, their skills and tools. These choices of characteristics are represented by High RRC when men will be idle at times awaiting work; Medium RRC when an occasional queue of defective items will develop; Low RRC when men will always be busy with plenty of work waiting for them.

The final choice open to the designer is simply stock out risk.

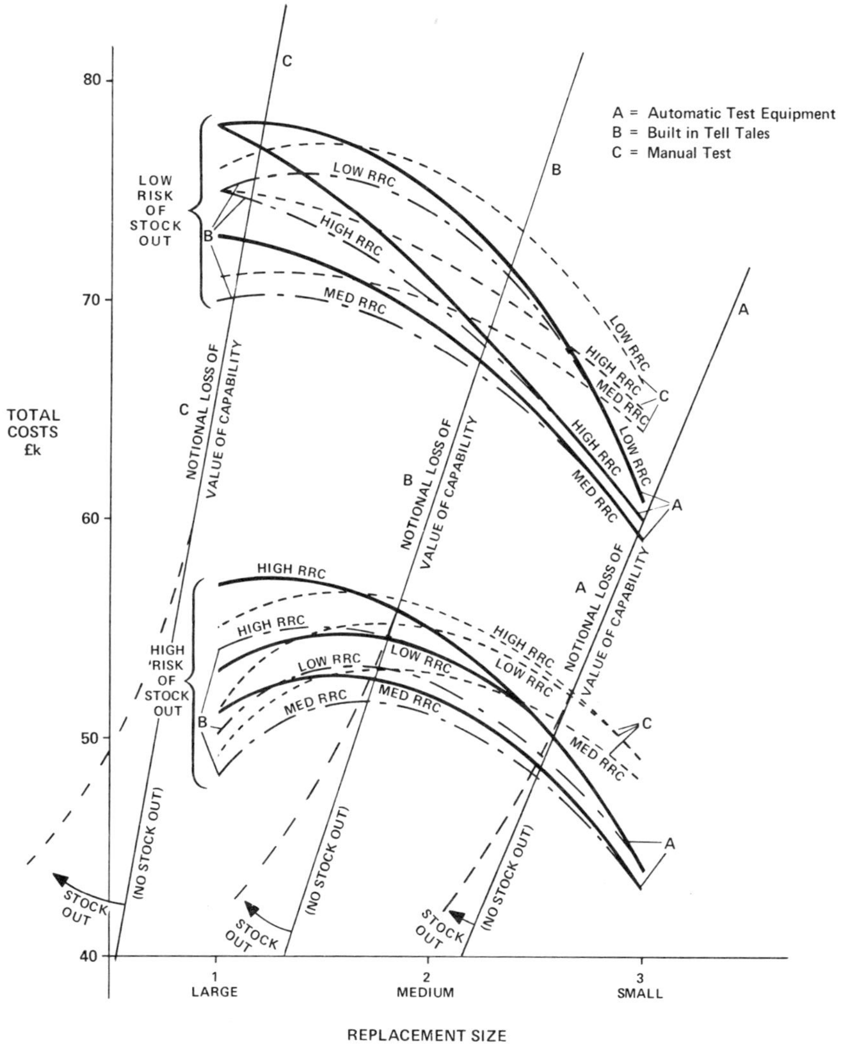

Figure 5

Fig. 4(i) shows the costs of maintenance on board derived from test equipment capital and upkeep costs, and equipment servicing diagnosis and replacement labour costs. Also provided are the downtimes of that equipment due to maintenance. These clearly depend upon the means of diagnosis and size of replaceable element.

Fig.4(ii) gives the labour and overhead costs at the repair resource, depending on its capacity.

Fig.4(iii) shows the cost of spares and their storage, on board, at a depot and awaiting repair at the repair resource. These stocks clearly depend upon stock out risk and Fig.(iv) has been produced similarly for a higher risk of stock out.

Total system costs derived from the four elements of Fig.4 are shown in Fig.5 as the two groups of curves. All variables have been retained, although it is now apparent that some of them are not very significant.

Also shown in Fig.5 are notional cost lines denoting a loss of value of the equipment's capability. These are derived from the downtime curves of Fig.4(i), presuming a financial value for time. These are drawn full on the presumption that, while there has been a risk of stock out, it has not actually occurred and so there has been no downtime waiting for spares. If stock out does occur, the wait is likely to be longer for the larger items because of more prolonged production and transport. The loss of value lines will therefore curve upwards as shown dotted. Fig.5 forms the basic tool by which the manager would make his judgement. In more sophisticated treatments, it would be an optimization. It would also be more numerate. Before making his decision, a manager would undoubtedly require some sensitivity analysis to be carried out, in particular, he may well require the assumed failure rate to form another variable. Of critical importance in this process is the financial value of fighting capability which increasing equipment complexity and stricter national economy renders increasingly pervasive. Many find the question repugnant. However, the question must be faced and figures allocated if a rational material state of naval forces is to be achieved. The Government, after all, places a financial value on defence forces in terms of a percentage of the gross national product.

These procedures do not always rely upon this final trade off between real cost and notional value. An actual application of these principles has been effected following Young [7]. The results of the computer print out are produced here as Fig.6. In this case, the size of the replaceable item is represented by its mean time between failures (MTBF) and constitutes a complete printed circuit board or a group of such boards forming a replaceable module. The policy decisions facing the designer were:

Policy A : repair on board ship
Policy B : replacement on board ship, repair effected at base.

Total costs of these options, together with some of the assumptions on which they are based, are given in Fig.6 showing an advantage to the smaller module and repair ashore; the manager also knows that Policy A leads to a higher downtime and therefore a greater loss of value.

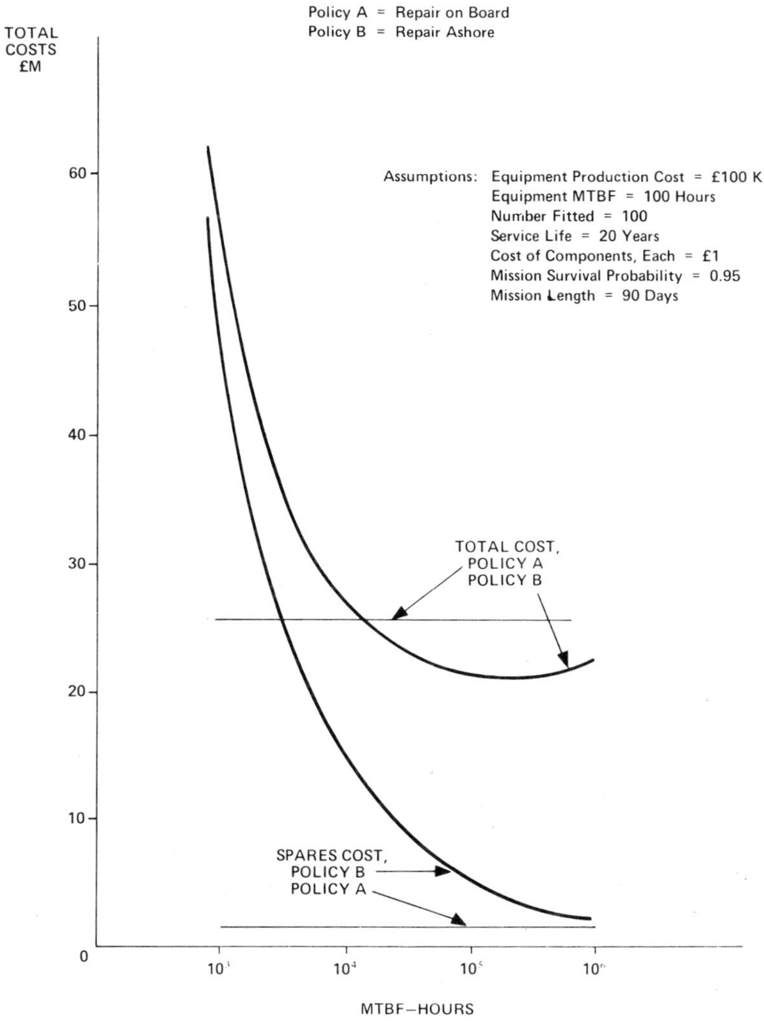

Figure 6 Printed circuit board maintenance policy

5. DEVELOPMENTS

Each of the networks presented imposes both a mental discipline and an organizational discipline. They provide the basis for the definition of task and its allocation to a manager either in existing organization or in a revised organization. By defining the communication amongst managers, they encourage the formalisation of these communications so that activities performed at any stage, be they engineering judgements or mathematical results are:-

(a) recorded, for subsequent updating,

(b) used for consequential activities,

(c) incorporated into logistics, performance and upkeep documentation.

The whole of the activities concerned with maintenance assessment for example, could probably be effected on a single form which could be transmitted to subsequent managers.

There are certain areas of crucial importance. Prime amongst these is undoubtedly the need for a data retrieval system structure from outputs required for the management of material on the basis of whole life effectiveness. In this, the Royal Navy is fortunate in having under development the Ship Upkeep Information System, a brief description of which is given by Button and Coates [9]. While not yet accompanied by a change in material policy or a declaration of intent, it clearly heralds such a change.

An exceedingly interesting application of the whole life costs of certain electrical equipments is given by Stokes and Stehle [10]. Phases of life similar to those of this paper receive retrospect cost allocations through a reporting system. These are then related by multiple regression analysis to characteristics of the equipments so that the costs formulae may be used for cost prediction. These predictions may then provide the basis for optimization as described herein. Such an approach, if found to be widely applicable, would avoid much of the painstaking synthesis of the whole life costs described in the equipment network. Predictive methods of this sort need more investigation.

A great deal of work yet requires to be done on the storage of numerate reliability data in accordance with the intensity of operating environment. Rawson [11] lists over 50 such ship environmental factors contributing to failure which is other than fair wear and tear and whose intensities vitally affect 'random' failure rates. While the application of numerate reliability techniques has advanced to a point where they are often part of the normal design process, the storage of failure rate data with environment has received curiously little attention. Such is the expense of reliability testing and of recording in the field that sharing of data is clearly desirable. With the critical position of numerate reliability in these predictive processes research in these areas is urgently needed.

The choice of the range of possible test equipments must depend upon the equipment or system under consideration. For electronics systems an excellent discussion of the degree to which automatic test equipment may be justified is given by Short [12]. Considerations of this type must precede the judgement of what range to include in the optimization process although it should not remove the possibility of choice altogether.

Perhaps the most conspicuous of the constraints upon change is that produced by the huge capital investment in upkeep resources—in the base and dockyard facilities, in the supply echelons, in the training and in the research and design processes. Each year, the further introduction of computers will render these upkeep system resources less conducive to change because of the great investments in structuring, coding and programming. A long term plan of evolution is therefore of the most profound importance to maintain cohesion amongst these resources so that they may continue to communicate in the interests of overall economy.

6. CONCLUDING REMARKS

The networks presented are only a beginning. They represent a model about which organizations that have grown up to a tradition of technological excellence can evolve to serve a policy of cost effectiveness. Before development can effectively take place,. there must be:-

(a) top management involvement; an exclusive material staff in concert with them should produce workable long term plans for each resource,

(b) a declaration of overall material policy,

(c) recognition of the human and capital constraints upon rapid change; change, however, needs to be positive and in accordance with a plan,

(d) recognition that logistics, like the other upkeep resources should not be treated in isolation,

Development of the model must then embrace:-

(e) development of common factors; these include ship models and files, codes, financial and material elements, language, methods of communication, manpower evaluation, reliability data storage, predictive techniques,

(f) structuring of organization, management systems, accountable units, management information systems, data retrieval systems and methods of training.

Some of these developments in the Royal Navy's upkeep system are now taking place.

ACKNOWLEDGEMENTS

This paper is presented with the permission of the Ministry of Defence; opinions expressed, however, are those of the author.

The author would like to express his gratitude to friends who have made many helpful suggestions. In particular, he is indebted for Fig.6 to Commander H.W. Young R.N. of whose development work this figure is a fragment.

REFERENCES

1. Frosch, Robert A., The Emerging Shape of Policies for the Acquisition of Major Systems. *Marine Technology,* July 1969.

2. Chestnut H., *Systems Engineering Methods.* Wiley 1967.

3. Sandler, G.H., *Systems Reliability Engineering* Prentice-Hall 1963.

4. Rawson K.J. and Tupper E.C., *Basic Ship Theory.* Longmans, 1968.

5. Bousquet R., Applications possibles de la gestion automatisée dans un chantier de construction et de réparation navales. *Association Technique Maritime et Aéronautique*, 1968.

6. Gurney, E.R., Un système de contrôle de production par ordinateur dans les arsenaux de la Royal Navy. Ibid 1969.

7. Young, H.W., Maintenance Philosophies for the 1970's. Symposium held by the Society of Electronics and Radio Technicians, July 1969.

8. Hadley G. and Whitin T., A model for procurement, allocation and redistribution for low demand items. *Naval Research Logistics Quarterly,* Dec.1961.

9. Button M.J. and Coates J.F., Naval material: advice on requirements and the feedback of service data to industry. Services/Industry conference on reliability 1968. Published by the Institution of Mechanical Engineers 1969.

10. Stokes R.G. and Stehle F.N., Some life cycle cost estimates for electronics equipments: methods and results. Annual Symposium on Reliability. IEEE.1968.

11. Rawson K.J., The Navy's need for reliability. Ibid. (9)

12. Short R.L., Shipborne Automatic Test Equipment in the Royal Navy. Cambridge conference on Electronics in the 1970's. Institution of Electrical and Radio Engineers 1968.

DISCUSSION

Buck: Do you have the same problem others have in getting supply and maintenance to work together?

Rawson: There will always be clashes of personality between supply and maintenance and, indeed, between designer and those responsible for support, those who train and those who allocate money and men. The better the organisation, the less the clashes matter. The crux of the matter is, of course, the statement of a common aim and the structuring of objectives for each contributor, in accordance with the aim. To this end, one must describe a large system in some form comprehensible to diverse managers so that they may evolve their organisations in accordance with an overall plan.

The Application of an Advanced Simulation System for Logistics and Transportation Planning.

D.F. Blumberg

1. INTRODUCTION

Many logistics managers are responsible for the development and implementation of complex logistics and maintenance support systems that defy analysis by any of the existing analytical techniques and inventory models. These logistics systems are characterized by having a mix of high-value expendable parts with relatively low demand over time. In addition, such complex support systems are generally composed of several locations where the repair and stock locations may not be the same. It is necessary to specify inventory levels at the several locations that will maximize the total system effectiveness, as well as to specify optimal maintenance and logistics operations policies and procedures. The achievement of this goal is complicated by the fact that any one location can, in time, draw upon the stock and maintenance capabilities of any other location. Because of this, it is impossible to maximize the system effectiveness by optimizing the stock and maintenance activity at each location independently of that at the other locations.

Using existing inventory models, (such as the Black and Proschan model), it is possible to optimize the stock levels for any one location. However, such models do not consider the interaction of the entire system, nor do they consider other dynamic conditions that affect the long range logistics and maintenance support systems performance. Existing models also do not consider such factors as:

. Variable repair capability by locations.

. Personnel levels, skills and capabilities available and utilized.

. Interaction of low-valve stock policy and high-value stock policy.

. Fixed and variable order cycles.

. Stock redistribution.

. Length of stock review period.

The ability to evaluate such interacting questions in the light of cost implications and ultimate logistics systems effectiveness is most desirable. The Logistics Analysis Simulation System (LASS) is directed towards this objective.

Simulation and related techniques in the field of Operations Research can be applied to the management and organization of logistical systems of any complexity and size. These techniques provide a flexible and analytic tool that can be used to determine the most effective support for a given logistics system budget.

Increased effectiveness will result from:

. Reduction in spares procurement costs for high-cost, low-demand spare parts.

. Maintenance or improvement in performance regardless of uncertainties of forecasting, obsolescence, ordering, holding, distributing, and activity phase-in/phase-out.

. Ability to obtain maximum system performance at any given level of system investment.

. Ability to continuously evaluate alternative logistics policies and procedures in light of experience.

Experience in utilizing logistics simulation techniques indicates that sizable reductions in logistics costs can be achieved at the same time that systems effectiveness is maintained or improved. In general, the simulation model approach offers these advantages:

. Allows analysis of systems having complex interrelationships and variables.

. Makes it easy to communicate the effects of alternative operational doctrines and system processes to management and command.

. Is flexible and adaptable to continuing system changes.

. Allows examination of total system dynamics and costs/budgets under varying assumptions.

Particular emphasis has also been given to the need for "self-learning" in the design of such an advanced analytical tool. The model design provides for continuous feedback from previous simulation runs which, in conjunction with data derived from the field, allows the analysis system to refine its own results. Thus, the simulation incorporates the capability to become increasingly accurate through use, evolving a realistic provisioning and maintenance policy.

In summary, LASS is an advanced analytical tool which can serve logistics analysts and managers. It reduces computational time, increases the number of alternative policies and concepts which can be efficiently evaluated, and serves as an aid to studying extremely complex situations using an orderly and logical model. However, the system is only a tool. It is designed only as an aid to, rather than a replacement for, competent logistics judgment and experience.

The system has also been designed to form the basis for a logistics management information system for continuous control and monitoring of Logistics Operations.

A flow chart of LASS showing the various submodels is shown in Figure 1.

2. METHODS OF LOGISTICS SYSTEMS MODELING

There are two kinds of simulation modeling techniques which are of interest from the standpoint of incorporating either directly or indirectly, relevant logistics factors into the environment. These are: (1) a Markov process model approximation to an inventory problem which has characteristics, particularly relevant to less expensive nonrepairable parts*; and (2) a model relevant to higher priced repairable spare parts*, developed by the Rand Corporation Logistics Department, involving the estimating of system effectiveness from a given overall spares budget allocated in a most efficient way. In both models, particularly the first, the value of each spare part to the system and the decline in this value over time through obsolescence or system phase-out is regarded as an important factor.

The reasons for selecting these particular formulations as components of the overall model should be briefly noted. For less expensive items, well-known and widely used policies lead to ordering whenever stock levels fall below a certain level, and enough stock is ordered to bring the level up to another prespecified level. Disposal of specified quantities also takes place when stock levels exceed certain quantities. Models based on this operating hypothesis (sometimes identified as (s,S) policies) have been consistently improved over the last decade, giving increasing ability to users to comprehend such factors.as: changing and improved estimates of demand rates; items becoming obsolete over time at a predictable rate; and various types of logistics costs. These have, generally, been described as Markov models. Computations have recently been developed which substantially reduce the computer time required for solution. With appropriate modification, this model implementing (s,S) policy is an attractive version for incorporation into a larger, total logistics system analysis framework.

Expensive (hi-value) spares, however, require significantly more complex management because of their repairable nature. In research on hi-value parts, significant advances have also been made in recent years, a major step being taken in the development of the $(s\text{-}1,s)$ policy by Feeney and Sherbrooke of the RAND Corporation.

Using the $(s\text{-}1,s)$ policy, the RAND Corporation developed two models. The first one, called an aggregate base stockage policy for recovereble items, is simply a model that applies the $(s\text{-}1,s)$ policy to a single location. This corresponds to the high value model of LASS' The second model is called METRIC—Multi-Echelon Technique for Recover-

* John Y. Lu, and R. J. Wolfson, *Dynamic Modeling of Inventories Subject to Obsolescence*, C-E-I-R, Inc. (Los Angeles Center) June 14, 1962; R. J. Wolfson, John Y. Lu, George W. Brown, "Dynamic Modeling of Inventories Subject to Obsolescence," *Management Science*, Sept. 1964, Vol. XI, No. 1, pp. 51-63. This modeling activity was originally conducted as research performed for the office of Advanced Logistics Research, Bureau of Supply and Accounts, U.S. Navy.

† G. J. Feeney and C. C. Sherbrooke, "Systems Analysis and Supply Management," The RAND Corporation, RAND Memorandum RM-4054-PR, April 1964; "The $(s\text{-}1,s)$ Inventory Policy under Compound Poisson Demand: A Theory of Recoverable Item Stockage," RAND Memorandum RM-4176-PR, Sept. 1964.

able Item Control.* METRIC deals with the situation where there is one depot and a number of bases. It attempts to optimize the stock levels at the bases and depot to minimize the total number of back orders. METRIC also has an algorithm for redistribution of stock between the bases and the depot, and for evaluation of performance for a given set of stock levels at the bases and depot. A few comments on the comparison between LASS and METRIC are given below.

2.1 Comparison between LASS and METRIC

1. Both models deal with the multi-location and multi-echelon inventory problem. However, METRIC is limited to one central depot supplying a number of locations while LASS can have the flexibility to model several depots supplying several locations.

2. The objective function in METRIC is fixed. This is the total number of back orders in the entire system. LASS, on the other hand, offers a choice to the user of four different objective functions. These four are ready rate, fill rate, service rate, and operational rate. In LASS, the user can choose any one of these for optimization and yet be able to evaluate the remaining three.

3. Both models assume a stuttering Poisson demand function which is estimated from historical data and then use Bayesian inference to project the demand in the future period.

4. METRIC assumes that all resupply is from the depot to the bases. This means that METRIC cannot evaluate the possibility of one base supplying another base. In LASS, on the other hand, in emergency situations, an item is obtained from the nearest base where it is available. The LASS system, therefore, offers more flexibility in this area.

5. METRIC assumes that all units have a probability r of being repaired at the base, and a probability $(1\text{-}r)$ of being returned to the depot for repair. LASS, on the other hand, specifies for each item, the echelon at which it can be repaired. Thus, in LASS, different units can be repaired at different locations.

6. METRIC assumes the same maintenance and repair capabilities at each location, while in LASS, the maintenance and repair capabilities at different locations could be quite different.

7. LASS allows evaluation of a much greater detail than METRIC in the area of maintenance capabilities, since in LASS we can model up to nine different skill levels for maintenance and specify for each high-value part, the minimum skill level required for repair.

8. LASS has options for facility and equipment phase-in and phase-out while METRIC does not.

* C.C. Sherbrooke, "METRIC–A Multi-echelon Technique for Recoverable Item Control", RAND report RM-5078-PR, November, 1966.

9. LASS can also evaluate more than two echelons while METRIC is limited to two echelons.

10. LASS can also evaluate the investment in test and repair equipment, while METRIC cannot.

11. Since LASS uses a simulation model, a history of the simulation process can be displayed in the output, thus giving the user an understanding of the operation of the system under different assumptions. METRIC does not have this capability.

In summary, LASS retains the basic analytical features of the METRIC model and combines these with a simulation model to provide flexibilities for evaluation of a total logistics support network. The use of simulation allows LASS to get around some of the simplifying assumptions made in METRIC, and also allows the evaluation of the impact of changes in some of these assumptions. LASS thus provides a logistics manager with a unique tool for evaluation of inter-related policy decisions in a complex logistics support system. The logistics manager, with the help of LASS, has the capability for evaluating not only the optimal stock levels by locations, but also other policy decisions as such as:

. Proportion of investment to be allocated to high-value parts, low-value parts.

. Repair capability requirement

. Number and type of personnel

. Investment in test and repair equipment

. Operation and maintenance procedures

. Echelons of stockage and repair between sites and source manufacturers.

. Transportation shipping and routing procedures.

. Reorder periods (low and high priority cycles)

. Stock review periods

. Amount of buffered parts (expressed as a percent of total parts)

. Optimal timing of site facility and equipment Phase-In and Phase-Out.

Another major advantage of LASS over METRIC is that LASS forms the basis for a logistic management information system in a real-time environment for continuous control and monitoring of logistics operations.

3. GENERAL DESCRIPTION OF LASS

The Logistics Analysis Simulation System consists of a set of mathematical models linked together by a simulation. The structure of the LASS system makes it flexible to be able to analyze complex logistics networks under varying assumptions and parameters.

The LASS system consists of four distinct submodels:

. The Low Value (*s*,*S*) submodel

. The High Value (*s*-1,*s*) submodel

. The Transportation/Distribution submodel

. The System Operation (SOS) submodel

A flow chart of the system is shown in Figure 1. A brief description of the submodels is given below:

3.1 The Low Value (*s*,*S*) Submodel:

The Low Value (*s*,*S*) submodel is a stochastic model useful for less expensive non-repairable parts with high demand. The significant costs in this submodel are holding (or inventory carrying) cost, stockout cost, and liquidation cost.

The "stock-out-cost" is the cost incurred whenever a part is demanded and is not in stock. In a commercial environment, this amounts to the profit lost because of missing a sale, as well as any prestige lost for not servicing the customer. In the military sense, a stock-out condition can have serious consequences in terms of causing a unit to operate ineffectively, thus imperiling an entire operation. The stock-out cost should be related to the effect of the part on the systems performance where low-effect parts have a small stock-out cost and high-effect parts have a high stock-out cost.

"Inventory carrying costs" or "holding cost" is the cost incurred when parts are held in stock. This includes the alternative cost of money invested in the inventory, as well as insurance, space, heat, and warehouse maintenance costs.

"Liquidation cost" refers to the loss that will be taken on the part when it becomes obsolete.

The total expected cost of inventory can then be determined for any given inventory level, if the probability distribution of demand is known. The model uses an iterative algorithm to determine the inventory level which minimizes total expected costs. A detailed description of the algorithm is given in [14].

3.2 High-Value (s-1,s) Submodel

The (*s*-1,*s*) submodel used in LASS is a model developed at the RAND Corporation for high-cost, repairable parts. These parts have such a high cost that the order cost is insignificant in considering the total system cost. Furthermore, the demand for these parts is very low so that it is difficult to predict the usage of these parts over a given time. In this case, a part is ordered every time one is taken from stock, and the failed part is not repaired and replaced in the stock. The inventory is used to keep the system

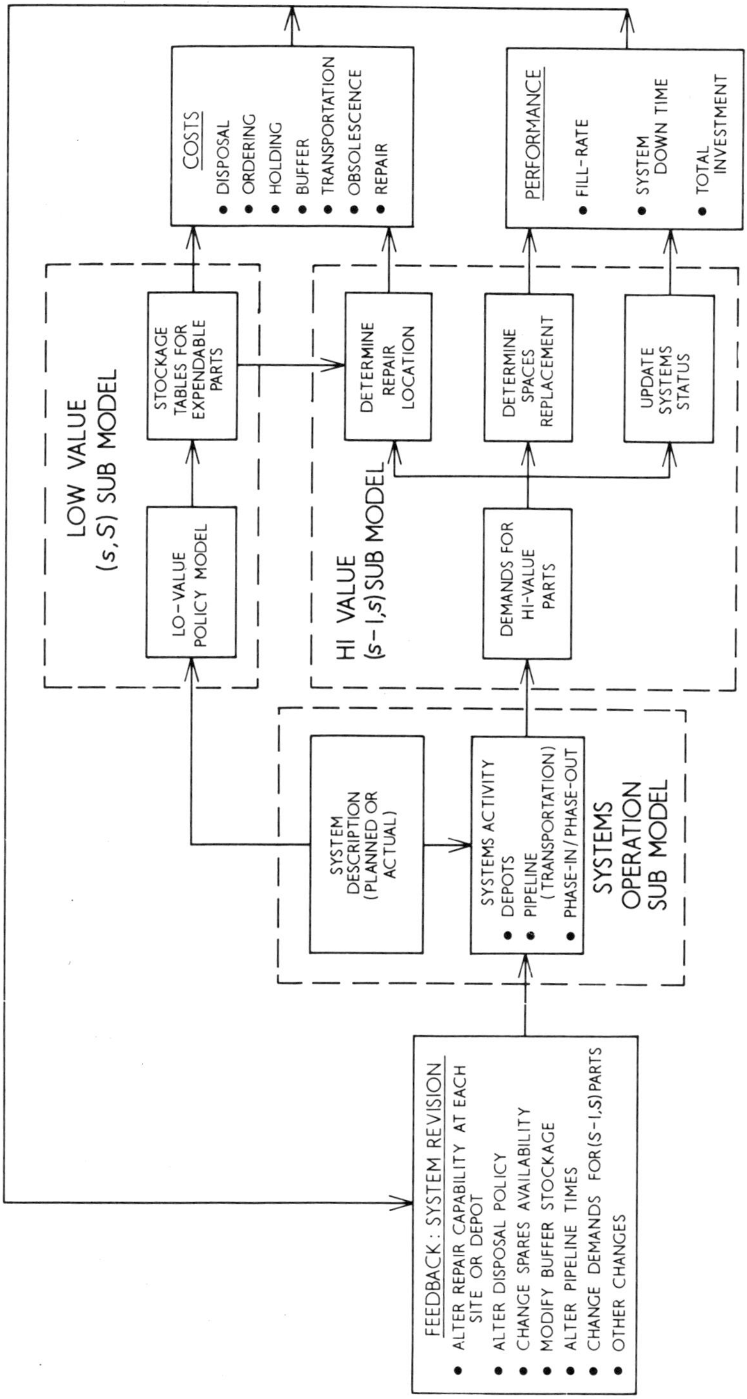

Figure 1 Logistics Analysis Simulation System model

up while the failed part is being repaired. Two policies can be followed with high-value parts. The first is to allocate a total amount of money for spare parts, and to stock enough parts to achieve the maximum amount of protection while parts are being repaired. The second policy is to stock enough spare parts to achieve a given amount of protection independent of investment.

The RAND model will calculate the optimum stock levels under either policy. The method used is to add one item at a time to stock until either the proper fill rate or the total investment is achieved. An optimum stock level is achieved by the addition of these parts to inventory that give the maximum expected fill-rate increment per each dollar of investment. An example is the case where one of two parts can be added to the inventory. One part increments the fill rate by ,002% and costs $500.00, while the other part increments the fill rate by .01% and costs $5,000.00. In this case, the $500.00 part will be added because it gives twice as much protection per dollar of investment. Proceeding in this manner, relatively low-cost, high-demand parts are stocked first, while the high-cost, low-demand items are stocked last.

A detailed description of the (*s*-1,*s*) model is given in [14].

3.3 Transportation/Distribution Submodel

The transportation/distribution submodel is used to redistribute parts after the (*s*-1,*s*) model has determined the new stock levels for each base. For each item a list of excesses and requirements is determined by comparing the old and new stock levels at each location. The requirements vector (excesses being treated as negative requirements) and a transportation cost matrix are then used as input into a transportation algorithm which determines the redistribution for minimum shipping costs. Vogel's method is used for solution.

A detailed description of this submodel is given in [14].

3.4 System Operation Submodel

The system operation submodel (SOS) represents the operation of the real logistics system. It is a simulation model which is applied to high-value, repairable parts. The simulation model replicates, over various time periods, the physical system of items failing, being replaced and repaired, and transported between various depots and sites, for different parametric inputs. At intervals pre-set by the model analyst or decision-maker, stock levels are reviewed in the light of past demand, then reset to reflect changes in demand. It is a "multi-echelon" model in which demand, repair time, and transportation are all treated as probabilistic events. Site phase-in and phase-out, as well as unit costs, are treated as known quantities with little or no uncertainty. The simulation permits the determination, for each different logistics budget, of the cost and fill-rate implications spread over a realistic time period.

The system operation submodel is made as flexible as possible so that it can be used to evaluate most logistics systems that have multi-location, multi-echelon characteristics. As a result, this submodel has the ability to process demands at 15 locations, stock parts at any of 15 locations, and repair at any of 15 locations. The numbers and skills of maintenance personnel at each location must be considered when dealing with repair capabilities. The skills are considered in up to nine levels of maintenance ability.

Another consideration is the amount of maintenance equipment at each location required for equipment analysis and repair. In this way, the model has the ability to consider systems where some locations can specialize in the repair of certain parts. The latter is particularly true of the military logistics organizations where one repair depot will repair engines only and another depot will repair communications equipment only. The 15 locations are not necessarily fixed, but can be increased if a computer with a larger memory is used.

Additional capabilities of the systems operations include the ability to:

. Simulate opening and closing of locations.

. Simulate increase and decrease in maintenance personnel.

. Simulate obsolete parts in the system.

. Vary the review period.

. Compute an inventory status index.

. Compute the system effectiveness.

. Process linked demands (a failure caused by a prior failure).

. Prepare output reports at varying time intervals.

. Provide a history of the simulation.

A detailed description of the various events in this submodel is given in [14].

3.5 Assumptions in LASS

Two types of assumptions are made in LASS. The first type are those regarding how the system will operate when complete knowledge of its operation is not available. Examples of this type of assumption are as follows.

. When an item is taken from stock at a particular location, the replacement eventually goes back into stock at the same location unless location is phasing out.

. Final determination of the repairability of a part is made at the factory level.

. Obsolescence is of two types. The first is due to the development of improved parts. The second occurs when a location closes and all parts at that location become obsolete in the sense that the cost of redistributing them makes them less valuable when stocking is considered.

. The redistribution of parts is determined by total least cost shipping route.

. The logistic system objective is either to maximize the system fill rate under a given investment constraint or to minimize the investment under a given fill-rate constraint.

- In an emergency situation, a part will be resupplied from any location that can supply it in the least time.
- A part requiring a specific skill level for repair must be repaired by maintenance personnel of that skill level or higher.
- Maintenance is at the same level twenty-four hours a day seven days a week.

The second type of assumptions are the simplifying assumptions that are made so that the model is computationally feasible. Examples of this type of assumption are

- Only two resupply times are significant in determining stock levels at each location in the $(s\text{-}1,s)$ model. These are normal repair time and normal resupply time for a location.
- The mean of the repair times follows a normal distribution and the repair times follow an Erlang distribution.
- Failures are randomly generated, i.e., follow a Poisson process.
- Order cost, stock-out cost, and obsolescence cost are constant for any one run, but can differ by location.
- Repairability, echelon of repair, and skill level required for repair are determined for each failure by generating these charactericts from probability distributions.
- The factory is represented by one location in the model, but it is assumed that this represents all factories that supply parts to the system.
- Shipping times follow an Erlang distribution.

4. DESCRIPTION OF INPUT AND OUTPUT

4.1 Input Requirements

For the low value model, the main inputs required are as follows:

- Stock Number
- Probability distribution of demand (which can be specified in terms of MTBF or failure rate)
- Unit cost of part
- Inventory carrying cost or holding cost
- Liquidation or obsolescence cost

- Stock-out cost
- Number of units per pack

For the high-value submodel, the main inputs required are:

- Stock number
- The number of failures in the base data period
- The cost of the high-value item
- The stock level at the beginning
- The repair code
- The level of complexity of replacement
- The replenishment time–normal and high priority

In addition, the total investment constraint or the total fill rate constraint must be specified at each location.

For the system operation submodel, the input required are:

- A description of the logistics network
- A description of the maintenance capability at each location in terms of the number of maintenance men and their skill levels.
- The mean time to repair and the repair time distribution for each high-value part.
- The lowest echelon of repair for each high-value part.
- The lowest skill level for repair of each high-value part.
- The degradation in effectiveness due to the failure of each high-value part.
- The minimum operational level of each location, that is, the value of effectiveness of the system below which an emergency situation arises.
- The loss of system effectiveness which causes a high priority classification on a failure.
- Control parameters for the simulation.

For the transportation/distribution submodel, the input requirement is a cost matrix describing the cost of shipping between the various locations.

4.2 Output Description

The output from the low-value submodel consists of a list of stock levels for each low-value part. These stock levels are optimum with respect to total costs of inventory in low-value parts.

The output of the high-value submodel, when run by itself, will be the stock levels in high-value parts which will optimize the total inventory investment with respect to a given fill rate constraint or which will optimize the total fill rate subject to a specified inventory investment.

The total LASS system combines the outputs of the low-value submodel and the high-value submodel with the system operation submodel and the transportation/distribution submodel to provide various other outputs.

By running the LASS simulation several times, a large number of different outputs can be obtained. Some of these results from the simulation are described below:

Figure 2 shows, for example, the relationship between the system fill rate and inventory investment for a given policy on location and maintenance.

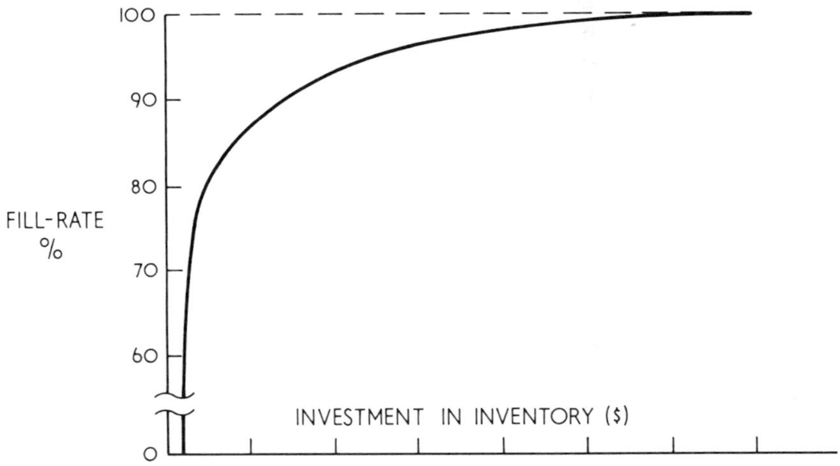

Figure 2 Figure rate versus investment in inventory

Figure 3 shows the relationship between shipping cost and maintenance costs for a variety of inventory investments. Maintenance costs can be reduced by repairing all spares centrally. However, shipping costs increase under this policy.

A large inventory investment will generally lower shipping and maintenance costs as fewer emergency shipments and repairs are required. These along with many other model outputs are considered in establishing the final inventory policy.

As maintenance costs increase, fill-rate goes up, as shown in Figure 4, reflecting the fact that items are repaired faster and are returned to stock sooner, all other factors being equal.

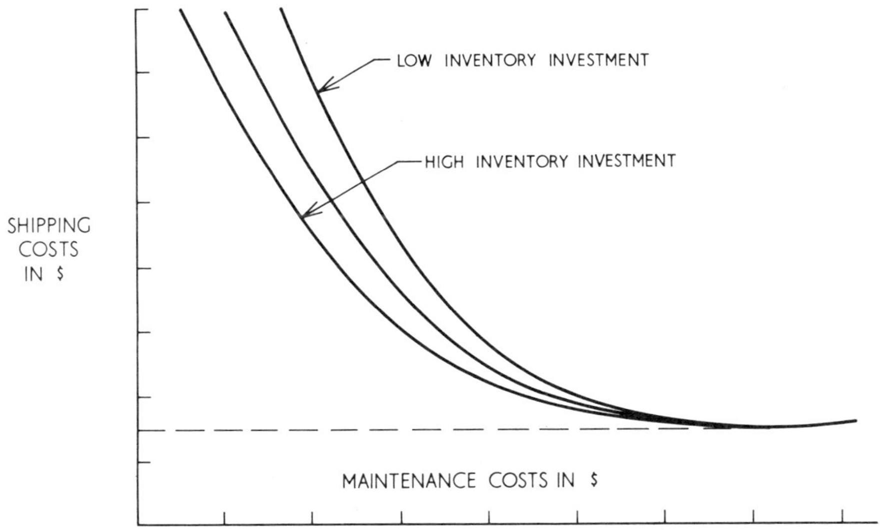

Figure 3 Shipping versus maintenance costs for various levels of inventory investment

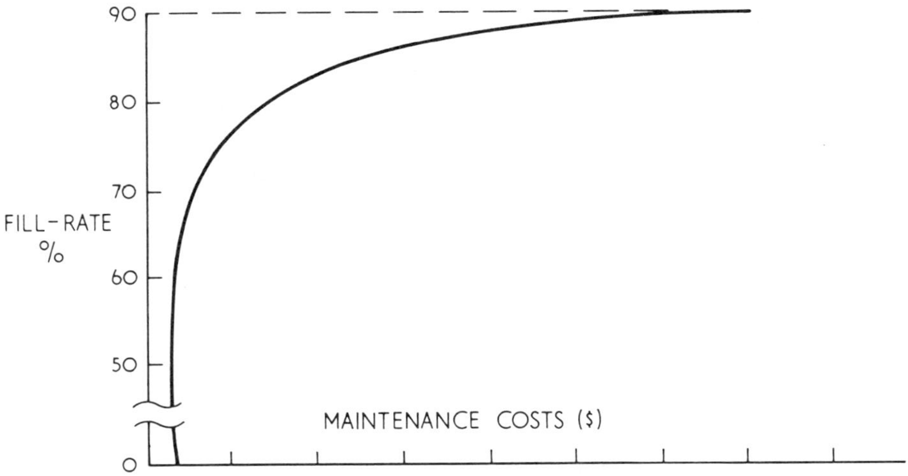

Figure 4 Fill rate versus maintenance costs

Figure 5 illustrates that regardless of whether stockage policies are oriented to base, or centralized, there is a consistent trade-off between inventory and maintenance. However, the figure shows consistent economies with more centralized stockage policy (risk held constant). This is so because with centralized stockage, more efficient maintenance routines and facilities can be adhered to; backlog can be more effectively handled; better test facilities are available; and there is more effective skill-distribution of personnel. In base repair, there are fewer personnel available; their range of skill is

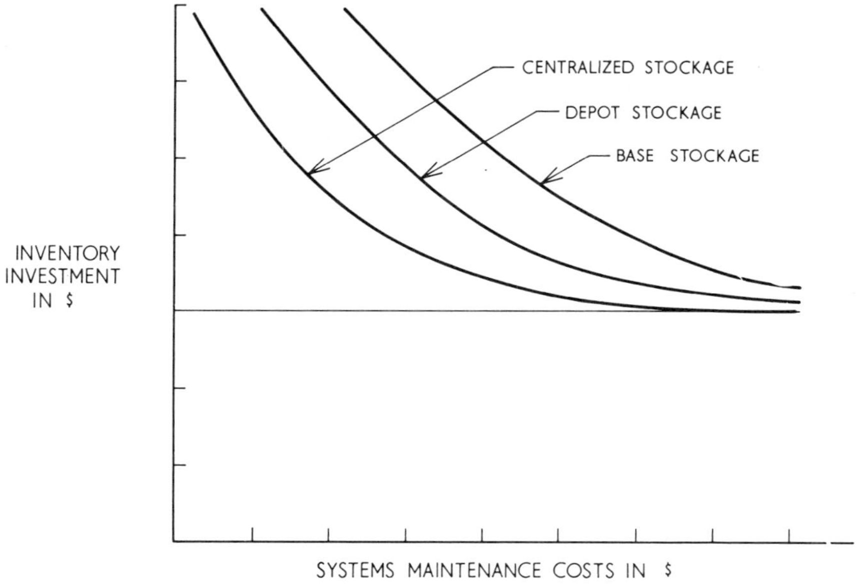

Figure 5 Maintenance costs versus inventory investment for various distributions, with risk held constant

more broad, although uneven, with less specialization and expertise. Consequently, maintenance costs are higher for a given level of inventory. In other words, for a given risk, or fill-rate, appreciably more parts sparing and/or maintenance capability must be provided for policies that concentrate spares/maintenance investment at a base level rather than a centralized point. However, the optimal mix of policies is clearly one which considers fill-rate (risk) and shipping costs, as well as inventory levels, the initial investment, and operating costs of the logistic support system. A comprehensive evaluation of logistics requirements trade-offs can also be made by the simulation system.

The simulation can compute the relationship between the fill-rate (or systems risk) and the total logistic systems budget consisting of initial investment in inventories and related capital expenditures such as test equipment, and operating costs such as personnel, transportation, etc. This can be done for any given logistics support concept at various total budget levels (Figure 6a). The simulation can also compute the relationship between logistics system budget and the cost of refilling the system in the event of a stock-out (i.e., those situations in which requests are not filled–or the "non-fill" rate*). This can be done for the same given logistics support concept (Figure 6b).

These curves can be computed for a wide range of logistic support concepts. Figure 7 depicts a family of relationships (between the minimum operating cost concept, and minimum risk concept) for various logistics support concepts at increasing systems

* The "non-fill" rate is, in fact the systems risk. It is equivalent to 1–fill-rate.

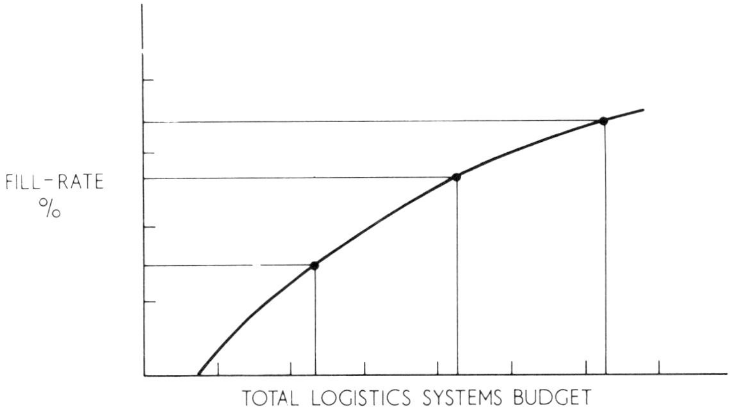

Figure 6(a) Fill rate versus total logistics systems budget

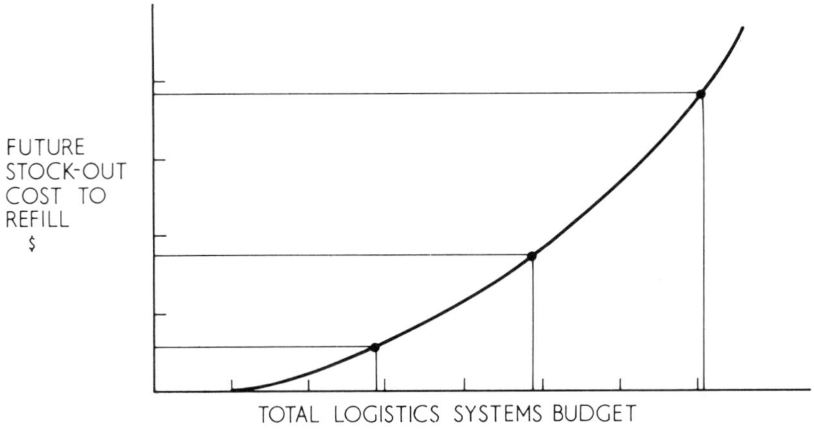

Figure 6(b) Future stock-out cost refill system versus logistics systems budget

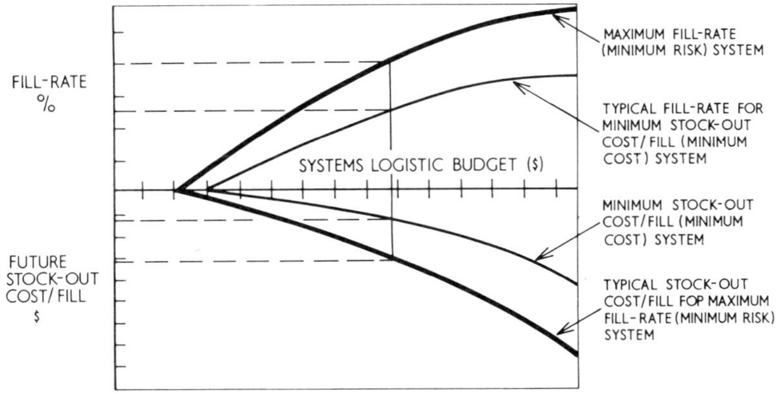

Figure 7 Trade-off of systems fillrate, stock-out, cost/fill and the systems logistics budget

budgets. This summary output can be used to evaluate the optimal logistics support concept, given either defined objectives for:

- Actual fill-rate (risk) and future stock/out cost to fill, or
- A fixed logistics budget and weight or values relating fill-rate and future stock-out cost/fill.

5. A TYPICAL APPLICATION OF LASS

The techniques and methodology of the system discussed in this report are intended for application in a broad range of logistical support systems. However, to give a feeling for the character and scope of such logistical applications we present in this chapter a discussion of a particular system for which logistical support was supplied, and the problems and requirements of the logistical manager in fulfilling system specifications and guaranteeing system performance. The typical system is outlined in Figure 8. The system involves a large number of operational sites (involving complex electronics, data processing and related equipments) and the total support system. Logistics Support is the responsibility of a single Logistics Operations and Maintenance (O&M) Manager. The responsibilities of the Logistic Manager include:

- Operation of a technical supply warehouse and stockroom at each site to provide receipt, issue, storage and protection of Government and contractor furnished supplies. Provide warehousing for parts, materials and supplies for replenishment of the site supply levels.
- Maintain a material control activity to account for the disposition of contractor-furnished parts, materials and supplies, and Government-furnished property. Accountability for Government property provided to the contractor for use in performance of Operations and Maintenance effort in accordance with Appendix "B", Armed Services Procurement Regulation.
- Establish a transportation control activity to assure expeditious dispatch of parts, materials and supplies to site locations. Provide transportation arrangements to insure timely arrival of required materials at each site.
- Recommend and establish realistic levels of operations and maintenance supplies for each site utilizing a central re-supply warehouse for storing and distributing high cost or critical items not economically justifiable for site stockage. Provide for initial receipt, inventory, and warehousing of operational materials, and maintenance spare parts at or for later issue to sites.
- Procure and/or package modification kits to comply with directed engineering change requirements.
- Maintain a special purpose tool and test equipment inventory, as required, for use by special maintenance support personnel.
- Establish a procurement activity for parts, materials and supplies and for special engineering, repair or test equipment calibration services which may be required.

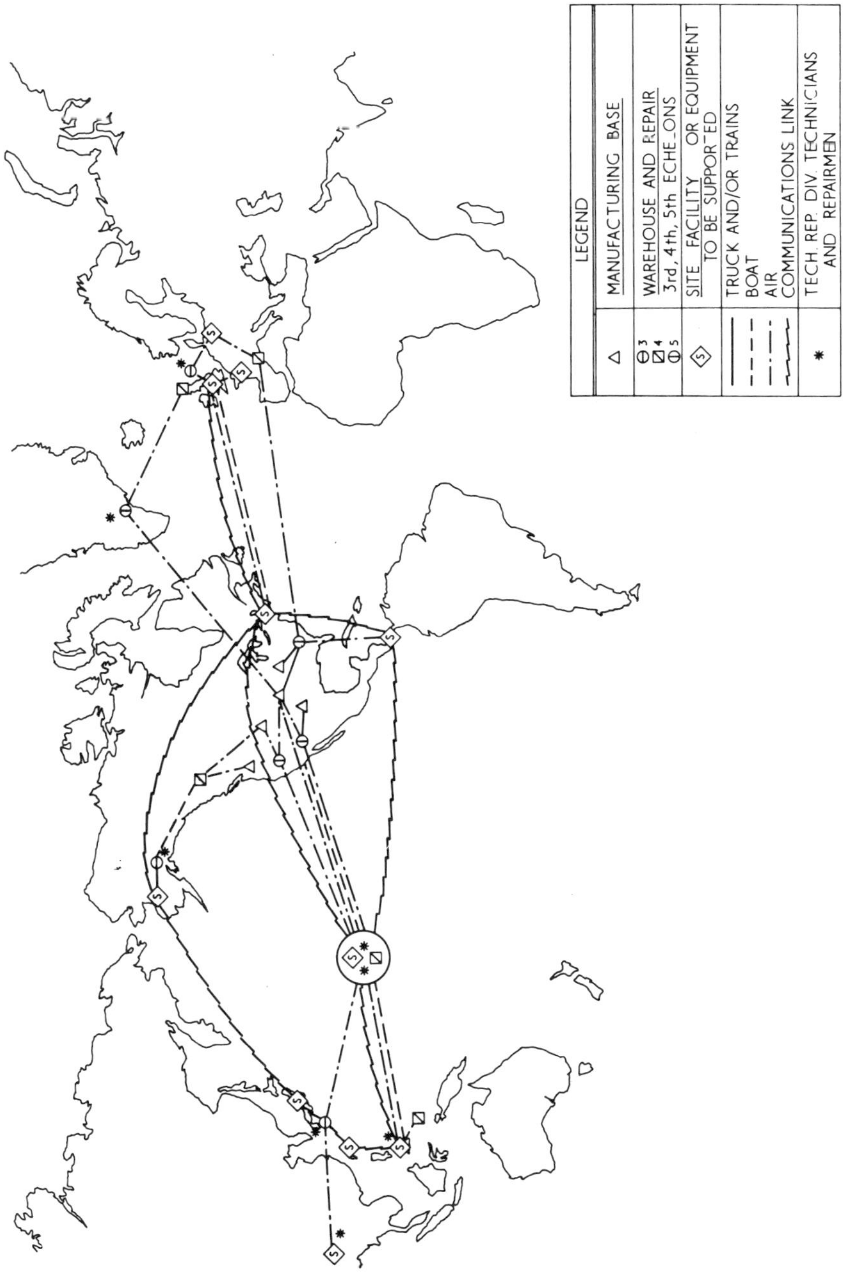

Figure 8 A typical world-wide logistics support system

Provide for procuring, requisitioning, and local purchasing of operations materials, maintenance parts, and maintenance service required to support operations and maintenance functions at all sites.

The Logistics Analysis Simulation Program can be used by this Logistics Manager to define and evaluate logistic policy for his system. Through this analytical tool the Logistics Manager can achieve these objectives:

. Reduce logistic system costs for high-cost, low-demand spare parts.

. Maintain/improve system effectiveness regardless of uncertainties of forecasting, obsolescense, ordering, holding, distributing, and site phase-in/phase-out.

. Obtain most acceptable trade-off between cost and effectiveness.

. Obtain maximum system performance for given level of system investment.

. Evaluate inter-related policy decisions in an aggregate manner.

These objectives are obtained by application of a simulation program, as a working tool used by Logistics Manager, to analyze logistic support policies such as:

. Stock levels (high/low value items)

. Maintenance abilities and person per facility or site

. Investment in test and repair equipment

. Use of local purchase

. Shipping and routing policy

. Effectiveness of roving maintenance teams

. Optimal policy for phase-in/phase-out of facilities

. Stock review periods

. Echelons of repair

. Echelons of stockage

. Amount of parts buffered by manufacturer.

The Logistics Manager can use the simulation system as a planning tool; he can evaluate policy and procedural alternatives before setting up the logistic support system. He can also use the simulation model as a management control technique after the logistic support system has been implemented. He is able to continually review and evaluate policies, stock levels, etc., using data from the field. The use of the simulation in an on-line M&O Management Information System, is illustrated in Figure 9.

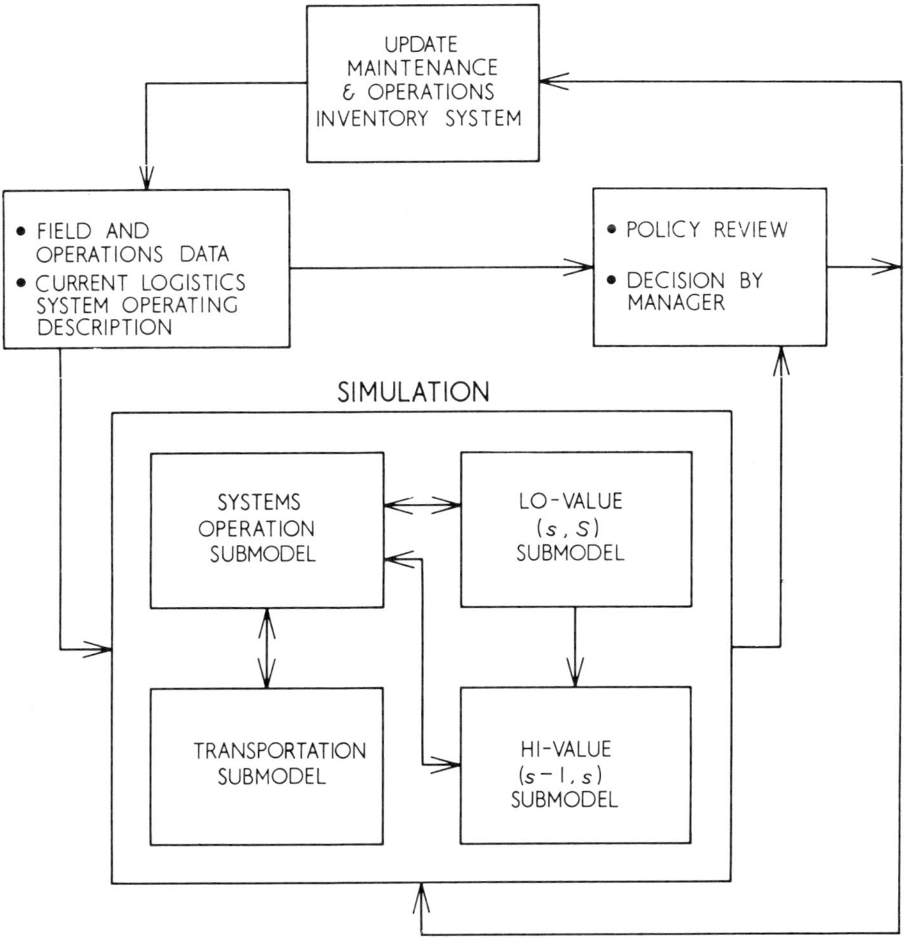

Figure 9 Logistics support (M & O) inventory system using simulation

6. BIBLIOGRAPHY

1. G. J. Feeney, J. W. Peterson, and C. C. Sherbrooke, *An Aggregate Stockage Policy for Recoverable Spare Parts,* The RAND Corporation, RM-3644-PR, June 1963.

2. Churchman, Akoff and Arnoff, *Introduction to Operations Research,* John Wiley & Sons, New York, 1957.

3. G. J. Feeney and C. C. Sherbrooke, *The (s-1,s) Inventory Policy Under Compound Poisson Demand,* The RAND Corporation, RM-4176-PR, August 1964.

4. W. Feller, *An Introduction to Probability Theory and Its Application,* Volume I, 2nd edition; New York; John Wiley & Sons, June 1957.

5. A. S. Goldman, T. B. Slattery, S. Firstman, and J. Regney, *Maintainability,* New York, John Wiley & Sons, 1964.

6. G. Hadley and T. M. Whitin, *Analysis of Inventory Systems,* Englewood Cliffs, N. J., Prentice-Hall, Inc., 1963.

7. Krasnow and Merikallio, "The Past, Present, and Future of General Simulation Languages", *Management Science*, November 1964, pp 236-267.

8. John Lu and Gabriele Michels, *Description of the Computer Program for Aggregate Base Stockage Policy of Recoverable Items*, The RAND Corporation, RM-4395-PR, April 1965.

9. John Y. Lu, and R. J. Wolfson, *Dynamic Modeling of Inventories Subject to Obsolescense*, C-E-I-R, Inc. (Los Angeles Center) June 14, 1962.

10. R. J. Wolfson, John Y. Lu, George W. Brown, "Dynamic Modeling of Inventories Subject to Obsolescense", *Management Science*, September 1964, Vol. XI, No. 1, pp. 51-63.

11. G. J. Feeney and C. C. Sherbrooke, "*Systems Analysis and Supply Management*", The RAND Corporation, RAND Memorandum RM-4054-PR, April 1964.

12. C. C. Sherbrooke, *A Management Perspective on METRIC–Multi-echelon Technique for Recoverable Item Control*; The RAND Corporation, RM-5078-1-PR, January 1968.

13. C. C. Sherbrooke, *METRIC; A Multi-echelon Technique for Recoverable Item Control*, The RAND Corporation, RM-5078-PR, November 1966.

14. Decision Sciences Corporation, *LASS–The Logistics Analysis Simulation System*, May 1970.

DISCUSSION

Tyler: What is your definition of "logistic support" in the context which you quoted of 60% of "down" time for this cause as opposed to 1% for repair? In what situation did these percentages apply?

Blumberg: In terms of the definitions presented in this paper, logistics support applies to the overall life cycle of equipment being supported. The particular figures used apply to a number of different types of situations with particular orientation to missile-type weapons.

Morris: Have you done any work on the cost of stock-outs?

Have you considered the proposition that the cost of a stock-out might be different in a high fill-rate system and a low fill-rate system?

Blumberg: Our firm has done work on the cost of stock-outs measured in terms of either the down-time of the system being supported or the cost of filling a demand in the event that a stock-out occurs at the point of use. Our studies and analysis clearly indicate that the stock-out cost is decidedly different in a high fill-rate system as opposed to a low fill-rate system. The high fill-rate system generally minimized down time of the system being supported, but has an extremely high price. In terms of the total logistics support budget factor, analysis indicates it is clearly better in many systems to pay a higher cost to improve the inherent reliability of the system than to move to a high fill-rate-oriented logistics support system.

Young: Does the simulation *systematically* study the options open to the equipment designer, such as the effect of module or sub-unit size and reliability, against various policy options, or is it essentially a model to optimise support for equipments already designed or contemplated?

Blumberg: The simulation described can be used to examine options open to the equipment designers such as the effect of module or sub-unit size and reliability against various logistics support policy options, as well as the design of a total logistics support system for equipments which already exist or are contemplated. However, it should be clearly noted that the simulation does not *systematically* evaluate alternatives, rather the analyst or designer using the system must create and impose these alternatives in a comprehensive manner. The simulation will merely evaluate each alternative in terms of cost and effectiveness.

Landau:

1. The speaker quoted the introduction of specialized maintenance equipment to reduce personnel costs. Did it need a large simulation to show this?

2. Are economic and social costs considered when siting maintenance facilities?

Blumberg: While the total logistics analysis simulation system would not be necessary to evaluate the introduction of specialized maintenance equipment and personnel allocation, this is one feature of the general model which can be utilized. The full simulation system, however, can be used to test these assumptions on other areas of logistics system as they apply to maintenance and test equipment and personnel allocations.

The logistics analysis simulation system model at the present time does not consider economic or social costs in siting maintenance facilities at the present moment. In general, these types of issues would not be best handled by the LASS except as introduced through constraints or costs on the ability to locate a site at a certain location.

Landau: To follow up previous question, the economic impact of, say, opening or closing a base must therefore be fed in by someone else, other than the OA team. Does this mean that the scope of the simulation is at the limits of OA?

Blumberg: The Logistics Analysis Simulation System as it is currently designed, considers the impact of opening and closing a base or a site only in terms of the total logistics support costs and not in terms of the social-economic impact on the area surrounding the base or site. This could be added to the simulation by means of having additional events or sub-systems in the systems operations simulations. Decision Sciences Corporation is currently involved in a study of the economic impact of closing bases or sites on the local region or metropolitan area in which the site is located. Such analysis is clearly within the scope of the general technique of simulation. However, it has not been included in the LASS system merely because it represents a special case of the more general problem of logistics support planning.

O'Connor:

1. Outline of the systems operational simulation.

2. What data feedback is used in operating the simulation?

Blumberg:

1. The systems operation simulation is described in some detail in the technical paper provided to the conference attendees. Additional information on detailed elements of the Logistics Analysis Simulation System can be obtained by writing directly to Decision Sciences Corporation, 814 Fox Pavilion, Jenkintown, Pennsylvania, U.S.A.

2. The primary data feedback used in operating the simulation are costs of maintenance and logistics support, effectiveness of this support, and its observed or predicted MTTR's and MTBF's in the case in which the simulation is used as an on-line evaluation of actual data from field use.

Developing a Logistics Management Information and Control System (LMICS).

Kjell Høgberg

1. SUMMARY

Faced with the termination of U.S. MAP and the future procurement and the operation of several new weapon systems, the Royal Norwegian Air Force, Air Material Command decided in 1965 it needed a modernized logistic system. As a consequence, systems R & D was undertaken in 1966. Now, three years later, the main parts of the LMICS are being implemented.

To develop the new LMICS, a project group was established in 1966 with the task of "... developing a Logistic Management Information and Control System that should:

- work equally well with different weapon systems,
- provide effective provisioning procedures and an orderly move from the introductory stage to normal operation,
- facilitate different support organizations for different type weapon systems if needed,
- work equally well with a system–oriented or functional organization,
- span the logistic support structure by means of an effective horizontal integration of all related functions and by integrating vertically all tasks within the same management area,
- utilize modern management methods as effectively as possible."

An important guide to systems development was the early recognition of the similarity between the maintenance activity within the logistic support structure and the production activity of a manufacturing industry. This led to the definition of a set of main tasks which has to be performed within the management cycle, irrespective of weapon system and support structure.

Describing the management cycle, the main management tasks are defined; the use of some of the important parameters explained; the interrelationship with personnel, comptroller and facilities management systems illustrated; and is demonstrated the close relationship between the operational system (to be supported) and the logistic support system.

Central to the management tasks are what we define as the Management Objects. These are the Equipment Configuration, Inventory, Work Shops and the Budget.

Associated with the Management Objects are the Management Objectives which in general terms are:

- to realize certain expected qualities of each component and a stipulated servicibility for weapon systems,
- to realize stipulated fillrates for each inventory item and stocking location as well as stipulated weapon systems NORS (Not Operationally Ready due to Supply) rates,
- to realize expected resource utilization factors and jobhrs. at each workshop and stipulated NORM (Not Operationally Ready due to Maintenance) rates for weapon systems.

Based on the tasks and functions which are performed in the management cycle, the LMICS is divided into the Managed System, the Decision System, and the Data/Information System.

The Managed System is further broken down into the Management Objects, the Management Areas, and the Organization (support structure). The Management Areas being closely associated with the Management Objects, are: Engineering, Maintenance (Production), Supply and Economy.

It is interesting to note that although the present organization is functional and task oriented, the organizational structure which is presently discussed nevertheless, takes care of the systems concept by a System Overlay-Structure. Provisioning it seems, will readily be performed within the support structure by standard routines.

The Decision System includes the task-oriented functional routines and decision models necessary to perform Resource Management, Objectives Management, and Operational Control, the meaning of which are described in later sections.

The Data/Information System includes the tasks of data collection, data transmission, data processing, storing, and data retrieval. Management oriented data systems have been developed, viz. the Supply, Economy, and Engineering and Maintenance subsystems. Being modular in design, they are nevertheless highly integrated.

The LMICS described is part of a larger Defence Management System. It will be shown how the LMICS fits in with the higher echelon systems, in particular it is well suited to PPBS techniques.

The particular logistic system to be managed by the described LMICS include approx. 200.000 items, approx. 30 stations, and approx. 4500 people. Although a relatively small system, it is sufficiently large to present all the complexities of a large logistic support system.

2. WORK METHODOLOGY

Developing a Logistic Support System requires a work methodology which must facilitate economic development. As it turns out, this requirement is best met by an efficient modularization of the system and the consequent system development work.

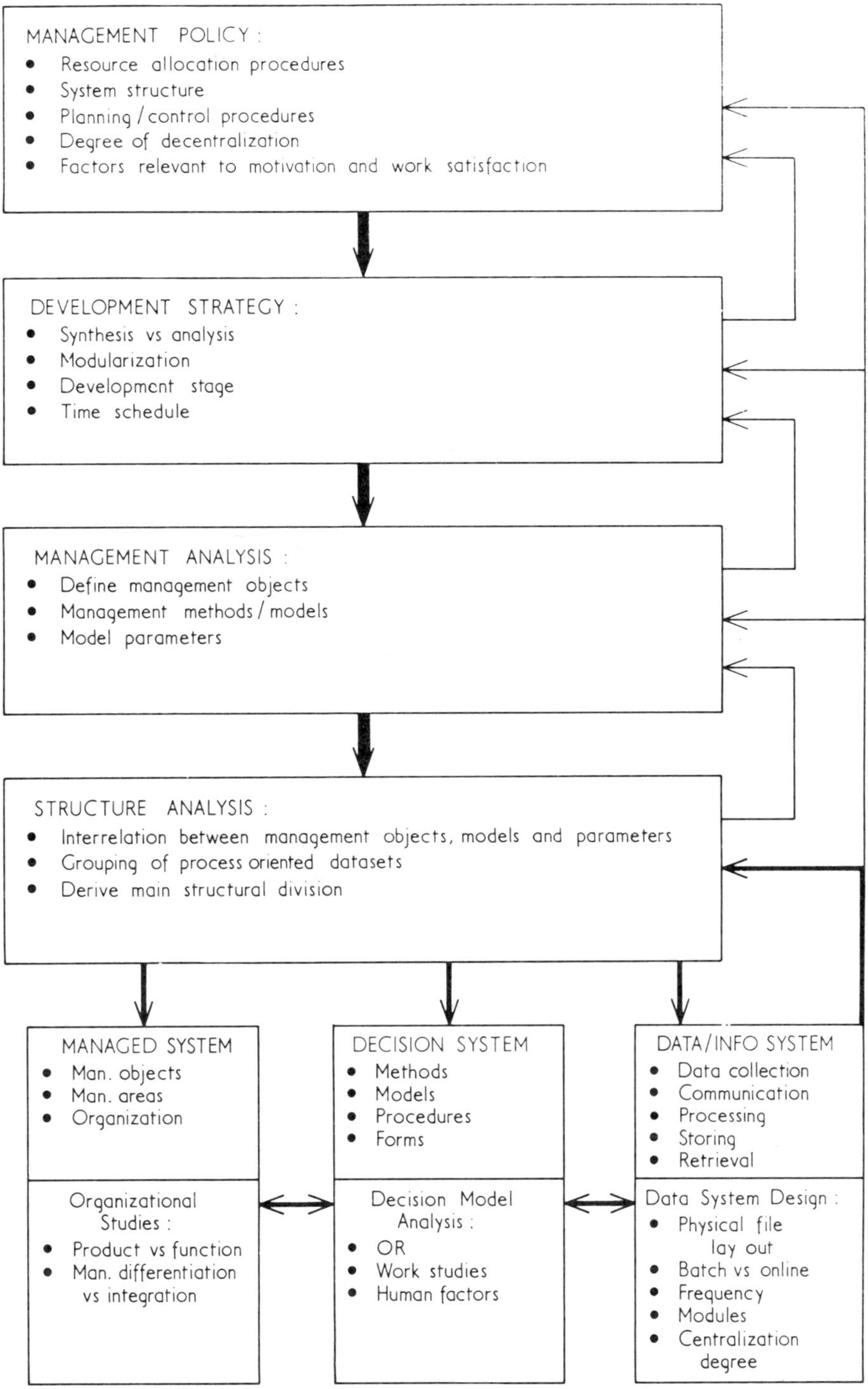

Figure 1 Management system development

The work methodology which has been used agrees well with that suggested for example by Langefors [1].
The methodology is shown in figure 1 and covers:

- establishing a Management Policy
- working out a Development Strategy
- performing a Management Analysis
- undertaking a Structure Analysis
- undertaking Organizational Studies, Decision Model Design, and Data/Information System Design.

3. THE INFLUENCE OF HIGHER ECHELONS' RESOURCE ALLOCATION PROCEDURES ON LOGISTIC MANAGEMENT POLICY

The resource allocation procedures' influence on a management system, which has been well described in [2], is illustrated by the very simplified representation in figure 2. The reason this is discussed is the fact that no single decision has stronger effect, than the adopted allocation procedures, on system structure, planning and control procedures, degree of centralization, and factors relevant to motivation and work satisfaction.

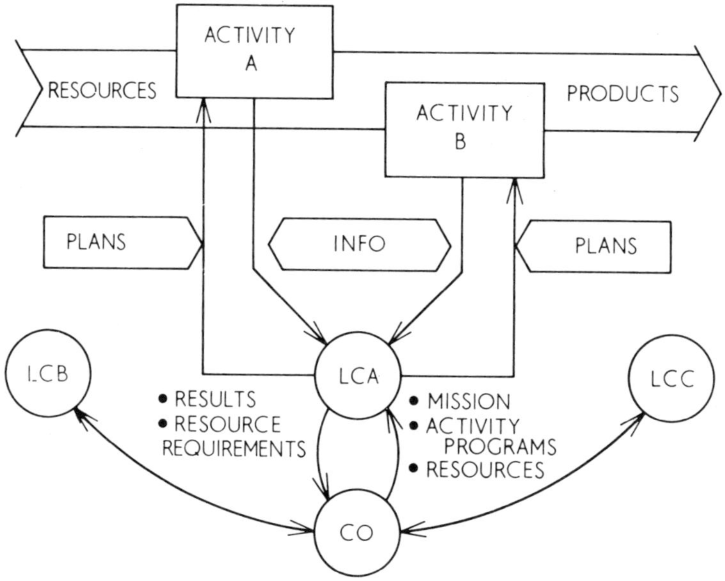

Figure 2 Effect on the organisation of method of allocation of resources

The Central Organization, CO, has the responsibility for optimizing the structure of the Armed Forces and for allocation of resources to several activities.

Activity A and B are logistic support activities managed by Logistic Command, LC A.

The CO could manage the two activities directly if it had perfect information about the activity processes and their effect on the value-function which is utilized for the calculation and subsequent allocation of resources. Utilizing a marginal allocation scheme, based on the principle of diminishing returns, the CO could allocate and reallocate resources until optimal allocation had been achieved.

At activities A and B there would really not be any management functions since they would only receive production orders, to be obeyed, and the resources calculated to be sufficient. Logistic Command A would be non existent.

This centralized management scheme has the weakness that the central information system necessarily must be enormous, that it is inflexible to changes in the activity environment and that it has a hostile effect on local motivation.

On the other hand, the CO could inform LC A about its mission in terms of deployment, response characteristics and annual operational programs for the weapon systems to be logistically supported by LC A. On this basis, LC A could calculate the minimum resources required to fulfil its mission, and if no changes in mission occurred, those resources would be allocated and administered with few restrictions by Logistic Command A. The CO would optimize its total resource allocations by marginally reallocating resources to those commands with activities showing the highest return rate in terms of the total value-function monitored by the CO. Achievements are supposedly reported to the CO from each Log. Command.

The last example is one with a high degree of decentralization and should consequently be highly motivational. Furthermore, the information system necessary at the level of the CO is much less than in the former case. Such an allocation procedure could be called budget allocation.

A further decentralization may be achieved if all resources are given a price and all "products" produced by activities are given a value. Then the CO may monitor "prices" such that the logistic system managed by LC A uses resources in an optimal fashion in accordance with the principle of supply and demand stability.

We may call this method price allocation. The plus-features of budget allocation cited previously are accentuated by price allocation. However, pricing all resources and products is no small task. In practice, as we shall see, many of the benefits of budget- and price allocation may be obtained if the Logistic Command breaks down its mission into objectives for each activity under its auspices, and for each functional Management Area; delegates the executive authority to the activities; and marginally allocates resources to them by systematically monitoring the value of each activity.

It should be clear that the relationship between the CO and the Logistic Command relative to the allocation procedures has a clear parallel between the Logistic Command and the activities under its auspices. Therefore, it is equally clear that several allocation procedures are possible and that the allocation procedure structures the logistic support system for which Logistic Command A is responsible.

In the case of the RNoAFAMC, the AMC participates in the resource requirement calculation and would be given mission objectives in terms of deployment, expected conflict response characteristics, and annual activity programs for the weapon systems to be supported logistically.

4. DEVELOPMENT STRATEGY

Any management system may be broken down into subsystems until the elementary system-level is reached. Each subsystem usually facilitates several processes. A particular event, identified by some predefined means, is managed by one particular process. This means that each *subsystem must contain a system* for process selection based on event identification and an executive system which executes the selected process. Thus, the subsystem with which we started may be broken down into two subsystems. This process terminates when no further breakdown is meaningful, and the elementary level has been reached. This methodology is development by analysis.

The opposite procedure starts with elementary systems. By combining these to subsystems at a higher system level, we are developing the system by synthesis.

Which procedure should we adopt?

Management and control automation requires undoubtely intimate knowledge of the smallest details. This could lead us to believe that development by synthesis is the best procedure to follow. However, to start development at the elementary level requires full knowledge of the environment surrounding the elementary system in order to avoid interface problems and redundancy. In other words, a certain degree of total insight into the position is necessary and this points towards development by analysis.

Our Development Strategy was to analize the entire problem area confronting us in order to modularize the system and the work. When the analysis was finished it was possible to develop and implement the modules in descending order of utility. Thus, it was possible gradually to define jobs and skill requirements, estimate job duration and derive a time-schedule for development and implementation.

5. MANAGEMENT ANALYSIS

The Management Analysis must start with the system to be managed. The RNoAF logistic system supports operational units at 30-40 stations (airbases and EW sites). Those units operate A/C or missiles, or air artillery, or EW stations. Number of inventory items are approx. 200,000.

The system is not large but sufficiently so to present all the complexities of a Large Logistic Support System.

In order to structure the logistic support system to facilitate management systems development, a planning and control flow-analysis was undertaken. The system requirements quoted in section 1, led us to concentrate on the tasks which have to be performed in the logistic support process. Thus, system development is made independent of the present organization and when development is finished, tasks may be assigned to organizational units.

In starting out with the tasks confronting us, we found it necessary to familiarize ourselves with the subject matter. In studying the literature on management systems, we found that very often was there mention of integration across functional boundaries, integration between organizational units, and integration between organizational levels. Seldom was there any mention, however, of the fact that it is necessary to integrate in time the management and control activities within a management cycle. It is in our opinion important to realize this necessity for integration in time. In our Management Analysis, it became our aim to present a picture of an integrated plan and control process which starts with long term planning and ends with job-execution on the workshop floor. For the purpose of this presentation, the different phases of the management and control cycle are depicted as

– Long Term Requirements Forecasting

– Annual Master Scheduling, Objectives Calculation, and Resource Allocation

– Short Term Scheduling

– Dispatching and Reporting

in figures 3-6 respectively.

5.1. Long Term Requirements Forecasting

Long Term Requirements Forecasting is depicted in figure 3. The Strategic Planning at higher command echelons has resulted in long term contingency plans and a mission for the particular logistic support structure we are dealing with.

Based on those plans and input from ongoing system procurement activities, the logistic support organization starts its long term planning by establishing characteristics and technical specifications for new equipment.

Prior to this event, there has probably been considerable activity, viz., engineering, reliability, and cost studies; and contracting and delivery negotiations.

On the basis of preliminary procurement documentation, as well as a technical decomposition of new equipment; preliminary engineering, maintenance, and supply parameters are established as part of the procurement documentation routine.

Preliminary data are correlated with current information on engineering standards, work shop performance, work force levels, anticipated maintenance schedules, and inventory status. On the basis of this information, aggregate maintenance requirements are generated per equipment and support activity location.

Then we may stipulate net resource requirements and draw up an anticipated Master Repair Schedule. In figure 3 is shown that the preliminary Long Term Resource Requirements Forecast includes.

– engineering documentation with part-break-down list, key parameters, explosive views, and technical orders upon which are based the further Engineering Management process;

– inventory plans covering annual demand forecasts, tool and support equipment

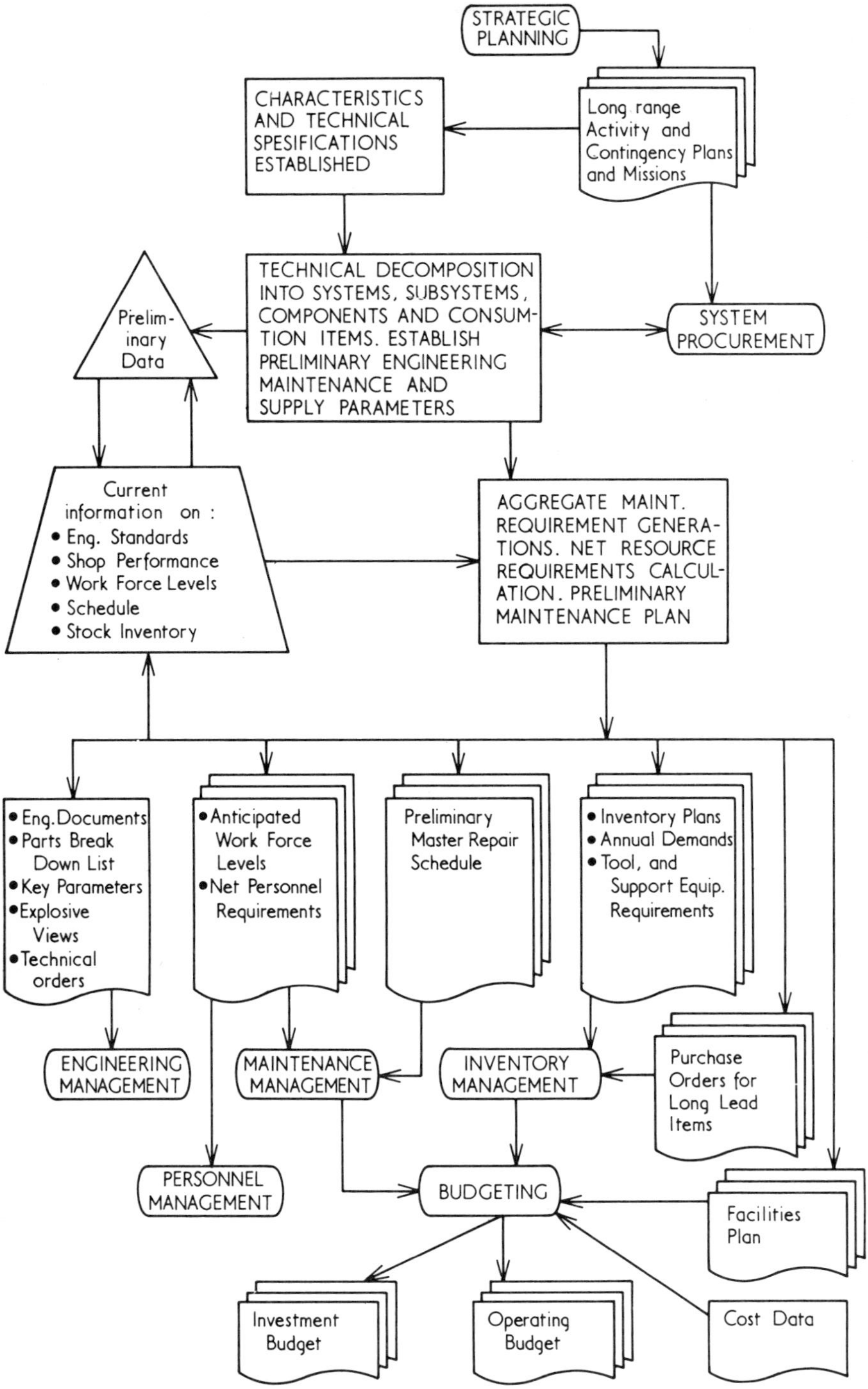

Figure 3 Long term requirements forecasting

authorization proposals, and purchase orders for long lead items, all of which are the basis for Supply Management action to follow;

– anticipated work-force levels is the basis of net personnel requirements which are forwarded to the agency responsible for Personnel Management.

Thus, the data and information collected and plans established are important inputs to the management activities in different functional Management Areas which we have named Engineering, Maintenance, and Supply Management. We shall have more to say about Management Areas in later paragraphs.

Mixing cost data with net resource requirements (except personnel), the budgeting process is undertaken resulting in an Investment and Operating Budget Proposal for the one-year period starting in one year, and Projected Program Cost for periods further into the future.

In performing Long Term Requirements Forecasting, the logistic organization is faced with a varying degree of uncertainty. Almost invariably, uncertainty will be greatest for the most distant years for which we plan. The reason for long term planning is our intent to make decisions now which hopefully give the logistic support structure flexible response characteristics commensurate with uncertainties regarding some of the future state variables.

Around the time that we perform our Long Term Requirements Forecasting, the logistic organization receives next years budget for which a proposal was forwarded a year ago according to procedures that have been explained above.

5.2. Annual Master Scheduling, Objectives Calculation, and Resource Allocation

Upon receipt of the authorized budget, Annual Master Scheduling, Objectives Calculation, and Resource Allocation, is undertaken as shown in figure 4.

Next year's activity and contingency program is firm, information about new and existing equipment have been updated; information is available on stock and work force inventory status, shop loads and capacity are known.

Thus, we may analize aggregate load generation, make up the Maintenance Master Schedule, calculate objectives for the different logistic functions to be performed, and allocate resources that are not inherent in the stipulated objectives.

Note that the engineering-, maintenance-, and supply data that were input to the master scheduling process came out of control processes which are named Configuration-, Shop-, and Inventory Control respectively. Note also that the consolidated objectives are the same, albeit not necessarily in magnitude, as the input parameters and are fed back to the control processes mentioned. Suffice it to say at this stage that the purpose of these control processes is to establish subobjectives, to compare performance with established objectives and report whenever maximum deviation–limits are exceeded, and to retrieve and report accumulated performance data when requested. The processes are EDP based. Of the more important objectives, I mention

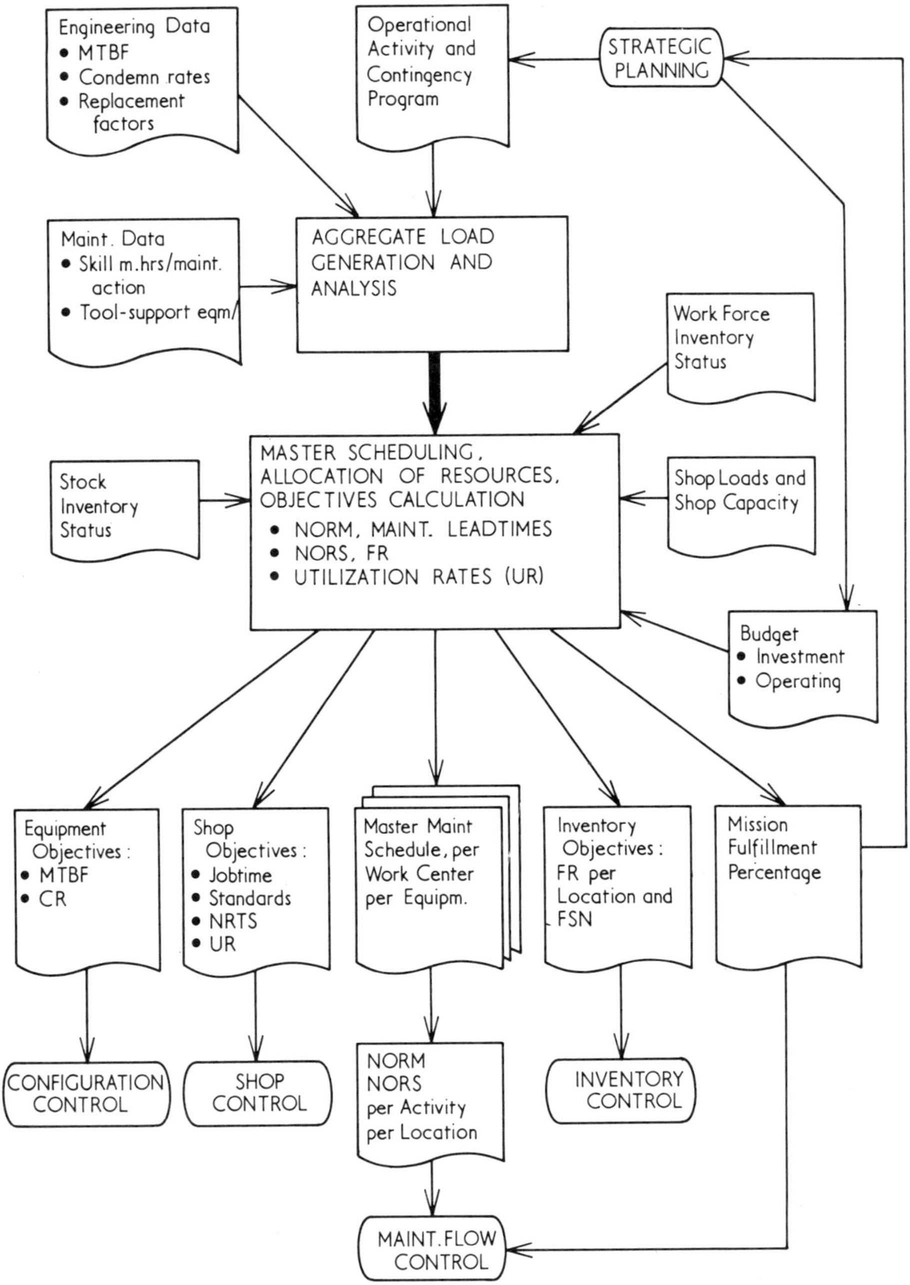

Figure 4 Annual master scheduling, objectives calculation and resource allocation

- Mean Time Between Failures (MTBF),
- Not Operationally Ready due to Maintenance (NORM),
- Inventory Item Fill Rate (FR),
- Not Operationally Ready due to Supply (NORS),
- Job Standards (JS),
- and Personnel Utilization Rate (UR).

It should be noted that in the Long Term Requirements Forecasting, we calculated required resources on the basis of status and activity and contingency programs. In Master Scheduling, it is necessary to calculate to what degree available resources are sufficient to meet activity and contingency programs. Resultant mission fulfillment percentage should be input to the Maintenance Flow Control and be the basis for a possible Mission/Resource Change Proposal to the agency responsible for Strategic Planning.

After completion of Annual Master Scheduling and on the basis of previous year's planning, material should be on its way into initial lay-in stocks, personnel should have been trained for and assigned to the work force, and maintenance facilities should have been readied. As we now move to lower logistic support echelons where the Short Term Scheduling takes place in cooperation with the local operational program manager, it is worth mentioning that the logistic support structure, due to earlier decisions, now operate under serious constraints. Furthermore, it should be noted that a change in objectives is also a change in resource allocation as mentioned in section 2. Thus, we conclude that the activities described with reference to figures 3 and 4 may be classed as Resource Management which we define as:

> proposing and obtaining resources, the determination and auditing of objectives for lower level activities, and marginal allocation of associated resources to those activities in an effort to realize the organization's mission.

5.3. Short Term Scheduling

Upon receipt of the Master Schedule and the objectives, the lower echelon logistic activities in cooperation with operational program managers perform their Short Term Scheduling as depicted in figure 5. After extensive explanation of figures 3 and 4, figure 5 should be self explanatory.

Note, however, that if resource availability is not sufficient to meet the local operational program, an Operational Program Change Proposal should be forwarded. If resources are consistently low, a Capacity/Mission Adjustment Proposal should be forwarded to the higher echelon activity responsible for objectives calculation and resource allocation.

Short Term Scheduling is primarily aimed at maintaining a satisfactory NOR rate while supporting the operational activity program satisfactorily. The planning horizon ranges from a week to several months. Of particular concern is the job generation process with regard to job spectrum and frequency since this must be compatible with shop infrastructure.

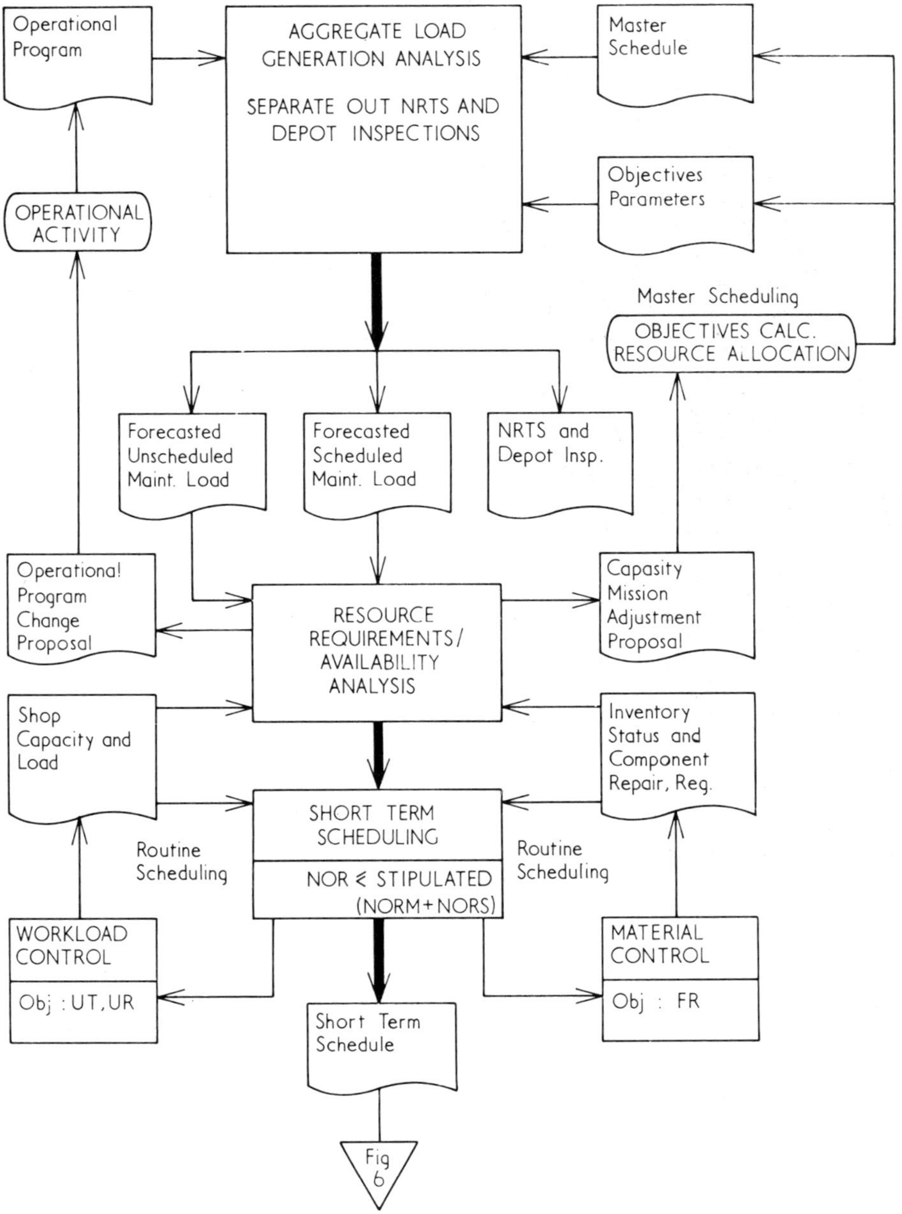

Figure 5 Short term scheduling

Workload Control is managing shop loading such that job standards and utilization rates are realized if at all possible. In the longer term, it should also be concerned with shop infrastructure with regard to work efficiency and ability to meet varying maintenance capacity demands.

Material Control is managing supply support such that stipulated fill rates are realized. This includes calculating anticipated supply support requirements to back up scheduled maintenance and to monitor local component repair requirements and the local supply points' ability to bring material to the user in a timely and efficient manner.

As we pass on from Short Term Scheduling to Dispatching and Reporting, we may say that Short Term Scheduling including Workload and Material Control is Objectives Management which we define as:

> a process of organizing and utilizing allocated resources to accomplish predetermined objectives.

5.4. Dispatching and Reporting

In figure 6 is illustrated Dispatching and Reporting.

The planning horizon is now at a minimum. Maintenance requirements are generated partly in conformity with the Short Term Schedule (planned maintenance) and partly in terms of unscheduled maintenance. The "Dispatcher" is confronted with a job–spectrum and–frequency that is a function of the actual equipment configuration and equipment usage. Furthermore, he is confronted with a shop capacity and stock situation which is the aggregated result of Workload and Material Control's short term management ability as well as the compounded effect of higher echelons' annual and long term planning. Thus, Dispatching is an activity of revealing truth.

It should be understood that problems encountered at this stage of the management cycle are most often caused by less than perfect management at higher echelons and in prior years (or periods) in the areas of Engineering, Maintenance, and Supply, rather than the result of poor dispatching techniques. (Although this may of course also be the reason).

The primary objective for Dispatching will be not to exceed stipulated NOR rates. In performing maintenance flow control, priorities are assigned and work orders dispatched with due respect to workload and availability of material. Dispatching thus includes very short term aspects of workload and material control. Upon dispatching a work order and having initiated material and equipment movement, the unscheduled event has transformed into a scheduled maintenance action. In this manner, the Short Term Schedule is continuously undergoing dynamic changes, On the basis of the above, Dispatching can also be classified as Objectives Management.

In the shop (or equivalent level), jobs are assigned to workers or operators by shop supervisors who also should be responsible for the quality of the work performed.

This necessitates both material quality control as well as operator performance control. We are in effect defining two new management activities:

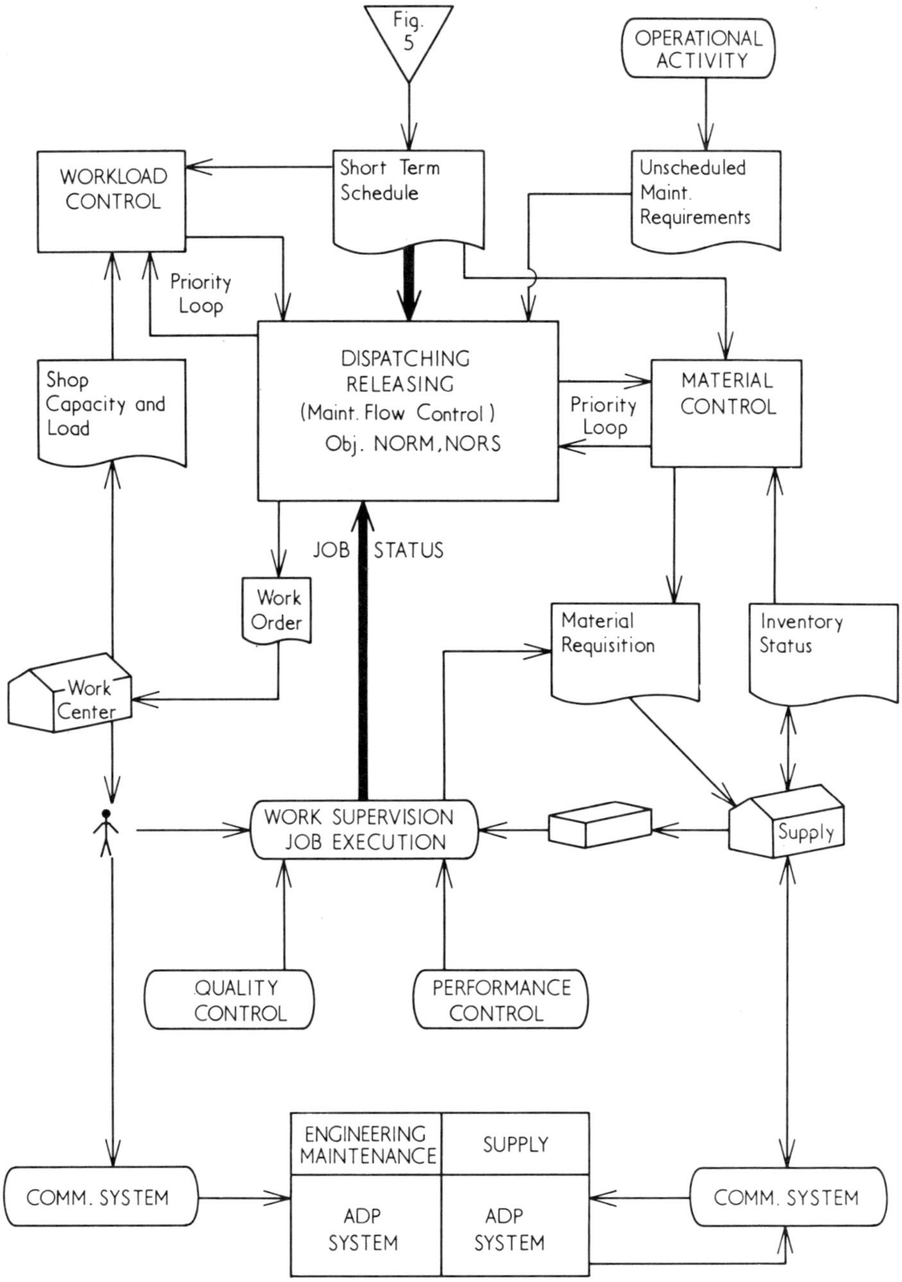

Figure 6 Dispatching & reporting

- Quality Control
- Training and Proficiency Control

Although very important, I shall not endeavour to detail those management activities since this in my opinion, would not significantly contribute to clarifying the structure of the management system.

We see in figure 6 that supply and maintenance actions are reported to the datasystem for processing.

Job assignment, physical quality control, job supervision, and proficiency control can suitably be labeled Operational Control which we define as:

> the process of assuring that assigned tasks are carried out effectively and in accordance with given orders.

5.5. Management Control

In previous paragraphs of section 5, I have shown the management cycle with emphasis on planning and allocation of resources. Control naturally is the other side of planning.

Suffice it to say that we consider it the responsibility of those who draw up plans, stipulate objectives, estimate performance parameters, establish basic data in the DP system etc. to compare the established parameters to real life results and take proper action to make real-life results conform to expected values.
This is not to say that we merely adjust expectancies to actual results. On the contrary, it is necessary to reevaluate, reassess, and replan before resetting parameters. Thus, planning and control are both important aspects of management.

The previously mentioned EDP based Configuration–, Inventory–, and Maintenance Flow Control System are important cornerstones of management control. They provide information for planning at the level of Master Scheduling and Short Term Scheduling, process logistic activity reports (transactions), compare performance with plans, and report deficiencies. Thus, those control systems provide the vertical integration within each Management Area between the two levels of management defined by Master Scheduling and Short Term Scheduling respectively.

5.6. The Management Cycle in Retrospect

In summary, we may reduce the management cycle in two steps as shown in figures 7 and 8. Shown in figure 7, are the levels of management including planning and control, the interaction between the activity programs and the logistic support system, and how resources are converted to products as a result of job execution.

In figure 8 the management cycle is simplified further to show the management levels defined as:

- Strategic Planning
- Resource Management
- Objectives Management
- Operational Control

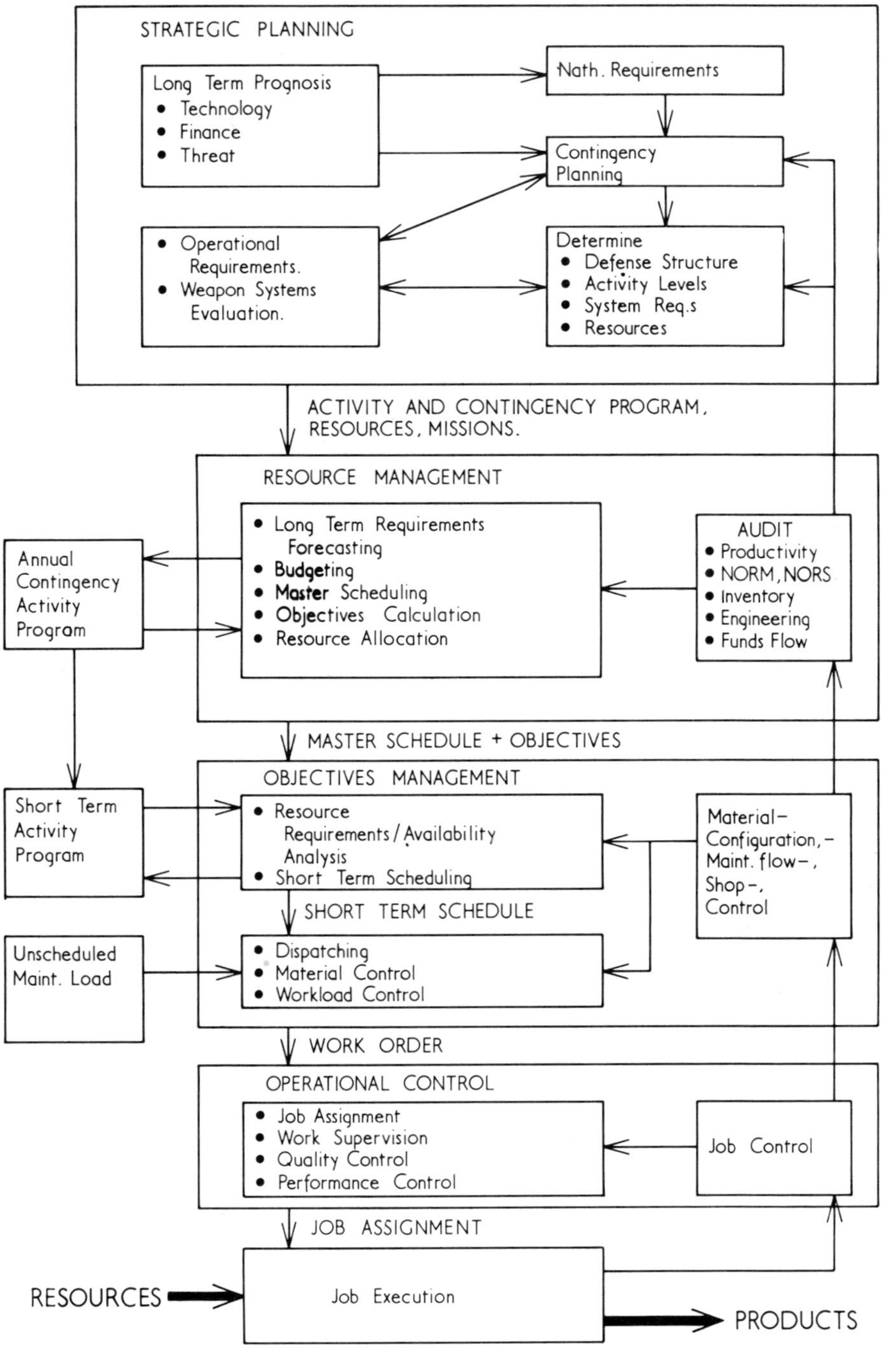

Figure 7 Consolidated management cycle

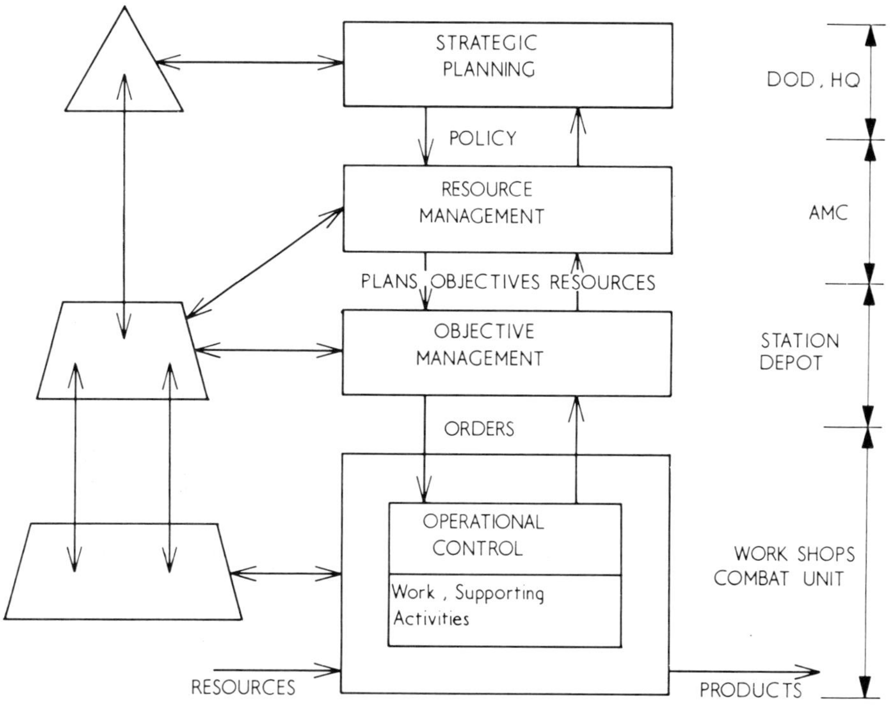

Figure 8 Management levels in a logistic support system

between which flow respectively

- Policy
- Plans & Resources
- Orders

For reference in figure 8 is also indicated how at RNoAF AMC the management levels may be referred to known organizational levels.

On the basis of the management flow-analysis, it is possible to perform the Structure Analysis. It should be noted, however, that as we continue at the same time with the detailed Management Analysis, we can also structure the system into finer segments.

As we penetrate the subject matter, our efforts should be concentrated on defining the Management Objects and the closely related procedures and decision models, all of which are necessary inputs to the Structure Analysis.

It is beyond the scope of this presentation to go into detail as to how the Management Analysis has been performed. However, the results as they apply to the structure of the system, will gradually become evident as we progress.

6. STRUCTURE ANALYSIS

In an effort to arrive at a main structural division, Structure Analyses of the interrelation between different management procedures is undertaken as well as a grouping of process-oriented datasets. As we progress with a continuous feedback to the Management Analysis and new input from it to Structure Analysis, the contours of the system structure gradually become visible. The details of our work are again beyond the scope of this presentation. Therefore, only the results are presented.

There are of course many ways to structure a management system. We found that for our purpose the structural division shown at the bottom of figure 1 was efficient. Namely, to divide the system into:

- the Managed System
- the Decision System
- the Data/Information System

For those being enthused by analogies, I suggest that the structure may be compared to a biological system composed of the anatomy, physiology, and the nervous system.

Substructure of each main structural element as well as the work division with regard to development and implementation are shown also in figure 1.

6.1. The Managed System

In my opinion, the most important aspect of any management system study is to define the Management Objects: what the system is supposed to manage. We found three primary logistic objects which we think it is vital to manage:

- Equipment Configuration
- Work Shops
- Inventory

6.*1.1. The Equipment Configuration*

This should be managed with the objective to meet the activity programs and response criteria with the most cost-effective equipment configuration. This would include, for example,

- the establishment and control of a set of engineering parameters inherent to each component, subsystem, system, and equipment both with respect to their quality, maintainability and effectiveness in relation to cost.

- establishment of max. NOR rates and NORM/NORS scales.

- establishment and control of maintenance instructions and skill-requirements for workers.

On the basis of the results of the Management and Structure Analyses, it is possible to separate out a Management Area that we designate Engineering Management, which should be responsible for the management of Equipment Configurations as outlined in very general terms above.

An important structure parameter with regard to Equipment Configuration Control is the Equipment Configuration Code, ECC. How this parameter is constructed is illustrated in figure 9.

The ECC facilities contract cost control, component discrepancy control, and item malfunction control, equipment labour consumption, activity program cost control, and segments of weapon systems cost control. This type of dataelement is used extensively in most production oriented support systems. To my knowledge, all three Services in the USA utilizes such a parameter. (see f.ex. USAFM 66-1).

With respect to the Decision System, ref. para. 6.2., Engineering Management should develop and maintain efficient procedures to facilitate their primary tasks. As a functional area of management, it will be concerned with design and configuration studies, product improvement vs. procurement, reliability studies, quality control, and maintenance procedures.

6.*1.2. Work Shops*

The accumulated efficiency of the work shops define the maintenance system configuration. The work shops should be managed with the objective of attaining established NORM and response criteria with minimum total resources. In the scheduling and dispatching phase of maintenance flow control, the objective is to realize established NOR rates. This would include:

- the establishment and control, per maintenance action, of responsible work centers, job standard times, and tool and support equipment requirements

- the establishment and control of work center utilization rates.

- maintenance load generation analysis, preparation of Long Term Requirement Forecasts, preparation of Master and Short Term Maintenance Schedules, dispatching and workload control.

Again it is possible to establish a Management Area which we label Maintenance Management. It should be concerned with the management of the accumulated and singular capability of the workshops as outlined in very general terms above.

The Work Center Code, WCC, is a dataelement with structuring significance, particularly if the logistic activity is organized in regional and local groups. It is important that similar work be designated to functionally similar work centers in order to facilitate Master and Short Term Scheduling as well as equipment oriented workload control.

The WCC is also important as a location code in connection with Material Allowance Authorization and Supply Management.

A possible work center structure is shown in figure 10.

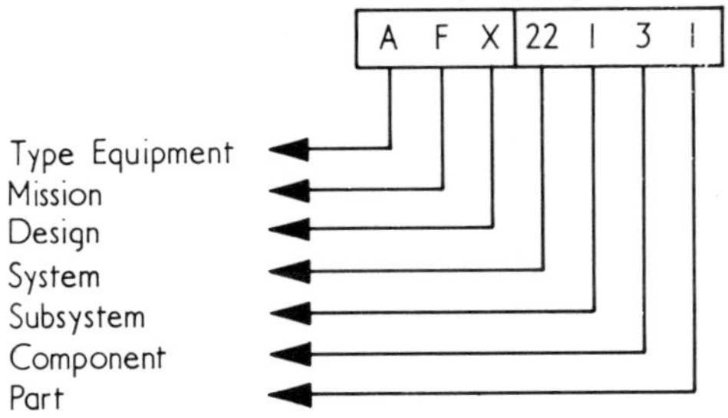

Figure 9 Equipment classification code

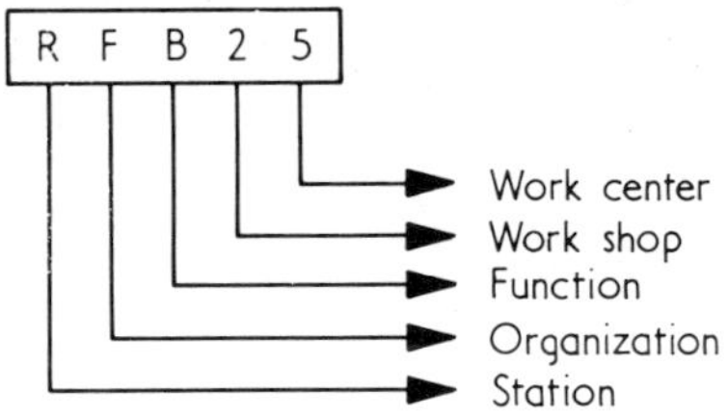

Figure 10 Work centre code

As regards the Decision System, Maintenance Management ought to be concerned for example with single-vs multiskill of workers, shift policy, manning criteria, dispatch techniques, priority procedures, scheduling methods and procedures, centralized vs decentralized maintenance as a function of equipment infrastructure and tactical mobility and selfsufficiency requirements, effect on production by methods which improve work satisfaction and motivation.

6.*1.3. Inventory*

This should be managed with the aim to furnish the user with supplies in the right condition, at the right place, in the right quantity, and at the right time. In short, Supply Management should endeavour to realize stipulated fill rates for each inventory item and supply issue point. Thus, we have defined a third Management Area. Typical activities associated with Supply Management are in my opinion:

- vendor selection
- freight and transportation media selection
- ordering, storage, distribution, issuing, and disposal
- cataloging
- requirements determination
- stock balance and consumption accounting
- establishment of criteria for efficient stocking procedures (MAX, MIN, EOQ etc.)

With regard to the Decision System, Supply Management should establish ordering, requisitioning, distribution, storage, issue, and disposal procedures; be engaged in studies aimed at improving transportation, inventory control, and issue and storage procedures.

6.*1.4. Economic and Systems Management*

In preparing and controlling the use of resources in the Budget, Economic Management spans such activities as

- cash flow control (accounting)
- contracting and contract cost control
- program, activity, weapon system and project budgeting and accounting (facilities interface with Planned Program Budgeting at higher echelons)

The previously mentioned equipment Configuration and Work Center Codes are important identifiers in program, activity, and weapon system budgeting and accounting.

With regard to the Decision System, Economic Management should be concerned with developing and maintaining efficient procedures for the above mentioned activities. Contracting and contract cost control, in particular, seem to have growing importance.

I have indicated as my opinion that after the management system structure has been defined and constraints have been established for each Management Area, they ought to have considerable freedom in developing management procedures and methods within their area of responsibility. Such a delegation of authority necessitates coordination. This coordination, as well as initial and follow-on general systems development, should be the responsibility of Systems Management. Since this presentation deals with that activity in particular, I think no further explicit mention is necessary as to the responsibilities of Systems Management.

6.*1.5.* As I have defined them in previous paragraphs, *the Management Areas,* and in particular those labeled Engineering, Maintenance, and Supply, spans the activity from Resource Management to Operational Control. We could indeed redraw figure 8 with three downward flows, rather than one, representing Engineering, Maintenance, and Supply.

This would indicate that there is a vertical integration between each management level which is functional and based on management needs. Within each management level, on the other hand, there is a horizontal integration between Management Areas facilitated by crossfunctional processes, which is necessary to fulfil missions, objectives, and tasks respectively.

6.*1.6. The organizational aspects of management* are dictated by a desire to manage effectively. Often foremost in the discussion, is the question of product vs. functionally slanted organizations.

A product or systems oriented organization will facilitate work-differentiation and ascertain a continuous monitoring of results vs. mission (systems) accomplishment. But if at the Objectives Management level, there are several systems being managed by one support activity and utilizing common resources, then this activity will take orders from several systems oriented management groups. This is an unwanted situation.

A functional organization will facilitate work specialization and the monitoring of the Objectives Management level. But at each management level, the functional organization must be made "systems inclined" by some means. I personally think that a System Overlay Structure at the Resource Management level would be an effective solution in an otherwise functional organization. This may be accomplished for each system by assigning systems responsibility to representatives for each of the three main Management Areas (Engineering, Maintenance, Supply). For each weapon system there would then exist a "group of three" with each member collectively responsible for the realization of program and mission efficiency.

At the level of Objectives Management, the NOR-objective for each weapon system will probably cause sufficient emphasis to be put on the systems concept when carrying out Short Term Scheduling and Dispatching.

6.*1.7.* I believe that *systems design* should be distinguished from studies in organization. The systems designer should be concerned with realizing a system that facilities several organization structures rather than working on methods of organization. For this reason, and with reference to figure 11, I shall give some indication–by examples–of how different organizations may fit the Management System Structure suggested in previous paragraphs.

Figure 11 is a reflection of figure 8. Shown is an Organizational Structure vs. the Management Structure. The full lines indicate command. The dotted lines indicate managerial integration within each Management Area. The presentation does not necessarily depict the structure of the RNoAF Logistic System.

In the RNoAF, the logistic structure is traditionally thought of as consisting of the depot–, field–, and organizational maintenance levels. I should like to make you think of those levels as category A, B, and C respectively, and associate them with maintenance complexity rather than a chain of command. The most complex type maintenance, and perhaps B category maintenance overload, is taken care of at the category A (depot) level. Remaining inspections, and fixing of major snags are taken care of at the category B (field) level. Pre- and Post-use inspections and fixing of minor snags are taken care of at category C level.

In terms of management, however, category A and B may well be mixed at the same location or work center complex. If work have been scheduled and resources made available, physical maintenance levels are reduced to two, A/B and C, under the auspices of Objectives Management and Operational Control respectively. Other factors may indicate however, that category A work should be centralized, perhaps for each type equipment.

Inventory Control is centralized at the RNoAFAMC. The Inventory Control System is of the push-type; i.e. stock balance is accounted centrally for all issue-points; max and min leveling, purchase initiation, and distribution initiation are centralized.

Buffer stock, traditionally though of as depot-stock, need not be stocked centrally. Rather, depot-stock may be distributed to issue-points in anticipation of usage if demand characteristics and the geography of locations are taken into account. Sometimes it may be advantageous to designate issue-points to carry depot-stock for classes of inventory items. If there is also A-category Component Repair Centers, CRC, those should be established at the location where the depot-stock is carried. The degree of decentralization of depot-stock is dependent on cost of transportation, stocking receipt control etc.

Thus, stock may also be managed in terms of two echelons, the issue-point and the user level

Further differentiation is possible.
For instance:

the maintenance resources of the tactical unit of figure 12 may be incorporated organizationally into the work center complex.

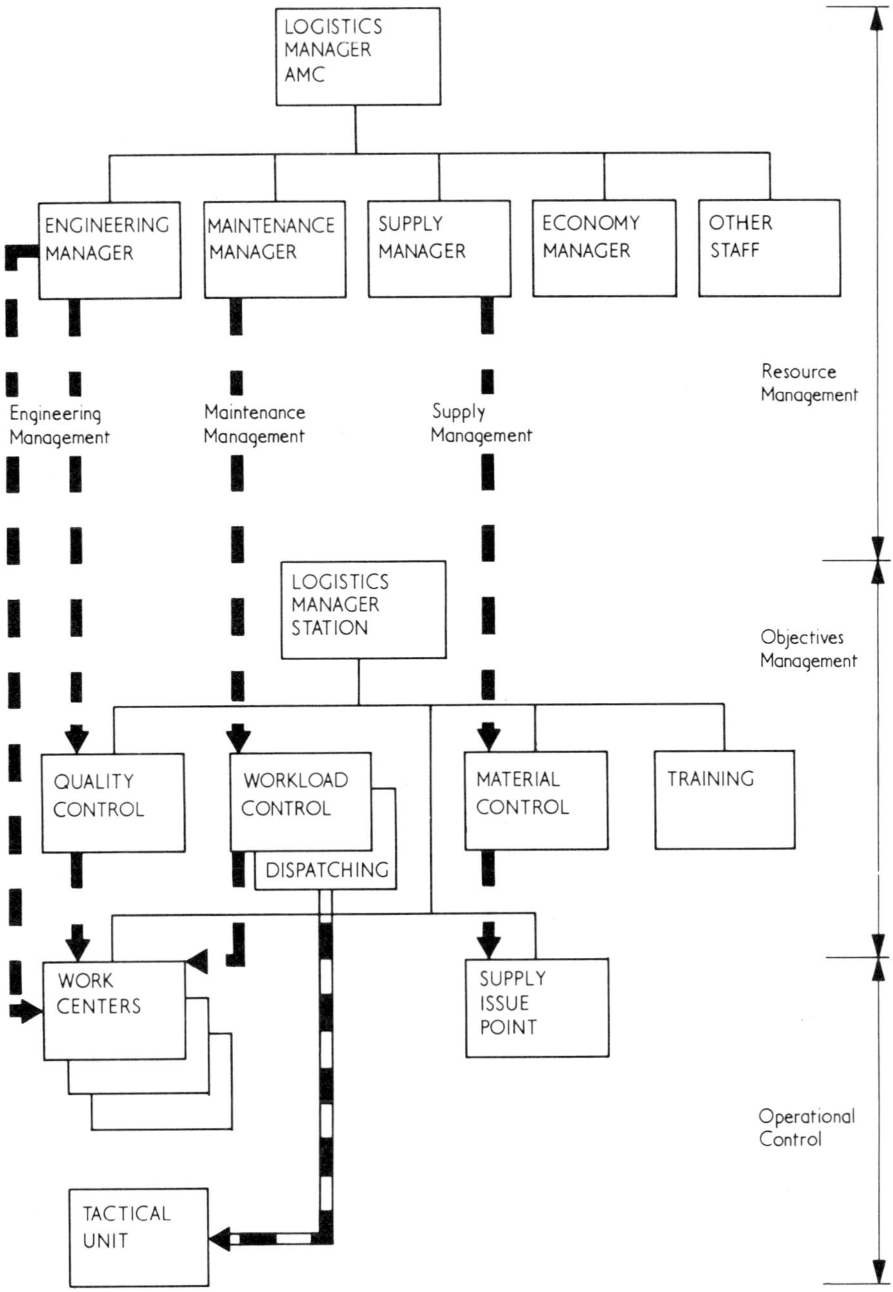

Figure 11 Organisation and management

- contrarily, consider the tactical unit having such a degree of self-sufficiency and be so large that it within itself incorporates the Objectives Management activity.

- the logistic management of EW sites may perhaps best be solved by establishing category C maintenance only on-site, merging organizationally Objectives Management and Resource Management at AMC, and dispatching specialists from category A and B work centers to site as need occurs.

May this suffice to show that although the Organization may change in various directions, the Structure of the management system stays unchanged.

6.2. The Decision System

The Decision System contains the models, methods, procedures, and processes necessary to carry out Resource and Objectives Management and Operational Control as they have been defined in sections 5 and 6. (see 6.1.1. – 6.1.4.). To reiterate, it is very important to have efficient procedures to facilitate the vertical integration, within each Management Area, between different management levels; and to have effective processes to facilitate the crossfunctional integration, within each management level, between the different Management Areas. This is a formidable task to achieve. The work of developing, documenting in manuals, training personnel, and implement the procedures and processes should not be underestimated.

This presentation is not concerned with decision methods; nevertheless, I should like to give you some feeling for the problems confronting us in designing the Decision System.

Figure 12 is intended to show the supply flow in the logistic system. It shows the flow of consumable items as well as the flow of recoverable components which are either useable, repairable or being condemned. The work shops both use and repair components.

Our goal is to realize stipulated fillrates at C-C with minimum cost while operating within known constraints. This means we have to develop inventory policies for procurement, A-A, distribution, B-B, and issue, C-C. I have indicated the type of policy we will use at the RNoAFAMC. Furthermore, as stated previously, the Inventory Control System is a push-system with regard to the cross-sections at A-A and B-B.

At RNoAFAMC we aim to use two inventory models to realize stipulated fillrates. One is homemade and used for consumables. The other, METRIC [3], is used for recoverable type items. Both models use Bayesian demand forecasting techniques. METRIC uses a marginal allocation algoritm to calculate levels for each supply issue point and depot stock-level within budget or efficiency constraints.

METRIC it turns out also renders near optimum NORS rates. Furthermore, it may be used as a basis for a distribution and repair priority model. That is to say, it may be used to pinpoint which of many supply issue-points should have the first component that becomes available after repair; and secondly it may be used to select which of

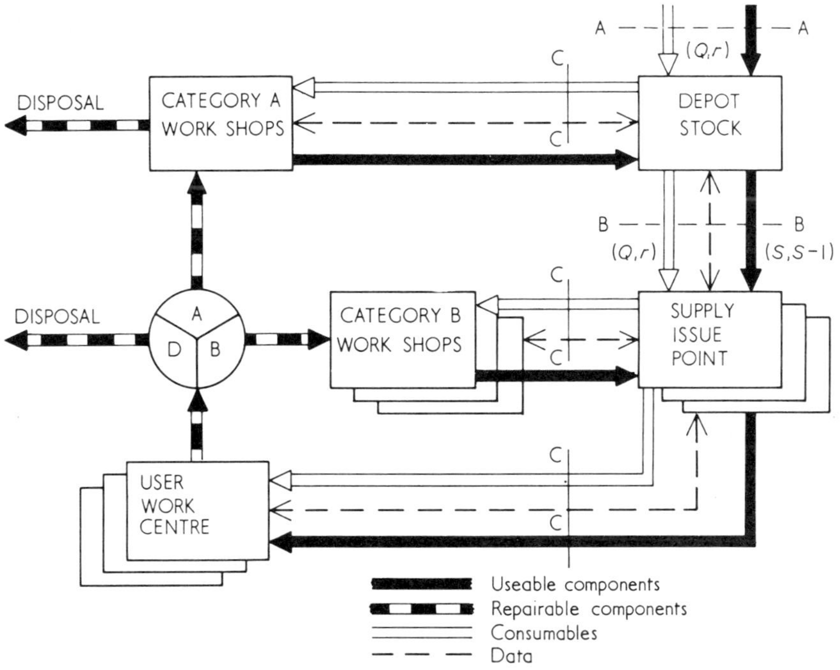

Figure 12 Inventory control

several different components should be repaired first. Thus, METRIC is an aid to Short Term Scheduling and Dispatching with regard to component repair.

In figure 13 is shown a very simplified illustration of scheduled and unscheduled maintenance. Maintenance flow control is performed by controlling the input, A-A, and the output, B-B, to and from category A and B shops respectively. The aggregate result is the relationship between the queue of unservicable equipment at A-A, and the pool of servicable equipment at B-B which gives the NORM rate (or NOR rate if NORS rate is zero).

Within the work shop illustration is depicted a workload diagram. With a slowly changing number of personnel and limited possibilities for shift work and overtime, Workload Control is trying to maintain an even load while realizing job standard times and utilization rates. There are some tricky scheduling and dispatch problems which await an optimum solution. I am somewhat excited by the possibilities which seem to exist if one thinks of personnel with a particular skill as inventory items and try to imply some inventory control techniques on the manning, and dispatch problems. (see remarks on the use of METRIC).

In Dispatching, one will frequently be faced with unforeseen, if not unexpected, problems. In order to avoid those problems, it is possible to use simulation techniques to identify future bottlenecks. We have not come this far yet, but I can see several possibilities for large pay-offs by introducing a reasonable degree of mechanization to aid Short Term Scheduling and Dispatching.

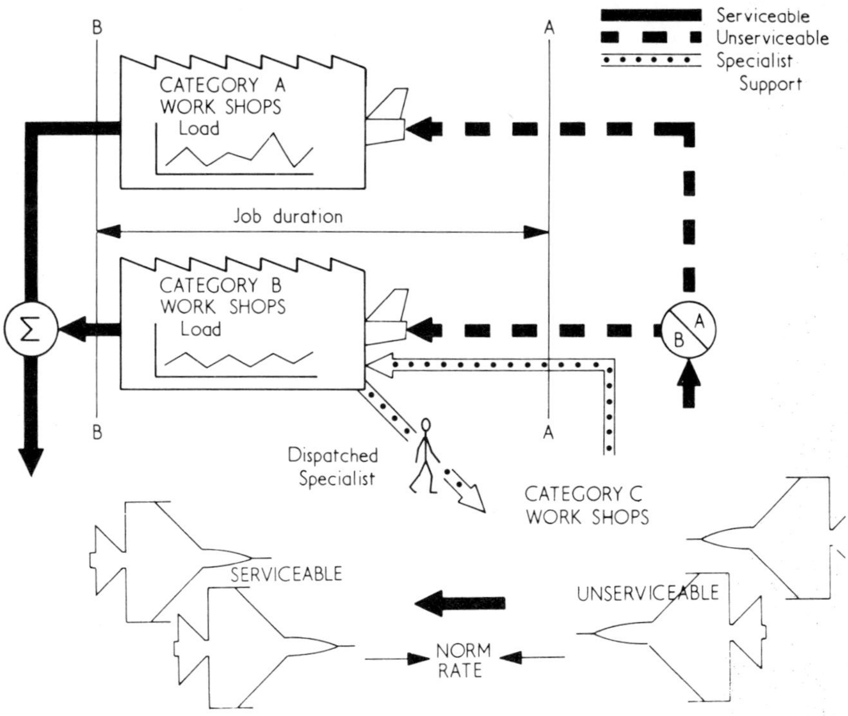

Figure 13 Maintenance flow control–workload control

The most important and difficult decisions are related to Master Scheduling, Namely, the calculation of objectives and allocation of an efficient job mix to different work center complexes.

So far, we can see several years of stumbling along with a mixture of manual and mechanized iterative procedures aimed at deriving reasonably good objectives and job mixes.

If we now superimpose figure 12 on figure 13, I think you will have a feeling for the complex problems confronting the Logistics Manager at the level of Objectives and Resource Management. Those problems should find their solution in a dynamic and adaptive Decision System.

6.3. The Data/Information System

On the basis of the conceptual framework for management which we have defined in previous paragraphs, it should be possible to start the development of the Data/Information System.

Information is defined as:

output from an intential and relevant data transformation process.

Data is defined as:

input to the transformation process which produces information.

If these definitions are valid, there are levels of information because output, i.e. information, from one process may be input, i.e. data, to another process.

The Data/Information System should support Management and Operational Control. Consequently, it should help to decide, plan, operate, and control. Therefore, the Data/Information System must include future, present, and historical data as well as processing routines, in order to furnish management with information, and be capable of:

- identifying events and/or actions representing resources and relate those to event-associated Objectives,

- present these events as data, the formal definition of which is inherent in predefined datasets,

- transforming and storing data such that it may be identified and retrieved,

- reporting information for decision and control.

The dataelements of such a system must be structured so that data collection, data transmission, data storage, processing, and retrieval are facilitated. We should differentiate now between the logical datastructure which is application oriented and the physical datastructure which is designed to make efficient use of data-processing machines and methods.

In order to develop an efficient Data/Information system, the most important single condition is that there exists a management policy or strategy for its use upon which development and implementation may be based.

The development of a Data/Information System is in itself a project of considerable magnitude. Different decisions that have to be made include

- decentralization vs centralization of Hardware and Databank

- accessrights to Databank

- batch vs on-line

- Hardware and Software selection

Each of those items are topics for a lecture. In this presentation I can only give you an idea of the structure of the RNoAFAMC Data/Information System which is depicted in figures 14–16, and consists of

- the Supply ADP-System

- the Engineering and Maintenance ADP-System

– the Economy ADP-System

– the ADP File Maintenance System

– the ADP Communications System

6.3.1. *The Supply ADP System*

This is illustrated in figure 14 and consists of 4 main system modules

– the I/O control
– the Distribution Control
– the Procurement Control
– the Inventory Control

The tasks performed by each module should be relatively well explained by the module task description and the file names shown in figure 14.

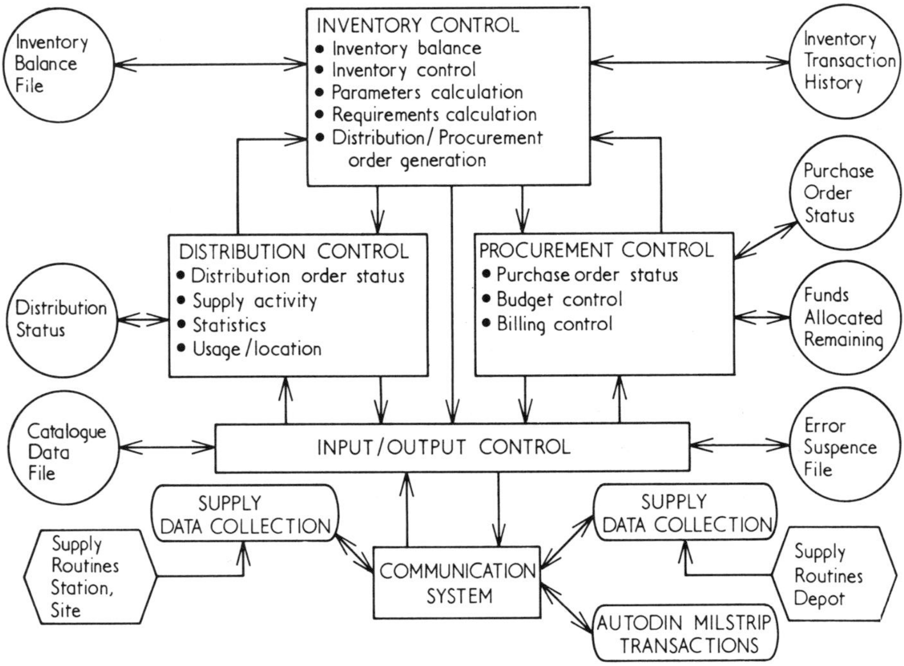

Figure 14 Supply ADP system

Input from the supply routines are collected via datacollection equipment, communications network, and the Communication ADP-System. Output is desseminated in the opposite direction.

The Distribution and Procurement Systems operate with Document number as the most significant identification data. Federal Stock Number, FSN, is the secondary identification element in those systems.

The Inventory Control System operates with FSN and Location as primary identification elements. Inventory control is carried out on the basis of Interchangeability–and–Substitution–Grouping, ISG. The inventory item structure is such that we presently favour daily batch updating.

The CPU used in the Communications System is also used as an in-house computer at the AMC. In this role the CPU provides simultaneous access from several terminals within the AMC area to local files containing inventory balance data as well as due-in and due-out data.

6.3.2. The Engineering and Maintenance ADP System

This is illustrated in figure 15 and consists of five modules:

- I/O Control
- Maintenance Flow-, Configuration-, and Shop Control
- Material Control
- Maintenance Load Generator and Scheduler
- TCTO

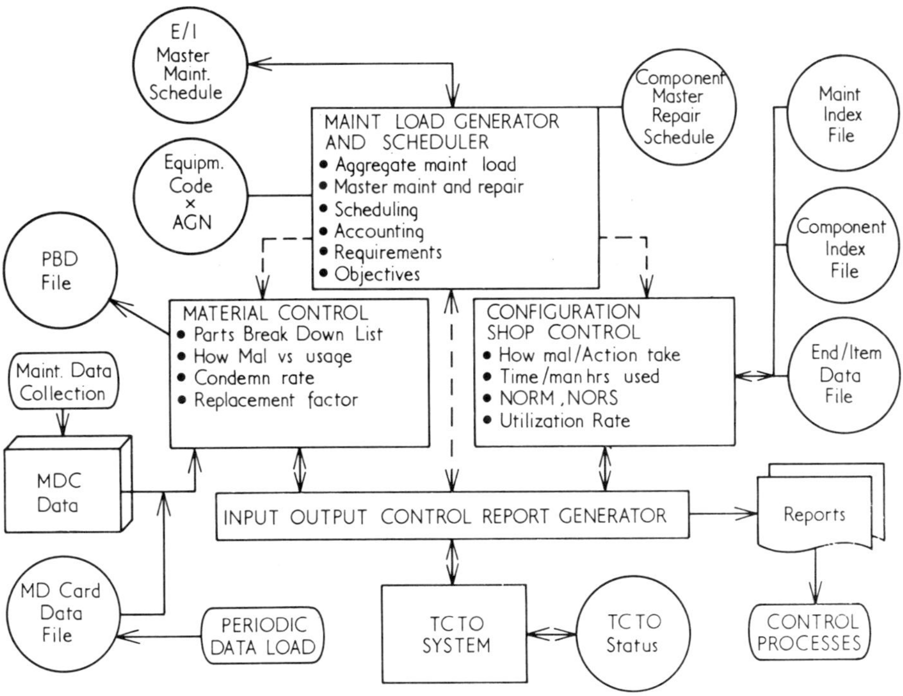

Figure 15 Engineering maintenance ADP system

The primary identification data in the Engineering and Maintenance ADP System are the Equipment Classification and Work Center Codes. The system works on data collected in conjunction with maintenance actions, viz., job duration, how malfunctioned, and action taken. The tasks performed by each module is exemplified in figure 15 by the filenames and the task description within each module.

Given an activity program and consumption-based parameters or engineering estimates, the Material Control ADP-System calculates expected annual demand for all ECC coded items. Crossreferencing ECC to FSN and taking into account ISG data, annual demand rates may be input to the Supply ADP System for max. and min. inventory level calculation. This method will be used in conjunction with provisioning and in case of large changes in Equipment Configurations and/or activity programs.

As demand history becomes available in the Supply and Material Control ADP Systems, it is integrated with initial estimates. Thus, a smooth transition from initial operation to steady state is facilitated as regards inventory control.

The further integration between Supply and Engineering/Maintenance ADP Systems is facilitated by a parameter reflecting the chosen job mix. This parameter, Percent Base Repair, is an important input to the inventory models contained in the Inventory Control ADP System.

6.3.3. *The ADP File Maintenance System*

As shown in figure 16, this periodically updates all files simultaneously in response to new or changed basic data.

The integration between daily and periodic ADP systems is also indicated in figure 16. File magnitudes are given in terms of million characters.

6.3.4. *The Economy ADP System*

This is not shown on any illustration. This is illustrative of the fact that there does not exist a particular Economy ADP System. The reason is that all information such as stock consumption, outstanding orders, stock on hand, received material, manhrs. pr. equipment usehr., labour consumption summary, labour backlog etc. are all inherently economic information. Costing such information as mentioned gives economic information.

Thus, the Economy ADP System is nowhere and everywhere. In the Supply, Engineering/Maintenance, and ADP File Maintenance ADP Systems, there are segments of the Economy ADP System.

6.3.5. *The ADP Communications System*

This includes, according to my definition, I/O equipment, Terminals, other Communications Hardware, and Communications Software. The Communications CPU at the AMC performs message switching in a store-and-forward mode, and facilitates on-line data retrieval in a multiprogramming mode from local files, as mentioned in para. 6.3.1.

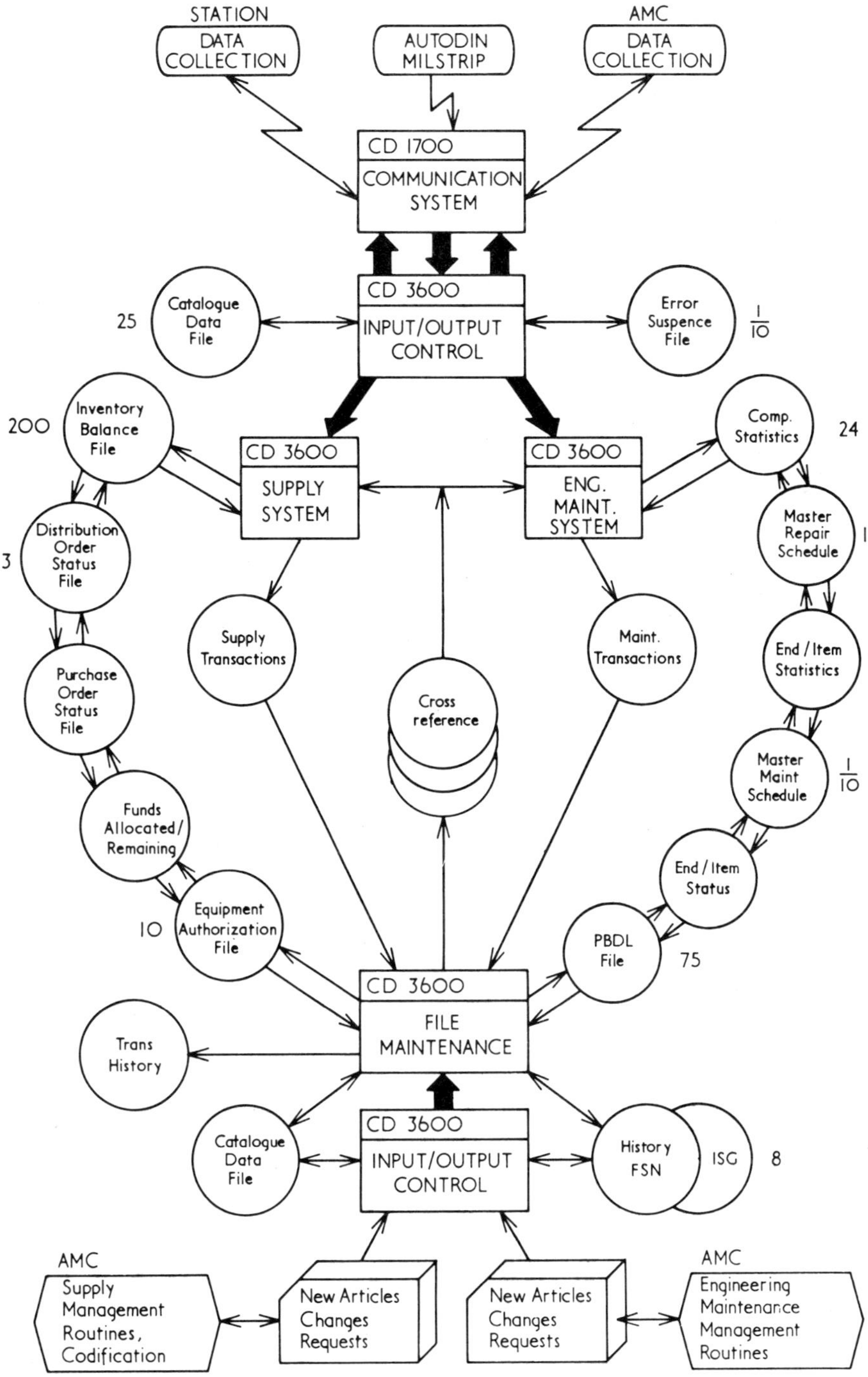

Figure 16 Daily/periodic ADP system interface

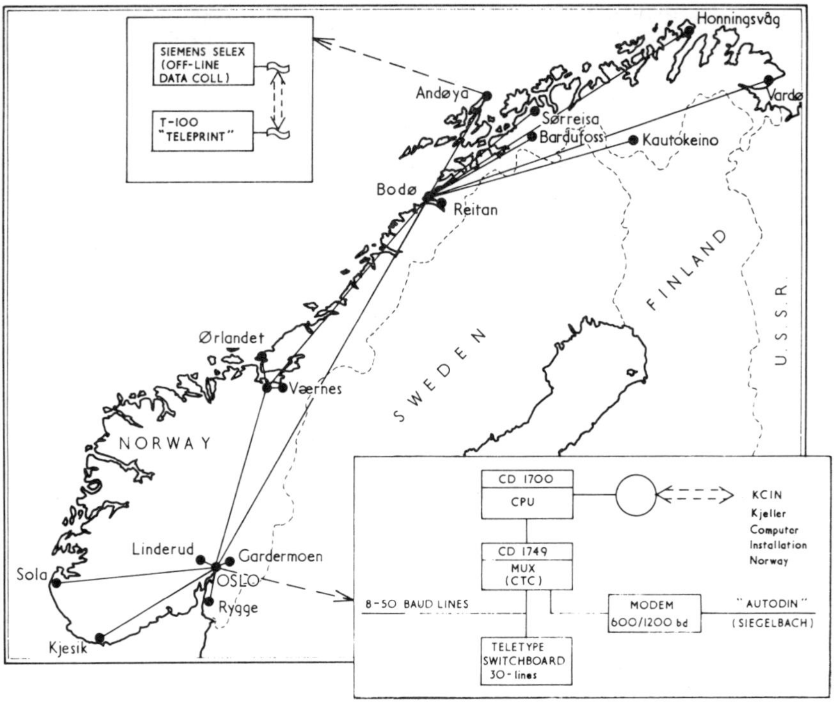

Figure 17 Logistics communication network

The Logistics Communications Network is depicted in figure 17. Presently all communication are low speed, except the Autodin connection. Present plans call for increased capacity and speed. In passing, it should be noted that the distance from south to north of Norway is approx. equal to the distance from Oslo to Rome. Presently, the network shown serves only the Supply ADP System. Future requirements will most probably result in some line communication in support of the Engineering/Maintenance ADP System.

The Communications ADP System is designed with improvement potential. The present on-line query-capability of the AMC may be rendered to all reporting units if communications capacity is sufficient. If it is cost effective, systems similar to the AMC in-house computer system may be distributed on a regional and/or local basis.

6.3.6. *The Data/Information System*

As a whole, this is perhaps best exemplified by the Shop-, Configuration-, and Inventory Control subsystems, should aid management to decide, plan, and control. This aim is accomplished by establishing sub-objectives, comparing performance with established objectives, reporting whenever maximum deviation-limits are exceeded, and by storing accumulated performance data which may be retrieved by request.

Thus, the Data/Information System should be much more than Hardware, Software, Analyses and Design. It should be a dynamic interweaving structure surrounding the organization so that management and the employees may function and work efficiently. The system must be flexible and have growth potential.

7. DEVELOPMENT AND IMPLEMENTATION

A Logistic Support System similar to that described in previous sections is under implementation in the RNoAF.

In order to relate to you manhr. consumption figures for development and implementation, I shall first give a resume of project status.

We have finished the work depicted in figure 1. The Management Flow Analysis illustrated in figures 3–6 is finished. The ADP systems depicted in figures 14–16 are ninety percent finished. The last ten percent contains models for optimal allocation of resources and job mix, and the implementation to full range use of the programmed inventory models. The percentages refers to programvolume. The associated work may well amount to as much as fifty percent of presently consumed labour.

User manuals for Supply, Engineering, and Maintenance Management have been distributed. Formal training of personnel has been undertaken as regards new procedures and processes.

Eighty percent of the Supply Management System is implemented.

The Engineering and Maintenance Management Systems are developed and at the stage of beginning implementation.

The ADP Systems are being implemented parallel to other segments of the LMICS. Final documentation is carried out as implementation progresses. Large portions of the formal documentation remain to finish.

The approximate amount of work in terms of manmonths which has been consumed in development and implementation is shown in table 1. Some think the manmonths expended on development and implementation are few. One reason for this may be the working procedures adopted.

We have approached the task as one would do if designing and fabricating a sophisticated weapon system. After the establishment of a management strategy, the project group defined the tactical and operational requirements to the system on the basis of initial studies. Then followed Management Flow and Structure Analyses upon which were based the system definition consisting of initial Systems Specifications and Characteristics. After having established performance and system baselines, followed system design, development, "fabrication", assembly, and testing.

Throughout development, contact with the user organization was kept to the necessary minimum and concentrated to motivational briefings and performance and specifications reviews. However, in the test phase and beginning implementation, the contact with the user is increased extensively as the system is being fitted to the customers' needs.

Work	Analysis	Progr. Testing	Inform Brief	Documentation Manuals etc
1. Strategy, Concept Structure Figure 1.	24	–	4	4
2. OR –	50	6	–	2
3. Decision and Data/ Information System				
a. Supply ADP System Fig 14.	80	30	20	30
b. Eng/Maint. ADP System Fig 15.	80	44	6	30
c. ADP File Maint. System Fig 16.	80	50	30	30
d. ADP Communication System	30	20	30	4
	344	150	90	100

Table 1 Labour Consumption in man months for Developement and Implementation

8. CONCLUSION

I have tried to show how necessary it is to realize the penetrating effect on lower echelons by higher echelons' planning and resource allocation methods; and furthermore, tried to illustrate how the datasets and parameters of the Logistic Support System in fact are the necessary foundation upon which rests the datastructure of the top (Strategic) level management systems.

In presenting the structure of a management system under development and implementation, I have tried to illustrate the complexities of Management Systems Design and Development. It is my hope that I have also succeeded in showing that it is possible and necessary to modularize the system under development to promote job segmentation and planning, because this is the only approach to orderly development and implementation.

REFERENCES:

1. Langefors, B., Management Information System Design, *IAG Journal*, Vol 2, no 4, 1969.

2. Fórsvars. dep., Sweden, Planering och Programbudgetering inom Fórsvaret, SOU, Stockholm, 1969, (Swedish text).

3. Sherbrooke, C.C., *METRIC: a Multi Echelon Technique for Recoverable Item Control*, the Rand Corp., RM-5078-PR, November 1966.

DISCUSSION

Ferrier: Vous avez souligné, avec beaucoup de justesse, l'importance de l "intérêt de métier", en tant que substitut partiel de l "esprit de corps". Avez vous une expérience personnelle sur les façons d'y parvenir ou d 'y tendre, au niveau de "management policy", particulièrement au stade de l "introduction" du système?

Høgberg: In answering the question, I do not draw on personal experience, but on the results of current research in Norway co-sponsored by the Labour Union and Industry.

It seems that motivation to a certain degree is dependent upon individual participation with regard to job situation. This entails participation in establishing performance goals and necessary resource requirements.

With reference to Management System Design this seems to indicate a necessity for a certain degree of decentralization and the establishment of autonomous groups, very often not following the established organizational structure.

Secondly, it is important to note that the structure of objectives, subobjectives etc. must reflect this decentralization of the management function.

Thirdly, it hàs been demonstrated that the establishment of management participating autonomous groups increases the requirements to the Information System, both with respect to the input/output configuration file system and the architecture of the total system.

Collins: What is the upper limit of MIS integration? e.g. can you or should you combine a Logs system within a larger system embracing such things as Command & Control, Finance, Personnel etc?

Høgberg: I think yes. For instance, in the personnel area, I have indicated, in figure 3, input from the Logistics Management function to Personnel Management. The Systems level or Organizational level at which you integrate is a function of your management policy, viz. functional or product (weapon systems) oriented management policies. Note however that the frequency of interaction between systems will most often reduce the higher the management level.

Fogden: The speaker mentioned the time scale for the development of a Management Information System. Could he amplify this point in relation to the inventory size, force deployment and personnel establishment mentioned.

Høgberg: I have indicated in table 1, page 37 that approximately 50 manyears have been used for approximately 80% of Systems development, testing and initial implementations. This effort has been used over approximately 3 years.

It is important to note that the initial work as outlined in figure 1, p.4, is an effort concentrated to a few people. As one is able to modularize the system and assign jobs and define skill requirements and recruit the people you need, the detailed work gets under way. Thus the nature of the work is limiting your progress in calendar time to a certain degree.

Important to realize is also the fact that implementation time is a function of the organizations' ability to absorb new methods and procedures, which establishes another threshold.

Lastly you have the revisions, the redesign.

Before you have a smooth running large scale logistics system, I think 8–10 calendar years will elapse from conception to full implementation.

O'Connor: In the RAF we have found problems in collecting high quality maintenance data. Have you considered the possibility of relating the data collected on an equipment to its initial and in-service costs, so as to concentrate the effort and reduce the total data collection workload?

Høgberg: Maintenance Data Collection in our system is aimed at providing data for Engineering and Maintenance Management as I have defined them. As you get into steady state in the sense that you are satisfied that your planning factors are relevant and performance is reasonably effective, I think you can reduce the volume of Maintenance data to what is actually needed. This points out an important point, namely that your system must have a built-in flexibility to adopt to different situations whether they be Management-directed or governed by development.

Falch–with reference to figure 15 page 31:

1. Can you briefly describe the difficulties you have experienced in implementing your maintenance data collection system?

2. What emphasis have you placed on work force motivation and active training to insure a reasonable degree of *input* accuracy for your maintenance data collection?

Høgberg:

1. First, let me make it clear that the Mechanized Engineering and Maintenance (ADP) System is only 80% developed and in the initial stage of actual usage. The parts of the system being tried out for 3 weapon systems and 3 bases are working as expected. Our main difficulties are related to your second question, namely motivation and data discipline.

2. I think a measure of the importance we put on motivation as a prerequisite to–amongst other things–data discipline, is the effort we have put into informing and briefing the policy makers and the users. In table 1 page 37, you will note that approximately 15% of the project effort has been allocated to information and briefings, (this excludes formal training which also is quite extensive) and which is equivalent to 2/3 the effort for programming and testing.

A Method of Searching for the Optimum Structure

J. Combes

INTRODUCTION

1.1–Foreword

1.*1.1. Purpose*

We belong to the NATO-HAWK Production and Logistics Organisation and nearly eight years ago we participated in the studies for the definition and setting up of the 1.2.1. HAWK Support System.

This lead us to view the problem in a NATO environment. In other words, "the" Logistic Support System to be studied would concern one (or several) weapons system(s) in service in several NATO countries and at least one part of its activities would be common to all these countries and borne by one (or more) common organisation(s).

Experience has shown that the most important difficulties encountered by a common organisation for its setting up and its operation resulted from the constraints introduced by the participating countries.

In our opinion, these difficulties result from the two following main causes:

- The interests peculiar to each country become directly apparent in the search for solutions, which makes discussions passionate, biased and consequently difficult to conduct and to conclude;
- No method is adapted to the conduct of the discussions; hence, misunderstandings and long delays, sometimes even clashes, which put the common will to reach a solution to a hard test.

Our ambition is, if not to remove, at least to attempt and reduce these drawbacks by proposing an evaluation method which would:

- Provide a basis, a framework and guidelines for the discussions as well as promote a better understanding;
- Transfer the discussions to a level where interests would not be directly apparent, which would facilitate discussions and enable a solution to be reached more rapidly.

In other words, more generally, this is an attempt at rationalization in a field–group discussions–which is currently considered as an art essentially based on individual experience, intuition and intelligence.

The method submitted is more particularly adapted to the search for the best structures of a logistic support system. With some adjustments, it could be used for other purpose. It refers to the exposés made by the "Centre de Prospective et d'Evaluation du Ministère français des Armées" during seminars on French methodology for the increase of productivity.

1.*1.2. Support policy*

Countries deciding to tackle in common all or part of the logistic support of a Weapon system must, *prior to* any study, establish the "policy" which they want to apply for the "conduct" of the support.

Very broadly, a support policy should include the three following aspects:

(i) Definition of the missions and, for each one, the goals to be reached. The list of the support missions will be given in the main part of the study. As regards the goals, it is essential to determine them to enable the correct evaluation of the means to be implemented and also, during the execution, the measurement of the support effectiveness and the correction of the weak points.

(ii) On the military level, formulation of the principles and rules to be applied and the respective responsibilities of the common and national organisations for the carrying-out of the support missions.

In fact, these principles, rules and responsibilities will result from the degree of centralisation decided by the participating countries for the carrying-out of each support mission.

(iii) On the economic level, determination of the conditions governing the funding of the common part of the support and the laying out of the common expenses.

Although the former conditions, as a rule, can be established easily because they result directly from the services which each country expects, this is not the case for the latter conditions.

For the latter, we will merely indicate two limits of the choice, i.e.:

base them on "free competition", extended or not to all NATO countries;

base them on the "fair return" rule according to which each member-country of the Consortium would be returned, through supplies or services performed on its territory, an amount of credits in proportion with its participation in the common support expenses.

Obviously, there are, between these two limits (one of which may result in a de facto monopoly and the other one in excessive prices) variants the form of which will depend on the environment.

In addition, a support system having a rather long lifetime, even when it applies to rapidly changing weapons systems (e.g. approximately ten years for missiles), the "support policy" must take into account the evolutions to be expected both at common level (NATO) and at national levels, during the period in question. In any case, procedures would have to be devised to enable re-adjustments as required by new developments and to avoid any "seizing" of the machine during the operation.

The method submitted could be used to facilitate the definition of the support policy, which it is not necessary to formulate for the understanding of the exposé.

1.2 General

1.2.1 Definition

The *structure* of a logistic support system will be taken to consist of *Activities*, related by command and functional inter-connections, in charge of operating the means for the fulfilment of the general support missions, in consonance with the established policy.

An *Activity* will be taken to mean an organisation vested with directive or executive functions. The table below shows these Activities with their respective functions and status (NATO or national). The listing given is a mere summary and could, of course, be developed (for instance, several national services might be considered per country or, several NATO agencies might be retained, etc.).

The word "elements" will be taken to refer indifferently to a mission, a structure or an Activity.

Optimizing the structures means selecting the best structures of all those which can possible be envisaged.

ACTIVITY / FUNCTION	NATO Agency (A.O.)	National Service (S.N.)	NATO Establishment (E.O.)	National Establishment (E.N.)	Independent Industry (I.I.)
DIRECTION	X	X			
EXECUTION			X	X	X

1.2.2. Logic Network

For a clear delineation of the scope and limits of this study, the inter-relations between the "goals" and the "means" should be specified.

At the start, there is, or there should be, a clearly-stated policy for the support of the weapon system, as a necessary basis for the speedy and efficient achieving of the study and, later of the support. This policy leads to the formulation of the general support missions which are its direct expression. These missions may not all have the same importance, and they should be classified according to their relative weights in achieving support.

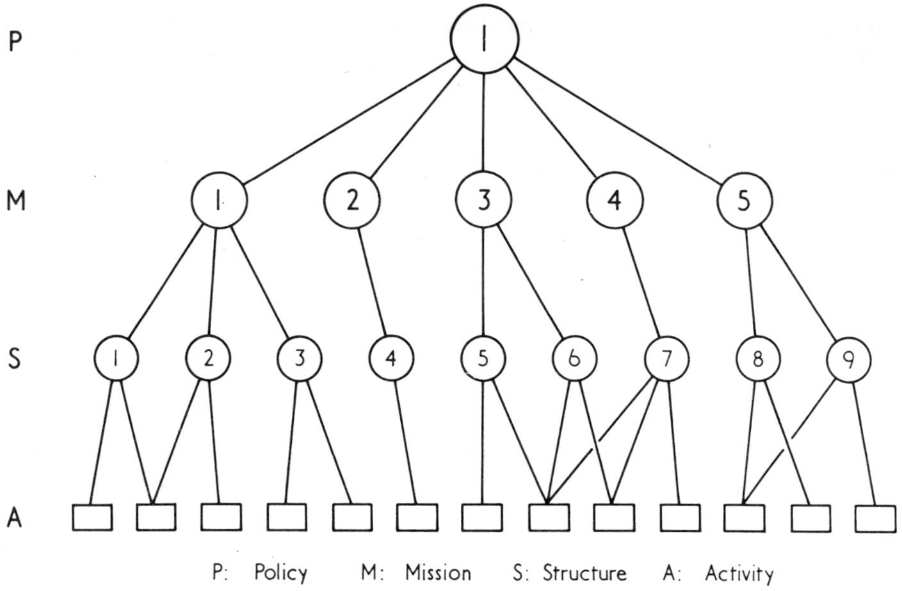

Figure 1 Logic network

M 1 _ CONFIGURATION CONTROL
M 2 _ CORRECTION AND MODIFICATION STUDY
M 3 _ 5th ECHELON
M 4 _ SUPPLY
M 5 _ TECHNICAL ASSISTANCE
M 6 _ TRAINING
M 7 _

Figure 2 Missions

Meeting the goals, i.e. the policy and missions implies "requirements", among which the "Structures". As each mission constitutes a goal, it is logical to envisage that structures suitable to a mission may not necessarily be suitable to other missions. On the other hand, as a rule, several structures may be considered for a single mission. The study should therefore lead to the optimization of one elementary structure per mission, the consolidating of these optimal elementary structures leading to a satisfactory solution for the overall system.

Lastly, the structures are based on Activities, which are the physical entities on which this study will bear for its greater part.

The inter-relations explained above can be set out in a *logic network*, a specimen of which is attached. It should be noted at this stage that there is a logic sub-network specific to each mission.[1] (see Fig. 1)

1.2.3. *Judgement Criteria*

A choice is a subjective matter to which the strictness of mathematical approach cannot usually be applied. It implies prior definition of "values" to be used as a basis for judgement and classification. In practice, it also requires a method for materializing the classification made in relation to each value and composing these various classifications into a single ordering.

The values retained, which are called "judgement criteria", are defined below. The method for making the classifications and composing them will be set out in the second part of this presentation.

(i) *Usefulness from the policy standpoint*

The elements to be classified are to be assessed as to their conformity with the established support policy. This criterion applies at all of the logic network levels.

(For instance, a structure suitable to the fulfilment of a mission may not be in conformity with the selected policy).

(ii) *Usefulness from the military standpoint*

Assessment will be made, in this case, on the basis of the importance of each element of a given level for the effectiveness of the next higher level.

(For instance, missions may be assessed from the stand-point of their relative contribution to the support; the structures considered for the fulfilment of a mission may be classified in terms of the resulting effectiveness in the fulfilment of this mission, etc.)

(iii) *Availability*

This criterion is to be taken into account only at Activity level. They will have to be classified according to the probability of their being available when required.

(iv) *Costs*

As the previous one, this criterion applies only at Activity level. The costs for the various Activities will have to be assessed, including setting-up costs as well as operating costs for the planned support duration.

[1] A logic sub-network therefore describes the interrelations specific to each mission.

2. DESCRIPTION OF THE METHOD

2.1 Influence of Intermediate Levels

An examination of the logic network, as already mentioned, shows very clearly that the assessed usefulness of a given element will depend on the judgements passed on the higher levels. Classifications are therefore to be carried out at each higher level on the basis of the applicable criteria and will be taken into account to assess each element in relation to the specific chain to which it belongs.

Quite obviously, comparison at each level can be made only between comparable elements; thus:

- Missions will be compared with one as other;
- Structures will be compared when they belong to the same logic sub-network, the same going for Activities.

2.2 Classification Rules

Classification of the elements according to a judgement criterion will be effected by assigning to each element an index number which will be all the higher as its importance is lower from the point of view of the criterion considered.

This requires prior assignment to each judgement criterion of index numbers characteristic of the classes into which the elements will be entered. These index numbers, i.e. classes, should not be too few so as not to result in too superficial a classification, nor too many, as fineness would not serve any purpose where subjective judgement is involved.

Four classes will be assigned herein to each of the first three criteria. The index numbers of these classes are as follows:

(i) *Usefulness from the Policy Standpoint*

Index	0	– In complete conformity
,,	2	– Satisfactory
,,	4	– Allowable
,,	∞	– Not in conformity

(ii) *Usefulness from the Military Standpoint*

Index	0	– Vital or the most useful
,,	2	– Satisfactory or useful
,,	4	– Allowable or less useful
,,	∞	– To be rejected or useless

(iii) *Availability*

$$\begin{array}{llcccc} \text{Index} & 0 - t_M & \leq & T & & \\ \text{,,} & 2 - t_e & \leq & T & < & t_M \\ \text{,,} & 4 - t_o & < & T & < & t_e \\ \text{,,} & \infty - & & T & \leq & t_o \end{array}$$

$$\text{with } t_e = \frac{t_o + 4t_m + t_M}{6}$$

where:

T = Leadtime at the end of which the element must be available*[2]

t_o = Optimistic leadtime for availability

t_m = Most probable leadtime for availability

t_M = Pessimistic leadtime for availability

t_e = Expected leadtime for availability

Classification according to Criterion No. 4 will be made in a different way (see Section 2.4.7), the costs being expressed in true values. This chapter and the following are not therefore applicable thereto.

2.3 Composition of Classifications

Once an element has been assessed in relation to given judgement criteria, it has to be examined globally on the basis of the combination of these criteria. In other words, the assessments made have to be "composed" into a single assessment, which requires the laying down of "composition" rules.

Besides, we have seen that for a comprehensive evaluation of an element, it had also to be considered from the standpoint of the chain to which it belonged and that the assessments made for the higher levels had to be taken into account for final assessment of each element. This amounts to considering the various levels of the logic network as many different criteria enabling to assess a lower level element according to the same classification and composition rules[3].

2 Formula drawn from the PERT Method.

3 We shall see hereafter that, as composition comes to an addition, an associate and commutative operation, it may be done:
- either vertically, according to the levels (taken as criteria), then horizontally, according to the established judgement.
- or horizontally, according to the criteria, then vertically, according to the levels.

2.3.1. *Composition of 2 Criteria of the same weight*

By definition, two criteria of the same weight will include the same number of classes with each the same index numbers (e.g. Criteria 1, 2 and 3 have been taken with the same weight).

Their composition is given by the square grid below, where C1 and C2 are the criteria, and 0, 2, 4, 6 the index numbers for the selected classes.

C2 \ C1	0	2	4	6
0	0	2	4	6
2	2	4	6	8
4	4	6	8	10
6	6	8	10	12

In this example, the result is a distribution of the elements into seven classes (0, 2, 4, 6, 8, 10, 12).

2.3.2. *Composition of 2 Criteria of different weights*

There are two possible ways of differentiating between two criteria of different weights:

(i) By assigning to the higher-weight criterion a number of classes all the larger as compared to the other criterion as the difference of weights is wider, or:

(ii) If the same number of classes has to be assigned to each criterion, by assigning to the higher-weight criterion classes with higher index numbers.

Two criteria of different weights are composed according to the two following grids (Where C.1 is more important than C.2):

C2 \ C1	0	2	4	6
0	0	2	4	6
2	2	4	6	8
4	4	6	8	10

C2 \ C1	0	2	4	6
0	0	2	4	6
1	1	3	5	7
2	2	4	6	8
3	3	5	7	9

3. Composition of more than two criteria

The rule is simply to compose two criteria, then to compose the result obtained with the third criterion, and so on.

The final result is independent of the composition order (associate and commutative rule of the addition).

For instance, the composition of two criteria C.1 and C.2 of a same weight (see paragraph 2.3.1 above) with a third criterion C.3 of a lower weight (comprising only three classes: 0, 2, 4), will be made according to the following grid:

C1/C2 \ C3	0	2	4	6	8	10	12
0	0	2	4	6	8	10	12
2	2	4	6	8	10	12	14
4	4	6	8	10	12	14	16

which gives 9 classes for a final distribution of the compared elements.

Remark 1 : The class index numbers should not be considered as marks, as the numerical difference between 2 indexes has no meaning whatsoever as such. In other words, it is an ordinal classification, merely to set the order of the elements.

Remark 2 : The above composition rules are based on axioms which may be set out during the oral presentation.

Remark 3 : The composition operations which have been described and will be used all come to additions, due to the fact the class index numbers retained for a criterion have been considered as expressing a constant weight independent of the other criteria.

However, it would be possible to consider compositions such as to give to a criterion a weight varying in relation to the weight of the criterion with which it is composed. It is clear that in such cases composition would no longer follow the addition rule and that a composition grid specific to each case would have to be set up.

2.4 Mode of Operation

2.4.1. Principle

The objective is to arrive, for each logic sub-network–i.e. for each mission–at an overall classification of the structures allowing final selection.

For the purpose of this classification, the logic network specific to the system under study has to be built up, which requires the identification of all possible structures.

Classification will then be carried out in two phases. During the first phase, overall classification of the structures according to Criteria 1, 2 and 3 will be performed, following which a first selection of the structures will be made by eliminating those classified last.

In the second phase, the structures retained will be classified according to Criterion No. 4 (Cost), following which final selection will be made.

Classification of structures in each of the two phases above will be achieved by composing the classification of the activities comprising them, the activities being the basic elements at the level of which the main part of classification work is done.

2.4.2. *Identification of the Structures*

(i) *Identification of the Activities*

As structures are combinations of Activities, the problem will be, in a first stage, to look for all activities which could take over the direction or execution functions required for the fulfilment of each mission.

To this end, each separate mission has to be broken down into tasks, each task relating to only one function (direction or execution) and one technical field (finance, contracting, procurement, planning, quality control, etc.).

All the Activities capable of assuming each of these tasks and set of tasks are considered.

(ii) *Combination of Activities*

Once they have been identified as indicated above, the Activities are grouped into combinations, in such a way as to have each task included in each of them, and once only. Each such combination constitutes a Structure.

2.4.3. *Horizontal Classification of the Activities according to Criteria 1 and 2*

(i) Classify the Activities according to each of Criteria 1 and 2 in succession,

(ii) Compose the two resulting classifications into a single classification of the Activities.

2.4.4. *Vertical and Horizontal Classification of the Activities*

(i) Classify the missions horizontally according to each of Criteria 1 and 2, then compose these two classifications into a single classification,

(ii) Classify the Structures horizontally (per logic sub-network) according to Criteria 1 and 2–then compose these two classifications into a single classification;

(iii) Compose vertically the classification of the missions (see 2.4.4 (i)), Structures (see 2.4.4 (ii)) and Activities (see 2.4.3 (ii)), to arrive at a single classification (horizontal and vertical) at Activities level.

2.4.5. Overall Classification of the Activities

(i) Classify the Activities according to Criterion No. 3;

(ii) Compose this classification with the classifications resulting from 2.4.4 (iii). The single classification obtained represents the overall classification of the Activities.

2.4.6. Classification of the Structures

(i) For each logic sub-network, classify the Structures by assigning to each of them a class index resulting from the composition of the class indexes of the activities comprising them (see 2.4.5 (ii));

(ii) Examine the classifications obtained for each logic sub-network and eliminate those structures which are classified last.

2.4.7. Classification of the Activities according to Criterion No. 4

Here, a mere ordinal classification would not be of interest, and it is preferable, whatever the amount of approximation, to evaluate the setting-up and operating costs of each Activity. The index assigned to each Activity following such evaluations will become in this case a true mark, indicative both of the classification of the Activity and of the evaluated difference between the Activities[4].

2.4.8. Classification of the Structures according to Criterion No. 4

Only the Structures selected (see 2.4.6) will be classified. The mark assigned to each Structure to allow their final classification will obviously result from the addition of the marks of the Activities comprising each Structure.

2.4.9. Structure Selection

For each logic sub-network, the Structure retained will be the one classified first, i.e. the least costly. Comparison of these Structures will lead to the definition of the final structure.

2.5 Utilisation of the Results

The results obtained will materialize in a Matrix of Tasks and an Organisation Matrix, which will make it possible to prepare the first drafter of the basic documents for the setting-up and operation of the selected Structures.

2.5.1. Matrix of Tasks

This Matrix of Tasks will consist of a double-entry table showing, for each task (as defined for the identification of the Activities–see 2.4.2 (i)), the Activity to which it

[4] Cardinal classification

will be all in the Structures finally retained. Conversely, the matric gives a summary of the tasks finally assigned to each Activity.

2.5.2 Organisation Matrix

From the above-defined Matrix of Tasks will be drawn up a double-entry Organisation Matrix showing the Activities involved in the execution of each mission and indicating the managing Activity responsible for the mission (Command Channel). Conversely, it will show the grouping of tasks relating to the same technical field and contributing to the fulfilment of the various missions, which will therefore allow to identify not only the Activity performing these tasks but also, within this Activity, the various departments of divisions directly responsible for their discharge (Functional Channel).

2.5.3. First Drafts of Basic Documents

The above-mentioned matrices can be drawn up in as much detail as required, to allow preparation of the basic documents.

A mere listing, not to be considered as exhaustive, is the only indication which can be given here, on a purely illustrative basis.

These documents would mainly consist in:

(i) Agreements between the member-countries for the making available of the Activities finally selected to the Community;

(ii) Agreements for the setting-up and operation of new NATO Activities (Charters) or for the utilisation of existing NATO Activities;

(iii) Agreements for the operation of the support system, including:
–Agreements for the funding of expenditure,
–General directives for execution,
–Carrying out of controls, particularly for financial matters.

3. APPLICATION

The above-outlined method is illustrated hereunder by a simplified example. The problem is define appropriate structures for an advanced weapon system support organization.

The policy will not be delineated but it is supposed to provide for centralized management by the Community (NATO). The general support missions retained are those indicated in the annexed table. Assessment will be made only for the first four missions.

The list of possible Activities will be limited, for clearness and simplicity's sake, to that given in the table in paragraph 1.2.1. of this presentation.

Still not to overload the text, classifications and classification compositions will not be explained in detail.

TASK ANALYSIS \ ACTIVITY	DIRECTION		EXECUTION			STRUCTURES
	A.O.	S.N.	E.O.	E.N.	I.I.	
	11	11	11	11	11	
a - Project management	x	x				
b - Management of ref. dossiers . dwgs proced. specs. . codification interch. . part list etc.		x	x	x	x	S.1 { A.O. 11, E.O. 11
c - Management of oper. card indexes . Items of supply . 5th ech items . test fac. and toolg etc.		x	x	x	x	S.2 { A.O. 11, I.I. 11
d - Technical specs. (procurement package)		x	x	x	x	S.3 { SN 11
e - Preparation of mock-ups military doc and chges etc.		x	x	x	x	S.4 { A.O. 11, E.N.11
f - Validation of results	x	x				
g - Execution control	x	x				

Figure 3 Configuration control (M_1)

ACTIVITY / TASK ANALYSIS	DIRECTION						EXECUTION			STRUCTURES
	A.O.			S.N.			E.O.	E.N.	I.I.	
	21	22	23	21	22	23	21	21	21	
a - Project management	X	X	X		X					S1 { AO 21, EO 21 }
b - Preliminary analyses. Defin. of goals	X	X	X		X					
c - Execution of study							X	X	X	S2 { AO 22, SN 21, II 21 }
d - Supervision of study execution	X			X	X	X				S3 { SN 22, EN 21 }
e - Examination of result selection of corrective actions	X	X	X		X					
f - Execution of modif. pilot phase pilot kitchgs to technical and military doc. etc.							X	X	X	S4 { AO 23, SN 23, II 21 }
g - Validation of results	X	X	X		X					
h - . Procurement of new facilities . modif. kits - Tool kits . new parts, etc.	X	X			X	X				

Figure 4 Correction and improvement studies (M_2)

ACTIVITY / TASK ANALYSIS	DIRECTION						EXECUTION						
	A.O.			S.N.			E.O.		E.N.			I.I.	
	31	32	33	31	32	33	31	32	31	32	33	31	32
a - Project management	x	x	x		x								
b - Administrative manag.- contracts work orders-disputes etc.	x	x	x		x								
c - Financial management budgets accountancy, payments.	x	x	x		x								
d - Industrial management programmes, facilities workload planning, progress.	x	x			x								x
e - Technical management, specs. documentation.	x	x			x								x
f - Execution							x		x			x	x
g - Supervision and technical acceptance	x			x	x								

Structures

- S1 { AO 31, EO 31
- S2 { AO 31, II 31
- S3 { AO 32, SN 31, EO 31
- S4 { AO32, SN 31, II 31
- S5 { SN 32, EN 31
- S6 { AO 33, SN 31, II 32

Figure 5 Fifth echelon (M_3)

TASK ANALYSIS \ ACTIVITY	DIRECTION					EXECUTION			STRUCTURES
	A.O.		S.N.			E.O.	E.N.	I.I.	
	41	42	41	42	43'	41	41	41	
a - Project management	x			x	x				S1 { A.O. 41, E.O. 41, S.N. 41
b - Management of stocks	x			x	x				
c - Storage						x	x	x	S2 { A.O. 41, S.N. 41, E.N. 41
d - Issue						x	x	x	S3 { A.O. 42, S.N. 42, E.N. 41
e - Tranpsort			x	x	x				S4 { S.N. 43, E.N. 41
f - Procurement	x	x			x				S5 { A.O. 41, S.N. 41, I.I. 41
g - Technical maintenance (Repairs modif.)						x	x	x	

Figure 6 Supplies (M_4)

3.1 Identification of the Structures

3.1.1. Identification of the Activities

The breaking down of the missions into tasks and the identification of the Activities capable of assuming them are dealt with in Figures 3, 4, 5 and 6.

3.1.2. Combination of the Activities

A list which, on purpose and for clearness's sake, does not include all the possible resulting combinations of Activities, is given in the same table in the column "Structures". These combinations constitute the structures from which selection will have to be made, mission per mission.

3.1.3. Logic Sub-Network

After having identified the Activities and combined them into structures, a logic sub-network can be constructed for each mission, making successive classifications easier.

These log sub-networks are shown in Figures 7, 8, 9 and 10. It is to be noted that, to facilitate assessments, the Activities have been placed each on a different line, with, for each of them, the mention of a serial number corresponding to the task assignments indicated in Figures 3, 4, 5 and 6.

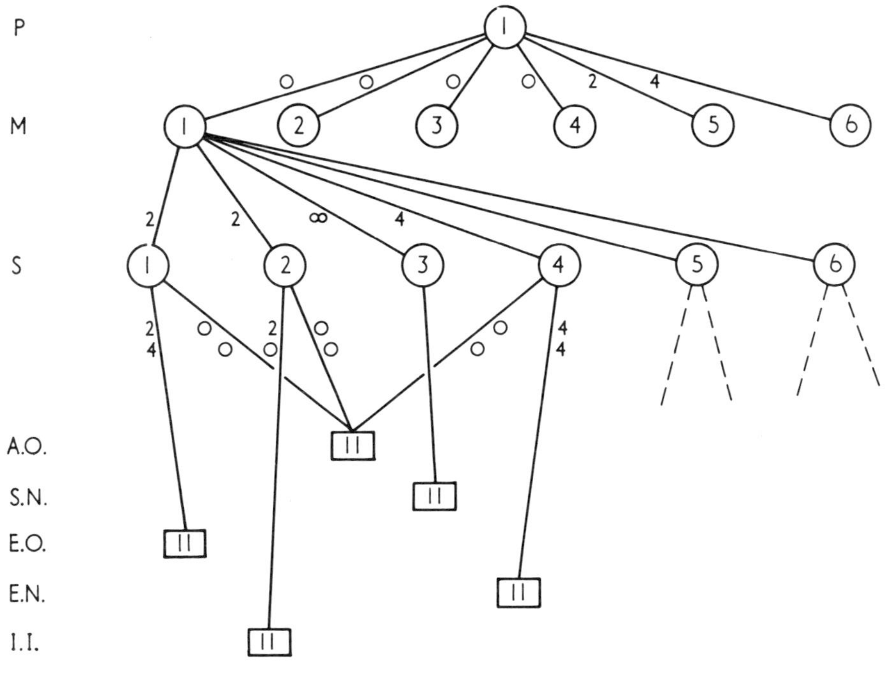

Figure 7 Network M_1

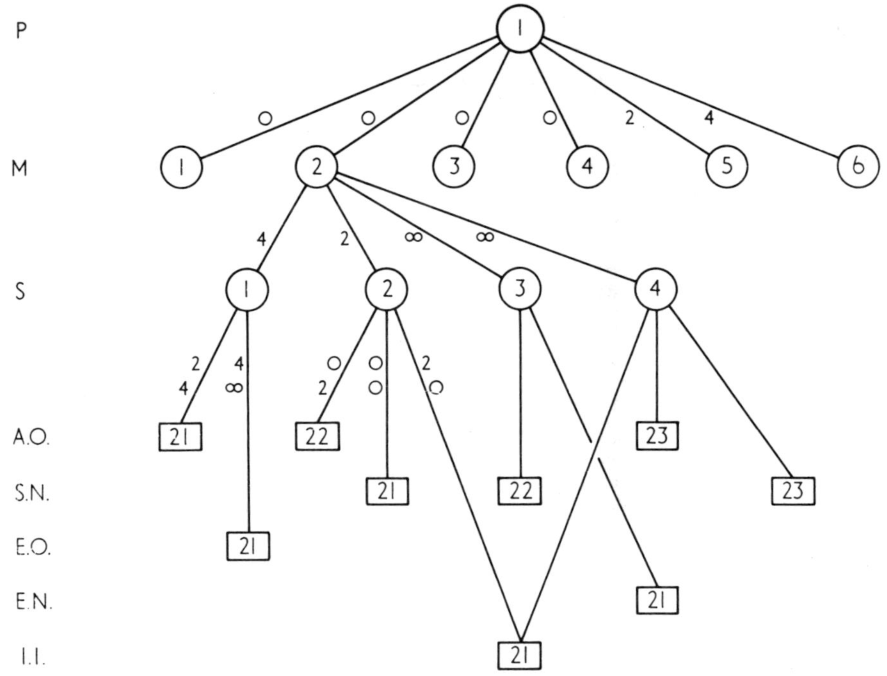

Figure 8 Network M_2

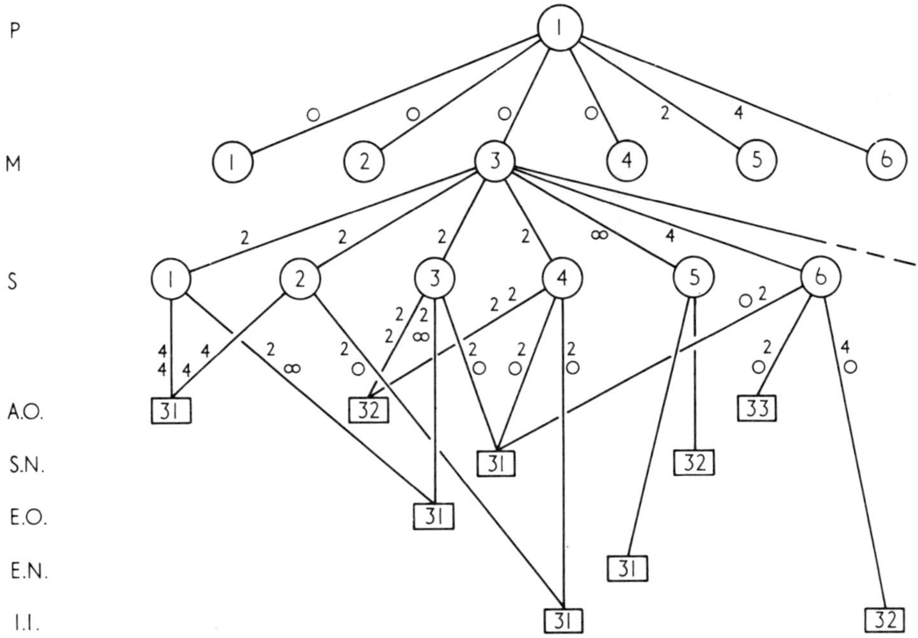

Figure 9 Network M_3

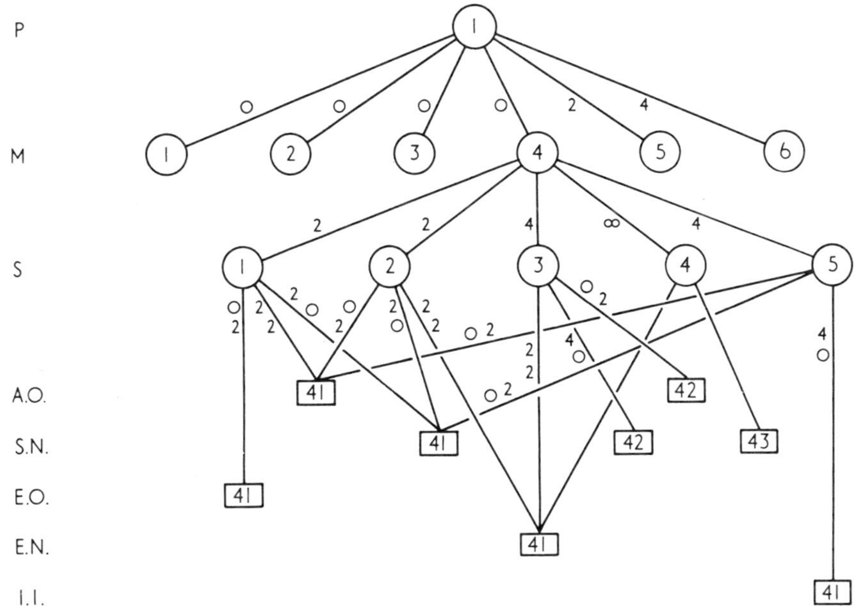

Figure 10 Network M_4

3.2 Classification of the Structures

3.2.1. Horizontal Classification of the Activities according to Criteria 1 and 2

The composition of the classifications of the Activities in relation to Criteria 1 and 2 is reflected, on each logic network, by the corresponding index number plotted along the line connecting the structure to the activity, nearer to the structure (for instance, for Network M1, the index number of Activitiy EO.11 is 2, plotted along the line S1-EO.11, nearer to S1).

3.2.2. Vertical and horizontal classification of the Activities

The result of the composition of the mission and structure classifications in relation to Criteria 1 and 2 respectively is reflected, on each logic network, by a classification index number plotted along the line connecting:

- for the Missions: each mission to the policy (for example: 0 for M1, plotted along the line M1-P1);

- for the structures: each structure to the related mission (for instance: 2 for structure S1, plotted along the line S1-M1).

The result of the vertical and horizontal composition of the classifications of each Activity is directly entered, for each Activity and related structure(s), in the column "Usefulness" of Figures 11, 12, 13 and 14.

Mission	ACTIVITY		Structure	ACTIVITY INDICES			STRUCTURE INDICES				
	NAME	CODE		Useful ness	Lead-Time	Compo-sition	1	2	3	4	
			1	2	0	2	2				
	AO	11 }	2	2	0	2		2			
			4	4	0	4				4	
M1	SN	11	3	∞		∞			∞		
	EO	11	1	4	4	8	8				
	EN	11	4	8	4	12				12	
	II	11	2	4	0	4		4			
							10	6	∞	16	

Figure 11 Policy $A_1 - M_1$

Mission	ACTIVITY		Structure	ACTIVITY INDICES			STRUCTURE INDICES				
	NAME	CODE		Use-fulness	Lead-Time	Compo-sition	1	2	3	4	
M 2	AO	21	1	6	4	10	10				
		22	2	2	2	4		4			
		23	4	∞	—	∞				∞	
	SN	21	2	2	0	2		2			
		22	3	∞	—	∞			∞		
		23	4	∞	—	∞				∞	
	EO	21	1	8	∞	∞	∞				
	EN	21	3	∞	—	∞			∞		
	II	21	2	4	0	4		4			
			4	∞	0	∞					
							∞	10	∞	∞	

Figure 12 Policy $A_1 - M_2$

Mission	ACTIVITY		Structure	ACTIVITY INDICES			STRUCTURE INDICES					
	NAME	CODE		Useful ness	Lead-Time	Compo sition	1	2	3	4	5	6
M3	AO	31	1	6	4	10	10					
			2	6	4	10		10				
		32	3	4	2	6			6			
			4	4	2	6				6		
		33	6	6	0	6						6
	SN	31	3	4	0	4			4			
			4	4	0	4				4		
			6	6	0	6						6
		32	5	∞	—	∞					∞	
	EO	31	1	4	∞	∞	∞					
			3	4	∞	∞			∞			
	EN	31	5	∞	—	∞					∞	
	II	31	2	4	0	4		4				
			4	4	0	4				4		8
		32	6	8	0	8						
							∞	14	∞	14	∞	20

Figure 13 Policy A_1-M_3

Mission	ACTIVITY NAME	ACTIVITY CODE	Structure	ACTIVITY INDICES Use-fulness	ACTIVITY INDICES Lead-Time	ACTIVITY INDICES Compo-sition	STRUCTURE INDICES 1	STRUCTURE INDICES 2	STRUCTURE INDICES 3	STRUCTURE INDICES 4	STRUCTURE INDICES 5	
	AO	41	1	4	2	6	6					
			2	2	2	4		4				
			5	6	0	6					6	
		42	3	4	2	6			6			
	SN	41	1	4	0	4	4					
			2	4	0	4		4				
			5	6	0	6					6	
		42	3	8	0	8			8			
		43	4	∞	—	∞				∞		
	EO	41	1	2	0	2	2					
	EN	41	2	4	2	6		6				
			3	6	2	8			8			
			4	∞	—	∞				∞		
	II	41	5	8	0	8					8	
							12	14	22	∞	20	

Figure 14 Policy $A_1 - M_4$

(For instance, in Network M1, Activity E0.11 is assigned index number 4, which results from the composition of the classification of Mission 1 (index 0) of the classification of Structure 1 (index 2) and of the horizontal classification of the Activity proper (index 2). This index is entered in Figure 11 against Activity E0.11 in Structure 1.

3.2.3. *Overall Classification of the Activities*

The Activities are classified in each logic sub-network in relation to Criterion No. 3 (Availability). As a result of the combination of these classifications, each Activity has been assigned a classification index plotted along the line connecting it to the structure, nearer to the Activity (For instance, the classification index of Activity E0.11 is 4, plotted along the line EO.11-S1 nearer to E0.11 in Sub-network M1).

The result of the composition of the latter classification with the one resulting from the previous paragraph is shown in Figures 11, 12, 13, 14, where the classification according to the Criterion No. 3 is given in the column "Lead-Time" and the classification resulting from the composition, in the column "Composition".

Remark: It will be noted that, if index ∞ is assigned in the process of the various classifications, it is useless to proceed with the classifications or compositions where such an index appears.

3.2.4. *Classification of the Structures*

The classification of the structures is shown in Figures 11, 12, 13 and 14. It is the result of the composition of the classifications of the various Activities constituting the structure. These final classifications for each logic sub-network are indicated at the bottom of the column "Structure Indices".

The structures finally retained are those with the lowest indices. They have been circled in the above-mentioned tables. From this stage on, the evaluation will bear on the structures only.

3.2.5. *Classification of the Activities according to Criterion No. 4 and Structure Selection*

As already mentioned, it may happen that a mere examination of the logic network obtained with only the structures retained allows to arrive directly at a final choice. Such is the case in the selected example shown in Figure 15.

For instance, it can easily be seen that, for Mission M3, Structure S4 is preferable, from the cost stand-point, to Structure S2, as these two structures differ in that the "Quality Control and Technical Acceptance" task would be performed by the NATO Activity in S2 and by a national service in S4. In Structure S2, a special Activity would necessarily have to be set up for this purpose, while in Structure S4 the task concerned can be ensured by an existing financial Activity at no cost for the Community.

Where such selection is not readily feasible, the cumulative expenditure has to be assessed for each activity with, in addition, determination of the cumulative expend-

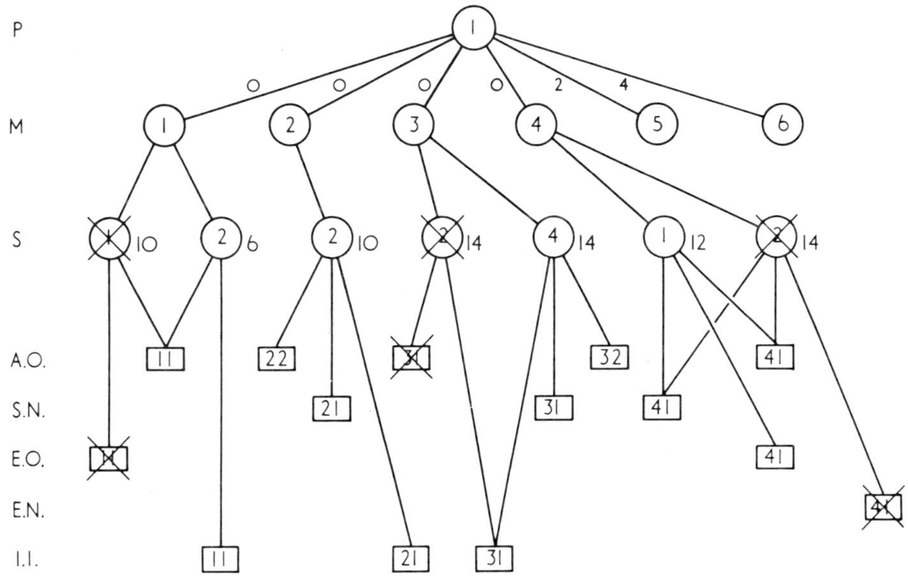

Figure 15 Network P_1

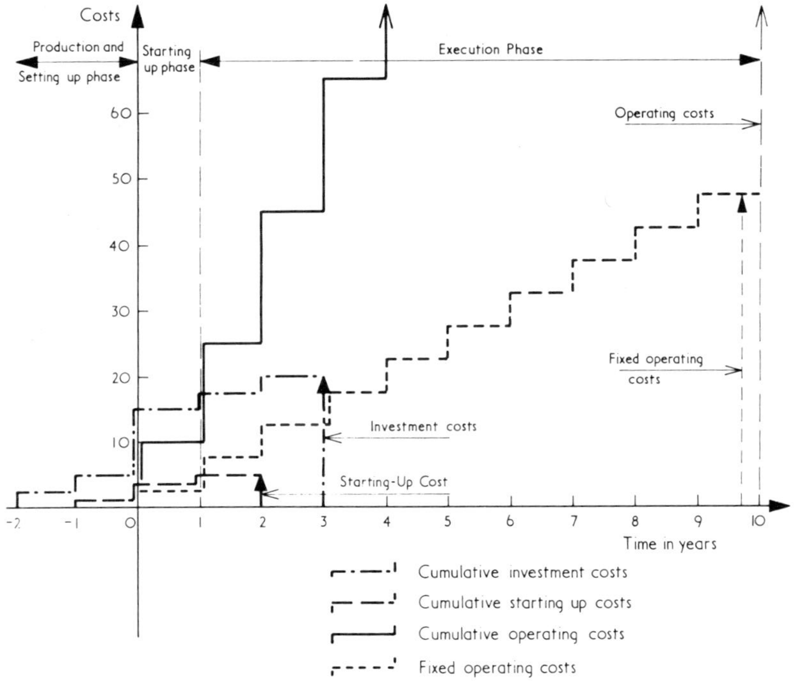

Figure 16 Activity

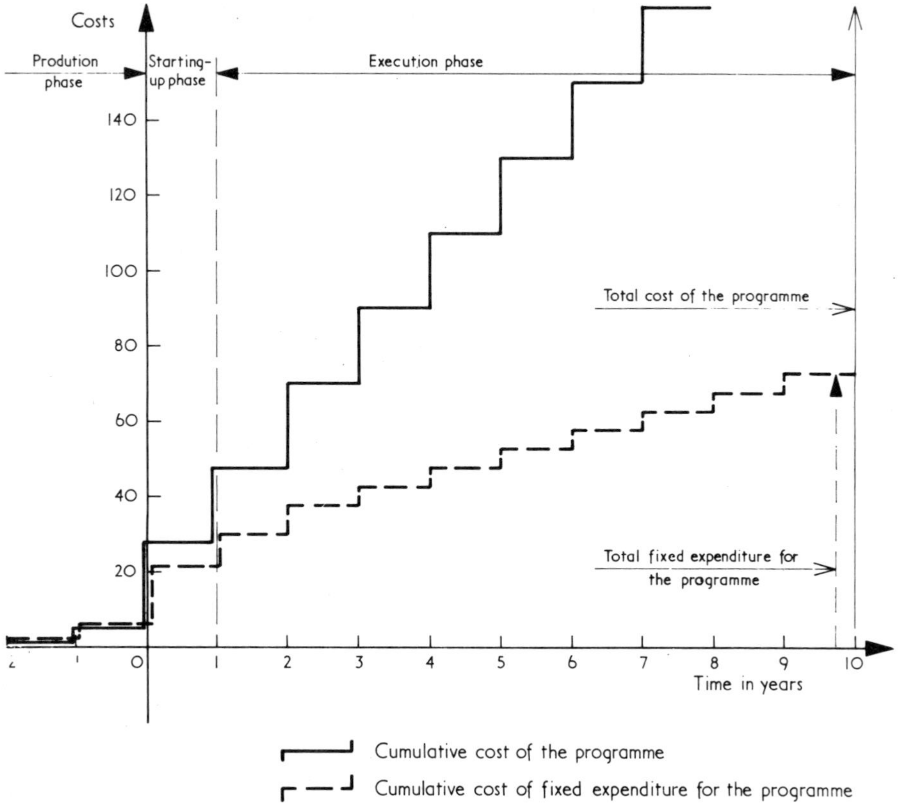

Figure 17 Activity

iture for each structure. Figures 16 and 17 give a presentation of these evaluations per Activity, and constitute at the same time a financial schedule.

One is a presentation of the separate costs: production and setting-up costs, starting-up costs and execution costs for the duration of the programme with, in the latter case, identification of the fixed costs. The other gives the cumulative amount of the various costs, with indication of the fixed costs.

Once such evaluations have been made, it is, in fact, easy to select the least costly structure to be finally retained for each mission, the grouping of these pre-selected structures permitting final structure selection.

3.3 Utilisation of the Results

The Matrix of Tasks and Organisation Matrix resulting from the above are shown overleaf.

MATRIX OF TASKS

MISSION		ACTIVITY / TASK ANALYSIS	AO	SN	PO	II	
I	a	General Management	x				
	b	Management of definition dossiers				x	
		i updating of dwgs, proced., specs.					
		ii Interch., codification					
		iii part lists failure rate					
	c	Management of the operation card indexes				x	
		i Items of supply					
		ii 5th echelon items					
		iii test fac. toolg etc.					
	d	Drawing-up of technical specs (procurement packages)				x	
	e	Preparation of military doc. final manuscripts and mock-ups printing, reprod. of all documents				x	
	f	Validation of results	x				
	g	Execution control. Financial management	x				
II	a	General management	x				
	b	Analyses	x				
		Definition of Goals					
	c	Execution of the study				x	
	d	Supervision of study execution		x			
	e	Examination of results, selection of corrective actions	x				
	f	Execution of pilot phase, corrective actions				x	
		i modif. pilot kit.					
		ii changes to doc.					
		iii modif. of spare parts stocks.					
		iiii modif of toolg. etc					
	g	validation of results	x				
	h	Procurement of new facilities	x				
		i modif. kits					
		ii tool kits					
		iii new parts, etc . . .					
III	a	General management	x				
	b	Administrative management	x				
		i contracts					
		ii work orders					
		iii disputes					
	c	Financial management	x				
		i budgets					
		ii payments					
		iii accountancy					
	d	industrial management	x				

Figure 18 Matrix of tasks

MATRIX OF TASKS (continued)

MISSION		TASK ANALYSIS / ACTIVITY	AO	SN	EO	II	
III		i programmes					
		ii industrial facilities					
		iii wordload planning					
		iiii progress					
	e	Technical management	x				
		i specifications derogations					
		ii documentation					
	f	Execution of work				x	
		quality control					
	g	Supervision of quality		x			
		technical acceptance					
IV	a	General management	x				
	b	Management of stocks	x				
	c	Storage			x		
	d	Issue			x		
	e	Transport		x			
	f	Procurement	x				
	g	Technical maint					
		i Repairs			x		
		ii Technical modifications					

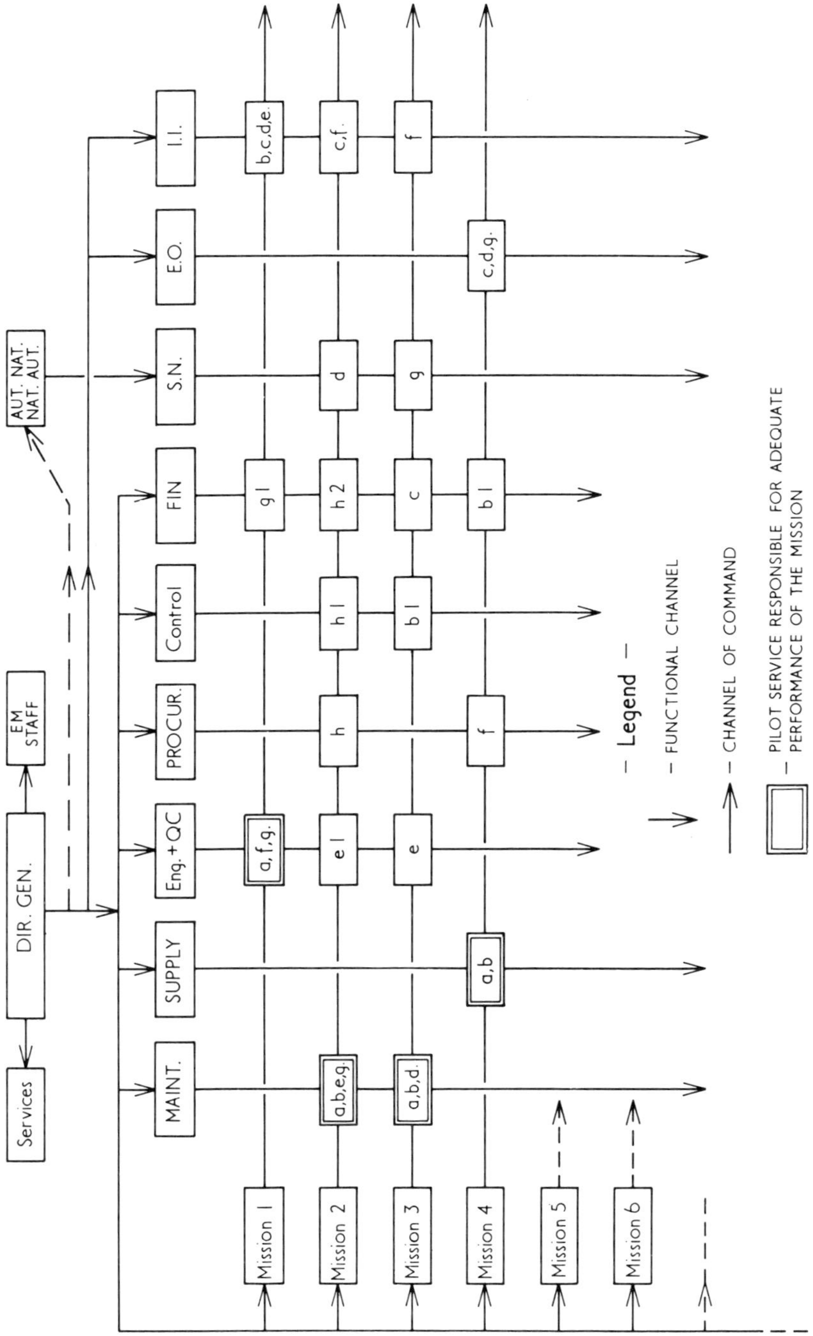

Figure 19 Organisation matrix

CONCLUSIONS

The method proposed herein could be used, if needed, as mentioned at the beginning, to facilitate the defining of a support policy; in such a case, the reasoning would have to start from the missions and no assessments would be made in relation to Criterion No. 1.

On the other hand, this method, which has been developed in a limited time, is to be considered as a first approach which would have to be corrected and complemented, more particularly by specialists in prospective and operational research, before full use could be made of it.

DISCUSSION

Bureau: En matière de politique de base, vous avez cité très justement la politique des Achats (concurrence–juste retour). Ne pensez-vous pas qu'il serait intéressant de se rapprocher de la concurrence mais en l'étendant à l'ensemble des programmes coordonnés?

Combes: La libre concurrence suppose, pour ne pas trop défavoriser certains pays, que ceux-ci soient en état, au moins dans certains domaines, de concurrencer les autres.

L'application stricte de ce principe peut amener à la limite à chaîner tous les marchés vers un même Etat.

Par ailleurs les retombées technologiques que procurent certaines fabrications ne sont pas à négliger et des pays peuvent accepter dans certains cas de payer plus cher pour avoir certaines commandes de ce type.

A notre avis et lorsque les dépenses concernées sont importantes la solution pratique serait à mis chemin entre la "libre concurrence" et le "juste retour":

– soit par l'intégration des systèmes de soutien qui pourrait, en laissant jouer la concurrence mais étant donné la variété des fabrications, assurer une certaine répartition entre pays.

– soit, dans le cadre d'un système de soutien, faire appel à la concurrence mais dans le but de rechercher le meilleur prix compatible avec une répartition des dépenses faite dans chaque pays.

Olivier: Votre exposé constate l'existence d'organismes de soutien logistique distincts des unités opérationnelles. Hier et aujourd'hui plusieurs orateurs ont évoqué les difficultés des communications entre la logistique et l'opérationnel.

Les recherches en matière d'organisation logistique ont elles examiné l'intérêt d'une intégration (totale ou partielle) des tâches logistiques dans le sein des responsabilités opérationnelles?

Combes: L'OTAN a intégré les responsabilités opérationnelles mais la logistique est restée une responsabilité nationale, malgré des essais de coordination et la création d'organismes OTAN pour le soutien de certains systèmes d'armes (A noter que ces organismes quoique OTAN ne sont pas intégrés–règle de l'unanimité pour la prise des décisions).

Les raisons de cet état de fait sont d'ordre politique, mais aussi économique et nous ne pouvons en débattre à cette conférence.

Cependant, au niveau des Systèmes d'Armes, un système de communication a été mis sur pied entre "opérationnel" et "logistique". Il s'agit essentiellement d'un "Système de Collecte et de Traitement d'informations" concernant la situation des matériels sur le terrain et dont des exemples font l'objet d'un ou même de deux des exposés qui seront faits à cette conférence. L'intérêt est d'assurer par ce moyen une liaison directe entre l'opérationnel national et le logisticien de L'OTAN en évitant les filtres parfois déformants que peuvent constituer les intermédiaires.

Britten: In your reference to "best price" in relation to tenders, do you mean only development and procurement price or is life cycle costing considered?

Combes: Il s'agit de l'ensemble. Le critère "Coût" concerne l'ensemble des coûts d'établissement (Recherche, Développement, Production) et des coûts de fonctionnement pour la durée de vie du système de soutien.

Solstrand: The speaker mentioned the 4 criteria on which the optimization is to be based. It seems to me that there is a great degree of interaction between them–that is: they are not independant. Could the speaker elaborate on this? Or to be non-specific: Isn't it essential to agree upon one criterion?

Combes: La réponse est non.

Les Critères doivent être établis en fonction du problème à traiter et ce serait là le premier travail du groupe d'experts qui voudrait utiliser cette méthode.

Par ailleurs il n'est pas absolument nécessaire que les critères retenus soient indépendants.

En effet il s'agit d'une méthode *subjective* et il suffit que chaque critère mette en valeur un aspect particulier.

Analytical Approaches to Large Scale Logistics Systems

Henry C. Alberts

During the period from 1967 to 1969, United States management consultants in Europe noted an increase in the emphasis placed on service improvements and cost reductions to inventory control and product distribution systems. This was especially true in Sweden where production efficiency had been pressed to the point where then extant distribution systems were hard put to effectively handle a smooth flow of products between diverse distribution points. The observable results of this condition were:

(i) A buildup of inventory at each distribution point with a concomitant decrease in service levels experienced there, and

(ii) An increase in the production costs (thus in the price per product) due to increased inventory carrying costs and because production forecasts in such situations could not be made to produce sufficient information to perform economic production planning.

Two cases of production and distribution problems were treated in detail during the period from October 1966 to June 1968. I will describe those problems and their solutions briefly. Then I will address an analytical means to investigate large logistics systems effectiveness based on the experience with these two specific problem areas.

The first case was begun in 1966. An organization operating within Sweden had a product line of 15,000 separate and unrelated items and 103 distribution points, all located within a land area of 1500 x 350 miles. The company questioned its then extant distribution system and had considered establishment of one or more central warehouses in one or more diverse locations to handle the distribution of some or all of the 15,000 items.

The problem characteristics were:

Number of items distributed	15,000
Number of related items	None
Number of suppliers of items	538-750
Number of distribution points (sales)	103
Transportation systems available	all commercial land, sea, air

The analysis revealed a quite irregular distribution of sales outlet distance from any point within or without the distribution area. Thus, calculation of distribution costs for alternative distribution methods required an iterated, programmed solution.

The distribution of supplier locations was also quite diverse. It ranged from locations within the distribution area to points over 5,000 miles distant from that area. Even when far distant supplier points were ignored, it was again the case that the geographical distribution of suppliers with respect to the outlet point distribution was irregular. Again, solution for distribution costs took the form of an iterative program.

During this analysis it became clear that when all distribution and inventory costs were systematically considered, the comparison between warehousing in one or more locations vs direct distribution of product mix could be easily treated, and a solution which provided for minimum cost was found. An analytical model of the total distribution system was constructed. Since there were, generally, no constraints on minimum or maximum quantities ordered (the minimum quantity acceptable to the distribution system was normally quite a bit above the point of price break), there was no need to consider the supplier's production cost in the system model.

This analytical model provided sufficient insight into the alternative cost structures to permit the company to continue to utilize its available combined office and warehouse facilities at reduced costs. Furthermore, through exercise of the model, the locations at which new warehouses should be built, and their propitious building times were related to sales experienced and predicted at each distribution point.

The second case was begun in 1967. This organization consisted of a single factory which produced 18,000 items, demands generated by nine wholly owned sales and distribution points, and a marketing organization (equivalent to a tenth distribution point). Each of the distribution points generated its own inventory policies, and each had been designated a separate profit center.

Since the production facility and each distribution point was managed as if it were a separate business, it was almost impossible to optimize the profitability for the total distribution and production organization. Inventories at each distribution point grew, and over the period of two years, reached alarming proportions. Inventory turnover dropped to an average of three times per year and the factory's poor response to demands caused even this low turnover rate to head downward at the close of the preceding fiscal year.

Poor response by the factory was the result of the management policy which failed to aggregate distribution point demands to provide a production forecast demand that could be programmed. Since each distribution point accounted independently, orders were generated without regard to factory scheduling. So, while the factory was always in a crisis situation, trying to produce particular products to a set of urgent but unaggregated demands, the distribution points were always in panic because the factory could not provide them with service. The distribution points then began to build inventory to compensate for poor factory performance and the system efficiency went down while the costs went up.

Systems analysis resulted in establishment of a two-tier distribution system. A systematic review of sales forecasting was undertaken and stock balancing at each distribution point achieved. Stock balancing alone reduced aggregate inventory costs by 25 percent of the value for the preceding fiscal year. Uniform sales forecasting rules were developed and computerized or costs produced monthly, quarterly and yearly. These forecasts were used to calculate inventory levels for each distribution point. The aggregate demand was serviced from a central warehouse established nearby. The central warehouse, in turn, placed its orders on the factory.

Because it had a single focus for system demand, the factory was able to establish a realistic "cost of production" program and to schedule production to the phased total system needs. Factory production costs decreased and lead times experienced by the distribution points decreased as well.

Central warehouse scheduling responsibilities also included transportation scheduling which was, in turn, based on the projected aggregate demand distribution. The central warehouse was able to respond very quickly to the distribution points' changing order quantities.

The total effect of the central warehouse operations was to permit even further lowering of the balanced inventories at each of the distribution points while at the same time providing a much better response to customer orders.

In the aggregate, the systematic approach to multiple distribution points operations fed from a single production point, resulted in increased sales through better response to customer orders, while at the same time it permitted considerably reduced production and distribution costs. The total profit picture was vastly improved.

There may be considerable learned from these two cases about systems treatments of military logistics problems. Following is a general statement of one possible application of the analytical techniques developed as the result of the two cases I have discussed.

The general analysis of logistic systems would be stepwise. The first step is to define a number of military "systems areas" within which a spectrum of equipment could be categorized. For instance, one might establish an "Infantry" systems area. This area would consist of all equipment specific to operations of self-contained infantry units.* A sample listing appears as Fig. 1. In addition to Infantry, three other systems areas are shown: artillery systems, aviation systems and aviation support equipment systems.

Within each "systems" area each component system is specified as consisting of "subsystems" and "parts." Subsystems are those system sub-units replaceable or to be replaced as discrete units of an individual system. Parts are limited to those pieces of an individual system, or subsystem, replaced as a single part. We can designate each "system" within each general systems or equipment area as System 1 to N. Further, each subsystem specific to each major or systems is assigned another descriptor number, say 1 through M. We also designate each part (whether it is within or without a subsystem) by a descriptor 1 through J. With these designators a parts list can be structured which specifies every system and subsystem, as consisting of individual parts and specified by a quadrad of numbers. For example: the trigger of an M16 rifle might be specified as Part No. A521. Where A denotes the Infantry systems area, 5 represents the M16 Rifle System, 2 represents the firing subsystem and the 1 represents the trigger element parts. A short notation for this would be $A\,[P_1]_2^5$.

Within each systems area, a general matrix can be formed with each part denoted by a number triad. Such a matrix appears as Fig. 2. Each system is a separate column heading, and each subsystem is a separate row. Within the resulting matrix cell every part belonging to that system and subsystem set is designated.

* Units operating at long distances from their parent organization, i.e., US LRRP or STINGRAY units.

CLOTHING	Number CARRIED	Number USED	Number DESIRED	MILITARY, INDIVIDUAL	Number CARRIED	Number USED	Number DESIRED	PERSONAL	Number CARRIED	Number USED	Number DESIRED	RATIONS	Number CARRIED	Number USED	Number DESIRED
Cap, bush				Gas mask				ID Tags				C-rations (meals)			
Cap, fatigue				Tent				Insect Repel-				Rice (lbs)			
Helmet, steel				Mosquito netting				lant (bottle)				Candy bars			
Helmet, liner				Flashlight				Foot powder				Chewing Gum			
Neckerchief				Cam. makeup				Wrist watch				Canteen, std.			
T-shirts				Machete				Pocket knife				Canteen, collapsible			
Shorts				Entrenching tool				Flashlight				Water purifiers			
Jungle jacket				Rucksacks				Plastic bags				LRP rations			
Jungle sweater				Combat pack				Tobacco				Other			
Jungle trousers				Air mattress				35mm camera							
Trouser belt				Ammo pouch				Razor							
Pancho				Grappling hooks				Soap							
Pancho liner				Electrical tape				Toothbrush							
Gloves				Wire				Air pillow							
Socks				Other				Writing material							
Jungle boots								Books							
Web belt								Radio personal							
Suspenders								Other							
Body Armor															
Other															

Figure 1 Typical individual equipment list for infantry systems area A

System No. & Nomenclature (rows) \ System No. & Nomenclature (columns)	1 Rifle M-16	2 Grenade Launcher M-79	3 Pistol Cal.45
Barrel 1	P_3, P_{20}, P_8 P_9, P_{15}, P_{17}	P_1, P_2, P_7, P_{15}	P_4, P_5
Stock 2	P_{28}	P_{34}, P_{39}	None
Trigger 3	P_{18}, P_{21}, P_{24} P_{30}, P_{57}	P_{18}, P_{22}, P_{47} P_{31}, P_{49}	P_{19}, P_{87} P_{43}, P_{28}
4			
5			
6			
7			
.			
.			
.			
.			

Figure 2 Equipment in infantry systems area A

The total number of parts and subsystems are found by analysis of the entire spectrum of Infantry equipment (for instance) to this matrix format. Parts or subsystems belonging to more than one system or subsystem are noted.

$$\left[P_1\right] \begin{matrix} 2,3,5,8.\dots \\ 1,3,7,10.\dots \end{matrix}$$

The total number of parts contained within each systems category can be ordered by this sort of analysis. The complete matrix set forms a "stock list" analogous to the stock list for any standard inventory system.

Expendable parts, for instance ammunition, food, POL, are calculated independently. When this is done, relationships between expendable items and the specific systems or subsystems which utilize them will be noted

$$A \left[P_i\right] \begin{matrix} 1\dots .N \\ 1\dots .M \end{matrix} \left[K_{MN}\right]$$

Such relationships might be derived on the basis of the number of rounds of M16 ammunition expended as a function of the number of M16 rifles expected to be operational.

The second step concerns specifying present distribution system outlets. The following definitions of levels of use are stipulated:

(i) User level: the actual point at which the system, subsystem or part is distributed for use.

(ii) Supply Level to User: the point of distribution responsible for providing systems, subsystems or parts to the actual user.

(iii) Coordinated Supply Level: the point of distribution which aggregates requirements of several user supply levels.

(iv) Major Supply Echelon: the point of distribution having responsibility for consolidated ordering for distribution to coordinated supply levels.

(v) Central Supply Level or Echelon of Purchase: the point of responsibility for procurement and distribution of the total supply requirement.

Once the present organizational structure has been diagrammed and the parts list obtained, the demand at all outlets can be calculated by aggregating the numbers of systems, subsystems and parts in use at each user or other distribution point. Expendables and non-expendables are denoted automatically by the matrix. Demands for both expendables and non-expandables are based on probabilities of system use defined by sets of "situations." A situation is a postulated level of activity at a user point. It is denoted by a descriptor (a_i). One or more use rates are assumed to associate with each situation level. The situation level also sets an implied priority or importance for each item, $[P_i]$. This priority designation is later used to assist in determining the service level for the item $[P_i]$ to be maintained at each distribution point.

For non-expendables the situation levels are used to stipulate use rates for systems, subsystems and parts. Use rates are the assumed rates of replacement for each system, subsystem and part based on operating failure rates experienced. Modes of failure in use or on the shelf failures and frequencies of failure experienced, are used as the basis for assumed replacement rates.

Again, a set of matrices are formed. These are illustrated in Fig. 3. Again, the systems are columnar, the subsystems are tiered, and the parts numbers are entered in each matrix cell. But there is in addition a level designator in the upper righthand corner of each cell to represent the specific situation (a_i) under which these systems and subsystems are employed. Thus, the situation analysis produces a parts list which consists of one matrix per situation per use site.

Site matrices are again aggregated to find use rates and thus distribution requirements for each system, subsystem and part along the total distribution system are found. At each higher level of aggregation a similar matrix will be produced. There will be one matrix per aggregate supply level per situation set.

When the aggregate flow requirement has been derived, weighted priority assignments can be compiled for all systems, subsystems and parts within the total logistic system. The complex of variable demands generated by the successive matrix analyses are then

System No., Nomenclature & Situation / Subsystem No. & Nomenclature	#1		#2				
	Rifle–M16	Rifle M–16	Grenade Launcher	Grenade Launcher	Grenade Launcher	Pistol Cal.45	
1 Barrel	1/ P_3, P_{20}, P_8 P_9, P_{15}, P_{17}	2/ $2P_3$, $5P_{20}$, $6P_8$ $20P_9$, $3P_{15}$, $5P_{17}$	1/ P_1, P_2, P_7, P_{15}	2/ $37P_1$, $20P_2$ $10P_7$, $10P_{15}$	3/ $53P_1$, $167P_2$ $15P_7$, $18P_{15}$	1/ P_4, P_5	2/ $47P_4, 10P_5$
2 Stock	1/ P_{28}	2/ $20P_{28}$	1/	2/	3/	1/ None	2/ None

Figure 3 Equipment in infantry systems area A by part numbers required for situation levels at user site

used to derive order quantities for the total parts list based on the service levels desired at each system node. These service levels in turn depend on the priority assigned to each stock list item at each node in each of the situation set conditions.

Lead times are calculated for each of the transportation systems available for each branch of the distribution system. Each transportation system is assumed to associate different elements of time delay. The time delays will be a function of the distribution point, the quantities to be shipped, the production and procurement time delays, and the transportation system used. Again, a matrix is structured which assigns each calculated variable time delay, for each transportation system, at each point of user distribution, for each discrete quantity of stock items used. A sample matrix appears in Fig. 4.

Having devised a set of possible aggregate demand loadings through the logistic system, having definitized requirements and priorities for each system, subsystem and part at each order point, having derived service levels, lead times and order quantities for alternative transportation methods, and having investigated and assigned costs to the total variabilities involved within the overall distribution alternatives, a computer model is constructed and exercised to determine for each case the characteristics of logistic systems, and costs, as a function of the following variabilities:-

(i) The anticipated range of situations at each distribution point.

(ii) The use rates for systems, subsystems and parts (expendable and non-expendable) implied by each situation.

(iii) The failure rates experienced at each use rate.

(iv) The transportation requirements generated by the complex demands implied by each situation set.

(v) The alternative transportation means available.

(vi) The lead times implied by the priority of transportation alternatives.

(vii) The variations in costs implied by each of the variables 1 through 7.

The analytical technique described thus far is almost exactly analogous to the distribution point analyses described in the second case above. It has also the capability to investigate the appropriateness of various kinds of logistics system configurations. It is possible within this analytical framework to investigate whether the established logistic system can be usefully modified. That exercise of the analytical structure is analogous to what was done in Case 1 above. There is, however, a great benefit that accrues because the complex demand can be assigned based on more structured sets of events which, in turn, can be stipulated in terms of a "probability of occurrence."

It may happen that changing the logistic system structure will also change the apparent aggregate demand as seen by the central ordering and processing facility. This too will be possible to examine. Such an examination would be analogous to the activity in Case 2 above, which indicated the benefits of establishing a central warehouse.

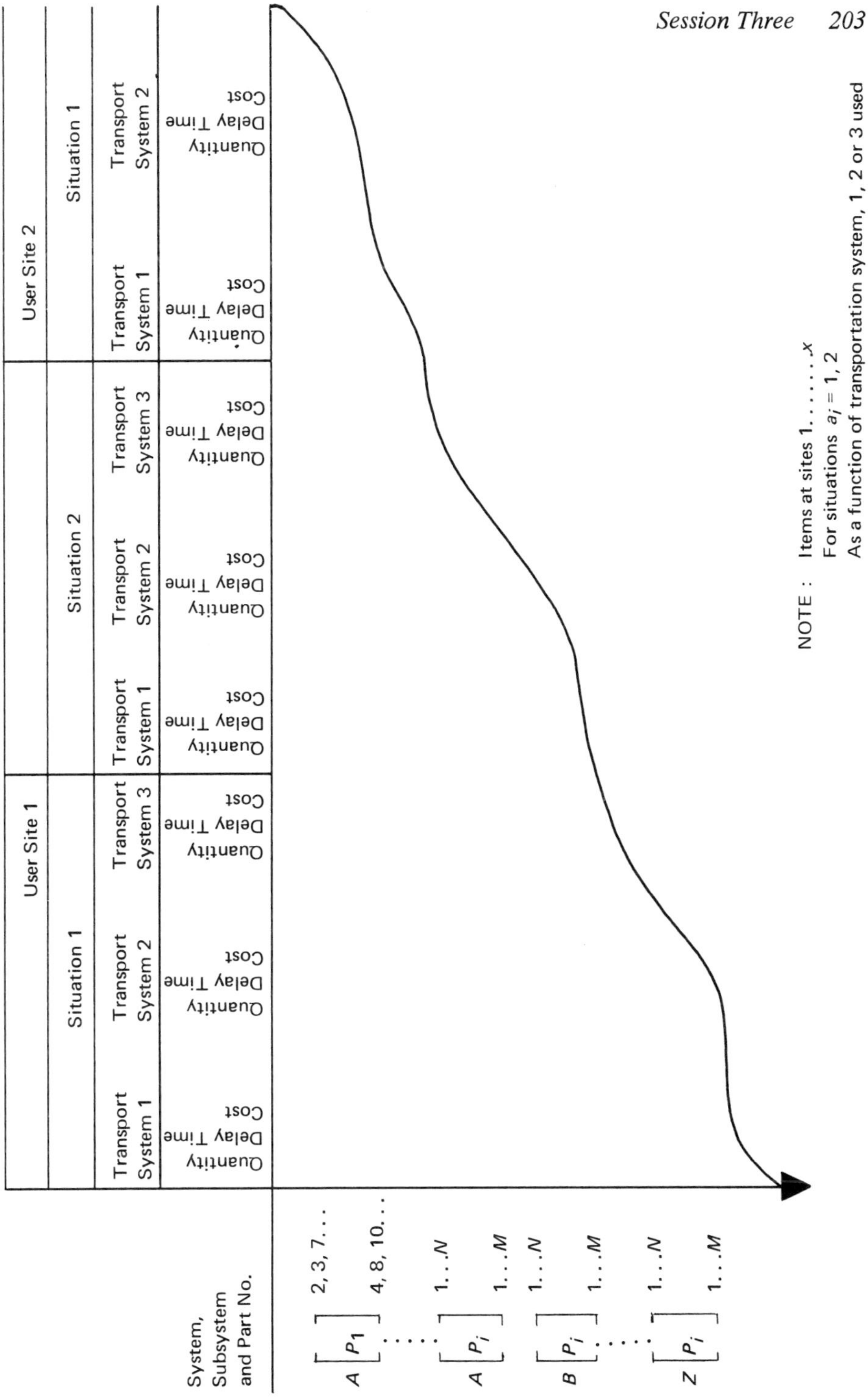

Figure 4 Pipeline time delays and quantities required for infantry and other equipment systems areas

Finally, given that overall sets of complex demands have been calculated for sets of variable situations, and that efficient distribution systems specific to each can have been developed, an analysis of production capabilities and costs analogous to Case 2 can be made. But since the complete production requirements may not be met by only one facility, two levels of analytical investigation are required here:

First, an investigation is required for each individual producer to determine production costs as a function of quantity produced per time period (this is no different than any other production cost control analysis performed for multi-product production lines); and

Second, based on the results of the cost for quality produced, analyses of each producer's capabilities, an aggregated multi-producer production analysis may be required to generate a level of production capability sufficient to satisfy the aggregate demand. This analysis is analogous to the exercise of the production and distribution system model described in Case 2 above with the assumption of multiple production sources.

The magnitude of the analytical effort necessary to scale the techniques used successfully for business operations to large logistics systems problem areas is unclear, but it does seem the kind of approach outlined in this paper will be costly, and will require development of considerable amounts of data which may not now be available at the required level of detail. But it might well be that the process of simply beginning such an analysis will produce considerable learning that would provide insight into which techniques of business analysis can be scaled for application to logistic systems. In itself this result might be so useful that the total analytical program described here might never need to be completed.

What has been described in this paper is an application of current analytical techniques scaled to permit an examination of large scale logistics systems. The techniques have all been used before with useful results. Their application to logistic systems at this time may point the way to similar beneficial results in that critical area of military operations.

DISCUSSION

Hirsch: You mentioned in your paper that you had discovered some relationships that were not intuitively obvious. I wonder if you would give an example of such a relationship?

Alberts: Again, I have to be careful about what I say. But let me mention one result which is quite clear. The number of rounds fired appears to correlate with the detail of the total combat situation and the day of the mission. Thus if one took the total number of rounds fired over all missions and tried to obtain an average for rounds fired per mission, the results might well be quite different than the data now indicate.

There are, of course, other relationships, but I do not feel free to discuss them here.

Rosenman: We did a study of parts demand for M-16 rifle parts in Vietnam as related to rifle population, rounds fired, type of mission, type of terrain, etc. The best we could find was a weak correlation between rounds fired and parts used in the aggregate (i.e. equal value)–almost no correlation betwen activity factors and demands for individual parts. At the time we did the study, we had our suspicions of the data base we were using. Was your experience based on your own data comparable?

Alberts: The problem in attempting to infer relationships such as you mention is familiar. There are two difficulties: First, the combat process and the interrelationships between use rates is complex. To understand what the data means required that one first have an understanding of the combat process and the situations that obtain when it occurs. We have performed a rather detailed study of the complex man-machine system represented by small groups of men performing patrol activities. And from that study we have been able to create a rather complex model of small unit combat. In that modelling process we investigated interactions between factors which could affect whether or not fire is exchanged between a patrol and an enemy. We also investigated the kinds of situations under which fire was exchanged, and the results of that fire. While we do not have a complete picture, we do feel that our data is real.

This brings me to the second difficulty: Since we had structured a model of the combat situation, we were able to obtain a rather broad spectrum of data. We attempted to get all of the information that the modelling structure indicated would be germane to the analytical determination of various interactions and relationships. The data continuum which resulted from our data collection program thus provided us the ability to apply multivariate analysis techniques to the problem of derivation of correlations. Through this process, I believe we were able to advance our knowledge considerably.

In short, I have no reason to mistrust our data base–and I would say further that the data base permits us to work in a much more enlightened manner. I think we have surmounted the obstacles to deriving correlations such as you mention; namely the less than complete understanding of the man-machine system involved, and the lack of a sufficiently broad data base to investigate that system analytically.

Høgberg: Is it not true that the data structure you define (system, subsystem, components, parts) is not limited to supply, but applies to Technical Configuration Control, R & D, Manufacturing, Maintenance, Contract Control etc.? Therefore that particular data set is one of the most significant structuring elements in a logistic system and great care must be taken to have a structure that reflects the *policy* of management?

Alberts: Certainly the structure described is useful in many ways. The data set necessary to construct a logistic system such as I have described will also infer a number of various kinds of other uses. Given the data set, for instance, one would want to develop use rates on that basis; to develop standarization rules on that basis, and so on. I agree with the comment that the data structure applies across a great deal of the board, and is quite significant in structuring logistic systems.

Landau: The speaker made references to the possible effect of changes in supplies on the scale of operation. To what extent does he think that usage rates depend on the quantity of, e.g., ammunition carried by the platoon and to what extent are objectives affected by type of weapon available?

As a corollary to the second part of that question, could this kind of analysis lessen the possibility of trying out new mission equipment?

Alberts: With regard to the first part of the question: I cannot go into detail here for obvious reasons. But I can make the statement that the data we have indicate no obvious relationship between the quantities of ammunition carried and the amount expended. There is considerable analysis of this point, and I do not wish to say there was never a case where the operating unit ran out of ammunition–there must be such cases. But the way in which the data was gathered; i.e., by interview of returning units, may skew the results considerably.

As to the objectives of missions and the effect upon them of the types of weapons available: I have not found that missions were assigned based on the kinds of weapons available to small units. In fact, one of the points that can be made is that units should be outfitted especially to perform missions. I personally believe this should be the case.

Finally, I do not believe analysis mitigates against the test of new kinds of weapons. Rather, I think quite the opposite is true. If one understands the relationships between mission performance and the variabilities inherent in different kinds of weapons, then the proper mixes can be found, or weapons can be created to fit desired patterns. Alternatively, mission operations can be structured to use fully capabilities of existing weapons. In either case tests of new weapons or equipment can be made with much more security given an analytical basis upon which to predict their usefulness in sets of situations.

Baker: I believe Mr. Alberts was lucky in having real data to study. Nevertheless he found unexpected relationships governing usage of parts. This suggests that the usage rates on which we plan for a European war may also be wrong. I therefore ask Mr. Alberts whether he has discovered any general principles which may help us in our prediction of European usage rates or whether he believes his findings are applicable only to the conditions and type of warfare in Vietnam.

Alberts: The carryover from Asia to Europe will depend upon the kind of combat situation we postulate. But even if one assumes that massive unit attacks such as those in World War II and Korea will be the major combat type, the usage rates may still be in error. I cannot say whether the relationships which appear in small unit activities will hold when those units operate interior to larger units (such as Divisions). But even given that kind of combat, and viewing the combat activities of forces as parts of a whole, I do believe that a review could profitably be made of the analytical basis used for predicting use rates. I think there might be more parallels between Asia and Europe than one might expect.

A Mathematical Model of Airborne Radio Servicing Systems*

B.E. Baker
Gp. Capt. R. Morris

1. INTRODUCTION

Airborne radio equipment is often common to several types of aircraft and some communication sets are very widely fitted indeed. In these circumstances the logistic support system has to be designed to cover a whole air force rather than a single weapon system.

The basic system for logistic support of airborne radio equipment within the Royal Air Force was set up in the 1950's and consists of a central, civilian manned, depot, supported by a group of vehicles which collect unserviceable equipment from each station once a week. Stations place a consolidated demand on the depot once weekly, stating the quantity and types of equipment requiring replacement. The delivery vehicle exchanges these equipments on the next week's run.

In addition to the main system, one or two specialised equipments have been serviced at one of a group of stations, since a more rapid replacement service can be provided in this way.

Until quite recently the central depot dealt mainly with sub-assemblies, and the procedure was that unserviceable Line Replaceable Units were removed from aircraft, tested at the base workshop and repaired by replacing sub-assemblies. The sub-assemblies (modules) were returned to the central depot for repair. With the increase in reliability of aircraft radio equipment, and the possibility of using automatic test equipment, which is too expensive to purchase in large numbers, this policy came to be questioned. It was decided to explore the possibility of centralising the repair of Line Replaceable Units.

As a first step, only those Line Replaceable Units with a predicted Mean Time Between Removals (MTBR) in excess of 500 hours were selected for central repair, but it was quickly realised that this was a difficult reliability target. A model was therefore devised to explore the factors which needed to be taken into account in deciding whether to service Line Replaceable Units at bases, or at the central depot. It can also be applied to examine the merits of using automatic or manual test equipment.

2. ASSUMPTIONS

2.1 Symbols

The symbols used in this paper are introduced in the text as they are required. A list of symbols is given at the beginning of Section 6.

2.2 Time

Throughout the paper, time is measured in working days. It is assumed that flying, maintenance work and deliveries occur only in working days, (and there are W such days in the year).

2.3 Servicing Systems

To facilitate calculation, the model assumes that a fleet of aircraft (F) is deployed in equal numbers on a number of bases (B) which can all be supplied from the same central depot. Under the present system, when the equipment is removed from the aircraft for repair it can be repaired at the base in a maximum time (R_S) days during which it occupies the test set for a mean time (R_T) days. Repair is by replacement of modules which are themselves repaired at the central depot. An order for replacement modules can be made once every f days. Delivery is also made once every f days, but when an unserviceability can take place immediately after the order has been sent, $2f$ days can elapse before a replacement is received, and bases must therefore hold sufficient stocks to cover them for $2f$ days. This is the present practice, but in the model, to cover the possibility of a change in practice the maximum time from an unserviceability to a delivery has been assumed to be R_m days.

Under the alternative system there are no test facilities at the base, but the same rule applies for obtaining replacement units as now applies to modules.

2.4 Effectiveness

At the base, the measure of effectiveness of the servicing system is taken as its ability to provide a serviceable spare when an arising occurs. Failure to do this must not occur more frequently than a given number of times (D) per year.

In addition, a central depot float is set up large enough to give a high probability ($S\%$) of meeting any base's requirements.

2.5 Manpower

It is assumed that the men required to repair main equipments can be deployed either at bases or central depots without change in cost. If not required at bases for servicing this particular equipment, fractional reductions in strength can be made. However, the basic model can be modified to take account of inflexible manpower, deployment, and this is shown in the section on the effects of changes in assumptions.

2.6 Test Sets

The numbers of test sets required are based on simple queuing theory as detailed in Annex A. All test equipment is assumed to be calibrated as a complete set at one interval (Z years). Cost of calibration (including repair) is given in manhours (H_T) and turnround time (R_C) is assumed to be a constant fraction of a year. For clarity of exposition, costs incurred in this way have not been discounted in the first instance. In practice, the results of any application of the model would have to be tested for sensitivity by discounting all running costs to present value, as in the numerical example.

2.7 Statistical Assumptions

The arising rate is generally assumed to be Poisson distributed, and calculations of spares requirements at bases are made on this assumption. Central depot floats are based on the Normal approximation to the Poisson distribution. Although observations of arisings have shown that their variability is greater than this, possibly due to the effects of weather and operational commitments on flying effort, it should be noted that:

(i) For base repair, the maximum repair time (R_s) has been used.

(ii) For central depot repairs, the maximum order interval (R_m) has been used. (Peak loadings will seldom occur over more than two or three days on a given base, and only one day in each service interval f can give rise to a demand which is unsatisfied for R_m days).

Numerical tests have shown that the above assumptions provide sufficient insurance to allow for wider variability in the arising rates.

3. CALCULATION OF SPARES HOLDINGS

3.1 Base Holdings of Main Equipments

In the worked examples which follow, it is assumed that all bases have the same number of aircraft, that the equipment has the same MTBR when installed in any of these aircraft, and that the sortie length is the same. In practice each of these parameters will vary from base to base, or even with groups of aircraft on the same base. However, the basic procedure is the same for calculating the spares requirements for the smallest flying unit for which all these parameters are the same. The difference occurs in the combination of these totals. Where all bases are uniform, total holdings will be the individual base holdings multiplied by the number of bases (B). Where stations are non-uniform, total holdings will be the sum of the individual base holdings. Where there are a variety of flying units on each base it is safest to sum the results of the individual unit calculations before summing over the bases.

3.2 Repair on base case.

3.2.1. *Assumptions*

The mean time between unscheduled removals (MTBR) of the main equipment is M_E flying hours. Mean sortie length is t hours and the daily flying task is T hours per aircraft.

3.2.2. *Arisings*

The expected number of arisings per sortie is $\frac{t}{M_E}$ and the probability that a main equipment requires removal after a sortie is given by:

$$p = 1 - e^{\frac{-t}{M_E}} \cong \frac{t}{M_E} \qquad (1)$$

(since this is normally small).

The expected number of removals per base per day is:

$$\frac{F}{B} \cdot \frac{T}{t} \cdot p \tag{2}$$

$$\simeq \frac{F}{B} \cdot \frac{T}{M_e} \tag{3}$$

3.2.3. Calculation of spares

Each base must hold at least one main equipment for immediate issue when an arising occurs. The defective equipment is then repaired (by module exchange) in R_S days. If a further arising occurs during this time it cannot be met unless the stock of equipments is greater than one. The risk of an inability r_{SO} during a repair interval R_S is acceptable if it leads to an annual inability rate less than D. The number of repair intervals during which this risk exists is the expected annual number of removals A where:

$$A = \frac{F}{B} \cdot \frac{T}{M_E} \cdot W \tag{4}$$

Thus r_{so} must be $\leqslant \frac{D}{A.}$ (5)

The procedure is therefore to assume $K + 1$ spare sets and by the use of cumulative Poisson graphs to calculate the probability r_{SO} of more than K arisings during a repair interval R_S, for which the expected number of arisings is:

$$\frac{F}{B} \cdot \frac{T}{M_E} \cdot R_s$$

The least value of $K + 1$ for which r_{SO} satisfies inequality (5) is the required stock.

3.2.4. Calculation of module stocks

The above calculation is not quite correct since if the base has run out of spare modules the set in work will not be repaired in time R_S. The probability of module inability $r_{SO}(\hat{m})$ must also be calculated in a similar manner and the spare module stock chosen so that:

$$r_{so} + r_{so(m)} \leqslant D/A \tag{6}$$

Since each station will normally hold at least one of each module it is usually sufficient to consider only the least reliable module. The contribution of the other modules to $r_{so(m)}$ is usually negligible.

3.3 Case of LRU repair at Central Depot

The procedure for calculating the number of spare equipments is exactly the same as that described above, except that the interval taken is R_m, the time for resupply from the central depot. There are still A risk intervals in a year so r_{SO} must still satisfy inequally (5) which for this case is correct, since spare modules are not held at the base.

3.4 Central Repair depot float

All deliveries go on the same day at each interval f so for the calculation of the central repair depot float, the mean arisings are calculated for the delivery interval f. For the main equipment this would be:

$$\frac{FT}{M_E} f \tag{7}$$

The normal approximation to the Poisson distribution is assumed and the insurance provided by adding sufficient standard deviations to this mean to provide a no-stockout probability of better than 5%. In the model s was taken as 90%, which requires 1.28 standard deviations to be added, giving an equation for central repair depot float:

$$K_3 = \frac{fFT}{M_E} + 1.28 \sqrt{\frac{fFT}{M_E}} \tag{8}$$

for main equipments.

For the i th module, the equation would be:

$$K_3 = \frac{fFT}{M_i} + 1.28 \sqrt{\frac{fFT}{M_i}} \tag{9}$$

3.5 Optimisation

These procedures do not, of course, optimise repair system effectiveness for a given cost of spares holdings, nor do they take into account possibly different national savings for different aircraft, where common equipment may have different operational implications. The model was devised for servicing policy decision-making, and not for optimisation of spares holdings.

3.6 Queuing Spares

For cases where the calculation of test set requirements reveals the need for a few spare equipments to form a queue (see Annex A) these have to be added either to base or central repair depot spares requirements.

4. COST–REPAIRS AT BASE

4.1 Capital Costs

4.*1.1. Spare Main Equipments*

Using the procedures of paragraph 3.2 the stock required for each base $(K_1 + 1)$ is calculated. The total stock required will be:

$$(K_1 + 1)B$$

Where K_1 is the number of equipments per base required to ensure that the probability of stockout during the repair interval R_S, combined with the expected frequency of base level repairs, gives no more than D stockouts a year, taking module stockout effects into account.

4.1.2. Number of Test Sets Required

Each base has a daily expectation of arisings

$$\frac{F}{B} \cdot \frac{T}{M_E}$$

The number of days of test time per working day will therefore be

$$I = R_T \cdot \frac{F}{B} \cdot \frac{T}{M_E} \tag{10}$$

the service intensity. If it is very much less than 1, as it usually will be in cases considered suitable for investigation, one test set can be allocated per base with reasonable certainty that it will cope with peak loadings. In this case, the number of in-use test sets required overall will be B. To cover the calibration float, an additional number of test sets will be required, and this is given by

$$\frac{B\ R_C}{Z} \tag{11}$$

rounded up to the nearest whole number of test sets. Thus the total test set requirement will normally be

$$B \left[1 + \frac{R_C}{Z} \right] \tag{12}$$

(Brackets indicate figures rounded up to nearest whole number).
Note. Where the service intensity I is not very much less than 1, it will be necessary to employ queuing theory to test whether a second test set is necessary to cover peak demands or, alternatively, whether it would be cheaper to provide a buffer stock of spare equipments to form the queue. This is examined in Annex A.

4.1.3. Spare Modules Required

The method of calculation is shown in para. 3.2.4. The numbers required are

$$\sum^{N} K_{2i}\ B + \sum^{N} K_{3i} \tag{13}$$

K_{2i} is the number of the ith module for each station which, combined with $(K_1 + 1)$ main equipments ensures that the number of main equipment stockouts at each base is not more than D per year. K_{3i} is the number of the ith module required for central repair depot float to ensure that all base requirements for that module can be met on 5% of occasions.

4.1.4. Capital Costs

The cost of spare main equipments is

$$C_E \quad (K_1 + 1)\, B \tag{14}$$

The cost of test sets is

$$C_T \left[B \; (1 + \frac{R_c}{Z}) \right] \tag{15}$$

The cost of modules is

$$\sum^{N} C_i \; (K_{2i} \;\; B + K_{3i}) \tag{16}$$

Thus the total capital cost is

$$C_o = C_E \; (K_1 + 1)\, B + C_T \left[B(1 + \frac{R_c}{Z}) \right] \tag{17}$$

$$+ \sum^{N} C_i \;\; (K_{2i} \; B + K_{3i})$$

4.2 Running Costs

4.2.1. Manpower

It has been assumed throughout that the manpower can be employed flexibly. If, however, it is necessary to have men specially provided for the equipment, who cannot be employed on servicing other equipment as well, then each base must have

$$\left[\frac{H_2 FT}{dBM_E} \right]$$

Where H_2 is the mean manpower cost of a repair to the main equipment and d is the average hours worked each day by one man.

Thus the total manpower required is

$$\sum^{B} \left[\frac{H_2 FT}{dBM_E} \right] \tag{18}$$

4.2.2. Test set Servicing

Calibration arisings $= \frac{N_T}{Z}$ per year $= \frac{B}{Z}$ per year (N_T = number of test equipments).

Manhour requirements per year are

$$\frac{BH_T}{Z} \tag{19}$$

$$\text{Men required} = \left[\frac{BH_T}{WdZ}\right] \tag{20}$$

4.2.3. Cost of Manpower Overall

Annual manpower costs

$$C_R = C_p \sum^{B} \left[\frac{H_2 FT}{d \ BM_E}\right] + C_p \left[\frac{BH_T}{WdZ}\right] \tag{21}$$

This would be discounted over the expected life of the equipment to give the present value of manpower costs. (C_p is the annual cost of a technician.)

4.3 Total Costs–Present System

Total discounted costs over Y years are obtained by adding C_O and $Y \Sigma C_j$ where C_j is the discounted annual running costs for the jth year. For simplicity we shall ignore the discounting giving a total cost of

$$\begin{aligned} &= C_E \ (K_1 + 1) B + C_T \left[B \left(1 + \frac{R_c}{Z}\right)\right] \\ &+ \sum^{N} C_i (K_{2i} \ B + K_{3i}) \\ &+ Y C_p \sum^{B} \left[\frac{H_2 \ FT}{d \ BM_E}\right] + Y \ C_p \left[\frac{BH_T}{WdZ}\right] \end{aligned} \tag{22}$$

5 COSTS–REPAIRS AT CENTRAL DEPOT

5.1 Capital Costs

5.1.1. Spare Main Equipment Required.

The method of calculation is shown in paragraph 3.3. The total stock required is

$$K'_2 \; B + K'_3 \tag{23}$$

Where K'_2 is the number of equipments required per base to ensure that inability to supply equipments at base level occurs less than D times per year and K'_3 is the number of equipments for central repair depot float to ensure ability to supply on at least $S\%$ of occasions.

5.1.2. *Number of Test Sets Required.*

The fleet generates $\frac{FT}{M_E}$ repairs per day.

The number of days of test time per day is therefore $\frac{R_T FT}{M_E} = I$ (24)

the service intensity. In this case, I is likely to be near unity or greater. In order to cater for peak loading, the methods shown in Annex A must be used to determine how many in-use test sets (N_T) are required. To provide for calibration float, a total of

$$N_T + \left[\frac{N_T R_C}{Z}\right] \tag{25}$$

test sets will be required.

5.1.3. *Spare Modules Required*

No spare modules are required at bases, or for a central repair float to meet delivery requirements. The float required within the central repair depot to support work in progress will be the same wherever LRU's are repaired.

5.1.4. *Capital Costs.*

The cost of spare main equipments will be

$$C_E \; (K'_2 \;\; B + K'_3) \tag{26}$$

The cost of test sets will be

$$C_T \left(N_T + \left[\frac{N_T \;\; R_c}{Z}\right]\right) \tag{27}$$

Total capital cost will therefore be

$$C'_o = C_E \;\; (K'_2 \;\; B + K'_3) + C_T \left(N_T + \left[\frac{N_T \;\; R_c}{Z}\right]\right) \tag{28}$$

5.2 Running Costs

5.2.1. Repair Manpower.

Manpower requirements excluding repair of modules is

$$\frac{H_2 FT}{dM_E} \tag{29}$$

5.2.2. Test Set Calibration

$$\text{Manhours} = \frac{H_T N_T}{Z} \quad \text{per year} \tag{30}$$

$$\text{Men required} = \left[\frac{H_T \, N_T}{ZWd}\right] \tag{31}$$

5.2.3. Total Manpower Costs

Annual manpower costs

$$C'_R = C_P \left(\left[\frac{H_2 \quad FT}{d \; M_E}\right] + \left[\frac{H_T \quad N_T}{Z \quad W_d}\right] \right) \tag{32}$$

5.3. Total Costs – Proposed System

As before (ignoring discounting) the total cost over Y years is

$$C'_o + YC'_R$$

$$= C_E \quad (K'_2 \quad B \;+\; K'_3) + C_T \quad (N_T + \left[\frac{N_T \, R}{Z}\right] \;)$$

$$+ \, Y \, C_p \left(\left[\frac{H_2 \, FT}{d \; M_E}\right] + \left[\frac{H_T \, N_T}{Z \quad W_d}\right] \right)$$

6. NUMERICAL EXAMPLE

The following parameters have been assumed.

F	=	fleet size	=	700
B	=	number of stations	=	35
T	=	daily task	=	1.5 hours
t	=	sortie duration	=	1 hour
N	=	number of modules	=	10

R_S	=	repair time main equipment	=	2 days
R_m	=	central depot order time	=	10 working days
f	=	central depot delivery frequency	=	5 working days
R_T	=	test time per repair	=	½ day
Y	=	life of equipment	=	10 years
$S\%$	=	central depot assurance factor	=	90%
H_T	=	manhours per calibration	=	100 MH
Z	=	calibration frequency	=	1 year
R_c	=	calibration turnround	=	2 months = 1/6 year
M_i	=	MTBR of *i*th Module—see Annex B		
C_E	=	cost of main equipment	=	£1000
C_i	=	cost of any module	=	£80
C_T	=	cost of test set	=	£3000
C_p	=	annual cost of technician (including overheads)	=	£4400
H_2	=	second line manhours per defect	=	8 MH
D	=	acceptable stockout frequency	=	2 times/year
M_E	=	MTBR of Main Equipment	=	100 hours
W	=	working days per year	=	200
d	=	working hours per day	=	8

Throughout the calculations manhour cost of repairs to main equipments has been assumed to be the same whether the repairs are carried out on the base or at a central repair depot.

6.1 Repair on Base

6.1.1 *Number of equipments required per Base.*

From (3), expected number of removals per base per day

$$= \frac{FT}{BM_E} = \frac{700}{35} \quad \frac{1.5}{100} = 0.3$$

Expected Annual removals (from (4))

$$A = \frac{F}{B} \cdot \frac{T}{M_E} \cdot W = 60$$

Thus from (5) $r_{so} \leqslant \frac{D}{A} = \frac{2}{60} = 0.033 = 3.3\%$

Expected Arisings in interval $R_S = 0.6$

which with $K_1 = 1$ gives 12% for r_{so}. It is therefore necessary to take $K_1 = 2$ giving r_{so} = 2.5%. Each base should hold $K_1 + 1 = 3$ equipments.

6.1.2. Number of modules required per base

Expected arisings in interval R_m for Module A

$$= \frac{F\ T\ R_m}{BM_i} = \frac{700}{35} \cdot \frac{1.5}{400} \cdot 10 = 0.75$$

The expected annual arisings are

$$\frac{F\ T}{BM_i} = 15$$

The number of inabilities per year for $K_1 = 2$ main equipments is $\frac{60\ .\ 2.5}{100} = 1.5$ leaving 0.5 possible inabilities for modules.

$$r_{so(m)} \leqslant \frac{0.5}{15} = 0.33 = 3.3\%$$

The probability of having 3 or more failures, given an expectation of 0.75 is 4%. Thus the base should hold 3 of Module A.

For Module B. $M_i = 800$.

Expected arisings in interval $R_m = 0.375$. Expected annual arisings = 7.5

$$r_{so(m)} \leqslant \frac{0.5}{7.5} = 6.6\%$$

The probability of having 2 or more failures, given an expectation of 0.375 is 5%. Thus the base need only hold 1 of all other modules and $\sum N\ K_{2i} = 12$ modules.

6.1.3. Number of modules for Central Repair Depot

Number of flying hours for fleet in interval $f = f\ T\ F = 5 \times 1.5 \times 700 = 5250$

Module	MTBR	Expected Arisings	Reserve	Total
A	400	13.125	4.63	18
B	800	6.5625	3.48	10
C	800	6.5625	3.28	10
D	800	6.5625	3.28	10
F	1600	3.28	1.92	6
G	1600	3.28	1.92	6
H	1600	3.28	1.92	6
I	1600	3.28	1.92	6
J	1600	3.28	1.92	6

$$\sum K_{3i} = 78$$

6.2 Repair at Central Repair Depot

6.2.1. *Number of equipments required per base*

Expected arisings in interval R_m

$$= R_m \frac{F}{B}\frac{T}{M_E} = \frac{10.700.1.5}{35\ 100} = 3$$

giving a minimum requirement of 3 equipments per station. To get $r_{so} \leqslant 3.3\%$ we need $K'2 = 6$, giving a requirement for 7 equipments and an expected 1.8 inabilities per year.

6.2.2. *Number of equipments required for Central Repair Depot Pool*

Expected requirement $= \frac{FTf}{M_E} = \frac{700.1.5.5}{100} = 52.5$

The reserve requirement = 9.3, so that $K'_3 = 62$ equipments.

6.3 Costs

6.3.1. *Repair on Bases*

(i) *Cost of Spare Main Equipments*
From para 6.1.1 $K_1 + 1 = 3$
Cost $= C_E\,(K_1 + 1)B = £105{,}000$

(ii) *Cost of Modules*
From para 6.1.2 $K_{2i} = 12$
$K_{3i} = 78$

Cost $= C_m B \;\sum K_{2i} + \sum K_{3i} = £39{,}840$

(iii) *Cost of Test Sets*
From Equation (10)

$$I = R_T \frac{F}{B} \cdot \frac{T}{M_E} = \tfrac{1}{2} \cdot \frac{700}{35} \cdot \frac{1.5}{100} = 0.15$$

so that no queuing is likely and each station has one test set.

From Equation (15) cost of test sets

$$= C_T \left[B \;\left(1 + \frac{R_c}{Z}\right)\right] = £123{,}000$$

(iv) *Calibration Costs*
From Equation (20)

$$= Y\,C_p \left[\frac{BH_T}{WdZ}\right] = £96{,}000$$

(v) *Total Costs*

	£
Spare Main Equipments	105,000
Modules	39,840
Test Sets	123,000
Calibration	96,000
	363,840

6.3.2. *Repair at Central Depot*

(i) *Cost of Spare Main Equipments*
From para 6.2:-
Number of spare main equipments

$$= \quad (K'_2 + 1)B + K'_3 \qquad = 245 + 62 \;=\; 307$$

$$\text{Cost} \;=\; C_E\,(K'_2 B + K'_3) \;=\; £307{,}000$$

(ii) *Cost of Test Sets*
From Equation (24)

$$I = R_T\,\frac{FT}{M_E} \qquad = \tfrac{1}{2}\quad \frac{700.1.5}{100} \;=\; 5.25$$

Intermediate calculations shown in Annex A lead to a requirement for 6 test sets and 7 equipments. The calibration float will add one more test set.

Total costs

	£
7 test sets @ £3,000	21,000
7 equipments @ £1,000	7,000
	28,000

(iii) *Cost of Calibration*
6 sets per year cost
6 x £275 x 10 years = £16,500

(iv) *Total Costs*

	£
Spare Main Equipments	307,000
Test sets	28,000
Calibration	16,500
Total =	351,500

Thus, in this example there is a clear advantage for the central repair case, of £12,340. Since this is less than the cost saving arising from lower calibration costs, calibration costs must be discounted. Using a discount rate of 10% the present value of the calibration costs are:

Repair at bases	£59,145
Repair at central depot	£10,115

The new totals are

Repair at bases	£326,985
Repair at central Depot	£345,115

so that the use of discounting changes the balance in favour of retention of repair facilities at the bases.

7. EFFECT OF VARYING PARAMETERS

7.1 Higher MTBR

The current servicing policy calls for the MTBR of an equipment to be at least 500 hours before dispensing with repair facilities at the base. The calculations of Sections 6.1, 6.2 and 6.3 have been repeated for this case, with the following results.

7.1.1. Repair at base

(i) Number of equipments per base

$K_1 + 1 = 2.$ Cost = £2,00 per base
Total £70,000

(ii) Number of modules per station

$\sum K_{2i} = 10.$ Cost = £800 per base
Total £28,000

(iii) Number of modules for repair depot pool

$\sum K_{3i} = 26.$ Cost = £2,080

(iv) Test sets required–unchanged at one per station–discounted calibration cost £59,145.

(v) Total costs

	£
Spare Main Equipments	70,000
Spare modules	30,080
Test sets	123,000
Calibration	59,145
	282,225

7.1.2. Repair at Central Depot

(i) Number of equipments per station.

$K_2' = 2$, at £1,000 each

$K_2' + 1 = 3$

Total Cost = £105,000

(ii) Central repair pool

K_3' = 15 at £1,000 each
Total Cost = £15,000

(iii) *Test Sets*. In this case I = 1.05 and two test sets will give a queue of just over one equipment. We require two in use test sets, and one spare, plus two main equipments, with a total cost of £11,000. The discounted calibration cost is £3,370.

(iv) *Total Costs.*

	£
Spare main equipments	120,000
Test Sets	11,000
Calibration	3,370
Total =	134,370

Thus, for this much higher MTBR, the central repair depot policy saves £182,855 after discounting.

In the subsequent paragraphs, the effects of changing parameters other than MTBR are considered. The main equipment MTBR remains 100 hours throughout.

7.2 Main Equipment with a number of LRU's

Throughout the previous discussion, it has been assumed, for the sake of simplicity, that the main equipment had only one LRU, or at least, the failure rate of one LRU greatly exceeded all others. If, however, there are a number of LRU's with MTBR's distributed in the manner of Annex B, then two stages of calculation are necessary, as follows.

7.2.1. Calculation of Base Holding of LRU's

(1) For the second line repair case Station holdings of LRU's would be based on R_s, second line repair time, and repair frequency. One set of LRU's would normally be held with the test bench and their cost should be added to the test bench cost. More than two of any individual LRU would only be required if the MTBR was below 500 hours.

(2) *For the "Central" Repair Depot Case*. Here the station holdings would be based on R_m, the third line order interval. There would be a requirement for three sets of LRU's per station.

7.2.2. Calculation of Third Line Pool of LRUs.

This would only be required in the "central" repair depot case, and would be directly analogous to the calculation given in para 6.1.3 ie 78 LRU's (say 8 sets) would be required.

7.2.3. Calculation of Module Holdings

Bases would hold one complete set of modules and the central repair depot servicing float would be about 3 times the cost of a set of modules.

7.2.4. Total Cost for Multi LRU Case

(i) LRU Repairs at Bases

	£
Spare LRU's (2 sets per station @ £800 a set)	56,000
Test sets (incl 1 set LRU's per station)	151,000
Spare Modules (1 set per station plus 3 sets equivalent at 3rd line)	30,400
Calibrated (discounted)	59,145
	286,545

(ii) *Repair of Central Depot*

	£
Spare LRU's (3 sets per station) plus	84,000
8 sets for 3rd line	6,400
Test sets (incl 1 set LRU's per in-use set)	27,000
Calibration	10,115
Total =	127,515

The cost advantage of the proposed policy therefore increases for the Multi LRU case.

7.3 Fleet Size.

If the fleet size is reduced to 350 without decreasing the number of bases, the following variations take place.

7.3.1 Repairs at Bases

(i) *Spare Main Equipments.* With the reduction in fleet size the daily flying per station falls to 15 hours, and the probability of more than two failures occurring in R_S = 2 days become 0.3 therefore each base can meet the risk conditions with only two spare main equipments, costing £70,000 overall.

(ii) Test Sets–Cost unchanged.

(iii) Spare Modules–the depot float would be approximately halved and only one module A held per station so that ΣK_{2i} would be 10, and ΣK_{3i} would be 43.

(iv) Calibration–Cost unchanged.

(v) Total Costs

	£
Main Equipments	70,000
Test Sets	123,000
Spare Modules	31,440
Calibration	59,145
Total	283,585

7.3.2. Repair at Central Depot

(i) Spare Main Equipments. $K'_2 + 1$ would be 4 per base, and K'_3 is 31, giving a cost of spare main equipments of £171,000

(ii) *Test Sets.* The number in use required would be 3, giving a requirement for 4 sets, costing *£12,000* plus 7 spare main equipments, total *£19,000*.

(iii) *Calibration.* Discounted costs for annually calibrating three sets would be £5,050.

(iv) *Total Costs*

	£
Main Equipments	171,000
Test Sets	19,000
Calibration	5,050
Total	195,050

It is demonstrated that a reduction in the fleet size, without a corresponding reduction in the number of bases, favours the centralisation of repair. A decrease in flying task with the same number of aircraft on each base has the same effect. Conversely increases in flying hours per day per base, whether by increasing the number of aircraft on a base, or by increasing the task, favours retention of a repair capability on the base.

7.4 Servicing Time for Main Equipments

Increased servicing time for main equipments favours centralisation since the number of main equipments has to be increased. Doubling R_s would call up one more equipment per base. Halving R_s might reduce the holdings by one, but it is a marginal case.

7.5 Servicing System–Central Depot

For administrative reasons, the RAF servicing system operates with $R_m = 2f$. This is not strictly necessary, and it was thought worthwhile to explore the effects of independent changes in these two parameters.

7.5.1. *Changing R_m*

a. *Repair at Bases.* Doubling R_m would increase the number of modules per station, module *A* requirements rising to 5. Modules *B, C, D* to 3 and increasing costs per base by £640, and overall by £22,400 or a 6.9% increase in costs. Halving R_m would reduce the number of modules by 2*A* modules per base, saving £5,600, a 1.7% decrease in cost.

b. *Repair at Central Depot.* Increasing R_m has a considerable effect on central depot repair system costs. Doubling R_m raises $K_2{}^1$ to 11 (ie 12 equipments per station) with a £175,000 increase in costs, or 50%. Halving R_m reduces $K_2{}^1$ to 4 (5 equipments per station) with a saving of £70,000 or 20%. There is clearly a good case for exploring the possibility of reducing R_m.

7.5.2. *Changing f*

a. *Repair at Bases.* This operates only on the module repair float. This is only £6,240 under this system, or 2% of the total cost, so that improvements in f can have little impact on total cost.

b. *Repair at Central Depot.* At £62,000 the central depot repair float represents 18% of total cost, so that a variation in f has a considerable effect, though not so much proportionally as a change in R_m.

c. *Effects of a change in f.* The effects of a change in f could, as well be seen from Annex A, lead to a change in the number of test sets at the central depot, and they will certainly cause a change in the transport arrangements for the depot. These effects would have to be carefully explored, lest their costs exceed the savings in equipments.

7.6 Cost of Main Equipment

The repair on bases system is not as sensitive to the cost of main equipment as the centralised system. Capital costs of equipments and modules account for only 44% of cost of the repair-on-bases system but 87% in the centralised system.

7.7 Test Time

If the main equipment test time was doubled, this would have no effect on the repair-on-bases system since the test set is still only 28% utilised. But in the centralised system, twice as many in-use test sets would be required, plus one extra for calibration, thus increasing the cost by £21,000, to which must be added additional calibration costs of £10,115. Doubling the test time thus works in favour of the repair on bases system. Similarly, halving the test time favours the centralised system, reducing the number of test sets from 7 to 4. This demonstrates the value of using automatic test equipment at the centralised depot.

7.8 Cost of Test Sets

Doubling the cost of a test set works in favour of the centralised system, since 38% of repair-on-bases system costs are absorbed in test gear, whereas in the centralised system only 8% of costs are absorbed. Halving the cost of a test set works in favour of repair on bases.

7.9 Life of Equipment

Except in the case where manpower is inflexible, the only cost dependent on the life of the equipment is the calibration cost. This represents 18% of the cost for repair on bases, but only 4.7% of the cost in the centralised system. The effect of discounting is that the advantages of a shorter equipment life to the servicing on base system are reduced, as are the advantages of a longer equipment life for the centralised system.

7.10 Number of Bases

If the number of bases is halved with the same fleet size–say 700 aircraft on 18 bases–not only are the test set and calibration costs decreased, for the repair on base case, but also the number of spare equipments and modules are increased, for both cases. The results are as follows.

7.10.1. Repair on Bases

(i) *Spare Main Equipments.* This is raised to 5 equipments per base costing £90,000.

(ii) *Spare Modules.* The protection against inability to supply afforded by 5 equipments per base is such that the number of possible inabilities for modules is rather more than that given in para 6.1.2. Nevertheless, the requirement per base rises to 4 *A* modules, and 2 each of modules *B, C* and *D*, giving a total of 16 modules per base, at a total cost of £23,040. The central repair depot pool could be slightly less, because the variability of demand reduces with the number of stations, but the quantity will be kept at 78, and the cost £6,240.

(iii) *Test Sets.* 18 in-use sets and 3 spare sets will be required, costing £63,000.

(iv) *Calibration.* Calibration costs for 18 in-use sets, discounted, over 10 years will be £30,417.

(v) *Total Costs*

	£
Main Equipments	90,000
Modules	29,280
Test Sets	63,000
Calibration	30,417
Total =	212,697

7.10.2. Repair at Central Depot

(i) *Cost of Spare Main Equipments.* Each base will require 12 main equipments making a total of 216. The central repair depot pool will remain at 62 making a total of 278 sets costing £278,000.

(ii) *Other Costs.* Other costs remain as in para 6.3, as amended for discounting.

(iii) *Total Costs*

	£
Main Equipments	278,000
Test Sets	28,000
Calibration	10,115
	316,115

7.10.3. Conclusion

Although concentration on fewer bases reduces both costs, the greater savings in Test Equipment and calibration favour the retention of repair facilities at bases.

8. CONCLUSIONS

The decision to eliminate repair facilities at bases, though very sensitive to MTBR, should not rest on MTBR alone. In the numerical example, an MTBR of 100 hours appears to be the break-even point, and, in some cases, rather lower MTBR's could be accepted.

Factors which tend to favour the elimination of repair facilities on bases are:

(a) Multi LRU equipment.

(b) Reducing the number of aircraft on each base, or reducing their task.

(c) Increased servicing time for main equipments.

(d) Reduced order interval from central depot (R_m).

(e) Increased frequency of delivery from central depot (smaller f). But there may be a penalty in increased transportation costs.

(f) Lower cost of main equipment.

(g) Lower test time for main equipment.

(h) Higher cost of a test set.

Factors which tend to cause retention of repair facilities at bases:

(a) Larger fleet size or higher task on same number of bases.

(b) Lower efficiency of central repair system.

(c) Higher cost of main equipment.

(d) Lower cost of a test set.

(e) Concentration of same fleet on fewer bases.

The key factors in determining the policy would appear to be:

(a) MTBR of main equipment.

(b) Ratio of test set cost to main equipment cost.

(c) Ratio of fleet size to number of bases.

(d) Whether main equipment has one or several LRU's.

ANNEX A: CALCULATION OF TEST SET REQUIREMENTS USING SIMPLE QUEUEING THEORY

The intensity $I = \dfrac{\text{arrival rate}}{\text{service rate}}$

is first calculated for the values of F, T, M_E, R_T, and B.
For each base this becomes

$$\frac{FT}{M_E}\,\frac{R_T}{B} \tag{A1}$$

For the third line case it becomes:

$$\frac{F\;T\;R_T}{M_E} \tag{A2}$$

Case 1: $\underline{I < 1}$

For $I < 1$ the Average Queue length for a single test position is given by:

$$L_Q = \frac{I^2}{1-I} \tag{A3}$$

and is tabulated below for values of I from 0.1 to 0.9

	L_Q
o.1	0.01
0.2	0.05
0.3	0.13
0.4	0.26
0.5	0.5
0.6	0.9
0.7	1.7
0.8	3.2
0.9	8.1

Case 2: $\underline{I > 1}$.

In case 2 the number of test sets required is first estimated by rounding up the results of equations (A1) or (A2) as appropriate, to the next highest whole number, then calculating the Average Queue length for this number of test sets until the queue is small. The equation used for the average queue length where $\frac{I}{N_T}$ is near unity is:

$$\frac{I}{N_T - I} \tag{A4}$$

(This excludes the equipments actually on test).

For the example given in paragraph 6.3.2.

$I = 5.25$

$N_T = 6$ $\qquad$ $N_T = 7$

$L_Q = \frac{5.25}{6-5.25} = 7$ $\qquad$ $L_Q = \frac{5.25}{7-5.25} = 3$

Economic Analysis. Having determined the average queue length, a trade-off must be made between additional equipments in the queue (4 in our example) and the whole life cost of an additional test set (suitably discounted). In our example, the additional test set would raise the in-use quantity to seven, thus increasing the calibration float to two, so that capital costs of two more test sets are offset against 4 more main equipments. Even without calibration costs, it is preferable to have only six test sets, and increase the queue length to 7.

Fortuitously, our numerical example has dealt with a relatively easy case. A comprehensive economic analysis would include:

(a) Queue length and cost of equipment.

(b) Number and cost of test sets.

(c) Size of central repair depot float.

(d) Effect of queueing on f, the delivery interval.

ANNEX B: ASSUMED DISTRIBUTION OF MODULE MTBR's

Module	*MTBR*
A	400
B	800
C	800
D	800
E	1600
F	1600
G	1600
H	1600
I	1600
J	1600

These module MTBR's were determined by empirical research on arisings of a typical equipment, from which the distribution of MTBR's among N modules was that the worst would have an MTBR of $\frac{(N-2)}{2} M_E$, $\frac{3N}{10}$ would have MTBR's of $(N-2)\, M_E$, and the rest MTBR's of $2(N-2)\, M_E$. In practice predicted MTBR's for each module would be used.

DISCUSSION

Volten: What is the effect of changing r_{SO} (risk of stockout during a repair interval)

a) on the choice betwen repair at base and centralised repair?

b) on total cost?

Baker: If we work with a higher acceptable risk of stockout we lower the total cost for either policy. Such a decision would favour the central repair policy since in this case we save in equipments whereas with the other policy the main saving is in modules. The test equipment and calibration costs do not change. It is not easy to decide at what risk level to work since the equipment is fitted in many types of aircraft with different rôles and different military priorities. However we considered that the equipment was essential for most missions and hence set a low risk level.

Ballow: How can you cost the risk of losing your one central depot?

Baker: Both systems are dependent on the central depot for the repair of modules so loss of the central depot will disrupt both systems. However, the base repair system may survive slightly longer provided that unusual demands for modules are not met.

The existence of the base facilities also permits cannibalisation of modules, which is a further safeguard against interruption of the supplies from the depot. These factors are qualitative rather than quantitative so they can not be costed.

Brackel: I don't see the influence of transport in both cases.

Baker: At the stage in the history of the organisation when the model was designed we were dealing with marginal cases, so that one additional equipment serviced centrally would not cause a change in the transportation system. We understand that now, as the policy of centralisation has developed, and, moreover, additional types of equipment are being serviced at the central depot (it was working only on radio equipment but now deals with electrical and instrument equipment also) additional transport is required. But it is difficult to separate the effects of these two policies. In para. 7.5.2. of the paper we point out that if a change in the delivery frequency was contemplated, we would have to perform a rather complicated trade-off between main equipments and transportation. At the time of the study, however, the vehicles being used had sufficient capacity to take the larger volume of L.R.V.'s as compared with modules. Nevertheless we think that the situation today would be a good subject for a further study.

Combes: Est-ce que vos évaluations ont tenu compte d'une réserve de matériels complets pour pallier les insuffisances du système de soutien?

Baker: The stocks calculated by the model form the reserves.

In addition, policy-makers would have to add special reserves for military requirements such as fly-away packs, and these could be used in emergencies.

Stephens: Why did you assume it would take the same total manpower at base or depot, as I would have thought with increasing labour costs this might be a significant factor if your assumption is false.

Baker: In principal the manpower bill is the same under either policy (except for the test equipment calibration cost). This is because the task is the same in each case, namely the replacement of unserviceable modules with serviceable ones. Cost differences can arise in two cases (a) if it is not possible to translate man-hour savings at the bases directly into equivalent personnel savings because fractions of a man's effort are involved or (b) if the cost of labour is different at bases and depots i.e. if service rather than civilian labour is involved.
Although for the single marginal cases explored by the model we would not expect economies of scale in going from base to central depot servicing, the cumulative effect of a number of decisions to centralise repair would be to realise economies of scale in overheads such as supervision.

Ferrier: Il est remarquable que les coûts totaux soient si voisins dans les deux hypothèses (réparations sur base ou en dépôt central). Il semble que les équipments d'essais et de réglage soient les seuls responsables de la préférence donnée aux réparations centralisées. La conclusion n'est-ille pas inversée si l'on considère qu'en toute hypothèse des réserves de tels équipements doivent être conservées au titre des stocks de guerre et qu'il est alors préférable de les utiliser que de n'en rien faire?

Baker: We were extremely lucky in our calculations to hit on the "break-even" point with the numerical example given in the text, without much iteration. The figures in the numerical example of para. 6 are typical of the past rather than actual, and, indeed practical tests of the model have shown that the much higher main equipment costs which are typical of the future, bias the decision towards repair at bases. We agree with the questioner that, if some reserve capacity of test equipment has to be purchased for war plans, it would be wise to use it, perhaps by introducing several central test points, and the original model showed how it would be used for the case of a few (4) bases, and multi-LRV aircraft equipment, to exploit additional reserve test capacity. However, we point out that by using test equipment you incur calibration costs–provided the calibration is, as is the case in the RAF, based on calendar periodicities. If, however, the calibration interval is based on running hours, then calibration costs are identical for both central and base servicing and in this case, for our numerical example, we would not need to discount and repair at bases would be a much cheaper alternative.

Irving: As the change from a centralised repair policy to a base repair policy would not be practicable to make, and as MTBR is not necessarily a constant, would it not be a more military solution always to select an initial base repair policy rather than to construct a major centralised repair depot right from the start of the equipment's life?

Baker: We do not believe it would ever be economic to make a change of policy during the life of an equipment since the total costs would be increased very considerably. We agree that MTBR is not necessarily a constant. It has been observed to increase considerably during the first year or two of the life of the equipment. However during this early period there are other factors operating which would relieve this strain on a central repair policy, at this time the fleet of aircraft is being built up so equipment utilisation is below its final level. Furthermore there may be extra equipments available at this time which are destined for the aircraft to be built at the end of the programme. However we agree that if costs are similar, base repair is to be prefered for a number of reasons, so it must be clear that substantial savings would be made before the centralised repair policy is adopted.

Solstrand: The speaker mentioned that the model was not directed at trading off personnel for sophisticated test equipment, but to find the best of *two* policies for repair.

It appears to me that there might be something to gain by such a trade-off; in other words, that it might be of *great* interest to study also some other policies, where various ways of doing this trade-off are included.

Baker: In order not to complicate an already over-long paper, we did not include trade-offs between manpower and automatic test equipment. We have, in fact, adapted the model for this purpose, and this version of it was presented at a conference on Automatic Test Equipment, held by the British Institution of Electronic and Radio Engineers at Birmingham in April 1970. In fact, we have found that, with automatic test equipment, such as TRACE, which is self-calibrating, we are trading off operating manpower plus calibration manpower against capital cost.

Logistics Trade-Off Model for Multi-Site Missile or Radar Systems

P. B. Buck

Long before large radar and missile systems reach the production stage it is necessary to know what resources will be required to support and maintain such systems. In selecting support resources one also has to select the best logistics support alternatives.

Figure 1 demonstrates the nature of the problem considered in this communication. There are numerous sites, each having complex missile and radar equipment. There is at least one depot performing major repair and supply functions. There is at least one factory and probably more with supply and repair capability. In some cases engineering dictates decisions as to what must be repaired at site, depot, and/or factory and how it must be transported. In many other cases, however, reparable components can be handled at any location. The major question, then, is which location is best for which items in terms of cost and effectiveness. Inherent in the repair location decision is the type of transport system to be used. Above all in selecting the right logistics alternative one must assume that enough resources are provided under each alternative considered so that when the radar/missile system finally reaches a "steady state" operating condition it will be continuously available or in a ready, operable, status the desired fraction of the time.

The problem of selecting logistics alternatives is also complicated, particularly in the case of the radars, by complex circuitry which may have thousands of reparable components and subsystems, a number of which are redundant or in parallel with each

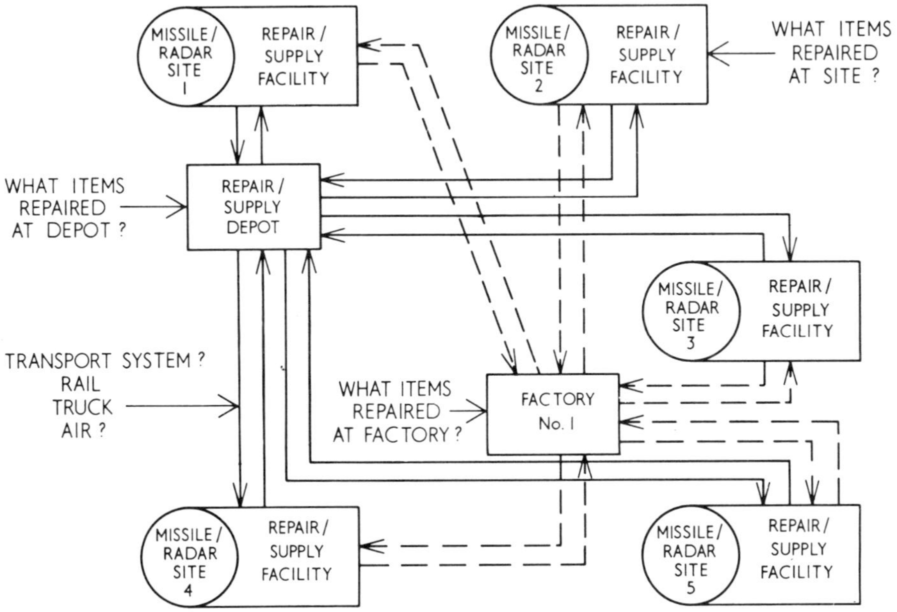

Figure 1 Missile/radar system and logistics questions

other to a varying extent. The degree of redundancy strongly affects support requirements.

We have developed a deterministic model for performing the trade-off analysis task whenever design information is available to give the inputs listed in Fig. 2. The model is not elaborate nor stochastic in treatment because in the pre-production stages of a new complex system detailed data are not available.

You may notice from the inputs that the model addresses the on-line replaceable unit, or ORU, as we call it. This is the smallest unit in the equipment breakdown where a failure can be detected by performance monitor equipment, and can be repaired by on-line replacement. Every failed ORU is removed and replaced.

I won't go over each of the model inputs and outputs on this chart. I might point out that the "off-site inventory" cost output refers to the cost of the off-site repair pipeline, if there is one, and the inventory required to replace those failures which are not reparable.

The POISSON calculation is required to develop the probability of having a replacement spare given a certain number of spares initially on hand. The Network Reliability Assessment Model, or NERAM, as it is called, is used to calculate system availability using the system block diagram input. The technique comes to us from the General Electric Corporation. I will describe it further later.

Let me define what is meant by system availability. System performance depends upon a number of sequential probabilities–the first is the probability of success when a weapon gets to a target; the second is the probability that when the button is pushed, the weapon performs its intended program in reaching the target area; the third is the system availability or the probability that all of the system is working and/or ready to function when the emergency is at hand. The probability of success and the weapon reliability after the button is pushed are primarily the responsibility of the designer and not the logistician. When we get to the system availability, both the system designer and the logistician get into the act.

Steady state system availability can be formulated as the uptime over the total time which is uptime plus downtime.

$$\text{SYSTEM AVAILABILITY} = \frac{\text{UPTIME}}{\text{UPTIME + DOWNTIME}}$$

The formula is equally valid for a single component and an entire system. The designer can choose components that have low failure rates so that the uptime goes up. He can also design good maintainability so that downtime is reduced. His concern for downtime generally relates only to reducing the time required for removal and replacement of a failed component. When he has reached the limit of these parameters in trying to attain an availability goal, the designer has one trick left in the bag–namely, redundancy. He develops parallel circuits as shown in Fig. 3 so that if one unit fails, another can carry on while the first one is being repaired. As long as some combination of darkened units work, as shown in the figure for a hypothetical circuit, the system is considered in an available condition. In his trade-off problem the

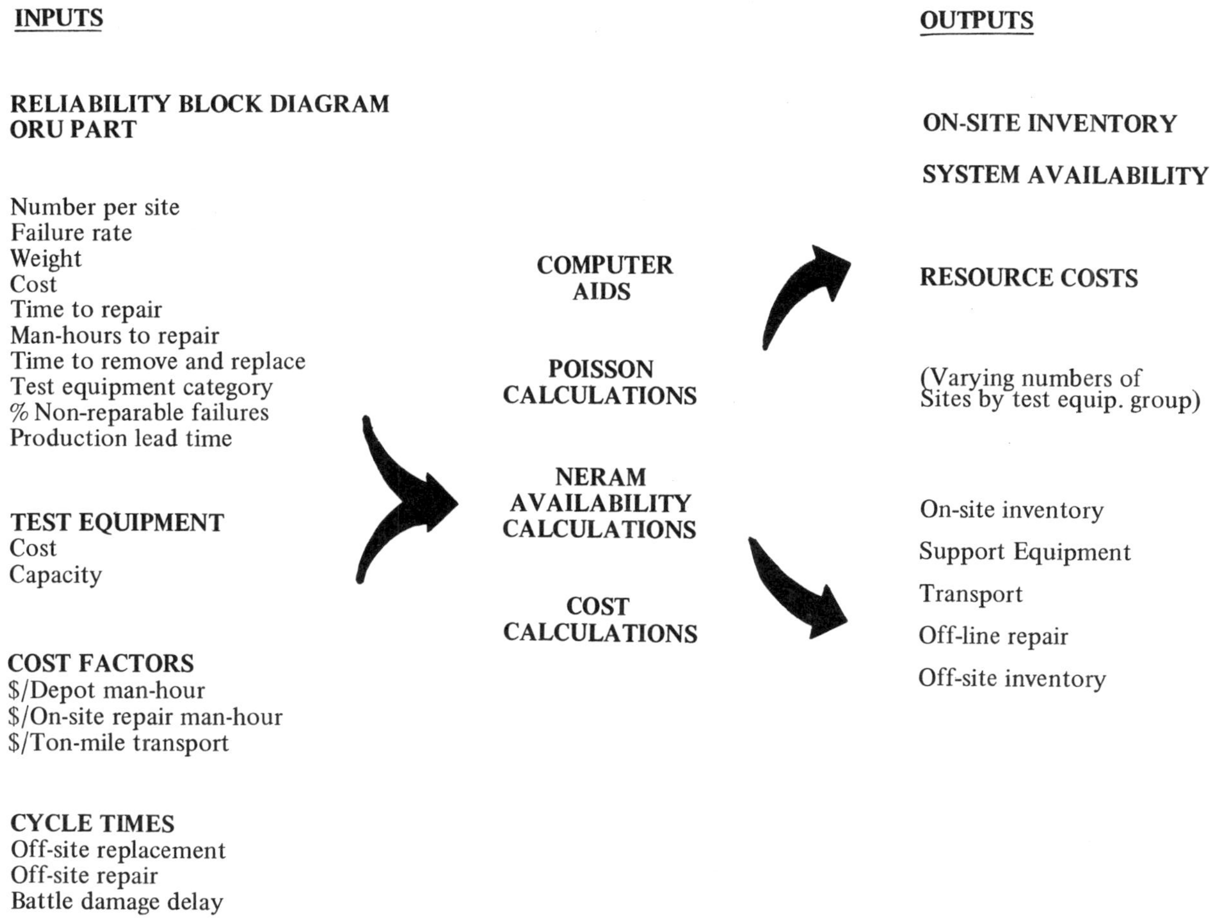

Figure 2 Model inputs and outputs

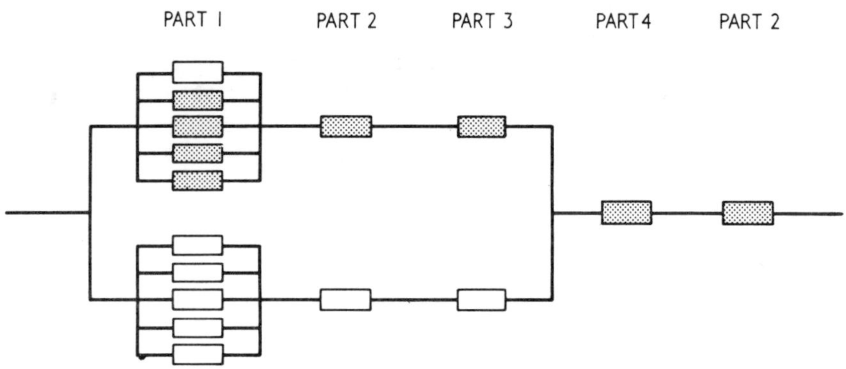

Figure 3 Hypothetical system with redundancy

designer may find redundancy less costly in some cases than higher unit reliability. Today the state of the art in connection with electronic component reliability is such that if the overall system goals assigned to the radar designers are high, they will be forced to use a lot of redundancy in their circuits. The missile design people get redundancy out of the number of missiles assigned to the missile farm to accomplish the same purpose.

Where does the logistician come into this? Remember the designer only concerned himself about the downtime of the sytem while a component was being removed and replaced. The logistician has got to produce the spare units for replacement. If a spare is not available for one of these non-redundant parts when it fails, then the whole system has to remain down until the removed component is repaired or until another one is obtained from another source. This affects the average downtime in our availability formula considerably. You can see, however, that the impact on availability of a failure without a spare may not be so bad in the case of our redundant parts.

Let's face it, the logistician can't always have a spare available for the remove and replace operation. As a result, system availability will be affected to the extent that logistics support is not perfect.

This is where our logistics trade-off model is necessary. If we change the logistics alternative as, for instance, from on-site repair to off-site repair at the depot or factory, we have to ensure that we are not degrading system availability below its intended goal. When we compare alternatives, we must compare the support resources necessary to achieve equivalent system availability levels.

How does the model get at these problems? I won't go into all of the formulae involved, but the next figure leads toward one of the most important ones. If you remember the downtime shown in the denominator of the availability formula, that is what we are dealing with here.

R is the remove and replace time; P_N is the probability (from a POISSON distribution) that a spare is available with N spares in stock and therefore $(1\text{-}P_N)$ is the probability that a spare is not available; T_R is the time for on-site repair of an item that has to be suffered if no spare is available. Beyond T_R is an even longer time, but unfortunately, much less probable. This is what happens if there are piece parts missing from the on-site inventory, and the repair is delayed until parts can be sent from the depot. This replacement time is indicated by $T_D + T_B$ where T_D is the expected unhampered transport time and T_B is the time delay which might be imposed at any time in war if the transport system is temporarily put out of action by enemy attack. If this is a probable situation we must throw this time T_B in as an ever-present impediment to supply delivery, one which has to be compensated for by inventory. Finally, of course, in the formula is the last resort situation which occurs when we go to the depot, and we can't find either the piece parts we want or a replacement spare. This, of course, is a very rare occurrence indeed, but it still has an effect.

$$\text{AVERAGE DOWNTIME FOR AN ORU (PART)} = R + (1-P_N)\left\{T_R + (1-p)\left[T_D + T_B + (1-g)\ T_L\right]\right\}$$

Figure 4 Downtime formula for on-site repair alternative

For off-site repair, this formula is changed. For the percentage of time we don't have a spare $(1\text{-}P_N)$, we must go off-site for a replacement and suffer at least the replacement plus battle damage time $(T_D + T_B)$ together with another last resort situation.

In providing resources the logistician can't do a thing about R, remove and replace time, or T_R, the average repair time. He can change the number of spares on-site which affects P_N; he may be able to affect the value of piece-part supply level, p, by providing a larger range and depth of piece part supply. He can also keep the replacement time down by using air transport rather than surface transport. Generally speaking, we have been given values for T_D, T_B, and p, however, we must obtain one for N. With the on-site inventory level N high enough we can reduce this whole term so that it has a trivial effect.

All this formula does, however, is give us a new downtime input for our availability formula as applied to a single item. We are concerned about the whole population of ORU parts that make up the system. Given values for each of the components, including the number N in inventory for each, we want to calculate system availability–just as the system designer does or should do–and also the associated cost.

The method for calculating system availability as developed by General Electric requires us to assume that failures of ORUs (parts) are entirely independent of each other. It uses standard binomial techniques to build up an availability for a group of components when given individual availabilities for components. In simplified form, the Network Reliability Assessment Model (NERAM) would see our hypothetical circuit in the next figure as having two types of relationships, the A type being essentially unlike items or groups of items in series, and the B type being like items or groups of items in parallel.

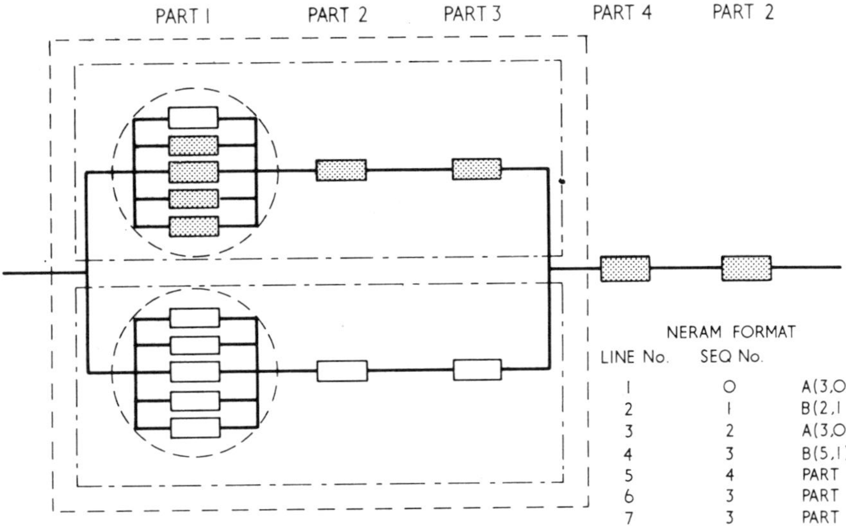

NERAM FORMAT

LINE No.	SEQ No.	
1	0	A(3,0)
2	1	B(2,1)
3	2	A(3,0)
4	3	B(5,1)
5	4	PART 1
6	3	PART 2
7	3	PART 3
8	1	PART 4
9	1	PART 2

Figure 5 NERAM analysis of hypothetical system

The program format would first see three items in series, none of which can fail, and would express this as, A(3,0). The first item in the series is the group of items within the largest rectangle, the second is a single part 4 and the third is part 2 in its second application. The program format then deals with the group of items within the largest rectangle and sees two smaller rectangles of like nature, in parallel, one of which can fail without putting the system down. This relationship is expressed by B(2,1). Within each of the smaller rectangles are three different items in series, none of which can fail, expressed as, A(3,0). Then within the circles are five like items, one of which can fail, expressed as, B(5,1). The B(5,1) situation refers to a single part, part 1. The sequence then calls in turn for the other two items in the smaller rectangles, parts 2 and 3 and then for the final two items under the first A(3,0) relationship. With design failure rates for each part providing mean uptime or mean time between failures and with downtimes inserted as directed by the model, availability figures can be calculated for each part and then for each of the group relationships or "nodes" shown in lines 1 through 4. The NERAM program knows by the letter and number index what type of binomial formula and what numerical inputs to use for each calculation. As you can see the same part may be used in different locations within the same circuit.

I have simplified things for demonstration purposes. The actual NERAM program can take care of many more complex circuit relationships than those indicated.

Our overall model enters this network analysis technique with the adjusted downtime for a single ORU part or for each ORU part in the system. The downtime adjustment reflects the average time losses incurred because of imperfections in the supply of spares.

For each logistic alternative the model develops on-site inventory requirements which meet the specific availability goal. The method used is, essentially, a trial and error process. We begin the process by looking first at each ORU (part) using the appropriate formula for downtime as previously described. When one is working with a single part the rest of the system is treated as if it were perfectly supported, that is–all other parts suffer downtimes equal only to their designed remove-and-replace times. Starting from zero we add spares to the ORU part in question and calculate the resulting system availability for each case. We continue the process until the resulting availability has just exceeded a pre-established arbitary intermediate availability figure lying somewhere between the higher figure for the perfectly supported system as designed and the lower figure representing the performance goal which one cannot drop below. The next figure shows the two availability levels and the results we are aiming for. If the system designers have not allowed some logistic margin between the

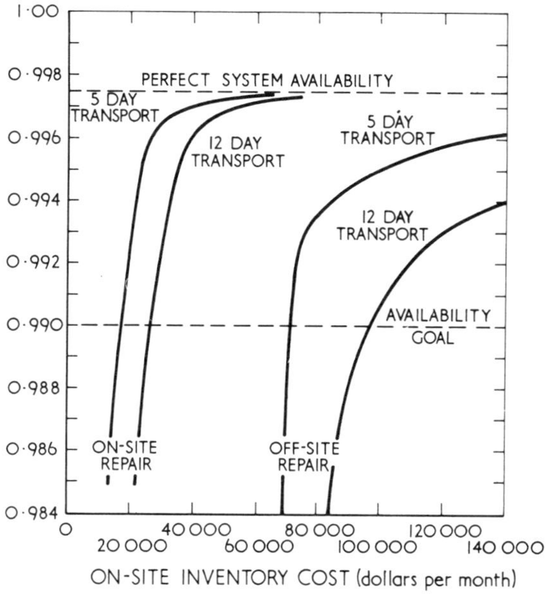

Figure 6 System availability versus on-site inventory

designed perfect system availability and the performance goal then there is nothing that the logistiician can do but demand a greater margin. After using the arbitrary intermediate limit to determine the spare inventory for each ORU part with others considered perfect, we then insert the resulting imperfect downtimes for all parts and calculate the system availability. If this is too high above the goal, we adjust our intermediate limit for each part downward and recalculate. Ultimately we can come as close as we wish to the desired goal, thanks to the computer which is doing all the work. We can also optimize our on-site inventory requirements by trading off high-cost, low-gain parts for low-cost, high-gain parts, but there is not time to go into this.

The table shows sample availability calculations for four different logistics alternatives covering two transport systems, on-site repair of everything, and off-site repair of everything. The dollar cost shown is a multi-site cost per month for the appropriate on-site inventory. As you might suspect, the off-site repair situation calls for a much higher on-site inventory to support the longer replacement cycle time. However, we must look at all of the other principal costs to determine the trade-off situation.

Rather than go through the cost formulae, the table below shows some sample cost figure results for a seven-site system. The number of sites can be varied as desired. The costs of each category are accumulated by test equipment category, because if one changes from an on-site to an off-site repair situation, the removal of a specific piece of test equipment affects all ORU parts assigned to it. All cases shown meet a specific availability requirement. The left-hand example shows a strong cost advantage for the on-site repair alternative. On the other hand, the right-hand example shows a cost advantage for the off-site repair alternative. In this case the cost of increased inventory requirements for the off-site repair situation does not overcome the increased cost of having a piece of test equipment at each site for the on-site repair situation. While the direct transport cost appears to be insignificant for all cases in this figure, our preceding figure indicates that a change in transport cycle can have drastic effects on inventory requirements.

	TEST EQ. 24		TEST EQ. 27	
	REPAIR ON-SITE	REPAIR OFF-SITE	REPAIR ON-SITE	REPAIR OFF-SITE
ON-SITE INVENTORY	$ 1,034	$25,586	$ 0	$1,012
SUPPORT EQUIPMENT	9,167	1,310	7,333	1,048
TRANSPORT	1	92	1	134
OFF-LINE REPAIR	1,280	2,057	453	485
OFF-SITE INVENTORY	40,078	63,088	2,632	3,414
TOTAL $/MO.	$51,560	$92,133	$10,419	$6,093

Seven Site Relative Support System Costs for On-Site Versus Off-Site Repair

There are any number of uses which can be made of the model:

1. Cost-effectiveness trade-offs. Off-site versus on-site repair, different transport systems, and other logistic alternatives

2. Stocking of spare ORUs to meet availability requirements and cost limitations

3. Evaluation of configuration changes affecting failure rate, redundancy, or maintainability

4. Analysis of support resource requirements for ORU repair—including repair personnel, test equipment quantity, repair task analysis, and transport requirements

5. Definition of overall availability goals

In addition to the indicated logistics alternative trade-off evaluations the model will be invaluable in assuring that the stock of spare ORUs is adequate to meet availability requirements. When dollar limitations are imposed on inventory quantities, the model should be able to indicate where savings can be made with the least impact on the availability level. The model can be used later as operational failure rate figures are generated so as to permit updating of spare stocks. While the model does not at present address provisioning of piece parts (components used for repair of ORU parts), a version can be designed which does address piece-part provisioning and ties it to availability.

The model also can be used to evaluate the impact of configuration changes. It can determine how much a potential improvement in the reliability or maintainability of a specific item will mean to system availability and system cost.

In the process of developing costs, the model can also generate certain support resource requirements in addition to inventory levels. For instance, it can generate repair man-hour requirements for either on-site or off-site repair. While the model can't tell the planner how to provide the main-hours—one shift, two shifts, round-the-clock, etc.—it can develop the average balance of different skill requirements so that men can be trained for the task to be performed. It also generates the workload assigned to any specific piece of test equipment and can thus determine whether more than one piece is required either on-site or off-site. When the on-site off-site decision has been made, the model will give the transport requirements in ton-miles per month per site.

Finally, the model can help to ensure that availability goals are realistic enough to take care of logistics support. Where system designers have used up all of the margin, the additional amount to be allowed for the logistics support system can be determined. The effects of changes in many of the support system parameters such as piece-part supply level, factory repair cycle, depot cycle, replacement time, etc., can be evaluated to see where the best areas for improvement lie.

Only a brief sketch of this modeling effort has been attemtped. The model was designed for the SAFEGUARD Anti-Ballistic Missile System and is being used by the U.S. Army in designing the SAFEGUARD logistic support system. The deterministic approach of this model is too gross for downstream system evaluation when time distributions and actual component failure rates are known. More detailed Monte Carlo-type models are being developed for that purpose.

However, this trade-off model is already having the very healthy effect of causing design contractors and logistics planners to discipline their efforts and outputs toward a uniform evaluation system.

ACKNOWLEDGEMENT

The work described in this communication was performed while the author was employed at Logistics Management Institute. Mr. T. R. O'Donnell, director of the LMI Safeguard logistics project was intimately involved in the effort and deserves co-author status. Significant help was received from Western Electric Corporation which performed all of the programming for the model.

We are especially grateful to Mssrs. Robert Pellitier and Joseph Carr of General Electric Corporation, Syracuse, New York, who actually provided the NERAM portion of the model.

Thanks are deserved by numerous other individuals from other Safeguard subcontractors and from the U.S. Army's Safeguard Logistics Command who have contributed valiantly to the effort.

DISCUSSION

Morris: Could you explain please the difference between off-site and on-site inventory costs?

Buck: The on-site inventory is what is required on site to support either the repair cycle for failed items when repaired on site or the replacement cycle (order and ship time) when items are repaired off-site. The off-site inventory considers the protection requirements for items which fail in a way that they cannot be repaired and also includes the pipeline of items through an off-site repair shop when off-site repair prevails. The great difference in off-site inventory cost between the two test equipment groups in the slide is due to the fact that the left-hand one has many parts attached to it while the right-hand one has very few.

Solstrand: To calculate the availability it was mentioned that it had to be assumed that the various types of failures which might occur, are entirely independent.

Is it the feeling of the speaker that this is an absolutely valid assumption? If not, is it possible to indicate anything about the type and size of the effect resulting from interdependence?

Buck: I agree with the questioner that failures are not always independent of each other. The assumption might be invalid for some systems. For the safeguard system, however, a great deal of effort and cost has been expended in designing extremely reliable components and in insulating them from each other. In view of the crudeness of design data on failures, I believe the assumption is valid enough for developing trade-offs concerning repair location. Indeed there is not enough information for any other assumption. We do recommend that as more information becomes available, a more sophisticated model is desirable—one which considers the possibility of multiple correlated failures.

Traitements de Systèmes Logistiques en Période de Paix et de Conflit

A. Rougerie

INTRODUCTION

Dans le cadre général du soutien logistique, différentes études ont été conduites par le Centre Interarmées de Recherche Opérationnelle. Elles ont permis d'étudier le comportement de plusieurs systèmes et d'éprouver différentes techniques de gestion des rechanges utilisant, en particulier, le concept d'essentialité militaire.

Il a paru intéressant de présenter une synthèse des moyens mis à notre disposition pour traiter avec efficacité le problème posé par le soutien logistique, dont l'importance ne fait pas de doute:

"en fait, le sort de la bataille est déjà décidé par les services du ravitaillement avant qu'un coup de feu ait été tiré" disait le Maréchal ROMMEL.

Dans la première partie de cet exposé, nous étudierons les critères de performances d'un système de soutien logistique. Différents modèles de gestion sont ensuite cités avec leurs mesures correspondantes,

–soit en régime transitoire, correspondant à une période d'isolement caractéristique d'une période de conflit,

–soit en régime permanent, c'est-à-dire en "période de paix".

Les modèles de gestion sont classés selon leurs domaines d'application les plus appropriés. Le problème de l'essentialité militaire est évoqué et un exemple permet d'en voir un traitement particulier.

Cet exposé ne serait cependant pas complet s'il ne présentait pas les moyens de réaliser l'adaptation des résultats du "régime permanent" à ceux que sont en droit de demander les utilisateurs en période de conflit (période d'autonomie). A cet effet, un exemple numérique est exposé à partir d'un modèle analytique conçu pour un régime permanent.

1. CRITERES DE PERFORMANCE

Pour concrétiser les mesures attachées au niveau d'un stock, nous considérerons, en figure 1, un système de soutien à structure élémentaire.

Ce système s'apparente à une chaîne de ravitaillement à deux échelons dont l'un est centralisateur. Dans cette structure, nous admettons que des ateliers de réparation peuvent être jumelés aux stocks mis en place auprès des utilisateurs. Selon l'option choisie, l'échelon centralisateur assure tout ou partie des réparations concernant les articles défaillants de niveau 1.

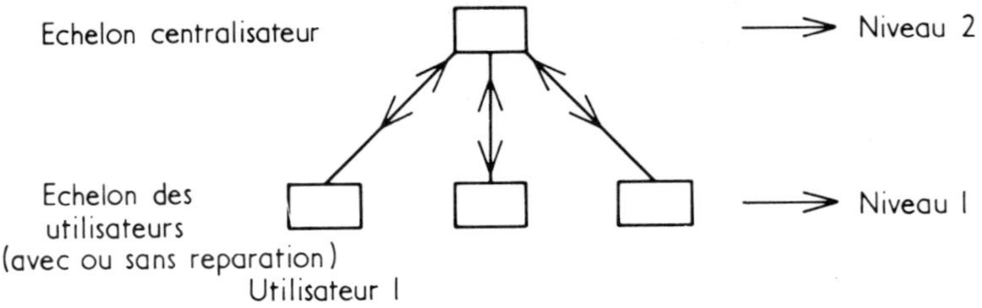

Figure 1

On peut constater que l'extension de ce système élémentaire de soutien à un système plus complexe à plusieurs niveaux devient facile si on sait dominer le système élémentaire.

Les paramètres d'états du système sont connus de tous. Nous ferons un rappel des principaux:

– Le taux de défaillance des matériels soutenus (ou taux d'emploi des biens consommés); ce taux est un des éléments indispensables pour établir la loi de prévision de la demande.

– Le coût de remplacement d'un article (ou bien) mis en stock.

– Le coût total du système ou coût de possession des stocks du système.

– Le temps de réparation par article et le niveau où s'effectue la réparation.

– Le coût de réparation par article.

– Les délais d'acheminement entre les niveaux.

– L'essentialité militaire de l'article; c'est un paramètre adapté à chaque système et dont la valeur procède d'un jugement souvent qualitatif guidé par un souci opérationnel: par exemple, pour un avion de chasse et en période de "conflit", l'essentialité d'une pièce de radar de tir aura une valeur supérieure à celle d'une pièce de la caméra-objectif et, inversement, si nous raisonnons sur un avion de reconnaissance.

– La politique de gestion suivie comprenant entre autres:

  la politique de rechanges à savoir: reparation ou mise en rebut systematique,

  – la méthode de remplacement préconisée à savoir:

    – soit un remplacement article par article dès l'apparition de la défaillance (ou de la demande),

    – soit un remplacement par groupe d'une quantité imposée ou à des époques précises.

Cette énumération ne peut être exhaustive. Bien sûr, d'autres paramètres existent. Nous avons simplement voulu vous en présenter les principaux auxquels il est fait appel dans toute politique de gestion.

1.1 Mesures des Niveaux D'Approvisionnement–Critères de Performances

Nous considérons un système de soutien doté d'un stock total (s) d'un article susceptible d'être renouvelé selon un temps moyen de réapprovisionnement T_m.

Par exemple, si à chaque panne au niveau de l'utilisateur, l'article défaillant est réparé sur place, avec la probabilité r et la durée de réparation R_p, ou remplacé par un matériel du stock central avec la probabilité 1-r, et la durée moyenne de réapprovisionnement R_C, incluant les transits et la réparation (cas où le stock central est nul), nous obtenons pour expression de ce temps moyen:

$$T_m = r R_p + (1-r) R_C$$

Différentes mesures du niveau (s) sont utilisées. Elles ont pour caractère commun d'être mesurables pratiquement sur une période d'observation assez grande, certaines à partir d'une information pauvre, par exemple le nombre d'articles délivrés, d'autres partir d'une information riche, par exemple le nombre de pièces manquantes et la durée de leur absence dans le stock.

Nous présenterons les plus couramment employées dans les modèles de gestion:

(i) *Le taux de disponsibilité de l'article ou sûreté individuelle.* (Ready rate)

Il indique à un instant quelconque la proportion de fois que l'article demandé a pu être fourni par le stock disponible. En terme probabiliste, c'est la probabilité que la demande soit satisfaite à un moment quelconque.

(ii) *La sûreté globale d'un stock composé d'articles différents*

Il est encore appelé "taux opérationnel" et s'exprime par le produit des sûretés individuelles.

Parfois, le terme "taux opérationnel" est employé pour désigner la proportion de matériels soutenus susceptibles d'être prêts opérationnellement à un instant quelconque. On peut constater que cette interprétation n'est pas éloignée de la définition donnée si l'absence d'un article entraîne la défaillance d'un matériel soutenu.

(iii) *Le taux d'indisponsibilité du stock*

Il désigne la fraction du stock indisponsible, cette indisponsibilité provient des articles en réparation et en transit vers (ou en retour de) l'échelon central.

(iv) *Le nombre de demandes non satisfaites.*

Il est encore appelé nombre moyen de ruptures ou espérance mathématique de rupture et nous le notons $E\{s_i$ pour un article de type i ayant s articles alloués dans le stock.

Par généralisation, pour un stock comprenant n types d'articles, on obtient un nombre moyen global de ruptures E:

$$E = \sum_i E\{s_i\}$$

(v) *Le taux de service du stock.*

Il a pour expression:

$$1 - \frac{\sum_i \text{demandes non satisfaites sur l'article } i \text{ avec un stock } s_i}{\sum_i \text{demandes non satisfaites sur l'article } i \text{ avec un stock nul}}$$

(vi) *Le taux de fonctionnement des matériels.*

Ce taux est employé aisément quand le stock(s) est constitué de matériels complets au niveau des utilisateurs. Dans ce cas, si m est le nombre de matériels que l'on veut en fonctionnement et T une période de temps donnée, ce taux H (s) indique le rapport de la durée totale de fonctionnement effectif sur la durée totale de fonctionnement désiré.

$$H(s) = \frac{\sum_i n_i t_i}{m\,T}$$

avec $m = \sum_i n_i$

et n_i le nombre de matériels ayant fonctionné seulement sur t_i.

Ce taux est un indicateur opérationnel car il mesure l'influence des différents états du stock sur le fonctionnement réel des materiels à soutenir. Il est lié à la notion de rupture et a ceci de particulier qu'il la personnalise.

1.2 Fonction d'Utilite du Systeme

Etant donné la mesure M (s) du niveau d'approvisionnement, il devient possible de se fixer une fonction d'utilité matérialisant une synthèse entre les besoins et les contraintes. Le plus souvent cette fonction d'utilité est de la forme:

$$U(s) = k\,M(s) - C(s)$$

où C (s) désigne le coût de possession du stock c'est-à-dire la contrainte liée au niveau s,

k, le coût des services rendus par l'unité de mesure choisie.

Ces coûts peuvent être d'origines diverses. Ils relèvent du même domaine que celui des contraintes liées au stock. Par exemple, si les contraintes sont d'ordre budgetaire, les coûts envisagés sont des coûts budgetaires; k exprime alors l'équivalence entre le plan financier et le plan opérationnel, d'une augmentation (ou diminution) d'une unité de mesure du niveau (s).

De même, si les contraintes sont liées au volume de stockage, les coûts sont exprimés en unité de volume; k exprime alors une équivalence entre un volume et un gain (ou perte) opérationnel relatif à un accroissement (ou diminution) d'une unité de mesure du niveau (s).

2. MODELES EN REGIME TRANSITOIRE "TEMPS DE CONFLIT"

2.1 Temps de Conflit–Temps de Paix

Le but cherché pour le responsable d'un soutien logistique est d'adapter les moyens disponibles aux besoins opérationnels des utilisateurs. En "temps de conflit", où l'isolement est toujours possible, le soutien logistique à prévoir revêt un caractère différent de celui du "temps de paix".

Il semble utile de voir en quoi peut différer l'approche des problèmes sous les régimes: "paix" et "conflit". Nous voyons deux différences fondamentales.

La première naît de l'idée même que le stock mis en place est là pour satisfaire au mieux la demande prévisible. Or, en régime "paix", cette demande n'est pas identifiable, même à un coefficient multiplicatif près, à celle d'un régime de "conflit". Pour s'en rendre compte, il suffit d'admettre que la demande "paix" vient de conditions normales d'utilisation en garnison ou en camp pour assurer:

- l'instruction ou l'entraînement militaire sous tous ses aspects,

- la participation militaire à la vie sociale du pays (il ne faut pas l'oublier).

Cependant, une prévision suffisamment sérieuse de la demande en régime de "conflit" peut être approchée par l'étude de la consommation observée au cours des manoeuvres d'unités constituées.

La seconde, vient de ce qu'en période de conflit, les impératifs opérationnels sont très restrictifs. Nous citerons pour exemple les impératifs inhérents aux unités terrestres mécanisées dont les qualités maîtresses de mobilité et de souplesse ne peuvent s'associer à des stocks démesurément gonflés qui peuvent, certes, satisfaire à la demande, mais qui gênent dans des proportions insupportables la liberté de manoeuvre.

Un compromis est donc nécessaire pour accorder les besoins aux impératifs opérationnels (nous ne retenons pas les procédures d'urgence d'approvisionnement qui ne fournissent pas une solution satisfaisante).

C'est par l'étude de l'essentialité militaire des rechanges (ou biens) que ce compromis peut être assuré avec toute l'efficacité voulue. Nous pensons qu'actuellement avec toute la complexité des matériels, nous avons là un moyen d'approcher la solution la plus satisfaisante qui soit.

Cette notion d'essentialité militaire n'est pas nouvelle. Elle a déjà été étudiée [1] et traitée dans certains problèmes [2] où l'environnement était cause de contraintes très sévères qui limitaient exagéremment les niveaux d'approvisionnement. Nous citerons, pour exemple, les rechanges électroniques embarqués à bord de sous-marins.

Les conclusions que l'on peut tirer des travaux effectués dans ce domaine sont qu'il n'existe pas de traitement a priori. C'est le matériel lui-même, que nous envisageons en tant que système d'arme, qui, par sa structure et sa finalité opérationnelle, oriente le traitement. Nous justifions ce point de vue d'après la définition même que l'on peut attacher à l'essentialité et à ses caractéristiques qui procèdent plus de propositions négatives qu'affirmatives. C'est sans doute là que résident toutes les erreurs d'interprétation.

2.2 L'Essentialité Militaire

Dans un effort de synthèse, nous avons cherché à définir l'essentialité militaire avec le moins d'ambiguité possible, ce qui nous a contraint à prendre pour référence le système d'arme soutenu.

Cette référence nous a semblé nécessaire car elle nous permet en raissonant à partir des missions dévolues au système d'évaluer les impacts opérationnels des défaillances des ensembles constitutifs.

2.3 Definition

Pour un système d'arme donné, l'essentialité militaire d'un des ensembles constitutifs indique l'effet de la défaillance de cet ensemble sur la capacité du système d'arme à réaliser sa mission.

Elle s'exprime par un indicateur, reflet du degré de dégradation des performances.

Pour plus de précision, il est nécessaire de l'assortir de certaines de ses caractéristiques:

(i) L'essentialité militaire est *différente de la fiabilité,* c'est-à-dire qu'elle est indépendante du taux de défaillance.

Ainsi un taux de défaillance élevé n'implique pas une haute essentialité.

Par exemple: un ensemble peut tomber 10 fois en panne par jour, mais si les performances du système d'arme ne sont pas affectées, alors cet ensemble ne peut se voir attribuer une essentialité élevée.

(ii) L'essentialité militaire *n'est pas unique* pour tous les ensembles d'un même système d'armes.

Par exemple: pour un char canon, les sieges des servants ne sont pas aussi essentiels que le canon lui-même.

(iii) *Le prix unitaire n'est pas*, en général, *un bon indicateur de l'essentialité* alors qu'il peut l'être pour marquer la complexité de l'ensemble considéré.

En annexe A.1., nous présentons à titre d'exemple de calcul d'essentialité, un questionnaire simple employé pour un hélicoptère envisagé en tant que système d'arme. Un autre exemple analogue peut être trouvé dans le texte référencé [3]. Les deux méthodes utilisées sont différentes en première analyse mais elles aboutissent, en fait, à un classement identique des pièces en fonctionnement (pièces à poste).

2.4 Modèles d'Allocation

Pour une période T d'isolement, le problème consiste à définir un stock de rechanges (ou de biens) permettant aux unités utilisatrices d'assurer au mieux leur mission. Le stock à définir n'est donc pas renouvelable sur T.

Différentes modèles analytiques existent. Nous présentons les plus adaptés actuellement: ce sont les *modèles analytiques à fonction coût-efficacité opérationnelle.*

Ces modèles ont pour qualité d'associer le niveau du stock à celui de l'efficacité opérationnelle correspondante. Il devient dès lors possible de choisir, parmi toutes les alternatives, celle réaliste qui présente le meilleur équilibre entre le coût du stock, exprimable en francs ou en poids ou en volume etc. . ., et l'efficacité opérationnelle du système d'arme; cette efficacité opérationnelle se définit comme le produit de la mesure $M(s)$ (cf. § 1.1.) du stock et de l'essentialité militaire attachée aux rechanges (ou biens) considérés.

Nous citerons pour exemple [4] le modèle où l'objectif serait de minimiser l'essentialité du nombre moyen de ruptures sous une contrainte du coût. Le programme mathématique correspondant a la forme:

$$\min \sum_i e_i E(s_i)$$

$$\text{avec} \sum_i c_i s_i \leqslant C$$

$$s_i \geqslant 0$$

c_i désignant le coût unitaire de l'article i

e_i étant l'essentialité militaire attachée à un rechange de type i

$E(s_i)$ étant le nombre moyen de ruptures à attendre sur la période T pour un stock comprenant s_i pièces de type i.

L'emploi d'un multiplicateur de LAGRANGE permet de résoudre simplement le programme et de le relier à une fonction d'utilité u_i:

$$u_i = k' e_i E(s_i) - c_i s_i$$

Le coefficient $(k' e_i)$ est à rapprocher du coefficient k indiqué au paragraphe 1.2.

La courbe coût-efficacité opérationnelle a l'allure générale indiquée sur la figure 2.

L'étude de la sensibilité du programme aux variations des paramètres que sont e_i, c_i et T permet de suivre leur influence sur le niveau du stock.

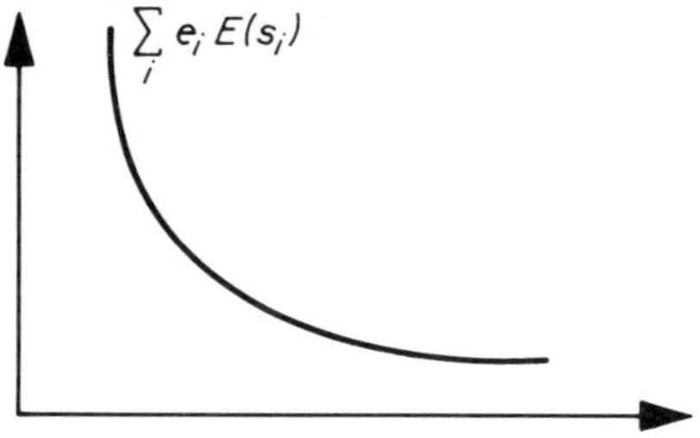

Figure 2

3. MODELES EN REGIME PERMANENT "TEMPS DE PAIX"

Précédemment (cf § 2.1.), nous avons insisté sur l'aspect restrictif des impératifs opérationnels en temps de conflit. Tout naturellement, écartant la solution des procédures d'urgence d'approvisionnement, nous avons été amenés à prendre en compte l'essentialité militaire des rechanges. La question se pose de savoir si cette essentialité est une notion à conserver dans le traitement des modèles de temps de paix.

Une brève analyse des méthodes actuellement utilisées en régime permanent semble indiquer que l'essentialité est laissée dans l'oubli.

En fait, ce n'est pas entièrement vrai car certaines procédures d'approvisionnement impliquent une idée d'essentialité: par exemple, les procédures différentes suivies pour des articles à coût élevé ou à forte consommation.

Les principaux modèles existants en régime permanent, que nous assimilons donc au temps de paix, peuvent classer en modèles analytiques et en modèles de simulation; nous laissons intentionnellement de côté ceux relevant d'un "arsenal" empirique, non justifiable théoriquement, bien qu'ils soient encore souvent employés.

3.1 Modèles Analytiques

Nous avons vu, en régime transitoire, que l'allocation était liée à la durée T de l'isolement et à l'essentialité militaire des rechanges.

En régime permanent, et pour la plupart des modèles analytiques, cette durée *T est* remplacée par la durée moyenne T_m de réapprovisionnement telle que nous l'avons définie au § 1.1. L'essentialité militaire est admise implicitement identique pour tous les rechanges assujettis à une même politique de gestion (c'est-à-dire e_i=1, quel que soit le type i de rechange).

(i) *Modèles analytiques à fonction économique*

Ces modèles ont constitué traditionnellement l'approche des problèmes d'inventaire. Tous reviennent à optimiser une fonction d'utilité de la forme déjà citée:

$$U\ (s)\ =\ k\ M\ (s)\ -\ C\ (s)$$

De sérieuses objections sont à opposer à l'emploi de ce genre de modèles. La plus fondamentale est qu'ils ne nous renseignent pas sur ce qui arrive au système quand il existe des contraintes limitatives; or, c'est le cas dans nos systèmes de soutien.

Que faire si les décisions conduisent à une mesure du niveau du stock de peu d'intérêt pour l'utilisateur?

(ii) *Modèles analytiques à fonction coût-efficacité*

Pour pallier les inconvénients précédents, des modèles relativement récents ont été exploités.
Ces modèles ont évolué dans le temps et ont été associés à des problèmes de plus en plus vastes. Les premiers [5] ont été employés pour la détermination de stocks au niveau de l'utilisateur et ne se souciant pas des liens éventuels entre les échelons d'apprivisionnement. Leur but est d'optimiser l'efficacité du stock de l'utilisateur pour un niveau donné de contraintes, le plus souvent ce niveau étant d'ordre budgétaire.

Nous citerons pour exemple les deux modèles classiques liés, chacun en ce qui les concerne à une structure particulière de la mesure M (s) et du temps moyen de réapprovisionnement T_m.

MODELE 1	*MODELE 2*
Minimiser $\sum_i M(s_i)$	Minimiser $\prod_i M(s_i)$
avec $\sum_i c_i\, s_i \leqslant C$	avec $\sum_i c_i\, s_i \leqslant C$
$s_i \geqslant 0$	$s_i \geqslant 0$

Où s_i désigne l'allocation de s pièces de type i, destinée à satisfaire la demande sur T_m

c_i le coût unitaire d'une pièce i.

Le modèle 1 est à rapprocher du modèle présenté au paragraphe 2 sous l'appellation coût-efficacité opérationnelle: il suffit en effet de prendre pour l'essentialité e_i, la valeur unitaire.

L'analyse de ces modèles montre que les résultats qu'ils fournissent sont pratiquement optimaux quelle que soit la nature de la mesure du niveau, à la condition que les articles considérés aient un coût unitaire et un temps T_m de réapprovisionnement sensiblement identiques.

Par contre, pour des articles à coûts unitaires très différenciés, ces deux modèles favorisent éxagérément les articles à coût unitaire faible. En général, ils ne peuvent être appliqués sans précaution à un ensemble d'articles très dissemblables dans leur coût, leur consommation et leur temps moyen de réapprovisionnement tel que nous l'avons défini au § 1.1.

Dans un effort de synthèse, nous avons essayé de classer les principaux types que l'on peut rencontrer avec la meilleure politique de gestion associée.

(i) *Articles à coût élevé ou à faible consommation*

La meilleure politique de gestion est d'assurer, unité par unité, le remplacement de l'article défaillant sans attendre de nouvelles défaillances pour lancer une "demande" et "envoyer les défaillants en réparation".

La meilleure mesure à employer pour le stock reste le nombre moyen de demandes non satisfaites (ou nombre moyen de ruptures). Le modèle 1 précédemment mentionné peut être utilisé sachant qu'il ne s'adresse qu'à un seul niveau, celui de l'utilisateur. Pour un système plus complet, c'est-à-dire comportant au moins deux niveaux (cf. figure 1), le modèle METRIC et ses dérivés, présentés dans le document référence [6], fournissent d'excellents résultats. Sous une contrainte de budget ou de performances, le modèle METRIC établit le niveau optimal des stocks entre les divers échelons et assure, si nécessaire, la redistribution d'un stock à période fixe ou non. Il faut cependant noter que se référant à un temps moyen T_m tel qu'il a été déjà défini au § 1.1, l'objectif de ce modèle qui est, rappelons-le, que le nombre de demandes pendant T_m ne dépasse pas le niveau du stock (s) présent dans le système, suppose que les utilisateurs ont à leur disposition (s) sur T_m ce qui n'est réellement pas assuré du fait des aléas des temps de réparation, des demandes et du moment de l'observation.

Citons aussi un autre modèle s'appliquant plus particulièrement à un circuit de matériels réparables, basé sur la théorie des files d'attente [7]. Ce modèle permet, pour une probabilité de fonctionnement souhaitée et sous certaines hypothèses, portant sur les lois des défaillances et les lois des temps de réparations, d'optimaliser le nombre (m) de matériels à mettre dans le circuit et le nombre (S) de bancs de réparation à envisager. La solution optimale n'est pas unique puisque (m) est relié à (S), mais il est toujours possible sur la courbe représentative de cette liaison (figure 3) de choisir un couple (m, S) convenable.

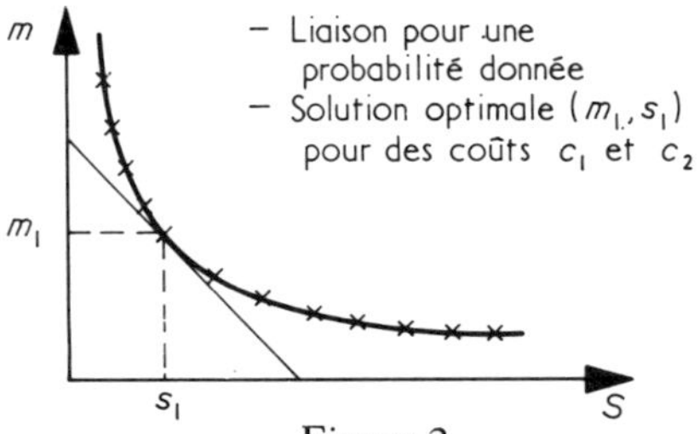

Figure 3

Par exemple si on veut, pour un probabilité donnée, minimiser la relation:

$$m\ c_1 + S\ c_2$$

où c_1 et c_2 sont respectivement les coûts d'un matériel et d'un banc de réparation, il est aisé de trouver le couple optimal (m_1, S_1).

(ii) *Articles à coût faible et à forte consommation*

Par opposition aux articles précédemment pris en compte, nécessitant une gestion de contrôle étroite en raison de leur coût élevé et de leur approvisionnement certainement insuffisant, les articles à faible coût, souvent non réparés n'ont pas besoin d'être approvisionné unité par unité* Dans la plupart des cas, les commandes sont groupées par lots ou à date fixe.

Cette politique de gestion se justifie du point de vue économique puisqu'il est remarqué que ces articles n'interviennent que faiblement dans le coût d'un système; en pratique, on peut observer que 70% environ du coût total actuel est imputable aux articles à coût unitaire élevé alors que ces derniers n'interviennent que pour 20% au plus dans le nombre total des articles alloués.

Le traitement de ces articles ne nécessite pas de prendre en compte l'ensemble du système. Un traitement, échelon par échelon, suffit en regard des faibles coûts budgétaires qu'ils mettent en jeu. Il suffit, en général, d'assurer une simple sommation des besoins par utilsateur et il devient facile de fixer une règle pour le niveau de déchlenchement des commandes. Nous citerons pour exemple le modèle E.O.Q. [8] et les modèles de gestion à calendrier.

3.2 Modèles de Simulation

Les modèles que nous avons précédemment mentionnés ont l'inconvénient de raisonner sur des moyennes, en particulier le temps moyen T_m de réapprovisionnement.

Ils ne fournissent donc pas *la mesure exacte* de l'efficacité. Appliqués à des régimes "temps de paix", donc à des régimes "permanents", intrinséquement de type dynamique, ces modèles ne donnent pas une solution réellement optimale Aussi, on peut constater pour certains problèmes relatifs à des matériels de coût élevé dont les stocks sont constitués de matériels complets au niveau des utilisateurs que les modèles analytiques, type METRIC par exemple, majorent assez considérablement la valeur des stocks.

En annexe A.2, nous présentons les résultats donnés par un modèle de simulation ayant pour mesure de performance le taux:

$$\frac{\text{temps réel de fonctionnement des matériels}}{\text{temps désiré de fonctionnement des matériels}}$$

que l'on peut qualifier d'opérationnel, sans aucune ambiguité. Nous avons tenu à les comparer (figure 4) avec ceux d'un modèle analytique classique ayant pour mesures de performances la sûreté globale (ou taux opérationnel défini au § 1.1.) du stock total mis dans le système. On peut constater l'avantage que procure, pour notre exemple, l'emploi du modèle de simulation plus réaliste mais non résoluble analytiquement.

On peut noter, en particulier, que nous pouvons être très souvent en rupture de stock pourvu qu'il manque peu de matériels et, pendant peu de temps, sans que cela influe sur le bon fonctionnement de l'ensemble.

Il n'est cependant pas dans notre intention de prouver que les modèles de simulation sont une panacée. Nous nous en garderons bien car les contraintes de réalisation

* à l'exception des articles dont l'absence peut provoquer l'arrêt du fonctionnement.

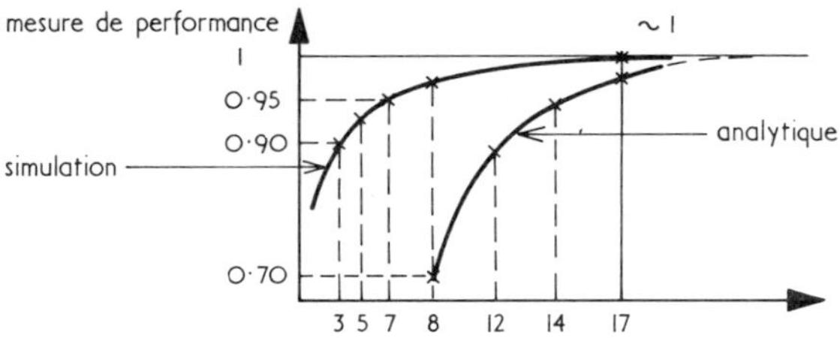

Figure 4

deviennent tout de suite prohibitives dès que le nombre combinatoire des solutions possibles à tester devient trop élevé et c'est ce qui arrive rapidement. Retenons cependant qu'il donne d'excellents résultats dès qu'il peut être utilisé.

4. ADAPTATION REGIME TRANSITOIRE – REGIME PERMANENT

Deux méthodes dissemblables existent pour assurer la transition entre un régime permanent (paix) et un régime transitoire (conflit). La première est de mettre en place un stock, dit de sécurité, dont le but est de s'ajouter aux stocks déjà existants mais dont l'usage n'est permis qu'en temps de conflit.

Le seconde est d'accroître le niveau des approvisionnements en ajoutant au niveau "paix", une quantité telle que, quel que soit le moment initial du "conflit", le niveau immédiatement disponible auprès de l'utilisateur puisse satisfaire la nouvelle demande sur la période T d'isolement prévue.

Ces deux méthodes ont des avantages et des inconvénients. La première a l'inconvénient, en figeant un stock de ne pas tenir compte tout ou partie du vieillissement éventuel des articles, des contraintes d'entretien et de stockage, des moyens réels en personnels etc . . .

La seconde satisfait sans doute les utilisateurs (ou consommateurs) car les niveaux d'approvisionnements sont plus élevés en temps de paix mais ils sont source de gaspillage.

Nous citerons comme exemple d'emploi de la méthode dite du "lot de sécurité", le problème traité par BROOKS and LU [9] relatif au soutien d'avions de combat.

A titre d'exemple, pour la seconde méthode nous indiquons la démarche suivie et les résultats obtenus sur le modèle mentionné au § 3. Il concerne le circuit de matériels réparables représenté schématiquement sur la figure 5.

Le but cherché est d'alimenter le circuit avec un nombre minimum de matériels pour que le stock de l'atelier de fonctionnement puisse satisfaire la demande, avec une probabilité donnée, sur une période T fixée pouvant survenir à un instant non priviligié.

La démarche suivie a été d'assurer une coupure dans l'alimentation de l'atelier de fonctionnement en aval de celui de réparation et de trouver le couple (*m*, *S*) permettant de satisfaire le but fixé.

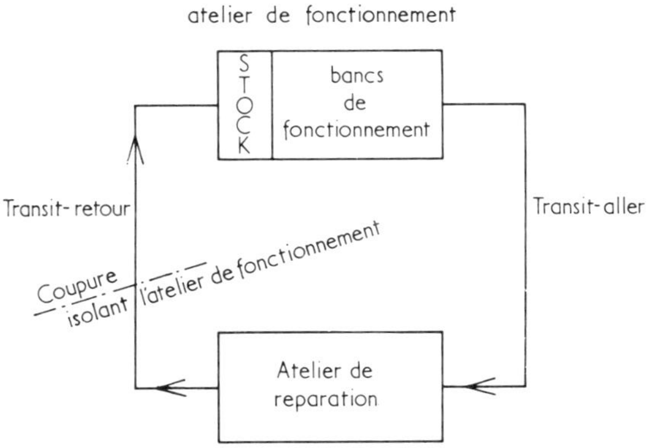

Figure 5

Des résultats obtenus font l'objet du tableau ci-dessous:

Probabilité	m paix	m paix et conflit	s (bancs de reparation) paix	s (bancs de reparation) paix et conflit
0,98	31	37	3	3
0,95	29	35	3	3
0.90	28	34	3	3
0.80	26	32	3	3

Nous constatons l'accroissement non négligeable du nombre de matériels dans le circuit sans pour autant changer la structure même de l'atelier de réparation.

Dans cet exposé, nous avons présenté un certain nombre de techniques de traitement de systèmes de soutien logistique en période de "paix" et de "conflit". Les plus satisfaisants, dans le contexte actuel de budgets décroissants, sont, sans nul doute, ceux à fonction coût-efficacité. Leur généralisation est à souhaiter et ce sera là notre conclusion.

ANNEXE A.1: EXEMPLE DE QUESTIONNAIRE D'ESSENTIALITE MILITAIRE.

L'objectif consiste à classer, selon leur importance opérationnelle, les pièces constitutitives d'un hélicoptère, élément d'un système d'arme.

Le procédé utilise la technique des enquêtes à partir d'un questionnaire d'essentialité établi à plusieurs niveaux selon le schéma ci-dessous:

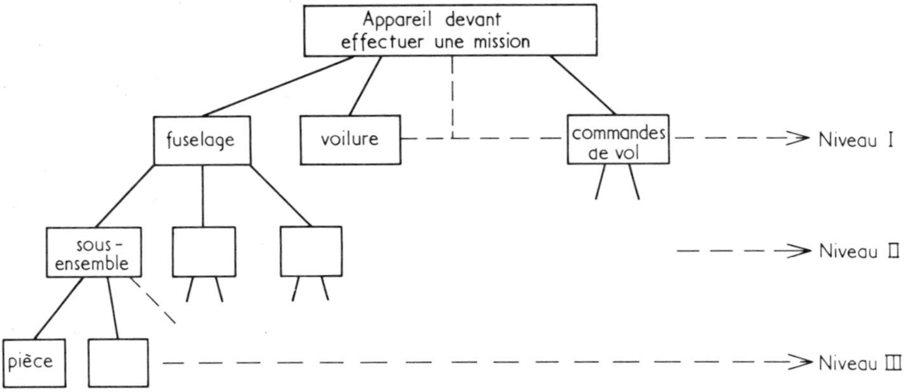

Figure A1

En pratique, le questionnaire est exploité seulement aux niveaux II et III, les ensembles du niveau I étant tous pourvus d'une même essentialité.

Questionnaire mis en place pour les niveaux II et III

– Impact, sur la mission, de la défaillance du sous-ensemble (ou de la pièce consommée):

– la mission est annulée : 1

– la mission peut avoir lieu, mais les performances sont notablement réduites : 2

– la mission a lieu sans réduction notables des performances : 3

– le sous-ensemble (ou la pièce consommée) est-il compensable?

– oui : 1 –
– non : 2

L'exploitation des réponses à ces questionnaires permet de situer en six classes d'essentialité les pièces en fonctionnement qui sont susceptibles d'être allouées en rechanges. Chacune des classes est affectée d'une importance chiffrée, en liaison directe avec les impératifs opérationnels.

Dans l'exemple considéré, le passage de l'essentialité des pièces en fonctionnement à celles des rechanges, est facilité du fait que chaque type de pièces n'est présent que dans une seule classe.

ANNEXE A.2: EXEMPLE DE MODELE DE SIMULATION

1. Description du Système Etudié

Un magasin de stockage et de réparation soutient un certain nombre d'unités utilisatrices.

Lorsqu'un matériel en fonctionnement dans une unité d'utilisation tombe en panne, il est remplacé par un matériel identique du stock de cette unité (s'il est approvisionné) et envoyé au magasin central.

L'arrivée au magasin central déclenche l'envoi d'un matériel de rechange au stock de la station d'utilisation concernée.

Après réparation, le matériel passe au stock central et y reste jusqu'à ce qu'on ait besoin de lui pour satisfaire les demandes des ateliers d'utilisation.

2. Objectif

Le nombre de matériels en fonctionnement varie aux instants τ_i qui sont, soit des pannes alors que les stocks d'utilisation sont vides, soit des retours aux stocks d'utilisation alors que l'atelier est en sous-utilisation:

Soit $t_n = \tau_n - \tau_{n-1}$

Soit $N_i \; (t_n)$ le nombre de matériels en fonctionnement pendant le durée t_n à la station i et D_i le nombre de matériels souhaités.

On définit la mesure liée à notre objectif comme la probabilité (P) de fonctionnement de l'ensemble du système à savoir:

$$P = \frac{\sum_n \left[\sum_i N_i(t_n) \cdot t_n \right]}{\sum_n \left[\sum_i D_i \cdot t_n \right]}$$

et on veut $P > \pi$, niveau désiré.

Cet objectif tient compte de tous les mouvements de matériels. C'est un objectif de "régime permanent".

On considère 3 unités numérotées A, B, C.

3. Valeurs des Parametres

– durée de vie moyenne des matériels : 4 mois

– durée de réparation : 1/4 mois

– durée transit–aller unité utilisatrice ↓ stock central 1/8 mois

– durée transit–retour stock central ↓ unité utilisatrice 1/8 mois

– nombre de matériels souhaités dans les 3 unités utilisatrices:

- unité A : 32 matériels
- unité B : 16 matériels
- unité C : 2 matériels

4. Resultats

Stock / Taux	A	B	C	Stock central	Total niveaux
0,928	2	1	0	0	3
0,944	2	1	0	1	4
0,956	2	1	0	2	5
0,960	3	1	0	2	6
0,977	3	2	0	2	7
0,984	4	2	0	2	8
~ 1	8	4	2	3	17

REFERENCES BIBLIOGRAPHIQUES

1. Brusco, Peter A., and Rosenman, Bernard B., Military Essentiality Coding. A.M.C. Inventory Research Office, Philadelphia, Pennsylvania. A.D. 677-387, July 1968

2. Denicoff, M. et al., The Polaris Military Essentiality System. *Naval Research Logistics Quarterly*, Vol. 11, n° 4, Dec. 1964

3. Denicoff, M., Military Essentiality of naval aviation repair parts. *Management Science* 13, n° 8, April 1967

4. Rougerie, A., Une méthode de calcul initial d'une allocation en rechanges électroniques sous des contraintes opérationnelles: volume disponible de stockage et essentialité. Conférence O.T.A.N. fiabilité, TURIN 30.6/4.7.1969

5. Feeney, G.J., and Sherbrooke, C.C., A system to base stockage of recoverable items. The Rand Corporation, R.M.-4 720-P.R., December 1965

6. Sherbrooke, C.C., Resource allocation techniques for logistics management. The Rand Corporation, P.-4 139, July 1969

7. Cruon, R., Rougerie, A. et van de Casteele, C., Etude d'un circuit de matériels réparables, détermination du nombre de bancs de réparation et du nombre d'équipements de rechanges nécessaires. Centre Interarmées de Recherche Opérationnelle, SAR 4/RT 1, 3/69

8. Lu, J.Y., and Brooks, R.B.S., An aggregate stockage policy for E.O.Q. items at base level. Rant Corporation, 5 678.P.R., June 1968.

9. Brooks, R.B.S., and Lu, J.Y., War reserve spares kits supplemented by normal operating assets. The Rand Corporation, R.M.-5 725.P.R., September 1968

DISCUSSION

Ferrier: Ceci est moins une question qu'une réflexion, susceptible peut-être d'orienter des recherches. L'exposé du Commandant Rougerie nous a fait plonger au coeur du problème de l'essentialité. C'est un concept capital, surtout dans le domaine de la logistique militaire. Malheureusement il est difficile de le cerner et bien plus difficile encore de le structurer par une relation d'ordre totale; et en admettant qu'une telle relation puisse être définie il est infiniment contestable de la transformer en une série de coefficients tout à fait arbitraires.

En 1967, à Athènes, le Professeur Haber, dans sa présentation du modèle POLARIS, s'était référé à un système de hiérarchisation et d'évaluation des essentialités qui répartissait celles-ci en 116 classes de M.E.C. (military essentiality classes). Le Commandant Rougerie a réduit ce nombre à 6 classes. Le nombre minimum compatible avec la survivance du concept est de 2 classes, respectivement E (matériel essentiel) et N.E. (matériel non essentiel). Il nous est arrivé de proposer la définition suivante pour le matériel essentiel, dans la Marine:

"article de matériel qui, s'il vient à faire complètement défaut, entraîne par son absence un dommage vraiment insupportable dans l'optique de la valeur opérationnelle du navire, de sa sécurité, ou du moral de l'équipage."

Cette définition, comme toute autre qui pourrait lui être substituée, a une orientation en grande partie subjective, mais on ne peut échapper à cela.

Si l'on parvient, de cette manière, à répartir les articles concurrents en deux classes, soit E et N.E, on devrait pouvoir aboutir, en traitant séparément les deux catégories, à deux indices d'efficacité globale foncièrement distincts que j'appellerais par exemple, pour la classe E la "sécurité opérationnelle", et pour la classe N.E.* le "confort logistique". A l'intérieur d'une contrainte commune (de budget, d'espace, de poids etc) le choix du décideur consisterait à échanger, "pour le mieux", selon ses propres inclinations, du "confort logistique" contre de la "sécurité opérationnelle", ou en d'autres termes à se placer "au mieux" sur la courbe d'échange correspondante.

* Pour cette classe N.E., les pénuries unitaires seraient avantageusement pondérées par un coefficient qui pourrait être soit le prix unitaire de l'article (indice commercial) soit son poids si la parade à la pénurie est un transport ou moins rapide.

Rougerie: Je suis entièrement d'accord avec M. le Commissaire Général Ferrier en ce qui concerne la difficulté de traitement dès que le nombre de groupes à même essentialité se trouve important. Nous avons été amenés, d'ailleurs, à réduire de 6 à 4 le nombre de groupes, constitué à l'issue du dépouillement des questionnaires. Descendre en-dessous de ce nombre reviendrait, je pense, à tronquer et à enlever toute substance au concept d'essentialité.

Dans la conduite du traitement de l'allocation, il est certain qu'elle doit aboutir, pour une même contrainte, à un choix du "décideur", consistant à se "mettre au mieux" sur les courbes d'échanges relatives aux différents groupes de même essentialité.

Olivier:
1. Quelle charge de travail représente la réalisation du questionnaire au niveau d'un hélicoptère?

2. Pouvez-vous nous donner une répartition en pourcentage des pièces dans le six catégories?

3. Peut-on envisager d'éliminer du système celles des pièces qui n'affectent pas les performances et qui en plus sont compensables?

Rougerie:
1. La réalisation du questionnaire a été facilitée par l'existence d'un catalogue-constructeur assurant une décomposition fonctionnelle de l'hélicoptère en 8 ensembles. La participation d'un "officier de marque" a sérieusement réduit les délais et la charge de travail.

2. Les six groupes de pièces ont été ramenés à 4 groupes lors de l'exploitation du modèle d'allocation. Le groupe, à plus forte essentialité, contenait 53% du total des types de pièces montés sur l'hélicoptère; les 3 autres groupes renfermaient chacun de 15 à 20% du total.

3. Je ne le pense pas, car à la limite, le matériel deviendrait une "caricature" de matériel dont l'absence de confort, et, à la limite, d'esthétisme nuirait au "bon moral" de l'utilisateur.

Morris: How do you take account of a weapon system with different missions for which different parts of the weapon system have different military essentialities?

Rougerie: Le type de questionnaire basé sur la décomposition fonctionnelle et une seule mission doit être modifié. Les différents ensembles du système d'arme se voient affecter plusieurs essentialités différentes dépendantes des différentes missions du système d'arme, auxquelles ils contribuent. L'application d'un facteur de pondération approprié sur les différentes essentialités d'une même ensemble permet alors d'obtenir, par ensemble, une seule essentialité.

O'Connor: The concept of fixed essentiality applied by you to a helicopter assumes that the helicopter itself is essential. This may be true if it is the only helicopter in the operating unit. Is not the marginal essentiality for a component variable dependent on such factors as numbers of alternative equipments available, cannibalisation, etc?

Rougerie: Je pense que la prise en compte de facteurs tels que l'existence et la quantité d'équipements compensables au sens "alternative" s'intègre sans mal dans la structure du questionnaire que nous avons vu. Elle a seulement pour effet d'accroître le nombre de groupes de pièces à même essentialité.

Par contre, pour la cannibalisation, je ne vois pas pourquoi on intégrerait ce facteur dans la mesure de l'essentialité puisqu'il a pour effet, si nécessaire se fait sentir, d'accroître le stock réellement en place, donc la mesure du niveau d'approvisionnement, paramètre lié à la seule demande et, de ce fait, indépendant de l'essentialité.

A Logistics System for Aviation-Electronics Material in a Counter-Insurgency Theatre

Col. James L. Burke
David W. Conrath

Preamble: Reflections on Some Behavioral Aspects of Logistics Systems (Conrath)

These initial remarks will be based on a behavioral science orientation. Initially my training was in economics and as I was oriented toward mathematics it was natural that I turned to operations research. Eventually, however, I became discouraged by the incompleteness in the assumptions upon which many operations research models were built. Hence, my training turned toward psychology and sociology. The persons I met in these fields, I must say, were extremely interesting, and they were great companions at cocktail parties. Nevertheless, I felt my first duty was to return to my former colleagues and attempt to convince them of the efficacy of a social science input into the general field of operations research. Today, then, I hope to expand my missionary activity by a demonstration for my logistics systems colleagues.

Two stories might be told to emphasize my point that models without a knowledge of human behavior may be of very little value. The first concerns a good friend of mine who was asked to devise a production scheduling and inventory model for a large paint manufacturer in North America. He completed his study and presented a model to management that would minimize total costs for expected demand on a yearly basis. Unfortunately, after implementation of the model profits appeared to go down rather than up. Production costs were not reduced significantly, and inventory costs were up due to the addition of a large warehouse to hold the tremendously long runs of white paint that were scheduled. Apparently the purging of the manufacturing process, especially for white paint, involved a considerable amount of down-time, and hence long runs of white paint were desirable in a total system sense.

After a number of re-evaluations of his parameters and a study of the model itself, my friend went to the plant, assured that he was correct. Many interviews later, he and the vice president, who first hired him, toured the plant and spoke with the scheduling supervisor. When asked if he was following the model, the supervisor said: "Yes, almost completely". "Almost", said my friend? "Yes, almost" was the reply. "I followed everything perfectly except for the run of white paint. No one in the world could want so much white paint, so I produced it as before. But I followed the instructions on all of the other pigments." This, of course was the cruel blow. The variation of one parameter made the entire schedule nonoptimal. The net result was that, after much education, the supervisor refused to change his thinking. But he was such a valued employee in all other aspects that he was kept on the job and my friend was released for evermore.

The second story, with perhaps a more relevant moral for this audience, concerns a logistics information system designed for a major command of the U.S. Army. It was a beautiful system from both a hardware and a software standpoint, and all logistics experts marveled at it. Unfortunately, a few significant behavioral facts were ignored. First, the general in command was not really included in the system. Is it surprising, then, that he hardly encouraged its use? Secondly, the input requirements, for the system to work as designed, were so great that most members of the Command would need to spend perhaps one third of their time doing nothing but gathering and inputting data. Thirdly, the output of information, though it was most comprehensive, was in such a form that it could not be readily understood. In fact, it would have taken considerable training just to understand it. Hence, most persons ignored the output and duplicated the system manually and by telephone, the way it was done before. About the only thing the system accomplished at the time was to increase the use of computers–reflecting the ability of computer salesmen–and to increase the employment of logistics systems programmers. While this may be desirable for those involved, it was hardly appropriate from a Command standpoint.

The above provides an introduction to the points I wish to stress with respect to the logistics system that Col. Burke and I have proposed. But here I wish to make a few remarks to provide the context for my following comments. Our study is conceptual– it is a proposal to the U.S. Army and is a long way from implementation. Hence, we cannot speak from experience regarding the system proposed. Hopefully, however, it will be clear from our paper that the system is designed with the knowledge of the problems of most existing logistics systems. Also, the system is as complete as possible. We envisage that it will entail not only the typical elements of production, inventory, maintenance and repair, but at the one end research and development, and at the other the disposal of materiel. Few systems include these extreme points,* but their inclusion may be critical to true system's effectiveness. Research and development should reflect field (operational) usage of equipment and the logistics problems that are encountered in handling it. At the other extreme, the disposal or surveying of equipment ought to reflect everything from that which currently is being developed for field use to a knowledge of what exists in the logistics pipeline. A system could not be called complete otherwise.

We are extremely conscious of aiming the system for the men in the field. This point was stressed by M. Bresson in his paper, and it cannot be stressed too strongly. All too frequently logistics systems reflect the desires of the designers rather than of the users. This must not be allowed to happen.

Most important from my personal standpoint, the logistics system we propose considers the organizational structure that is to operate it as the critical element. By doing so we believe can reduce the estimate of design to implementation made by Major Högberg of ten years, to something less than half of that, by not trying to marry a new system to an existing, and frequently unwilling, organizational structure.

The Bases For a Logistic System

The army has been the slowest of the services to develop "modern" logistics systems, primarily because their materiel requirements did not change radically over time, and partly because their operations remained reasonably similar over the short run. But all

* For example, SPLICE and VESPER as described by Col. Page.

of this has changed. The army of today not only has extremely sophisticated and costly equipment–as evidenced by the particular topic of our paper–that of aviation electronics,* but its operations are likely to be vastly different than they have been in the past. The concept of a mobile force which can conduct combat operations in virtually any theatre of the world means that the logistics support has to accomodate this. In particular, counter-insurgency operations cannot be tied down to a cumbersome logistics support system.

Today logistics systems for the army appear to require five elements: Mobility, Flexibility, Accountability, Maintainability and Integrability. Mobility is required because the locus of operations of field forces may change rapidly. Consistent support must exist no matter how great the change in the nature of field operations might be. Flexibility must be maintained so that the logistics structure can accommodate new equipment and new types of logistics problems that are likely to arise with the rapidly changing technology that we see about us. We can no longer afford the luxury of designing a logistics structure for each new technology that comes along. The system must be flexible enough so that it can adapt to new situations without destroying its capability of meeting the needs of the troops in the field.

Accountability is required because of the great increase in the value (essentially reflected in cost, but also in operational readiness) of the materiel being handled. This point has been stressed by other speakers, so there is little point in expanding it here. Nevertheless, we must have a system that requires a minimum inventory and redundancy, and still meet given requirements of operational readiness. Maintainability alludes to both the needs of the materiel that are being handled by the system, but are in reality a part of the system, and the needs of the system itself. The increased use of module units in electronic equipment has greatly increased their ability to be maintained, and is an example of what we are speaking of in regard to materiel. But the system itself must be easily maintained as an integral unit, or its effectiveness diminishes quickly. This can be insured by the creation of a separate organization which has this as its prime responsibility. We have done this because our past experience has indicated that when a system is under the jurisdiction of several commands, no one command sees it as its responsibility. Either one command always assumes that all of the others are in a better position to maintain the system, or all try to maintain and control it. In either case the system often fails to provide the service intended by its designers.

The final point of integrability is critical in the design of today's systems. Since a great many separate systems are being developed by various elements of the armed forces, each element ought to be designed so that it can be made compatible with all the others. Obviously this requires some basic preplanning so that this is feasible. We, for example, were most concerned with what other elements of the U.S. Army were thinking about doing with respect to their logistics system. Thus, if they developed what they said they would have liked, we could marry our system to theirs. This is not always possible, but if a common language can be developed, if common storage can be used, and the like, one can achieve greater total system efficiency. I realize that some of my colleagues question the desirability of such centralization, and I do not wish to exclude the possibility of decentralized memory or data banks, for example. But the extent to which a system ought to be integrated with another is a separate

* Also the topic of the paper by Mr. Baker and Group Capt. Morris.

research program in itself, and if integration did appear to be desirable, the systems ought to be so designed that the integration can, in fact, take place.

Let me go over the point again of just how important the organization is. First, let us note that the major impact of any organizational structure is the enhanced efficiency to be obtained by cooperative behavior. Thus, the organization has as its basic charge that of coordination. Without coordination each of the parts or elements of the organization tends to go its own way, and there is little chance of anything we might call "system effectiveness". For a logistics system, then, this means that the organization that is monitoring the system must not only be cognizant of the state of maintenance and repair of all items under its authority (such as aviation electronics, as in our paper), but the production and inventorying of all such items as well. As mentioned earlier, we do not intend to use a definition of logistics systems in a context so narrow, for we wish to include the research, development, introduction to field use, and the disposal and surveying of materiel as well. But these many diverse activities cannot be coordinated by as many commands. One command must be responsible for the development and operation of the required information system. This command must also have resupply capabilities, and hence must have transportation facilities available at all times, especially when supporting combat operations. We envisage a command structure that will comprehensively view the entire problem of logistics for the materiel for which it is responsible. Like commands may be structured for other items of materiel, and eventually it may be that these will all come under the authority of a single command structure, so divisionalized and decentralized that the handling of materiel of the armed forces of a nation or group of nations can be handled with true system's efficiency and effectiveness. Note, that while it is most desirable to coordinate these efforts at as high a level as possible, this does not imply that the most efficient thing to do is centralize. It may well be that extensive decentralization is the most effective means for handling the logistics problems of the armed services. But the assumption that this is the case because this is the way it is done is naive to say the least. Perhaps we ought really to be talking about logistics sub-systems for that is the level at which we are still working.

Input Data Reliability

There is one additional topic I wish to discuss briefly, that is the problem of the accuracy of the input into the system. All of us know the expression GIGO–garbage in, garbage out. It is nonsense to talk about a logistics system that is based on unreliable data. Since we are talking about all of the data input coming from the field, its reliability is subject to question.

Five factors may help to ensure data accuracy. First, the information system, from an input standpoint, must be kept as simple as possible. The use of prepunched cards would enhance this, but other factors ought to be considered as well. Perhaps the design of field unit consols ought to be a project by itself, with considerable use of our knowledge about man-machine interfaces.

Secondly, motivation to maintain accuracy is required. Both the field unit commander and the personnel assigned to input the data ought to be convinced, hopefully by demonstration, that their operational readiness is enhanced to the extent that they participate as an effective member of the team, making timely and accurate inputs into the system's date base. The unit commander needs to be convinced of this or he will encourage the establishment of larger than optimal inventories to maximize his own operational readiness as he sees it. He must be convinced that his real inventory lies within the system itself, and this requires transportation facilities that can meet his needs quickly. The man at the machine must have no other duties, and must be held accountable for accurate usage and inventory reporting. He cannot be given other duties and be expected to maintain an accurate accounting. Both he and his commanding officer must be aware of this.

Thirdly, motivation will be enhanced to the extent that transshipment from one unit to another can be minimized, occuring only as a court of last resort. If a commanding officer sees that materiel he believes he needs is always being shipped to other units, he will become convinced that he must disguise some of his inventory to enable him to keep up a "satisfactory" state of readiness. These fears must be alleviated in every way possible. Perhaps we should note here that we do not assume that there is a fixed number of items that must be in inventory at all times. If so, the system would be in continual action, and could oscillate if the information flows faster than the materiel. Rather, inventories would be kept within a range, and only when the range is exceeded or the number of items falls below its lower limit would action be taken. The range might be continually adjusted, but if this resulted in too much transshipment, for example, the inventory policy would be corrected on the basis of experience.

Fourthly, direct reporting of the data to the central memory bank will have the advantage that every command in the chain cannot adjust the data to suit its own needs. This is a necessity. There is always a tendency for each command to think that its needs rate first priority. If they had access to the data flow, lower commands would quickly learn how to adjust the data to give them the highest possible operational readiness–or so they would think. But such behavior would greatly reduce the system's total effectiveness, and in the long run each command would be worse off than they would be if they let the data alone. If they cannot get to the data at all, this is ensured without their having to learn via experience.

Finally random inspection would help to insure that actual inventory levels of materiel are consistent with those kept in the data bank. Updating must be done periodically, and this can only be done by an actual accounting. But such a counting is meaningful only if done when it is not expected; hence the random inspection. Most commands always look great to the inspector when they know he is coming, but what is more important is how they look when he is not around. Operational effectiveness is an every day thing.

In conclusion, we offer nothing new from a logistics systems standpoint. Virtually all of the pieces that we allude to exist, or are at least in a form such that a modification could be made to meet our requirements. What does not yet exist, however, is their integration. This is what must be accomplished to achieve true system's effectiveness, and this is what we propose to do.

1. INTRODUCTION

Any organization that continues to operate with a logistics system designed for one environment, but operating in another, is bound to incur econonomic inefficiencies and operational ineffectiveness. Most of today's military organizations face just such a situation. In an era of "brush fire" wars and insurgency and counter-insurgency efforts, the logistics operations supporting these activities are those designed when armies felt it necessary to take and hold large land masses. They have been altered only by a matter of scale. While the physical aspect of replenishment has been changed in some instances, the system that lies behind it has not. Even the use of computers has been to speed up record keeping rather than to change the basic concepts behind the records. "Systems programming" has usually resulted in little more than new software for more efficient and effective data retrieval.[1]

As the notions of warfare, at least those involved with respect to existing combat conditions, have been altered, they have had two fundamental impacts upon logistics system. Firstly, mobility of field forces has been a keystone. But mobility is greatly limited to the extent that the logistics system constrains it, particularly with the requirement of co-location. Mobile forces require flexible logistics systems. Secondly, the monetary value of field equipment, resulting from the introduction of electronics and advanced weaponry, has risen rapidly. Overstocking rifles and ammunition may not have been relatively expensive, but overstocking electronic direction finding equipment is. Hence, two prime requisites of today's logistics support are that it must enhance rather than detract from mobility and it must be accurate in its knowledge of what materiel is where.

1.1 Rational for the Proposed Aviation-Electronics Logistics System

This paper is essentially a proposal to establish a logistics system suitable to accomplish the above objectives for a specific Army requirement, that of handling aviation-electronics materiel. While the system suggested is structured to meet a given need, the concepts incorporated are sufficiently general to be applicable to a wide variety of military, and perhaps even non-military, applications. Furthermore, the prototype described is for a purpose that is critically involved in the requirements of modern warfare; it is structured for the expensive electronics systems required by highly mobile helicopter squadrons.

The specific proposal resulted from the experience of the United States Army in Vietnam. Two factors of both political and military importance have been instrumental in the development of the logistics system. The first is the existing high ratio of support to combat personnel currently stationed in Vietnam. This has impeded military operations, and the size of the total force has had negative international relations repercussions. The second is the lack of inventory control that currently exists, with its resultant waste in resources and lowering of combat effectiveness. All too frequently combat units have been operating at a level considerably below that for which they were designed because of a shortage of electronic equipment, when it was in surplus elsewhere within the overall comman structure. Given the fear that re-supply

[1] Initial efforts at a systems approach are usually involved with such problems. For examples seè Johnson, R. A., Kast, F. E. and Rosenzweig, J. E.: "The Theory and Management of Systems" (2nd ed.). New York: McGraw-Hill, 1967, pp. 243-245. Nevertheless, the potential benefits from computer oriented systems can be great; for examples see Diebold, J.; "Bad Decisions on Computer Use," *Harvard Business Review, 47*, Jan.-Feb. 1969, pp. 14ff.

could not be counted upon, combat units have the tendency to understate their actual inventory so that they have a reserve "just in case". Such a situation can be avoided only if field forces have confidence in the re-supply system, and its ability to meet their needs.

Aviation-electronics present logistics problems distinct from those of traditional Army materiel. These aircraft oriented systems have many common subsystems including modules. Materiel management must be done at *each* subsystem or module level if operational efficiency and effectiveness are to be maintained. In some cases, total systems differ by just a few subsystems so that interchangeability of components and modules between total systems is not only possible, but highly desirable. Each subsystem, equipment, or module, once removed from the aircraft weapons system for maintenance, enjoys an individuality of its own. The complexity of such a hierarchical materiel system virtually requires reliance on a computer oriented logistics system if the goals of reducing inventory, minimizing air craft downtime, minimizing on-site and incountry maintenance, and maximizing the responsiveness of the aviation-electronics organizations to the aircraft weapon system as a complete entity, are to be met. A well designed unitary organization under the authority and responsibility of a single commander could insure that these apparently inconsistent goals are not dealt with independently of each other.

1.2 Design Criteria

Our intention is to present a logistics concept not only broad in its own capabilities, but one which is also capable of being integrated with other similar logistic systems. The system is designed to take advantage of the latest developments in information systems and computer technology, especially as they relate to logistics efforts. Furthermore, the system is developed with knowledge of the work being done in organizational theory, so that the common problems of man-machine interface and organizational structure are minimized.

Computer oriented systems have demonstrated the ability to permit improved distribution of supply items among active Army elements. The use of such a system makes possible the operation of a centralized support organization with uniformity in supply planning and distribution, and provides improvement in the training and utilization of personnel. If given "real time" capability, it also provides complete visibility of all items, highly desirable in an environment of rapidly changing demands. In addition, improved logistics information ought to be used to assist in focusing research and development activities on the needs of the field operating units, and this aspect should not be ignored.

The concept proposed takes the process that makes up a logistics system and designs a management system around this process, an organization totally oriented to providing logistic support of materiel for a specified force on a theatre or world-wide basis. The concept is tied to user needs. It reflects not what the technical computer and logistics experts might want to deliver, but what the military functional specialist in logistics supply, transportation and maintenance need.

The complexity of modern day support suggests that the needs of the user will be best served by an integration of the functional specialists within a single support organization that has direct access to the most vital logistic information. This indicates that the creation of an integrated aviation-electronic logistics battalion could provide visibility of assets including: quantity, condition, authorized stock levels, location (including quantity and condition data), and items in transit. An aviation-electronics logistics specialist, provided with a single command structure, should be able, upon direction, to redistribute assets anywhere within his command on a priority basis within a very short time frame.

2. ORGANIZATIONAL STRUCTURE

The logistics system proposed suggests the establishment of an aviation-electronics (A-E) logistics battalion to supply and maintain all aviation-electronics materiel required by a field force, such as that operating today in Vietnam. The battalion would receive materiel from existing depots, deliver it to its organic divisional and non-divisional direct support units (DSUs) and perform all maintenance and replenishment functions. The unit would be responsible for predicting usage, requisitioning, distributing, and inventorying all end items, components and repair parts for elements of the force.

The battalion would consist of a specified number of companies for regional logistical administration (if required) and a number of direct support units, normally one per aviation company supported. Regional companies will be established whenever the span of control of the battalion commander becomes too large,[2] or when it appears highly desirable to have an administrative activity in the proximity of combat operations. The direct support units (which are now in operation) perform a supply and maintenance function. In addition, they provide the real-time input to the several

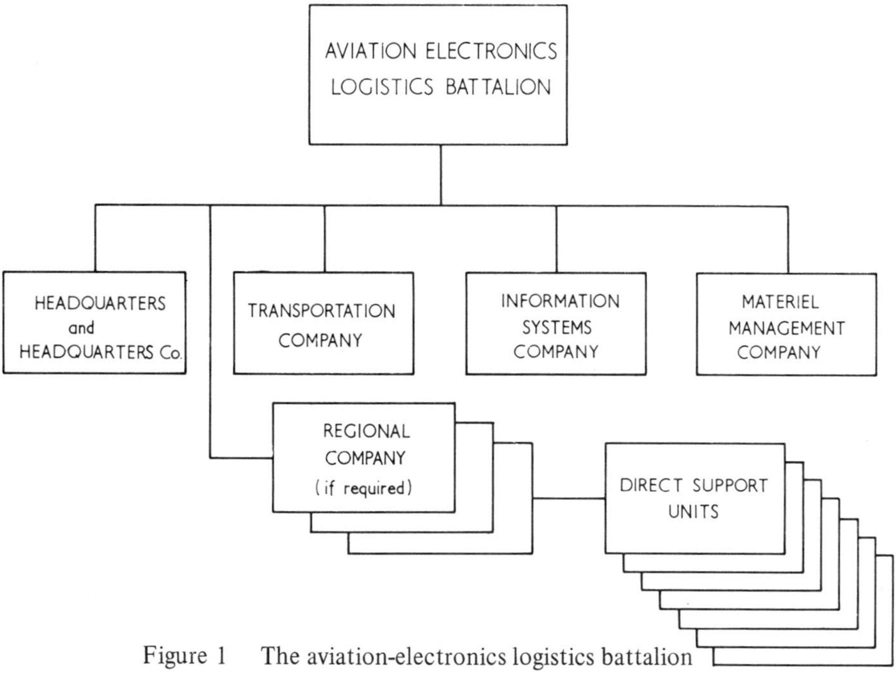

Figure 1 The aviation-electronics logistics battalion

logistics and inventory models maintained by the battalion. Predictive demands for the entire force, including shipping schedules, are prepared and transmitted by the battalion to a National Inventory Control Point (NICP) and/or to units within its command which maintain inventories. Custody of shipments is accepted at the depot or other source and delivery to the DSU is a function and responsibility of the battalion. Similarly, custody of reparable equipment and modules is exercised in the return cycle. This assures physical control and minimization of loss, a problem that plagues most commands.

To accomplish its task the aviation-electronics battalion would require a transportation company—to supervise and operate the means for the physical movement of materiel, a materiel management company—to conduct and control materiel maintenance and repair, and an information systems company—to operate the communication and data processing system necessary to achieve true systems effectiveness. Such a proposed organizational structure is shown in figure 1.

2.1 Aviation-Electronics Battalion

The battalion is to be a single operating command structure reporting directly to the US Army Electronics Command (ECOM), the command with direct responsibility for aviation electronics materiel. Broadly, the A-E battalion inventory management system can be divided into three subsystems: an information system; a transportation system; and a materiel management system.

The *information system* is required to maintain a complete file on the state and location of all aviation-electronics materiel under its cognizance. The information system will also retain the capability of real-time updating so that the state of materiel readiness can be known at any moment in time. Necessary physical verification procedures based on randon inspection of all inventory points will be required to ensure that the data on file reflects the actual inventory.

The information systems company will also have a programming capability. Programming will be required to develop procedures to ensure that demand information can be effectively related to inventory requirements. For example, changes in field requirements ought to be immediately reflected in the inventory levels necessary to ensure maximum operational readiness of field units. Also, programs will need to be devised that can integrate research and development efforts, production schedules, and repair and maintenance operations with the experience of combat units that must use the materiel.

We expect that, with only slight modifications, a central (Army wide) data bank (such as NICP data bank) can provide most of the data necessary for decision making at the battalion level. This would minimize the memory bank facilities required by the battalion. Further specifics of the information system are noted later.

The second component, the *transportation system*, is designed to deliver the require-

[2] For an example of a formalized means of determining the limits of span of control see Stieglitz, H.: "Optimizing Span of Control," *Management Record, 24*, 1962, pp. 25-29. Also see Udell, J. G.: "An Empirical Test of Hypotheses Relating to Span of Control," *Administrative Science Quarterly, 12*, 1967, pp. 420-439.

ments of the division and non-division support units. The A-E battalion will retain the necessary aircraft or airlift allocation to ensure materiel re-supply and to return that which needs maintenance and repair to permanent facilities not in the theatre of operations. Aircraft indigenous to the transportation unit are most desirable, and space available on a regular (daily) basis on other aircraft outside the command would be required. The aviation-electronics battalion should have the capability of 24-hour, or at most, 48-hour delivery to any point in the world. This would allow the integrated direct support units to maintain the minimum inventory possible consistent with their combat readiness posture. The transportation unit would also retain escort capabilities to ensure that each item delivered arrives at its destination in a readiness condition.

The third component, the *materiel management system*, involves maintenance capabilities. Stateside maintenance and repair will be required for aviation-electronics materiel that is being returned by the DSU's. Such activity will be done in facilities, or under contracts, administered by the Commanding General, US Army Electronics Command. The information system would have the necessary programming to be able to determine whether and when it is economic to repair items or to request that ECOM purchase new equipment. The maintenance unit would also have the capability of quality assessment of maintenance conducted within the battalion and at the direct support level.

The A-E battalion is to be based in CONUS (continental United States) or some off-shore area not related to a counter-insurgency theatre. Only regional companies, when required, and the DSU's will be stationed in combat zones. This is essential, for one of the purposes of the entire logistics system is to minimize the number of personnel that are not in a fighting status in the theatre of combat operations.

At his direction, the battalion commander may detach certain of his units if necessary to enhance operational efficiency. It may be desirable, for example, to have the transportation units located at one of the three major points of debarkation in CONUS, or co-located with field units, to expedite materiel support. Also regional companies might be required to coordinate the activities of the direct support units from a logistics standpoint. This could arise, for example, when the locus of combat operations changes rapidly. This is likely to result in rapidly changing inventory levels for each of the DSU's. These changes in demand would probably best be met by the transshipment of materiel from one DSU to another. It would probably be easier for a command located in the proximity of the counter-insurgency operations to coordinate and conduct these transshipments than it would be for the battalion itself.

2.2 Direct Support Unit

The direct support unit (DSU) is attached to a tactical operational aviation company. The unit consists of three elements: an inventory analysis and control section; a subsystem (black-box) repair section; and an aircraft system testing and repair section. Each DSU maintains its own inventory of repair and replacement equipment and parts.

The inventory analysis and control section is the major element of the logistics system in terms of the DSU's relationship with the A-E battalion. The prime function served will be that of inventory control and the management of mission stock. This section will maintain the inventory of aviation-electronics equipment consistent with the mission of the non-divisional aviation company or divisional support command to which it is attached. The inventory will be that pre-specified by ECOM, modified by recent demand data as processed by the A-E battalion, reflecting local combat conditions, repair data, and other programmed support factors. The unit will also provide the data source for all logistics requirements. All data concerning the replacement, repair, loss, return and stocking of aviation-electronics materiel will emanate from the DSU via a remote computer terminal in direct communication with the A-E battalion.

The other two elements of the DSU are reasonably self explanatory. The subsystem repair section is to repair those modules that require simple adjustments or sub-module replacement. All repairs of a complex nature that would require difficult to maintain test or repair equipment, or scarce highly-skilled personnel, will be returned to a maintenance operation outside of the combat theatre. The aircraft system testing and repair section will inspect aviation-electronics systems as a whole. Its personnel will be responsible for the testing, removal and replacement of malfunctioning equipment in the aircraft.

The organization of a typical direct support platoon is shown in Figure 2.

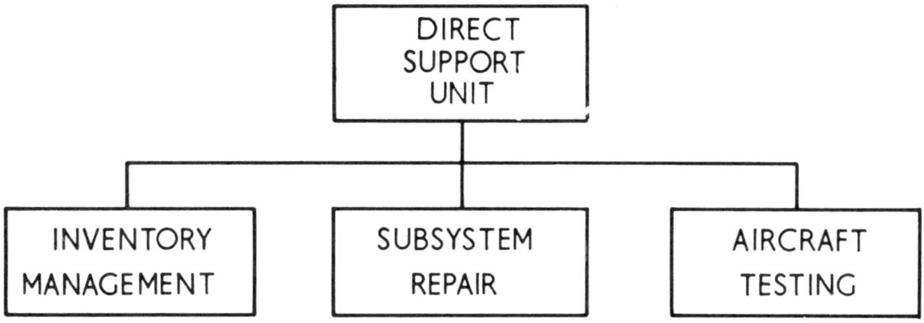

Figure 2 Organisation of a direct support unit

2.3 Logistics System

To obtain a better understanding of the logistics system proposed, the flow and custody of A-E materiel will be described. A detailing of the accompanying information system is contained in the next section.

Each DSU maintains an inventory of all A-E items deemed necessary to the success of the mission of the force being supported. The content and the level of the inventory is established by the Commanding General of the Electronics Command, but this is modified by the computer programmed inventory model reflecting current usage and combat conditions. Each piece of equipment that is lost, surveyed, or unserviceable is accounted for by a "change in status report' submitted to the A-E battalion by the DSU. These change in status reports make up the data base that is used to maintain appropriate inventory levels in each of the attached direct support units. The only other vital information coming into the battalion headquarters is the combat situation report submitted by the Field Force Commander, which is expected to include priority categories for the various field units. From these reports, together with the change in status reports, desired inventory levels are determined, and shortages and surpluses are calculated, both for specific DSUs and for aviation-electronics in general.

As change in status reports are received at the battalion, the information that is relevant to the central data banks is forwarded to them. Such data are now available for recall. That information unique to decision making at the battalion level is put in the battalion's memory bank for later use. The bulk of the storage presumably can be accomplished by central data files such as that maintained by the NICP.

As new data are communicated to the computer, it will continually ask itself to determine the appropriate inventory level for the item and unit in question. The computer will then determine, by checking with its memory file, whether or not the inventory now held by the unit is adequate. If so, no action is taken other than to record the fact that a surplus exists, if that is the case. If not, another program is activated to determine the "best" source of supply for the needed item.[3] Usually this will be one of the CONUS depots. If the equipment in question is not in a readily available inventory, its manufacturer will be queried (via ECON P&P) about his available supply not yet delivered to the Army. Another alternative source remains, that is the possible surpluses that might exist in other DSU's. If they exist, transshipment among the DSU's may be ordered. If an item is in overall short supply, the "desired inventory level" for each of the DSU's will be adjusted downward to reflect the general shortage, but this will be done only as long as the "shortage" exists.

The transportation company attached to the battalion will usually pick up the items to be shipped to the DSU and escort them to their destination. If a local transshipment is required, however, it may be more appropriate to utilize local transportation facilities (such as indigenous aircraft). If regional companies exist, they may conduct such trans-shipments.

The transportation company will also be called upon to return unserviceables to a non-combat zone maintenance and repair facility (NMP). The battalion information systems company will have the programming capability to determine an optimal repair. policy, including such factors as the location of maintenance, the optimal lot size for maintenance, and the desirability of maintenance and repair given the cost of purchasing the item, demand considerations, and the advent of more modern equipment to replace it. This information will be transmitted to the Commanding General of ECON for action.

The aviation-electronics logistics battalion has actual physical custody of materiel only when it is in transit. On the other hand, it has knowledge of the whereabouts of all A-E materiel from the point of production to the point of being surveyed. It is the battalion's ability to comprehend and process, via its software, the entire picture that

[3] For many items the "order point" will be significantly below the "desired level"—the base for determining the existence of a surplus. Such a gap, or "slack," will reduce the level of activity that might be required to react to each unit level of change. The establishment of these levels will depend upon the criticality of the item, its expected usage rate, and other aspects of the relative costs of combat theatre inventory vs. inventory in CONUS and elsewhere.

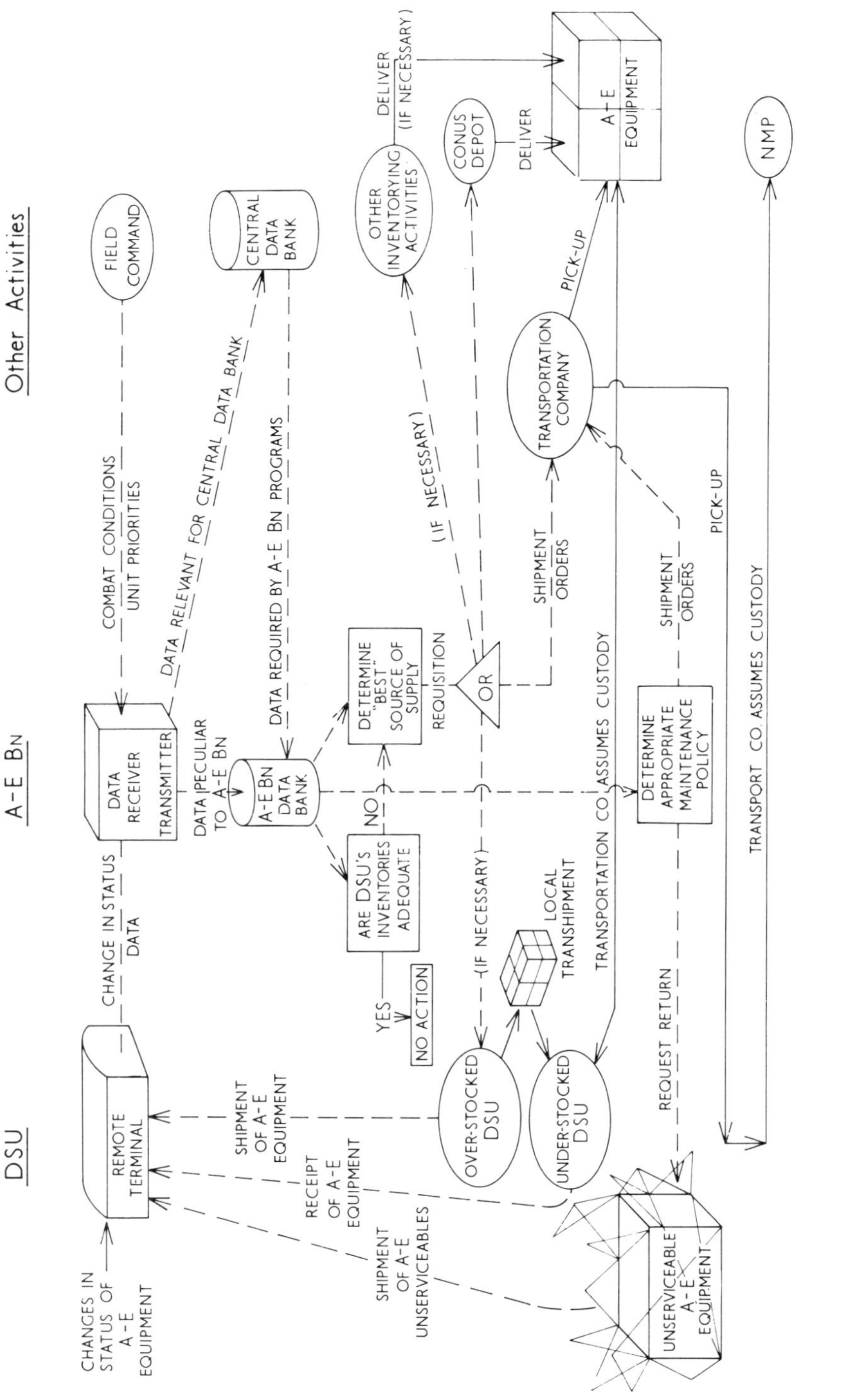

Figure 3 A–E logistics system

will greatly aid in the efficiency and effectiveness of aviation-electronics logistics. For a schematic of the system in its entirety see Figure 3.

3. THE INFORMATION SYSTEM

The development of a sophisticated data/information system is absolutely essential to the success of the mission of the A-E battalion. Such a system will require indigenous data processing and computer programming capabilities, with the necessary support personnel to sustain and operate them.

The core of the information system will be the several data banks to which the battalion will have access. Hopefully the various data banks, such as that retained by the NICP, will be compatible with the data processing equipment of the A-E battalion so that duplication of data will not be necessary. To the extent that the compatability is lacking, data will be retained in the memory banks of the logistics battalion.

The essential requirement for the information system is the accurate direct reporting of materiel changes of any kind by the DSU to the battalion. This will allow real-time information to be available to the battalion and will greatly enhance materiel readiness. Furthermore, the very fact that data will be transmitted directly to the memory bank will eliminate the considerable bias that currently exists via the modification of data at each command level. In addition, it must be demonstrated to the field commands that accurate and immediate reporting enhances the readiness capabilities of the DSU, by allowing more immediate response to its needs.

Proper motivation to encourage accurate reporting is a necessity. Furthermore, it is suggested that the DSU *not* be informed of its optimal inventory level as determined by the A-E battalion. This will help forestall attempts to maintain such a level. Nevertheless, verification of the DSU inventories will be made at random intervals to ensure that accuracy is maintained.

The reporting of all materiel changes (except local repairs accomplished by the DSU) will be monitored by each higher command level so that each command can pull off that data needed for its own decision making requirements. Modification of the data, however, cannot be accomplished by the monitoring activity. It can only be done by submitting exception data to the core memory via the A-E battalion commander. This is to minimize the possibility that each command level might introduce corrections it considers necessary for its operational effectiveness. For a system to operate as a complete system, unauthorized modifications must not be allowed. Individual adjustments for subsystem reasons will virtually always reduce the level of efficiency for the system as a whole[4] This must be fully appreciated.

3.1 Direct Support Unit System Components

The DSU will require a remote computer terminal. The specific equipment used will depend upon the equipment maintained at the battalion, ECOM, and central data bank

[4] The basis for this conclusion can be found in Lipsey, R. G. and Lancaster, K.; "The General Theory of Second Best," *Review of Economic Studies, 24*, 1956, pp. 11-32.

levels. The system at the DSU level is to be as simple as possible. This will involve sets of prepunched cards to reflect every item maintained in inventory by the DSU. The cards should be impervious to weather changes and to moderate handling. The sets of cards will not only contain the identification number (FSN) of each item or piece part of aviation-electronics materiel, but other sets of cards will indicate all possible statuses of equipment, cardinal numbers, unit location, cause of failure if determinable, and the like. This will allow the DSU to make all required change in status reports. Although the DSU will have the capability of making exception reports verbally, these are to be kept at an absolute minimum. None will be forwarded past the A-E battalion. All data not required by the central memory cores will be placed in the computer data bank of the battalion.

The remote terminal will also provide the means whereby the A-E battalion will communicate with the DSU. In particular, requests for more information will be transmitted via the terminal, and notification of shipments between DSUs are required, these requests too will be made via the terminal and authenticated in an appropriate manner. Such requests, however, ought to be kept to a minimum to prevent the negative motivational impact that is likely to result.

In addition to the typical complement of trained electronics technicians, the DSU will require the services of personnel who are especially trained to operate the remote computer terminal. To ensure the reliability of the performance of their tasks, they should have no other responsibilities, and the importance of their work must be impressed upon their commanding officer. They will be evaluated on the accuracy and timeliness of their data transmissions.

3.2 Aviation-Electronics Logistics Battalion System Components

The A-E battalion will require computer memory, data processing and programming capabilities. It will also possess the ability to call desired data from the various central memory banks for needed logistics calculations. The size of computer operations will be such as to require several skilled logistics programmers and several skilled computer operators. It would also be desirable if the A-E battalion commander had sufficient knowledge of computer information systems to be "snow proof".[5]

The A-E battalion, in conjunction with its data processing capability, will be able to perform the following functions: establish optimal inventories consistent with predicted operational usage, repair and other supply information, and capital and transportation constraints; compare optimal with actual inventories for all DSU's; determine "best" supply sources to meet DSU shortages; establish optimal transportation of aviation-electronic materiel consistent with transportation constraints; prepare predated requirements data for supply and maintenance scheduling by ECOM

[5] Many would say that this is an understatement, for example, Ackoff, R. L., "Management Misinformation Systems," *Management Science, 14*, 1967, pp. B-147-156, and Stern, H., "Information Systems in Management Science," *Management Science, 16*, 1969, pp. B-144-147. Also, it has been suggested that the manager not only understand the system, but actively participate in its design as well, Feigenbaum, D. S., "The Engineering and Management of an Effective System," *Management Science, 14*, 1968, pp. B-721-730.

consistent with desired inventory posture; establish optimal scheduling of maintenance; and develop the data base required for most effective logistics decision making.

The A-E battalion will be tied directly into the counterpart Army Materiel Command NICP such that optimal decisions for purchasing, for research and development effort and the like can be made without requiring separate computer facilities. This will require very close coordination, but will enable the NICP to make decisions with respect to aviation-electronics that reflect the most efficient and effective means for accomplishing a mission for the command as a whole. For example, decisions on whether and when to schedule depot or contract overhaul of a particular A-E item ought to be made with due regard to possible changes in the equipment in the near future, the price of new pieces of equipment, the relative time lags required by overhaul and production facilities, and the expected demand requirements by the field commands. Only when decisions can be made that involve all such considerations can true systems effectiveness and efficiency be achieved.

Another example can be related to R&D activities. While a piece of aviation-electronics materiel might not be meeting the desired specifications with respect to the mean time between repairs, the low cost of undertaking repairs in terms of time and dollars may make further R&D efforts to increase the MTR of much less value than equivalent research spent on other items of aviation-electronics. In essence, to achieve true systems capabilities, the A-E battalion has to be an integral part of the ECON command structure.

3.3 Programming Requirements

The proposed system will require a number of computer programs in the area of logistics. These programs will use a common data base. Therefore it is essential that the development of the programs and the data base be done in conjunction with one another. To the extent possible, standard inventory and logistics programs, especially those that have been tested and have been found to be reliable, are to be used.[6] Nevertheless, the requirements of the A-E battalion are such that considerable programming will need to be done within the battalion to reflect the unique aspects of counter-in-surgency support.

Stock programs should be made available for determining desired inventory levels for ECOM according to pre-specified demand requirements based on static models. These will be the basis fromwhich inventory levels at both the A-E battalion and the DSU levels will be modified in response to current demand data.

Inventory programs should be made available that not only include static responses to changes in demand, but dynamic responses as well. Demand data should not only be evaluated in terms of changes in the flows, but changes in the rate of flows. These second order changes can be used to make more accurate predictions of demand. The potential for counter-insurgency activities to vary rapidly over time requires that the logistics system be effectively responsive to these changes.

Programs will have to be developed to make maximum use of the transportation

[6] IBM's IMPACT package of programs might provide one such source.

facilities made available to the command. Such programs also ought to be of a dynamic nature. It is likely that availability of transportation will not be constant. Hence, at any moment in time the A-E Battalion should be capable of determining the optimum use of existing transportation facilities to perform its mission. Such programs will have to incorporate not only aircraft availability, but location of aircraft, location of available inventories for shipment, and available escorts.

Programs ought to be developed for the optimal scheduling of Stateside or off-shore maintenance facilities (NMPs). These programs will take expected demand, available inventory, cost and lead time of repair and overhaul, and cost and lead time of purchase into consideration. In addition, such programs should determine the economic lot size for either maintenance or purchase consistent with forecast future demand.

The above requirements suggest that two sets of programs ought to exist. One set will contain econometric type forecasting models that allow for second order characteristics of demand. The second set will incorporate given demand data, and will be consistent with reasonably standard inventory and logistics (in terms of transportation) models. Programs of both types currently exist. Their appropriate integration, however, has not yet been accomplished.

3.4 Data Requirements

The above programs will require a substantial data base. As has been noted, much of this can be obtained from an operational central data bank system. It is assumed that such a system will exist. If it does not, adequate memory facilities will have to be created.

All changes of status, whether into or out of inventory, will need to be reported. Status will involve both the physical status of aviation-electronics materiel and its location. Changes will be reported by FSN, number of items so changed, the old status of the item and the new status, elapsed time in the old status if relevant, reason for status change, aviation company or other command units involved in old and new statuses, and specifics of the change if requested by the A-E battalion.

Periodic inventory reports will be made by unit, status, FSN, number of items, and exceptions, if required. These reports will be correlated with the centralized memory and discrepancies will be subject to investigation as to cause.

The A-E battalion will maintain demand data on all items, cost per item, lead times for repair, overhaul and production, quantities available at every command location, and all data on transportation capabilities.

The asset posture indicated by the above data will not only apply to aviation-electronics materiel, but to all test and maintenance equipment assigned to the battalion and its subordinate activities as well.

A schematic of the data flow as described is presented in Figure 4.

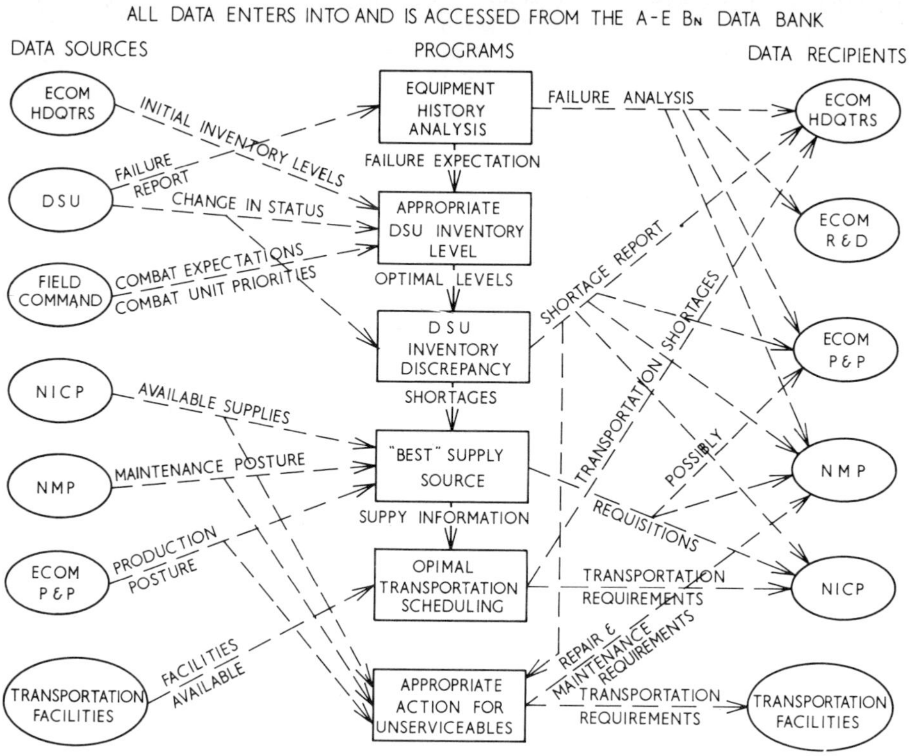

Figure 4 A–E battalion data flow

4. CONCLUSION

The most obvious conclusion is that much remains to be done. The system as described is just a skeleton, and before it could be put into operation it must be filed out. This requires that extensive programming be undertaken, that the system be debugged, that the necessary command support be obtained, and that it then be implemented. In addition, a new organization needs to be created, and appropriate means for motivation must be devised for the system to be put into operational readiness. The aspect of motivation for accurate data reporting cannot be overstated. Many a system that is technically beautiful has failed because of inadequate data input.

Assuming that the system in its entirety can be developed, we believe that there is little doubt that the two objectives upon which we based this proposal would be accomplished. A major proportion of the support personnel now required to be in or adjacent to the combat zone could be removed elsewhere, hence enhancing the mobility of the fighting units. Moreover, operational readiness would be significantly increased by better allocation of existing supplies to the units of "greatest need." In addition, supplies would be made available on the basis of a greater awareness of actual need rather than on preplanned programs, many of which are soon out of date in an ever changing combat environment.

ACKNOWLEDGMENTS

The authors wish to thank Professor Edward Silver for his helpful criticism of an earlier draft, and Mr. Theodore Conti for his many contributions as our work developed. Needless to say, the authors take sole responsibility for this final version.

DISCUSSION

Morris: You have stated a high standard of data input and accuracy of reporting. Is this attainable without serial numbering of stock items, especially with reference to modification states?

If serial numbering becomes necessary have you considered the data processing implications?

Conrath: I do think the system can be implemented without serial numbering, and that stock number and number of items on hand at each inventory point will be sufficient. The cost of running a serial-numbered system is admittedly too great, the data-processing requirements would be tremendous, that is why I would propose nothing more sophisticated than a stock numbering scheme.

Influence of Reliability on the Organisation of a Complex Electronic System of Logistic Support

Mrs. G. Piu

1. FOREWORD

The purpose of maintenance activities is to insure correct operation of the equipment within their performance parameters.

Efficient maintenance is even more an essential requirement in a missile test range where the equipment must operate with utmost reliability to allow achievement of mission goals and collection of required mission data, and to insure launch safety.

Data gathering and recording instrumentation must provide the data to be utilized after completion of the mission for evaluation of the missile behavior. The Range is used for missile and rockets qualification launches and the tracking equipment (cinetheodolites and radars) supplies the data which is subsequently utilized for trajectory, velocity and acceleration data reduction, and for determining the miss-distance.

Launch safety must be insured to prevent damage to operating people and to inhabited areas.

It is therefore necessary that the equipment provides the safety officer with the data which allows him to assess the behaviour of the missile in case of danger, determine to destroy it.

The requirements outlined above as well as cost considerations both as regards the equipment and the manpower, point out the importance of the Range instrumentation reliability and the need for an adequate maintenance organization and maintenance quality assurance program.

The effectiveness of a maintenance program cannot be evaluated by precise criteria since there are several parameters which cannot be easily assessed.

The approach in the case under analysis, has been to check that the range system characteristic parameters (in this specific case, the reliability and availability parameters) have kept within standards limits and have not been degraded by maintenance performed.

Experience has shown, in fact, that the best way to face the problem is to set up a logistic organization and subsequently adapt it to the requirements of a given weapon system until it becomes satisfactory.

2. PREVENTIVE MAINTENANCE PROGRAM

The Range consists of 13 sites which are linked one to the other and each to the Control Centre through a communications system. The instrumentation constitutes two tracking chains, one for the missile and one for the target.

A typical site includes one tracking radar, one coordinate converter, one cinetheodolite, data transmitting and receiving equipment, and other auxiliary equipment.

The number and variety of the equipment installed on each site has necessitated the organization of scheduled maintenance cycles and the establishment of a section responsible for collecting the documentation and analysing the actions taken by maintenance people.

Clearly, this program is concerned with preventive maintenance since off-schedule maintenance is not acceptable for the reasons we are going to discuss.

The Range activity is not always sceduled in a detailed and definite way, therefore, the equipment must be kept at a level of efficiency such as to allow it to be used at practically any time.

Furthermore, the equipment installed on the sites is diverse in nature: the complexity of both equipment and circuitry do not allow troubleshooting and corrective actions to be performed in a short time.

It is further to be noted that some of the sites are located at great distance from the place where maintenance people collect. Consequently, an off-schedule repair would be performed with a not acceptable delay considering that the average duration of a mission does not exceed one hour.

Preventive maintenance has been scheduled to meet Range operation requirements, and fit the specific characteristics of the equipment to be maintained. Information collected through failure reports on the behaviour of the equipment has been of great help, not only in evaluating the skill and efficiency of maintenance people, but in adjusting preventive maintenance schedules to suit the actual requirements of the equipment installed, with the ultimate purpose of getting the best possible results from both technical and cost viewpoints.

As a matter of fact, MTBF information has made it possible to know the number of random failures and hence the number of repair actions required; information on physical characteristics and aging of components has allowed a check of maintenance rate to ascertain that their periodicity was adequate to prevent degradation of equipment performance.

All these data and the estimation of the average time required for troubleshooting and repair have made it possible to establish maintenance schedules and manpower necessary to meet the actual Range requirements.

3. DATA COLLECTION ORGANIZATION

Failure repair actions were initially recorded on maintenance logs for each site. In a short time, however, this reporting system proved inadequate, because:

- the information recorded was in general approximate and incomplete,
- it was extremely difficult, for the above reason, for those who analysed the reports to judge on adequacy of the repair action,
- it was not possible to collect data in an organic and uniform manner.

In fact, some reports indicated as a breakdown what was merely to be considered a minor malfunction (as it did not make the equipment unserviceable or exceed tolerance limits). It was necessary to establish for each type of equipment what was to be considered a real breakdown and set up well-defined test procedures in order to avoid subjective judgements by maintenance people.

The setting up of proper failure report forms was required to assure that collected information be homogeneous and complete, and at the same time, to simplify paperwork for maintenance people. An example of failure report form is shown in Fig. 1 where the following information is recorded:

- equipment failed
- subunit, model, serial number
- symptom of failure
- description of failure

Recto

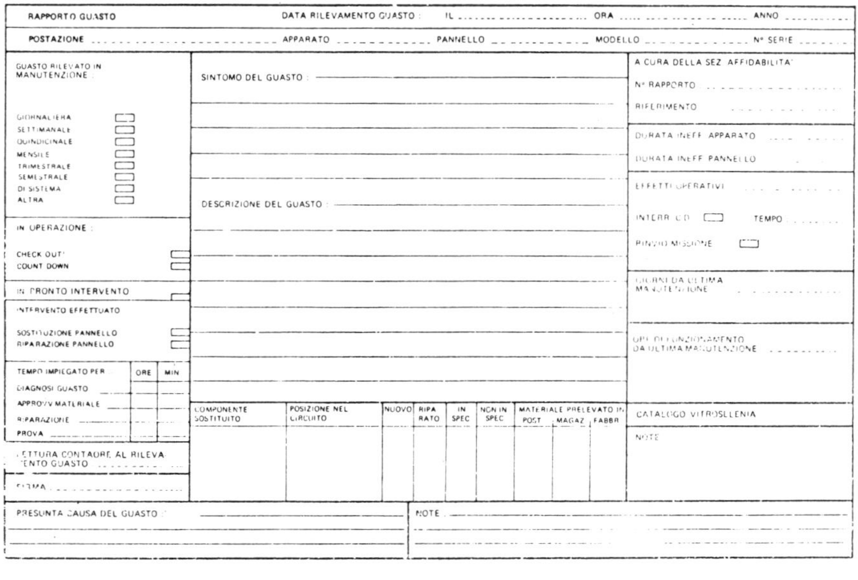

Figure 1 Failure report form

Verso

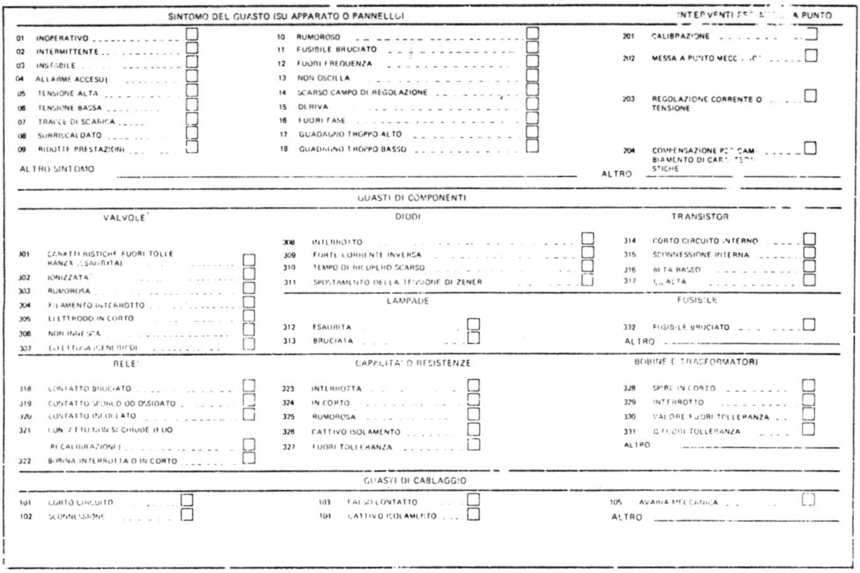

SINTOMO DEL GUASTO (SU APPARATO O PANNELLO)

01 INOPERATIVO
02 INTERMITTENTE
03 INSTABILE
04 ALLARME ACCESO(I)
05 TENSIONE ALTA
06 TENSIONE BASSA
07 TRACCE DI SCARICA
08 SURRISCALDATO
09 RIDOTTE PRESTAZIONI
10 RUMOROSO
11 FUSIBILE BRUCIATO
12 FUORI FREQUENZA
13 NON OSCILLA
14 SCARSO CAMPO DI REGOLAZIONE
15 DERIVA
16 FUORI FASE
17 GUADAGNO TROPPO ALTO
18 GUADAGNO TROPPO BASSO

ALTRO SINTOMO

INTERVENTI [illegible] A PUNTO

201 CALIBRAZIONE
202 MESSA A PUNTO MECC[illegible]
203 REGOLAZIONE CORRENTE O TENSIONE
204 COMPENSAZIONE P[illegible] CAMBIAMENTO DI CAR[illegible]STICHE

ALTRO

GUASTI DI COMPONENTI

VALVOLE

301 CARATTERISTICHE FUORI TOLLERANZA (ESAURITA)
302 IONIZZATA
303 RUMOROSA
304 FILAMENTO INTERROTTO
305 ELETTRODO IN CORTO
306 NON INNESCA
307 DIFETTOSA (GENERICO)

DIODI

308 INTERROTTO
309 FORTE CORRENTE INVERSA
310 TEMPO DI RICUPERO SCARSO
311 SPOSTAMENTO DELLA TENSIONE DI ZENER

TRANSISTOR

314 CORTO CIRCUITO INTERNO
315 SCONNESSIONE INTERNA
316 BETA BASSO
317 [illegible] ALTA

LAMPADE

312 ESAURITA
313 BRUCIATA

FUSIBILE

332 FUSIBILE BRUCIATO

ALTRO

RELE'

318 CONTATTO BRUCIATO
319 CONTATTO SPORCO OD OSSIDATO
320 CONTATTO INCOLLATO
321 CON[illegible]TTO NON SI CHIUDE (FUO[illegible] RICALIBRAZIONE)
322 BOBINA INTERROTTA O IN CORTO

CAPACITA' O RESISTENZE

323 INTERROTTA
324 IN CORTO
325 RUMOROSA
326 CATTIVO ISOLAMENTO
327 FUORI TOLLERANZA

BOBINE E TRASFORMATORI

328 SPIRE IN CORTO
329 INTERROTTO
330 VALORE FUORI TOLLERANZA
331 Q FUORI TOLLERANZA

ALTRO

GUASTI DI CABLAGGIO

101 CORTO CIRCUITO
102 SCONNESSIONE
103 FALSO CONTATTO
104 CATTIVO ISOLAMENTO
105 AVARIA MECCANICA

ALTRO

- type of remedial action
- time to repair
- equipment downtime and reason for it
- components used for repair.
- effects of failure on mission performance (countdown hold, mission postponement)
- time when failure occurred: (i.e. during maintenance or in service, premission checkout or countdown)
- hours of operation from last maintenance
- clock reading.

These forms have proved very useful particularly for information completeness and clearness. Failure reports are checked and if not exhaustive, further information is requested from the compilers.

As mentioned earlier, the reports indicate the period of occurrence of each failure (preventive maintenance, pre-mission check-out or countdown).

In fact, in order to assess equipment and Range System realiability, only count-down failure rate is of interest; this figure is based on hours of equipment operation during missions and failures occurred.

Failures during countdown are reported even if they have not jeopardized a mission.

However, for an overall evaluation of equipment behaviour and maintenance effectiveness, it is important that all remedial actions taken during maintenance be reported too.

4. PROCESSING OF DATA

Information collected through the failure reports, not only allows to evaluate the equipment reliability but also it provides a means of monitoring maintenance effectiveness and logistic support organization.

The number of equipments installed and the quantity of information collected, dictated the need for a computer.

All data were coded (site, equipment, sub-unit, type of failure, etc.) and written in tabulated form (an example is shown in Fig. 2) and subsequently punched on cards for processing in a computer.

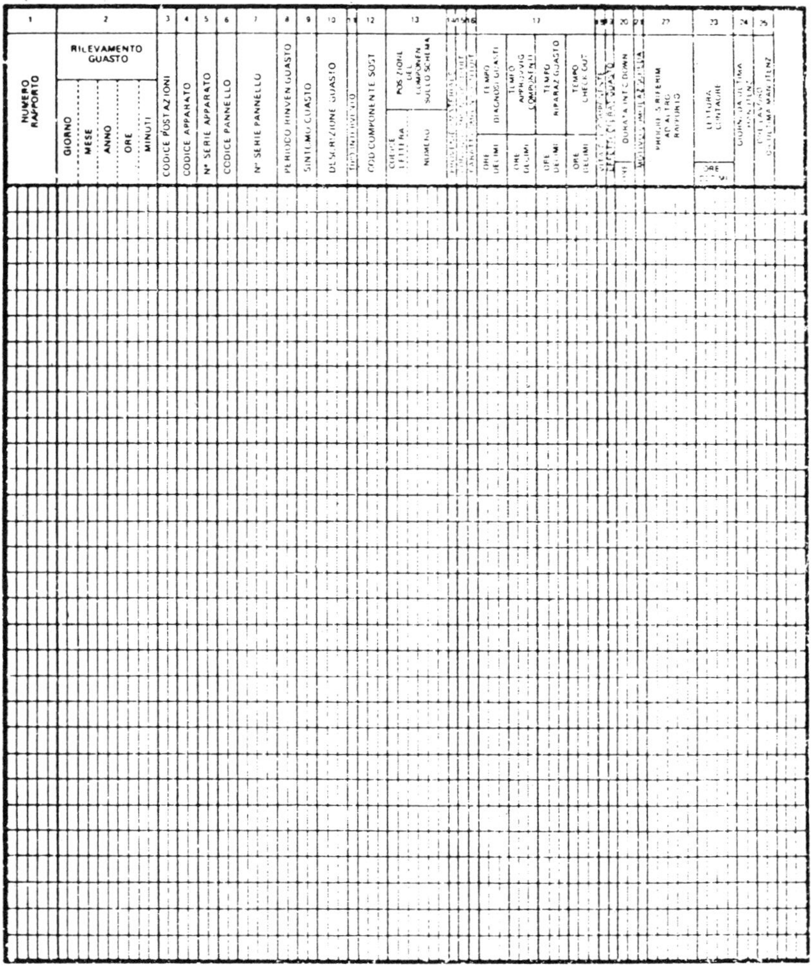

Figure 2 Tabulated failure report form

Data processing gives the following results:

– equipment failure rate

– range system reliability of each site and of the whole system

– statistics on types of failures for each piece of equipment

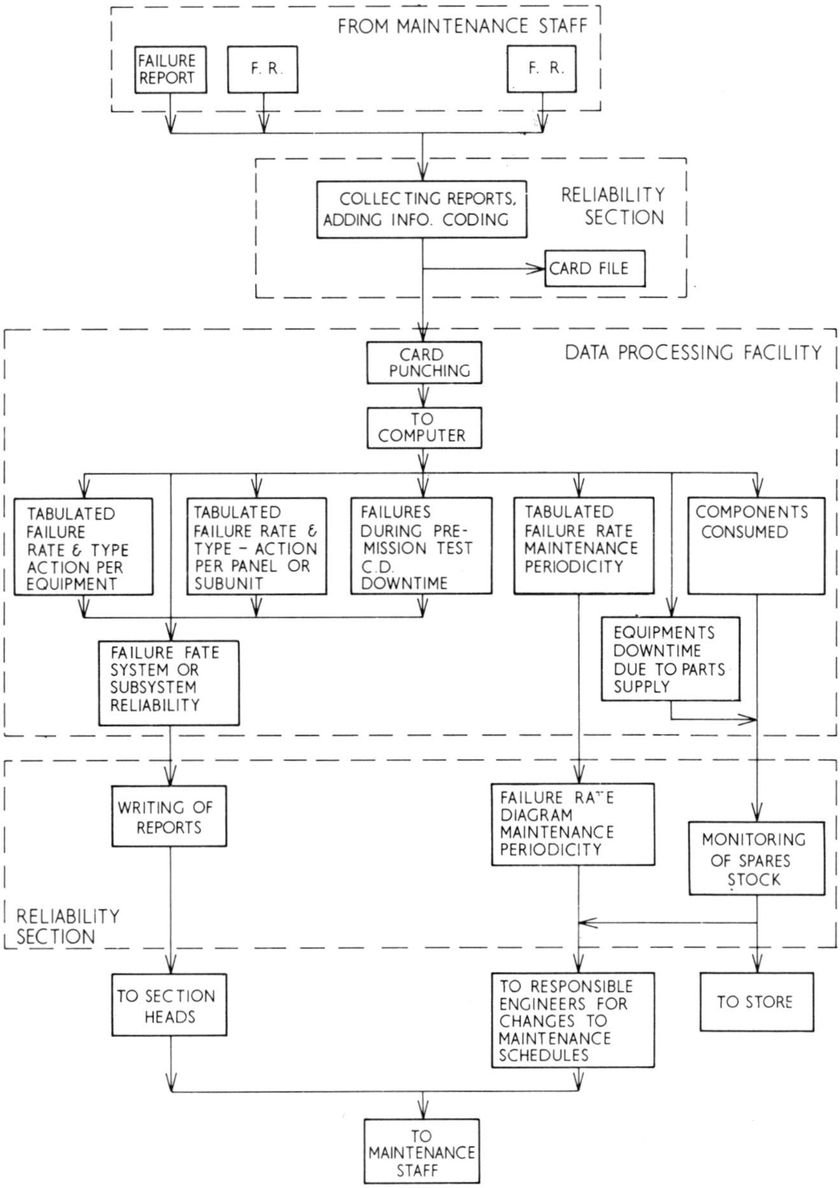

Figure 3 Processing of failure reports–flow chart

	4- 1-64 9-15-64	9-16-64 3-31-65	4- 1-65 12-31-65	1- 1-66 6-30-66	7- 1-66 12-31-66	1- 1-67 6-30-67	7- 1-67 12-31-67	1- 1-67 6-30-68	7- 1-68 12-31-68
Operation Time (hrs)	1396	700	1524	726	606	955	1059	1313	2386
Number of failures during operation	13	7	7		2	2	2	2	4
Failure rate	$9.3\ 10^{-3}$	$10\ 10^{-3}$	$4.6\ 10^{-3}$	$4.13\ 10^{-3}$	$3.3\ 10^{-3}$	$2.1\ 10^{-3}$	$1.9\ 10^{-3}$	$1.52\ 10^{-3}$	$1.68\ 10^{-3}$
MTBF (hrs)	107.4	100	217.7	24.2	303	477.5	529.5	676.5	596.5
Percent operativity	39.5%	35%	24.4%	18.3%	21.8%	36.4%	39.2%	47.7%	50%

Figure 4 Co-ordinate converter reliability summary

– statistics on types of failures for each site

– mean time to repair, allowing to estimate equipment availability

– statistics on maintenance actions taken on the equipment, allowing to check adherence to maintenance schedules

– consumption of components, allowing to make a check of the spares in stock.

The results obtained are recorded on reports which are distributed to the people concerned as shown in Fig. 3.

4.1 Evaluation of System Reliabiltity

Range subsystems MTBF is computed periodically, based on failures and hours of operation during countdown:

$$\text{MTBF} = \frac{\text{total hours of operation during count-down}}{\text{total number of failures during count-down}}$$

By combining the MTBF relating to the various subsystems, the reliability of the overall Range system can be evaluated.

According to Agree definition, reliability is the probability of a system performing its purpose adequately for the period of time intended under the operating conditions encountered.

In the evaluation of Range system reliability according to the formula

$$r = e^{-\lambda t}$$

the time introduced is the average duration of missions, whereas λ is the total failure rate.

The fairly high figures that have been obtained have not changed during the six years of operation of the Range.

Figures 4 and 5 show an example of life curve plotted on the basis of the failure rate

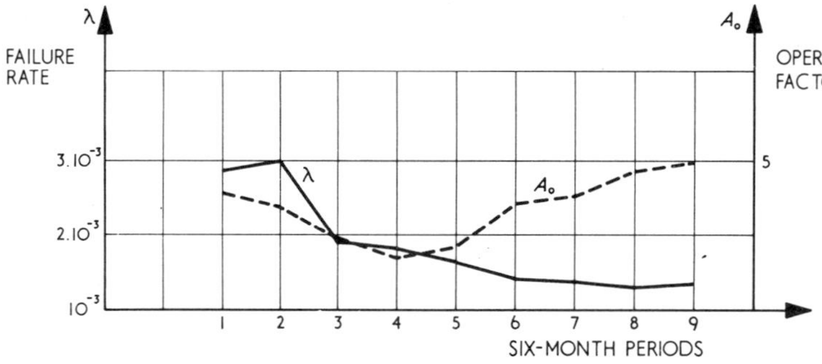

Figure 5 Co-ordinate converter life curve

computed at various periods, starting from initial activation of the equipment. In this way it is possible to check whether the failure rate, as computed at different set periods changes or if there are differences between one period and the other.

4.2 Evaluation of System Availability

More significative parameters than Range System reliability, are Range "operational availability" and "achieved availability".

The first parameter permits to evaluate the probability that the system will be available for operation for a given operating time; the second one provides the percentage of maintenance hours over the total number of hours of equipment operation.

Both figures are evaluated on the basis of the following parameters provided by the computation program for each subsystem:

– mean time to repair, defined by the formula:

$$MTTR = \frac{\text{sum of failure repair times}}{\text{number of failures}}$$

– mean time between maintenance:

$$MTBM = \frac{\text{operating time}}{\text{number of maintenance actions + number of failures}}$$

– mean active maintenance downtime:

$$\overline{M} = \frac{\text{time to repair + time to perform maintenance}}{\text{number of failures + number of maintenance actions}}$$

– total downtime, defined by:

$$MDT = \frac{\text{time to repair + time to perform maintenance + downtime}}{\text{number of failures + number of maintenance actions}}$$

The achieved availability is computed for each subsystem by means of the formula:

$$A_a = \frac{MTBM}{MTBM + \overline{M}}$$

The correlation of achieved availability curves and life curves indicates the behaviour of each subsystem and how it responds to maintenance.

This is a very important indication mainly for those periods of maximum activity of the Range, when maintenance schedules cannot be completely met. In fact it permits to select the specific equipment (like cinetheodolites where mechanical components exceed electronic components) for which maintenance is absolutely necessary to avoid degradation of equipment performance.

The subsystem operational availability is defined by the formula:

$$A_o = \frac{MTBM}{MTBM + MDT}$$

The overall system operational availability is computed as a product of the availability of each sub-system.

The "operational availability" parameter, therefore, takes into account operation and maintenance time in addition to equipment downtime. This parameter may be degraded by the equipment downtime due to performance of maintenance as well as by unavailability of spare parts.

Therefore, operational availability should be one of the parameters to be used for assessing the effectiveness of a logistic support system.

The probability of operational availability of each subsystem for a given operating time, is computed by the following formula:

$$A = 1 - e^{-ut} + e^{-(t + \lambda T)}$$

with the following meaning for the symbols:

u : maintenance rate $= \frac{1}{MTBM}$

t : mean active maintenance downtime ($\overline{M}$)

λ : failure rate ($1/MTBF$)

T : operating time over which probability is computed

4.3 Failure Analysis

The failures which occur in each equipment are classified by type in statistical form. An example of such statistics is shown in Figures 6 and 7 which are related to two subsystems in the Range: the analog computers and the cinetheodolites. These tables also indicate the units where the failures have occurred.

Analyses of these data have allowed in the most cases to identify the cause of the failure and to take appropriate actions. These statistics have in fact provided the means of identifying not only the failures merely due to chance, or those due to untimely replacement of aging components, but also, for some equipments, certain recurrent failures.

In some instances, failures have been recognized as having been caused by circuits working under critical conditions and remedy was found in making modifications to the circuitry.

PANNELLO	ESAUR.			AVAR.			NON NOTO			CHOP.			SCONN. F.C.e. C.C.			RESIS.			POTEN.			FUSIBILI			DIODI			CAPAC.			REGOL.			TOT.
	m	p	cd	m	p	cd	m	p	cd	m	p	cd	m	p	cd	m	p	cd	m	p	cd	m	p	cd	m	p	cd	m	p	cd	m	p	cd	
AMPLIFICATORE DC	4		1	4						5	2	1	4												1									22
ALIMENTATORE 10089		1	1	3											1								2											8
PANNELLO AZIMUTH													1																					1
PANNELLO ELEVAZIONE														1						1			1	1										4
SERVO AMPLIFIER																																		
ALIMENTATORE DOPPIO																																		
P1 REGULATOR AMPLIF.							1																											1
PANNELLO DISTANZA													1	1											1									3
SWITCH AMPLIFIER																																1		1
Z7																																		
Z6, Z5, Z4															1																			1
Z3, Z2, Z1																																		
Z9, Z10, Z11																												1						1
RANGE SER..																																		
RACK															1																			1
CONTROLLO																																		
PARALLASSE																																		
PANNELLO PROVA																																		
CARTOLINA D F G																																		
TOTALI	4	1	2	7			1			5	2	1	6	2	3					1			3	1	2			1				1		13

Figure 6 Statistics of analog computer failures per panel

CITE	N°	PANNELLI	VALV.			LAMP. GENE.51			FLASH			FUSIBILI			CON-DENS.			CON-NETT.			CAN-NOCC.			C.C.F.C. CONNES.		
			m	p	cd	m	p	cd	m	p	cd	m	p	cd	m	p	cd	m	p	cd	m	p	cd	m	p	cd
OTTICA	1	OTTICA MT.3																								
	2	OTTICA MT. 1,5																								
	3	CANNOCCH.PUNTAMENTO																			1					
	4	MICROSCOPI DI LETTURA																								
	5	DIAFRAMMI, FILTRI, ETC...																								
ASSERVIM. ALFA	6	AMPLIFICATORE	1																					1		
	7	ALIMENTATORE																								
	8	MOTORE ALFA																								
	9	GENERATORE ALFA																								
	10	SCATOLA INGRANAGGI α																								
	11	POTENZ.VOLANTINO α																								
	12	MOTOGENERAT.VOLAN. α																								
ASSERVIM. LAMBDA	13	AMPLIFICATORE LAMBDA	1																							
	14	ALIMENTATORE λ													1											
	15	MOTORE λ																								
	16	GENERATORE λ																								
	17	SCATOLA INGRAN. λ																								
	18	POTENZ.VOLANTINO λ																								
FLASH E VENTILAZ.	19	BULBO								1																
	20	PARTE ELETTRON.VENTIL.																								
CONTATORE	21	CONTATORE ELETTRONIC.																								
	22	COMANDO ELETTRONICO																								
TELEGUIDA	23	SINCRO ALFA																								
	24	SINCRO LAMBDA																								
	25	ERROR DETECTOR																							1	
	26	CABLAGGI																								
CAMERA FOTOGRAFICA	27	PARTE MECCANICA																								
	28	CONTENITORE CARICAT.					6	1			1														1	
	29	ELETTR.COM.CAMERA	1		1								2													
	30	MOTORE																								
	31	GENERATORE FASE OTT.																								
	32	CABLAGGIO																1								
	33	DISTRIB.ALIM.A.C.																								
PARTE ELETTRICA	34	LAMP.RETE CAMERA																								
	35	LAMP.RETE PRINCIP.																								
	36	LAMP.RETE BOLLA																								
	37	LAMP.RETE COLLIMAT.																								
	38	INT.POT.DEVIATORE																								
	39	VENTILATORI																								
ASTRODOMO	40	ASTRODOMO																								
		TOTALI PARZIALI	3		1		6	1		1	1		2		1			1			1			1	2	
			4			7			2			2			1			1			1			3		

Figure 7 Statistics of cinetheodolite failures per panel

In other instances, failures were due to inadvertent operation: that has been the case with the cinetheodolites: imperfect film loading and transport caused it to break and prevented it to be utilized for mission data processing.

Another cause of failures has been recognized in unfit ambient conditions which accelerated decay process of certain components making them unserviceable before the average life guaranteed by the manufacturer had expired.

This kind of failure was experienced in the radio link equipment. There, the inadequacy of the environmental control system of the air conditioning system caused certain tubes to wear out before they were due for replacement.

4.4 Monitoring of Spares Stock Level

Indication of the failure reports of the components used for repairs, allows a periodic check of the spares stock level and identification of the components to be re-purchased. This timely check is very important because many components are not available on the domestic market and take a long time to procure.

Furthermore, the mean life stated by the manufacturer can be checked for some components. In fact experience has shown that owing to the particular operating conditions to which certain components are subjected, the actual mean life is below the theoretical one. This makes it necessary to evaluate the mean life of the components on the basis of data collected during the past years in order that realistic predictions may be made of the quantities of spares to be kept in the Range stock for the year to come.

The most important results obtained with the above described philosophy are:

- saving of spare parts, to date the cost of spares has been only 5% of total cost of Range equipment.

- avoiding equipment downtime due to lack of spare components, thus preventing reduction of the "operational availability" parameter for the various subsystems and consequently for the overall range system, as indicated in paragraph 4.2.

5. PROCESSED DATA REPORTS

Results obtained from data gathering and processing are collected in a report which is distributed to engineers responsible for the range subsystems and to military agencies concerned. The report provides information on the Range operational activity, namely:

- missions performed

- equipment and sites involved

- manpower used for operational support.

In addition, the report indicates the preventive maintenance actions that have been carried out compatibly with the Range activity. Each subsystem is covered in terms of failure occurred during countdown, number of hours of operation of each equipment, failure rates during the period covered and the precėding period.

Analysis of the types of failure for each equipment and comparison with the situation relating to the preceding periods allows examination of the equipment behaviour and determination of any problems not directly arising from failures but, for instance, from reduced maintenance caused by intense Range activity.

This information cannot be drawn from the failure rate figure only since deterioration of equipment performance does not always result in a failure.

The recurrence of certain types of malfunctions has shown that for certain equipment containing more mechanical components than electronic ones, the scheduled maintenance carried out on the Range is no longer sufficient, but overhaul is necessary in order that equipment may be restored to the efficiency level required for its specific use.

6. CONCLUSIONS

It has been indicated in the above, the reasons which have determined the establishment of a preventive maintenance organization and of a section having the specific task of collection and processing the failure reports data provided by maintenance people.

It has been emphasized that the function of the Reliability Section does not end within itself but has a well-defined task covering efficiency of the maintenance activity, reliability of the system, and finally, proper use of both equipment and people. The Reliability Section is the only one, in fact, in a position to assess the effectiveness and adequacy of the logistic support system and, if necessary, propose modifications.

Conclusively, we may say that the organization of the maintenance activity (and of the reliability organization which controls it) combined with the redundancy of some of the equipment have contributed to the successful operation of the Range: in fact during the six years of Range activity, not one mission has been deferred because of equipment failure.

DISCUSSION

Bax:

1. Please explain the nature of identification code to identify the systems, subsystems and components (for example: work unit code, part number, etc).

2. What type and skill level of personnel fills out the data collection form (or sheet)? (for example: soldiers or Vitroselenia personnel, etc.)

3. Can you explain the system to control the different components during the whole life cycle? Especially the procuedures to control all the small parts (resistors, etc.) contained in black boxes?

Piu:

1. The identification code is made as follows:

   – there is a numeric code for the sites of the range

   – a numeric code for the equipment

   – a numeric code for each piece of equipment, which is the code given from the manufacturer.

2. Vitroselenia personnel fills out the data collection form.

3. I spoke about "life cycle" for equipment and not for components. Data about "mean life" of components are generally given from the manufacturer or are taken from RADC data.

Segerson: Do you know if it is planned to use the same techniques for Italian Air Force Systems?

Piu: The answer to this question has been given from Generale Matacotta and is "yes".

Ballow: Did you consider having technicians fill out reports in coded language as opposed to the form you have used? If so, why did you choose the system as you did?

Piu: No, because the failure report form has been studied in such a way as to simplify paperwork to technicians.

Towards a Theory of Maintenance

W.M.J. Geraerds

1. THE PHENOMENON OF MAINTENANCE

Since the early days of mankind man has used tools. They served as a means of extension of the capabilities of the human body in processes like winning raw material in nature, growing and harvesting, building a dwelling, bridges and canoes and hunting animals or each other.

Along with the use of tools went the experience that they showed a decline in suitability through wear and tear, while sometimes sudden breakdown appeared. One way to meet this situation was the manufacturing or procurement of a new tool as a replacement of the unsound one. In many a case, however, the unsound tool could be restored to its serviceable state. Though not familiar with todays economic criteria, experience showed that the effort and the time for this restorative process were often considerably smaller than the process of replacing the tool. Thus repair was born. Also it was discovered that sometimes the breakdown and the damage resulting from the age and the use of the tools could be prevented or reduced if attention was paid to their state during the period in which they still fulfilled their function. This is what we call preventive maintenance nowadays.

Observing the development through the ages, we find an increase in the number of tools and of their complexity, especially in the last decades, where we see that large and complex systems are designed, manufactured and put into operation in record time. Though a vast amount of scientific knowledge and of experience is included in such systems, deterioration and breakdown does not disappear. This is partially due to the fact that all materials show deterioration by aging and by use. Partially it is due to the fact that there is, in the continuous process of development of new systems, insufficient knowledge, at the time of design and manufacturing, of an economically justifiable way by which deterioration and breakdown can be postponed to the time that a system is obsolete.

This means that it is unthinkable that maintenance as such will ever become superfluous by elimination of its causes. To meet the maintenance task efficiently it is therefore necessary to direct the attention to the way in which maintenance should be organized and to the way in which the properties relevant to the maintenance task can be influenced.

2. THE MAINTENANCE FUNCTION WITHIN THE ORGANISATION

2.1 The impact of maintenance

Little is known about the magnitude of maintenance in the organisation. A detailed inquiry by means of a questionnaire was made by Mr. E. Turban [1], covering 309 valid answers out of the addressed 770 medium- to large-size plants across the U.S.A. The results show that the trend over the years 1958–1965 is a decrease in the number of maintenance men per 100 production workers and of the repair time per work order, an increase in the time period between serious breakdowns, but a rise in the cost per serious breakdown. Maintenance costs as a percentage of total manufacturing costs vary between 10.2% for process plants and 6.3% for non-process plants, averaging 8.2%. Figures in consumption–rather than production–organizations, like military organizations, show quite a different picture, probably owing to early introduction of new discoveries and operations close to performance limits. The British Navy, for example, allocates 40% of its 1967–1968 vote to upkeep [2]. The RAF shows the same impact of maintenance on its total Defence Budget: £200m out of £500m. The amount of money spent on new equipment is equal to the amount of money spent on maintenance [3]. These figures can be taken as characteristic of military organizations.

In both industrial and military organizations, the equipment in use has become much more expensive over the last decades. This went along with increase in size, complexity and performance. The–not quantified–aspect that is gaining importance is the loss of production capacity caused by down time, in particular the unexpected down time, of the equipment in use. To cope with this growing maintenance task, a need exists for a comprehensive theory of maintenance, which would serve as a guide in organizing maintenance within an organization and in indicating areas requiring research.

2.2 Definition of maintenance

The literature shows a variety of definitions of maintenance, which is not unusual in managerial subjects. In order to prevent ambiguity, it is necessary to define maintenance as it is understood in this paper. In an attempt to arrive at an all encompassing theoretical structure of maintenance, the approach must be based on the goals to be reached by maintenance. This means that maintenance cannot be characterized in terms of organizational departments, job descriptions, responsibilities and similar notions taken from the static, hierarchical approach of organizations. If objectives are to be achieved by having things happen, the attention has to be focussed on the collection of processes that lead to the realization of the objectives and, in particular, on the way in which the output of these processes can be influenced. This point of view, derived from the systems theory approach to organizations, leads to the following definition, which includes the goal element and the process element:

> Maintenance is the total of activities serving the purpose of retaining the production means in or restoring them to the state that is considered necessary for fulfilment of their production function [4].

It should be noticed that this definition implies repair: "restoring to" corresponds to corrective maintenance and "retaining in" corresponds to preventive maintenance. The definition does not imply that maintenance is efficient. Whether maintenance in a given case is efficient or inefficient is dependent on the "considering", which depends on the criterion applied for the evaluation of the efficiency.

2.3 Relation of maintenance to production

The production function manifests itself materially in the products that are manufactured. It is because of the use of means of production that maintenance is necessary, and consequently the costs of maintenance are to be attributed to the costs of the products manufactured. Optimization of maintenance must therefore be found in criteria which directly or indirectly minimize the cost of the products. This means that maintenance exists only by virtue of the fact that its function is derived from the production function by supplying its output to the production function. Seen in this way, it is possible to detach maintenance from the total systems of the organization and study it as a more or less independent subsystem (fig. 1).

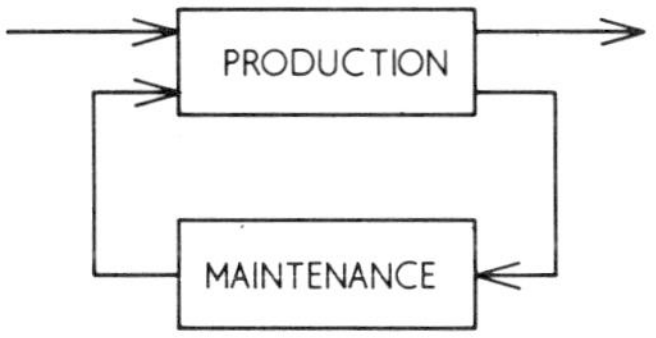

Figure 1 Relation between production and maintenance functions

3. MAINTENANCE AS AN INDEPENDENT SUBSYSTEM

3.1 Maintenance as a closed system

In the simplified flow diagram in fig. 2 the maintenance process is characterized by the

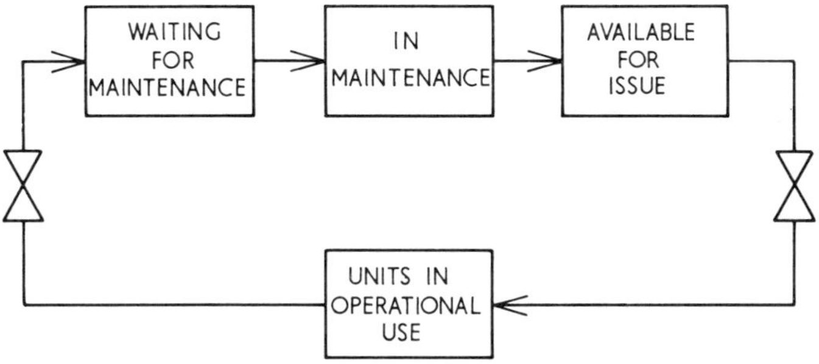

Figure 2 Maintenance as a closed system

change in states that units to be maintained undergo, starting from the state of operational use and returning to this state. In order to fulfil the task set by production, it is necessary to control this flow of units through the maintenance phases. By situating production in this flow-diagram as shown, the interrelated behavior of maintenance and production can be detailed further.

3.2 Maintenance echelon structure (fig. 3).

The execution of maintenance is by nature technological and asks for facilities and skills, capable of covering the diversity of objects to be maintained. Specialization leads to centralization for different levels of work, usually called echelons, including the appropriate facilities and manpower for each level.

The units in use in an organization have a certain useful lifetime, after which they are replaced by identical units, or by more or less similar units according to the operational requirements and to the state of art of design at that time. This renewal process is continuous, though implying in some cases a rather radical operation, like the introduction of a new weapon system together with the phasing out of the one to be replaced.

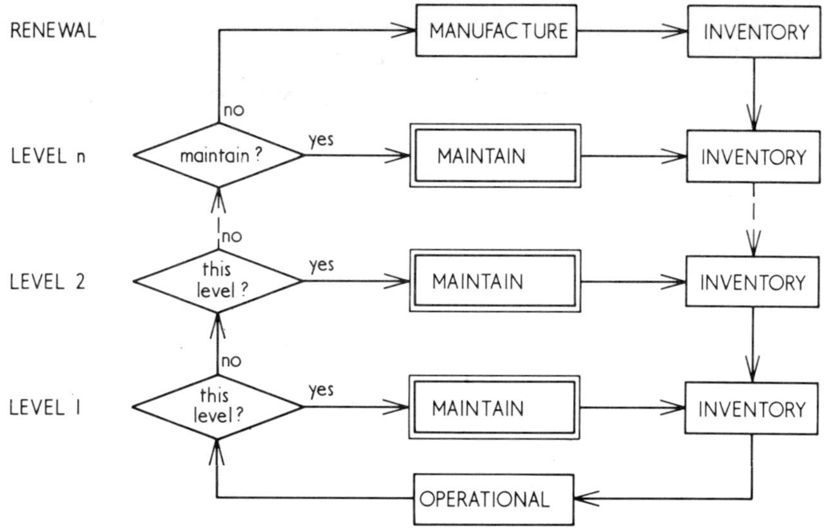

Figure 3 Maintenance echelon structure

The more permanent nature of the echelonized capacity exists at the times that new objects to be maintained enter the organization. The capacities in the maintenance echelon structure, given in fig. 3 will then have to be adjusted to the requirements of the new objects. The left part of this figure represents the decision-rules for the selection of the maintenance level, controlling the actual flow of an object to the levels to be considered. If maintenance is not possible or not efficient renewal takes place by purchase. Contract maintenance is included in the structure as it is regarded as a capacity available for selection. The total behavorial content of this structure is too large for direct theoretical development. That will be done by looking closer at the elementary maintenance loop in fig. 2.

4. ANALOGIES OF MAINTENANCE AND PRODUCTION SYSTEMS

4.1 Fundamentals of maintenance

An essential condition to arrive at control of a process is quantitative prediction and measurement. A tremendous amount of scientific work has been devoted to quantitative models for a large diversity of problems, often valid for more situations than those that originated the research. To prevent the superfluous effort of reinventing the wheel, theoretical elements already developed should be traced as far as they fit in logically. If we look at maintenance activities for that purpose, the following facts are noticed.

In the end maintenance consists of technological treatment of physical objects. In order to realize that, the work to be done must be assigned to the appropriate capacities. As capacities are scarce, there will be more than one job competing for assignment to the capacity. The theory that covers this assignment problem is known as the theory of *scheduling*. The resulting reservation of capacity is known as *loading*. To be able to schedule the work to be done, it should be prepared and presented in a shape that is adjusted to the scheduling process. This is known as *process-planning [5]*.

The capacity being called upon cannot be expected to be more or less according to the requirements if the qualitative and quantitative dimensions have not been predicted and adjusted accordingly in due time. This process is known as *long-term* and *mid-term planning*.

These fundamental maintenance activities clearly show far-reaching analogies with fundamental production activities, calling for a more systematic comparison of the maintenance process and the production process.

4.2 Analogous diagrams

To arrive at a more detailed comparison of maintenance and production the production process and the maintenance process are represented by a flow diagram in figure 4. The suggestion–coupled with analogous diagrams–that the behavior is identical, requires a comparison of the contents of the respective blocks in the diagrams, so as to distinguish between identical and non-identical behavorial aspects.

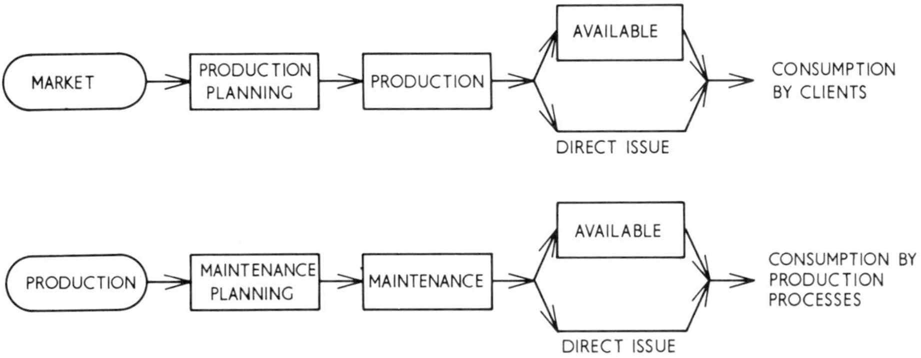

Figure 4 Analoguous flow diagrams of production and maintenance

4.3 Production and maintenance

4.3.1. The assignment of work

At lower levels of maintenance, the assignment of work concerns logically more diverse work than that which is done at higher levels, which is more specialized. The work at the lower levels shows the combination of diversity of work and universality in the capacities–in particular, in the skills. The majority of scheduling problems at these lower levels are therefore analogous to those in the job shop. From research in this field, done by Conway and Maxwell [6, 7, 8, 9] and Bakker [10] we know that different priority rules lead to quite different results. The state of the art at this moment does not furnish a complete methodology of selecting the optimal decision rules applicable in real situations. The greater part of research in this field concerns singular products. Maintenance objects usually are assemblies; this leads to a network structure of maintenance operations. The scheduling of orders of that type is known as multi-project scheduling, which is in the early stages of development.

The assignment of work at higher levels brings on specialization, which can lead to several process types–continuous, batch or single project–presenting relatively less scheduling problems than the job-shop type [9], [11].

In general it can be said, that a great store of knowledge on scheduling in production processes is available. This can be applied to maintenance processes because the problems to be solved are identical. Hence, independent development of techniques for application to maintenance is not required. However, areas exist where the state of the art of scheduling has not yet matured to the desired level. Maintenance processes are on a par with production processes as suitable objects of research into scheduling.

4.3.2. *Process planning* [5]

This results in the plan of execution, specifying route, operations and material needed. It is based on the design of the product. In maintenance, the same information has to be supplied, but in this case the rather well defined design fails as a basis. Because the technological state of each object differs from one to another, the operations and the material needed show a stochastic behavior. The actual picture for each case can only be determined by individual inspection. This leads to a splitting up of a maintenance job in two major phases, the inspection or diagnostic phase and the execution phase (fig. 5).

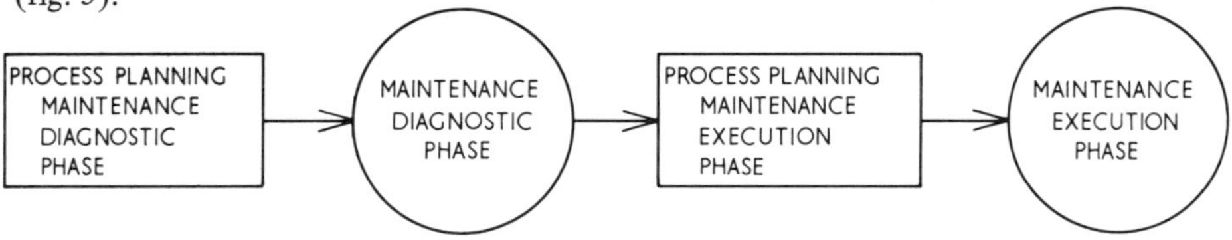

Figure 5 Maintenance job as a two stage process

The inspection or diagnostic phase replaces the design phase of production processes. If projects of a certain identity arrive more frequently, throughput times can be predicted more reliably. If the frequency is very low the reliability of the prediction becomes low, and this is also usual in job shop production.

Though the final content of process-planning and maintenance is similar, the information needed as a basis comes from non-comparable sources. That means that the information needed will have to be found by studying the behavior of the objects to be maintained. This is a typical maintenance problem for which there is no analogy in production. The provisioning of material will be treated separately.

4.4 Available versus direct issue

If clients' orders could be predicted perfectly, waiting for delivery could practically be eliminated. As prediction is not that accurate, a certain inventory is necessary to serve as a buffer between the fluctuating output of production and the fluctuations of the demand. Direct issue by manufacture on order is possible but it will have to be balanced against the capacity needed for replenishment of the inventory when promising delivery times. It is evident that, if there is no balancing mechanism, the risk of favoring direct issue production at the expense of the replenishing of the stock–with undesired results in the longer term–is very great. In comparing production processes, we find a very direct analogy in the stock of inventory, which serves the same purpose of buffer to disconnect the fluctuations of maintenance output and demand by the client–in this case the production process, or in the military, organization operations. A practical difference of great importance is the fact that the client, possibly wishing direct issue, is the same as the one who will suffer from lack of future availability if insufficient capacity is left for the replenishment of the buffer. This means that, in such a situation, the client must be willing to reduce his initial requirements or, putting it positively, he must tune his operational plans to the maintenance possibilities. Though the analogy of potentially conflicting interests is evident, the problem to be solved is substantially more simple in the maintenance case than in the production case. If the fluctuations between maintenance output and demand for maintenance lead to inacceptable waiting times, a buffer is to be built in between maintenance and consumption.

4.5 Maintenance planning and production planning

Long-term planning for the size and kind of facilities and manpower involves the same problems in maintenance as in production. Mid-term planning for maintenance, however, ought to be much more accurate as the future "clients" will all arrive, and there will be no possibility of refusing them. Though the planning processes are analogous, the mid-term planning in particular requires a relatively high quality.

4.6 Market

Market research as a basis for long- and mid-term planning in production supplies the forecast of the demand for a diversity of products within the potential production range. The products considered attractive for the producing organization are selected by policy decision. At the time of realization clients will turn up and the policy decision will prove right if the resulting demand is in balance with the forecast. In maintenance, however, all clients are known. The "market" in this case is the entire collection of facilities and equipment of the organization. The study of the future workload requires study of the behavior of these objects, which is not comparable with the methods used in market research. As the manufacturing processes consuming maintenance constitute the maintenance market, it is possible to represent the maintenance process as a closed loop, as illustrated in figures 1, 2 and 3. It stresses the possibility of direct observation of all clients and of direct control of the use of the production means in order to achieve the tuning in of maintenance and operations. These possibilities are practically unavailable in production. That means that the constant tuning of maintenance and production is a typical maintenance problem.

4.7 Conclusions

In comparing production with maintenance, the following conclusions can be summarized.

- Analogies between production and maintenance exist in terms of scheduling and process planning, long-term and mid-term planning. There is no need for an independent theoretical approach to these aspects in maintenance.

- The behavior of objects to be maintained has no analogy, and requires an independent theoretical approach.

- Relatively more simple than in production is the reconciliation of maintenance required by production and maintenance capability of the organization, as all objects to be maintained are known beforehand and production and maintenance can be controlled within the same organization. This, however, requires an independent theoretical approach if the control system is to be regarded as closed.

5. BEHAVIOR OF OBJECTS TO BE MAINTAINED

5.1 Relevant aspects of behavior

From the production point of view–disregarding optimization constraints–it would be appreciated if production means would:

- not fail
- if failing, show very few failures
- if failing, show failures that are somehow predictable
- if failing and if repairable, be returnable to the working state within a short time.

Requirements of this nature cannot be met if they are not quantifiable. Scientific development in the last decades has yielded the quantifiable properties that cover the aspects mentioned.

5.2 Quantifiable relevant behavorial properties

The requirements mentioned concern three behavorial aspects:

- failure behavior
- maintenance time behavior
- availability behavior.

5.2.1. *Failure behavior*

A failure is when a system, e.g. a production unit, changes from the state in which it can fulfil its productive function to the state in which it cannot fulfil that function. It does not matter if restoration of the unit to the ready or up state is possible or not. Failure analysis is of importance for the prediction of failure. The probability of failure is quantifiable by measuring the failure intensity. This property, the *failure rate*, is expressed as the ratio of the number of failures and some relevant measure of productive use, measured over a certain period of time. For instance, it could be the number of punctures of a tyre in 10,000 miles, or the number of times a lighter does not work out of a hundred times it is pressed to be lighted. The failure rate as a function of time (or practical equivalents like mileage, flying-hours, number of operations, etc.) shows whether the failure intensity is constant, decreasing or increasing. An often quoted characteristic, which is sometimes supposed to have general validity, is the so called "bathtub curve" (fig. 6). Research in real situations has shown that the curve is not generally valid. This is of importance because the validity of the models in which the failure rate is used is not independent of the failure rate gradient [14], [15].

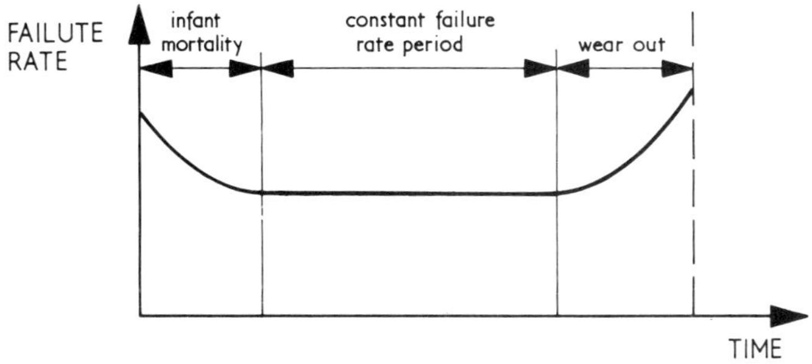

Figure 6 The bath-tub curve

As the times to failure will show variance, the failure rate as such does not give an answer to the primary question as to whether a mission that requires a certain failure-free operation time can be accomplished. If a number of failures is known and arranged in a histogram (fig. 7) [16], it appears that the answer can be given in

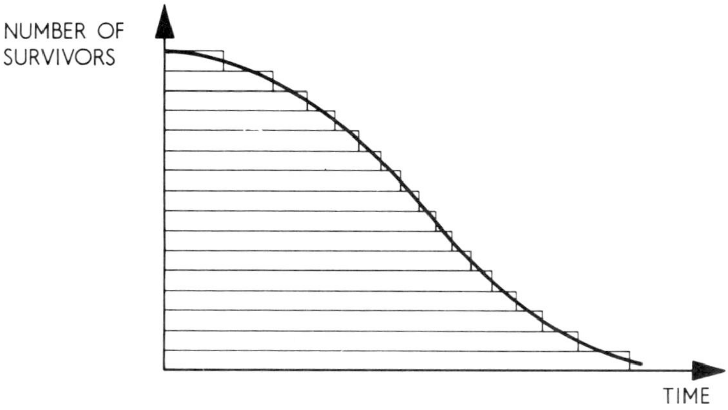

Figure 7 Survival characteristic

probability terms. There is a diversity of characteristics that can be found, as illustrated by MORSE [16] (fig. 8). The property characterized this way is known as *reliability*, defined as the probability that a system will accomplish a mission of given mission time successfully, under stated operational and environ-

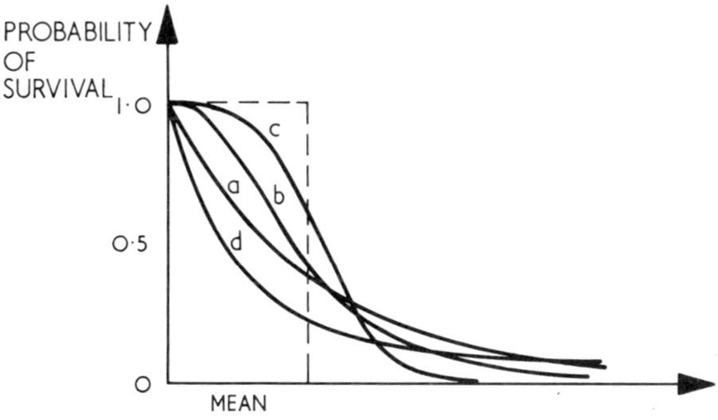

Figure 8 Reliability characteristics

mental conditions, and if the system starts the mission new or at the beginning of an up-interval.

Reliability theory has developed out of the necessity for prediction of system reliability at the time of design, in order to be able if necessary, to increase the inherent system reliability by redesign. There is a vast amount of theory on reliability and a growing number of practical performance requirements for a diversity of components.

The majority of investigations in real situations concerns electronic components. The structure of reliability knowledge is influenced by its use in the process of designing and by the method of measurement of the promised or required reliability.

5.2.2. Maintenance time behavior

If a failed system is repairable and repair is executed, the repair process takes time. The interval of time from the moment that the failure is known until the moment that the system is restored to the ready state, is called time to repair. The property representing the behavior of the time to repair is a probability, known as *maintainability*; this is analogous to the time between failures (fig. 8). It is defined as the probability that repair will have been executed successfully within a certain time of the failure becoming known and under stated conditions of the maintenance organization. Figures 7 and 8 represent maintainability by substituting "maintenance time" for "time to failure" and "maintenance" for "survival". Studying maintainability leads to splitting up the property into elements, of which preventive maintainability and corrective maintainability are mentioned here. The knowledge of maintainability is far less developed than that of reliability. A rather broad and systematic approach has been made by Goldman and Slattery [17]. Practical applications are reflected in particular in Military Standards. As with reliability, the structure of maintainability knowledge is influenced by its use, primarily in the design phase.

5.2.3. Availability behavior

A system can be in either of two states, the *up state* and the *down state*. It is in the down state when the system cannot operate because of preventive or corrective maintenance being performed. It is in the up or ready state if it is operating or if it can start operating when called upon.

The property characterising the availability behavior is *availability*, defined as the ratio of the sum of the up times and the sum of the up- plus down times, measured over a period of time. Availability has been known under several names for quite some time, e.g. serviceability and readiness. Studying availability requires the examination of each sub-element. Availability knowledge is less developed than the knowledge of reliability and maintainability. In practical design applications, one can make use of reliability and maintainability models and standards, as these implicitly implement availability to a high degreee. This will be shown in the next paragraph.

5.2.4. Relation between reliability, maintainability and availability

If the number of failures over a period of time is known and the mean of the maintenance times in that period is known, the product of these two represents the sum of the down times. The total period of time being known, this means that availability is also known. The expression of availability as a product of number of failures and maintenance time, illustrated in figure 9, [17], shows that a specified

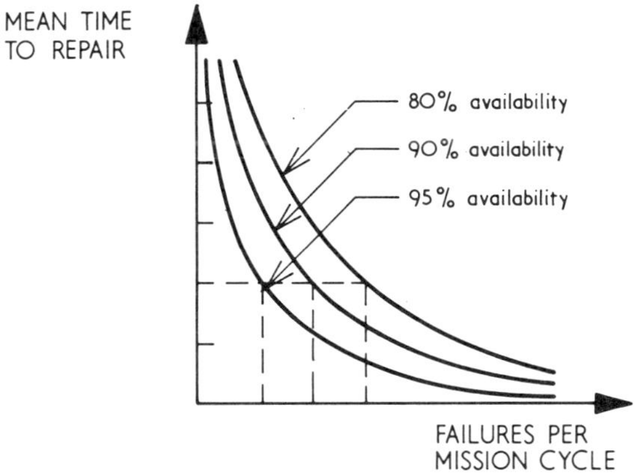

Figure 9 Graphical relation between reliability, maintainability and availability

availability can be achieved by means of several combinations of reliability and maintenance times. Actually, the iso-availability curves plotted in the plane are projections of a–hypothetical-availability surface, illustrated in fig. 10, [17], in which

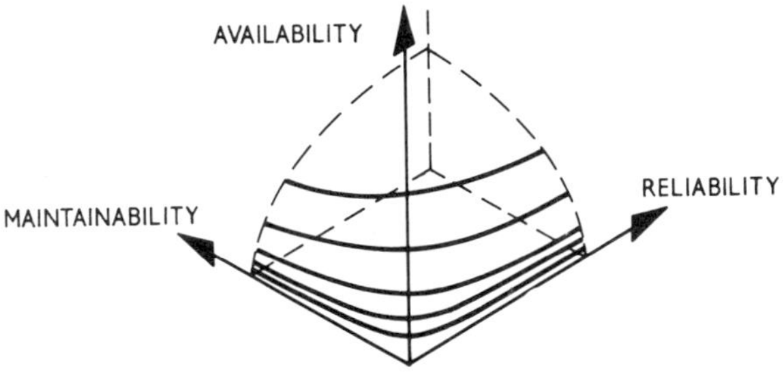

Figure 10 Hypothetical availability surface

the availability axis has been added. It is to be noted that reliability and maintainability in this illustration do not stand for these properties in their strict sense of probabilities but as averages over the same period of time for all three properties. Goldman and Slattery [17] develop this idea further to a direct relation with cost. In the case of design, reliability or maintainability can be changed: in the case of a system in use and with an inherent reliability, availability can be influenced only partially by influencing the length of maintenance times. A combination of reliability and availability turns up if conditional probabilities are considered: for instance, the probability that a system will perform a mission of a given length of time if it is called upon at any random instant of time, or at a random instant within the interval in which the system is in the up state.

5.2.5. Practical use

The practical use of failure rate, reliability, maintainability and availability appears to be great in the process of design of new, complicated and expensive systems. There is not much evidence of application and systematic use in the design of relatively cheaper systems. The majority of experience is related to electronic systems. It is likely that the application in other technological areas and concerning simpler systems will gradually increase, as inductive experience and methods become available from large scale research activities.

6. MAINTENANCE CONCEPTION

6.1 Maintenance policies

When maintenance has to be executed because of failure or because of preventive maintenance being due, a number of activities have to be developed. The structure of these activities for one maintenance level can be represented as shown in fig. 11.

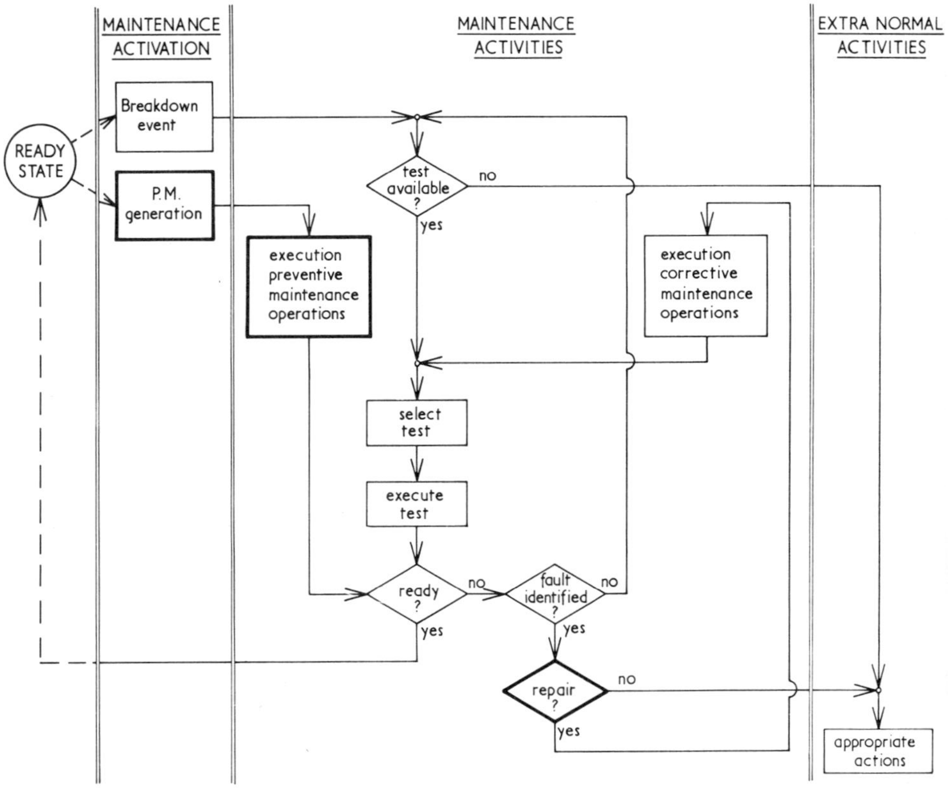

Figure 11 Item oriented maintenance activities structure

This diagram illustrates the need for decision-rules in the mechanism generating preventive maintenance, concerning the actions to be taken when restoring to the up state will not be performed at the same level of maintenance. This of course has to be laid down for each type of object requiring maintenance. The total set of determined directives dealing with the way an object is expected to be maintained in the using organization is called the *maintenance conception* [18].

For the choice of the different primary possibilities roughly the following strategies can be distinguished:

- *"Wait and see,"* which means that no preventive maintenance is performed and that no maintenance will be considered until the event of failure.

- *Opportunistic* policy, which means that maintenance does not start until the event of failure, but at that time preventive maintenance is executed in addition to corrective maintenance.

- *Preventive policy*, which means that part of the maintenance is executed in the non-failed state with the objective of preventing or reducing failures.

To arrive at a maintenance conception, a choice has to be made out of a diversity of alternative possibilities. To have a closer look at these possibilities and their effects, the following three aspects are distinguished:

- the behavior of the object to be maintained

- requirements, wishes and constraints from production

- the echelonized structure.

6.2 Behavior of objects to be maintained

6.2.1. Inherent versus organizational properties

From the point of view of production, which maintenance has to serve, failures should be scarce and if failure turns up, restoration to the up state should not take much time. The failure behavior can be measured by the failure rate–for the intensity of the failure events–and reliability–for the probability of undisturbed production. The maintenance time behavior is measurable by means of the maintainability. An object has received a reliability and a maintainability through design and manufacturing, independent of the use of the maintaining organization. As such, these are *object-inherent* properties. In real situations these properties will be influenced by the way the objects are used, by their environment and by the maintenance organization. The practical problems encountered, if the inherent properties are to be distinguished quantitatively from the *organizational* properties, cannot be neglected. However, they will not be discussed here. It may suffice to stress the fact that the properties are, to a very large extent, functions of the objects chosen (the inherent value). It is on this basis that a maintenance conception is to be constructed; the importance of implementing requirements that are derived from the maintenance process in the specifications for procurement is, of course, beyond doubt.

6.2.2. Policies and assumptions

In general there are two assumptions underlying maintenance policies. The first concerns the assumed relationship between the amount of maintenance and failure behavior; the second concerns the assumed conditions for application of preventive policies. The assumed relationship between the amount of maintenance and failure behavior is usually visualized in graphs of the type of figure 12 [19].

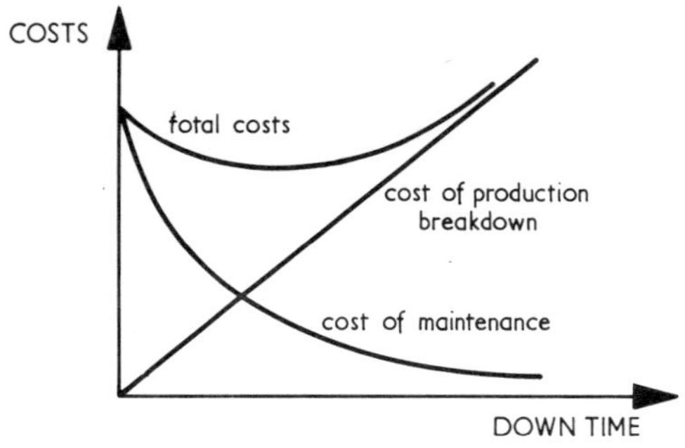

Figure 12 Relation between maintenance and down time costs

In this model, the cost of maintenance curve implies that down time decreases continuously to zero if maintenance costs are raised. This implies that the failure-reducing effects of preventive maintenance are unlimited. Practical experience shows, however, that there is a limit to the down time reduction that can be reached. What happens in reality is shown in fig. 13. If the minimum of the attainable down time is

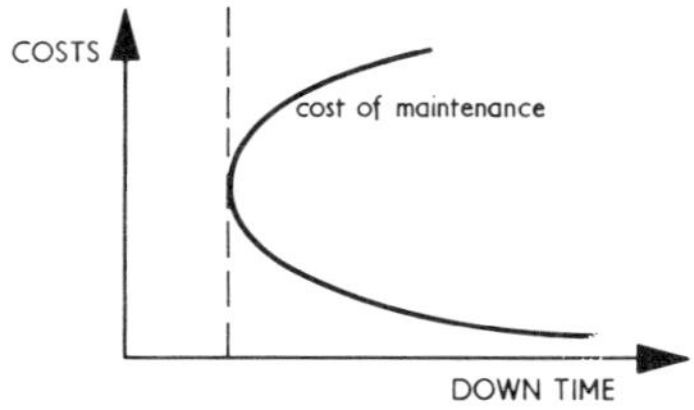

Figure 13 Relation between maintenance and down time costs

reached, further increase of maintenance will result in an increase of down time. Figure 14 shows the effect of this on the total cost curve. This aspect, which is shown more explicitly in figure 15 demonstrates that increasing preventive maintenance does not necessarily lead to a decrease of down times; an increase of down times may result.

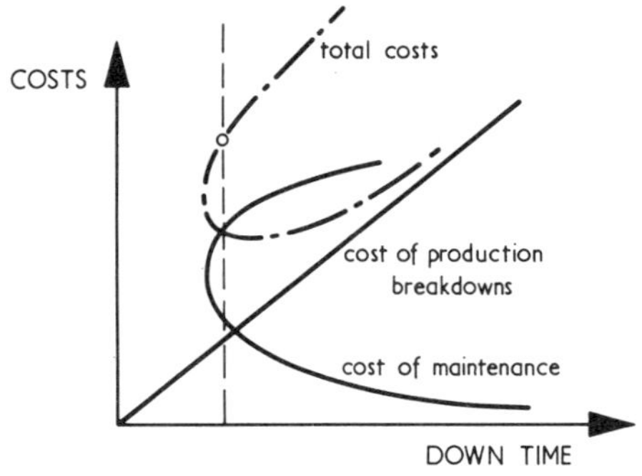

Figure 14 Relation between maintenance and down time cost (corrected)

The effect, also known as "overmaintenance", has been investigated rather thoroughly by HUSSEY et al [14] for an air line, in which it was found that the majority of the components investigated were over-maintained.

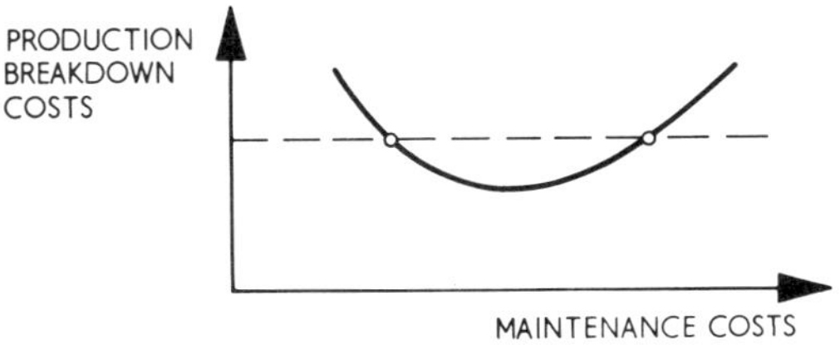

Figure 15 Relation between maintenance and production breakdown costs

The second assumption sometimes found in the literature (e.g. [15] page 71), is that the failure rate of the parts subject to preventive maintenance must be strictly increasing. The rationale for this assumption is the consideration that constant failure rate implies that the probability of failure is independent of the time in use; consequently, replacement does not increase the reliability. In figure 8 curve a, representing the negative exponential case having constant failure rate, preventive maintenance yields no benefit. In the hypergeometric case represented by curve d, preventive maintenance would decrease the reliability. The curves b and c and the ideal case represented by the dotted line hold possibilities of increasing the reliability by means of preventive maintenance. As will be shown later, the assumption mentioned can be held only with the addition: "if the policy depends on information that is limited to the times to failure (or times between failure) only".

6.2.3. *Problems in using behavioral properties*

The failure rate, reliability, maintainability and availability have great potential possibilities in arriving at a structural frame for a theory of maintenance because of the combination of quantifiability, predictability per object and relevance for the possible alternative maintenance policies and maintenance conceptions.

According to the present state of the art however, there are several problems hampering application, namely:

- the composite nature of maintenance objects
- the ambiguity of function and failure
- the problem of data collection
- the difficulty of determining optimal preventive maintenance policies.

The *composite nature* of maintenance objects leads to a collection of failure rates and reliabilities, because the objects consist of various parts. The failure rate of an object is therefore the result of the superposition of the individual failure rates of all parts. Reliability characteristics are not related to one technological type of failure any more. It is likely that, in the end, failure behavior of a system is always random. That makes analytical determination of a preventive policy for a composite piece of equipment much more difficult than the elementary approach suggested.

The *assumed relationship* between *function* and *failure* suggested that there is an unambiguous connection between the two. In studying the function of a system we usually find that the function consists of a number of capabilities. Some of these are not always needed, some are of more importance than others. Function appears to have a composite structure, consisting of a number of subfunctions, which do not all have the same importance and which sometimes can be combined in a different way, corresponding to different operational modes. This leads to rather severe complications when trying to analyse the behavior of objects and the results of the maintenance performed on them.

The problem of *data collection* is or will partly be solved by means of modern automatic data processing and data transmission equipment. This does not solve the problem of how to collect the relevant data at the place of origin. The careful registration of the state at the time of installment, state at the time of failure, failure type, corrective action, etc. requires an expensive amount of work, while the accuracy of the collected information is limited by human constraints [20], [21]. An investigation by the Royal Netherlands Air Force of a sampling system to determine reliability, maintainability, availability and failure rate for aircraft showed that the variance of the deviation from the completely registered figures was far too large to be usable. The majority of data available to-day concerns electronic components. For a wide application of quantitative methods, more figures will have to be gathered originating from tests and real situations, eventually leading to figures that are generally valid for defined conditions and material. Then statistical sampling and testing methods will undoubtedly enable reduction of the workload of registration.

The difficulty of *determining optimal preventive maintenance policies* was mentioned before. If failure times analysis were the only basis for preventive maintenance, the number of possibilities would in all probability be limited. If failure behavior is stochastic, optimal policies can be determined for several situations. Jorgension et al [15] present the analytical theory for this case. In the usual analytical approach, the time to failure is the only information to start from. As this period is regarded as a black box, only the statistical probability of failure can be determined. This does not, however, enable prediction of difference in failure behavior from one individual failure to the other. In many a case there is still a possibility that this can be opened up by looking into the black box period, if individual indications can be found that failure is to be expected. This is illustrated in figure 16. It is based on the assumption that an

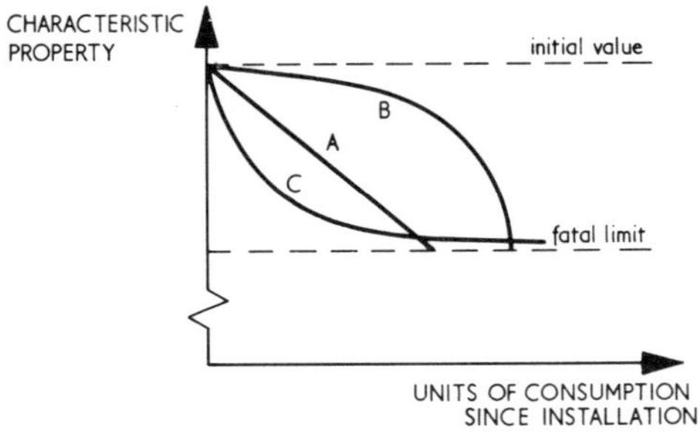

Figure 16 Endurance curves

object subject to failure shows properties that change gradually from an initial value to a fatal limit, being the value of the property at which the object will enter the failed state. If such a property exists, which is manifest if found, the quantitative drift can show different behavior, characterized by the curves A, B and C in figure 16. Curves A and B will be treated separately. Curve C is clearly rather insensitive, as the point at which the fatal limit will be crossed varies much on the horizontal axis, if the value of the characteristics property measured varies a little. This means that it is incorrect to use this type of property characteristic. It could be something like measuring the quality of the paint of a car in order to quantify the operational time left before overhaul of the vehicle will become necessary. Other properties, showing the behavior of the curves A and B will have to be found in such a case. If we look at the distribution of the times to failure, figure 17 illustrates that this concerns the values on the horizontal axis only.

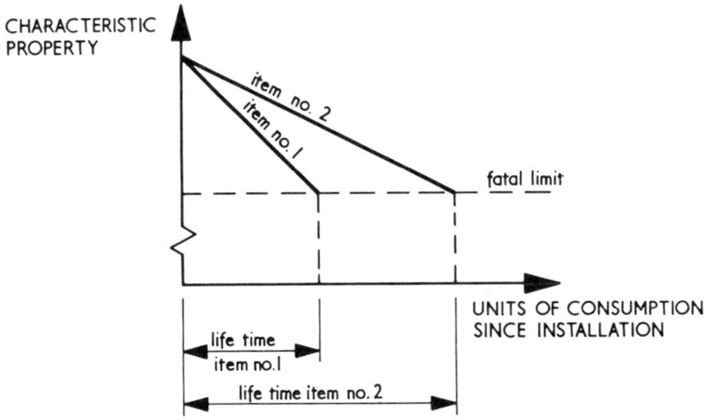

Figure 17 Endurance variation

The same goes for the curves of the B-type. In the case of the A-type curve, the drift of the characteristic property is a linear function of time (or some other time-related unit of consumption).

As figure 18 illustrates, periodic sampling of the value of the characteristic property enables one to decide at what time preventive maintenance will have to be accomplished in order to prevent failure. The sampling constitutes *preventive inspection*, meaning a check of some property after which taking or not taking corrective action is the result of the comparison of the value found and the appropriate control value. In this way it is possible to select individual objects out of a population, in which the distribution of the times to failure may be of any of the possible types. The control limit is termed the *absolute control limit* because the value measured in the sample is compared with a control limit that is directly derived from the value which is fatal. The determination of the length of the periodic inspection interval and of the value of the control limit resembles the inventory control problem of determining the order cycle interval for a system with a possibility of ordering at fixed intervals only and the determination of the reorder control limit. Expected delivery time in that case in analogous to the time estimated to be necessary to plan and execute the appropriate maintenance in order to bring the characteristic property back to its initial level (fig. 16). In the case of curve B the same method, using an absolute control limit, is possible. The model of this curve, however, opens up another possibility that reduces the risk of being too late if the curve has a steep descent in the last phase. In this case (fig. 19), the value of the interval difference could be measured and compared with a control limit–*the relative control limit*.

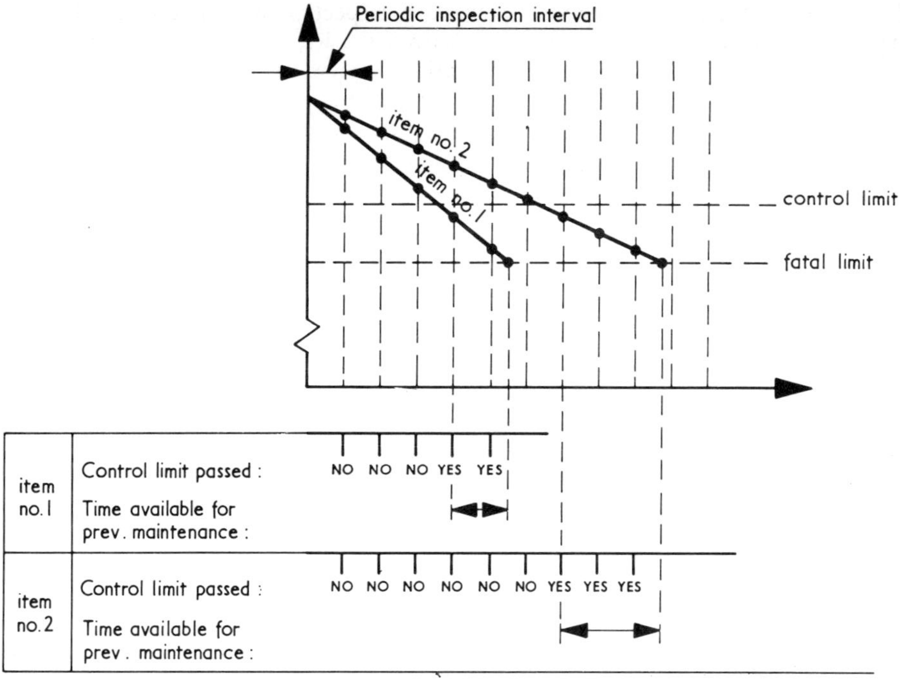

Figure 18 Preventive inspection by absolute control limit

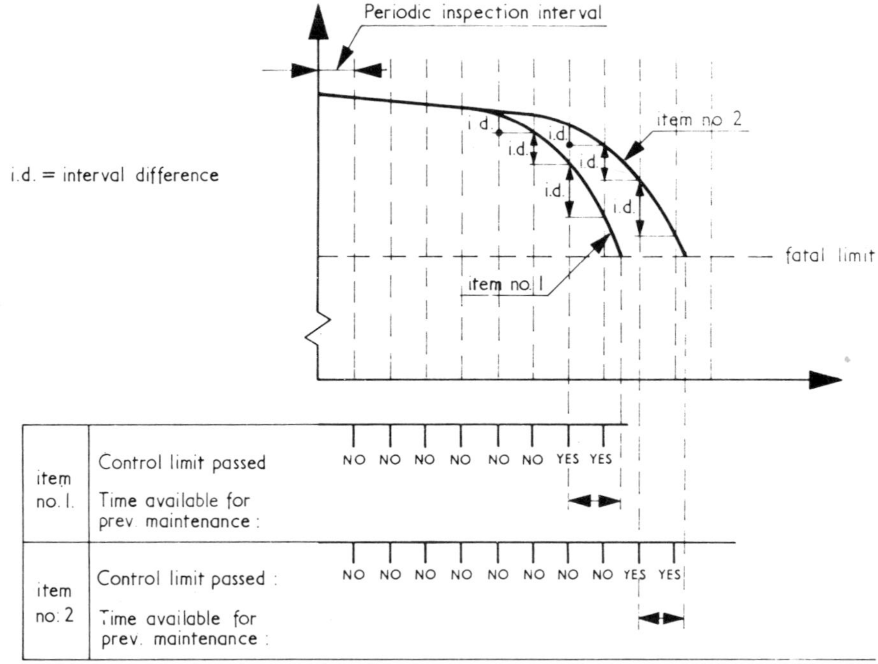

Figure 19 Preventive inspection by relative control limit

The theoretical aspects underlying preventive inspection have not received much attention, though it is not entirely absent from the literature [22]. The systematic research into the possibilities of the application of preventive inspections and how to assemble them into periodic packages in a maintenance conception, seems of primary value for the development of maintenance. The new ideas about designing out maintenance, which lead to more systematic application of modular structures, will undoubtedly offer more profitable opportunities for this method. The measurement of the depth of profile of tyres is a rather well known example of preventive inspection by absolute control limits. A much more sophisticated application, indicating the new possibilities of preventive inspections, is worth mentioning. It is a method to predict imminent failure of aero engines by periodic analysis of engine oil samples, the so called SOAP–Spectrometric Oil Analysis Program [23]. By spectrometric analysis of the oil sample, the concentration of a number of metals–considered characteristic–are measured. By logging these values of an engine and by studying their development, abnormal increases are detected. Analysis of the underlying technological behavior of the values, individually and in combination, shows what components in the engine assembly are expected to fail. A new development in data collection is AIDS–Aircraft Integrated Data System–opening up the possibility of continuous valve registration during operation [24].

It is evident that the application of the preventive inspection method does not depend on a theoretical approach to the maintenance aspects only. Also the systematic search for characteristic properties of the technological processes deserves special attention. A working group of the Netherlands Society for Efficient Maintenance is studying practical aspects of the preventive inspection method and is expected to report on their findings early next year.

6.3 Requirements, wishes and constraints from production

6.3.1. Primary aspects

The primary wish of production is continuity of the production process. In some cases production will be uninterrupted. This is to be regarded as a special case of the production pattern. Usually, the production pattern consists of a series of productive and non-productive intervals. The maintenance conception has to take into account the opportunities for maintenance in the non-productive intervals in the production pattern, as well the reliability and availability requirements in the productive intervals. The alternative policies are determined by the object to be maintained. This leads to distinguish between the following types of objects:

- large, *singular* systems (such as the energy supply system in a factory, a radar installation, a paper manufacturing machine, etc.)
- *groups* of identical rather *simple* and cheap objects (such as lamps)
- *groups* of identical rather *complex* and expensive objects (such as motor vehicles, aeroplanes, lathes etc.).

6.3.2. Large, singular systems

The most important property of large, singular objects is the fact that an important operation cannot be fulfilled if the system fails, because there is no alternative capability available. That means that reliability and availability should have high values. The reliability is given, being an inherent, design property; it can be high by virtue of the use of components with high reliability and also by using redundancy. The availability may be influenced by the using organization by minimizing maintenance times and by searching for preventive inspection possibilities, thus increasing the organizational reliability. The maintenance times may be reduced by having components available for replacement, so as to eliminate the maintenance time of the component itself from the critical path. For the lower levels of maintenance, the schedule will make maximum use of the non-operational periods in the production pattern if it is progressively structured. This *progressive maintenance* policy is a preventive policy in which the collection of preventive maintenance activities that have to be performed periodically has been split up into packages of activities that require the same length of maintenance time. In this way the periodic maintenance can be dispersed over the whole period, so that execution can be fitted into the non-operational periods in the production pattern or can be combined with non-operational periods because of corrective maintenance (opportunistic policy).

For higher levels of maintenance requiring a relatively long period of down time, the appropriate method to minimise down time is process-planning and control by means of network-planning techniques. The availability results are limited by the inherent availability built into the design, which in some cases can be relatively low. The maintenance policies to be considered for systems of this type need fundamental study to determine the maintenance conception and the achievable availability to be accepted as a constraint of the existing design.

6.3.3. *Groups of identical, simple objects*

If groups of identical, simple, and hence usually cheap, objects are in use, renewal instead of repair is more economic in many cases. If the effort to renew the units hardly depends on the number to be renewed and if the renewal effort is not to be neglected compared with the price of the item, it can be more efficient to renew the whole group at a time, when one or more have failed or when a characteristic measurable property has reached a certain value. In the case of lamps, this could be the rule for renewal of a group based upon a certain number failed or after the decrease of the light intensity below a certain level. This is a preventive inspection method in cases where the failure rate increases with age. It is known as "group replacement" or "block replacement" and has been studied analytically, giving models for practical application [26].

6.3.4. *Groups of identical, complex objects.*

If an object is used in quantity (e.g. aircraft, vehicles, lathes, etc.), the availability manifests itself to production as the number of units in the ready state. One can influence this quantity by having an additional number that are allowed to be inoperative for maintenance. Optimization then concerns the investment in this additional number and the cost of achieving a certain maintenance throughput time. Equipment of this type usually requires several levels of maintenance. If the number available with the user varies too much, this may require a more systematic preventive structure of the maintenance conception. It is then posssible to determine the average number that will be in the maintenance part of the loop (fig. 2 and fig. 3). This average has to be increased by an additional amount because of fluctuations in the arrival pattern for maintenance and in the maintenance throughput times.

The final decision on the number of items that will be used as buffer, called turnaround, should be based primarily on the mission effect [27] of the object and on the investment in procuring additional unit over and above the number needed for direct use. This decision implicity includes the determination of average throughput times for preventive maintenance and the quantified non-availability to be accepted at the user's level.

In controlling this number, the so called "staggerline" (fig. 20 and 21) enables anticipation of the arrivals for maintenance. The staggerline is a graphic representation of the behavior of the property characteristic between two preventive maintenance treatments, e.g. hours operated or mileage in an overhaul policy. Fig. 20 shows a case

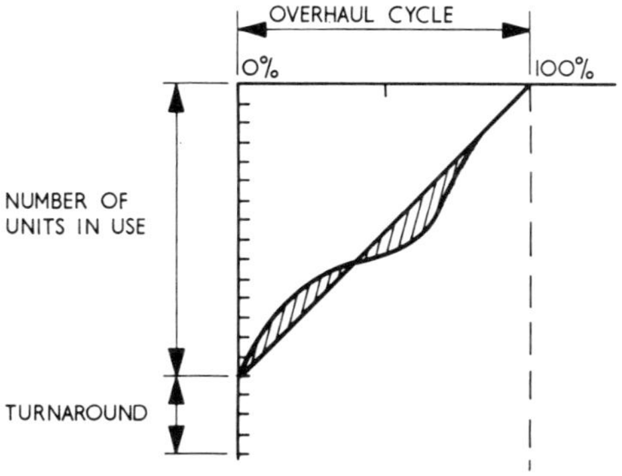

Figure 20 Example of HSO Status

where the arrivals in the near future will most probably be about the average. Fig. 21 shows a case where arrivals in the near future will, in all probability, be lower than normal, but will be followed by a period in which a larger number than normal will become due for maintenance. This may be prevented by taking in units at an earlier moment than the nominal end of the cycle or by influencing the use of the units in use by indicating which units have to be used more or less intensely. In the case where a

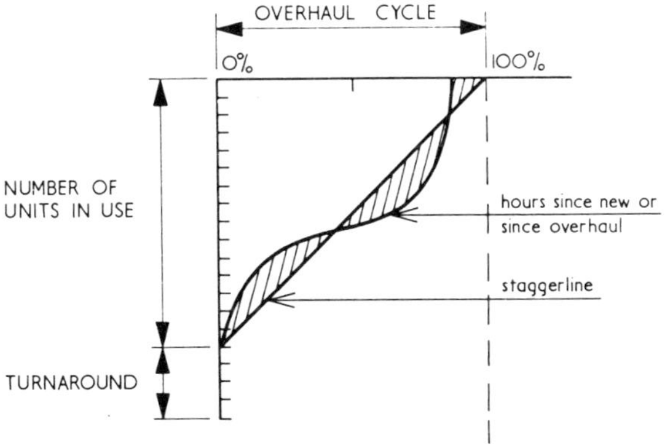

Figure 21 Example of HSO status

group of identical units is introduced at the same time, serious danger exists that the arrival pattern will show a high peak, causing the number of units not available for production to become greater than the number calculated as turnaround. This effect is illustrated in fig. 22. To prevent this peak, it is necessary to start the maintenance

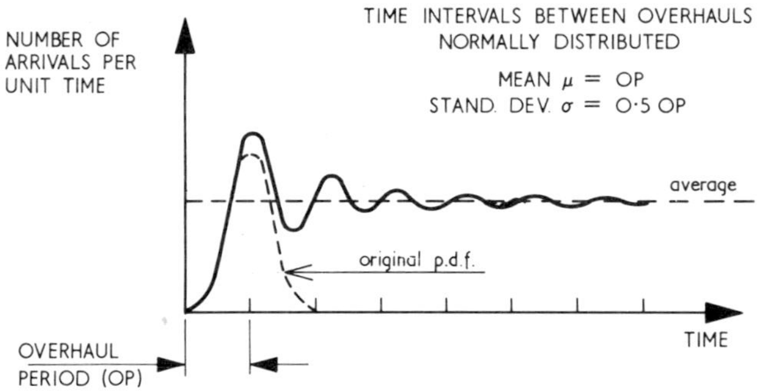

Figure 22 Arrival pattern for overhaul

earlier in the first period for a number of the units. The effect of this is shown in figure 23.

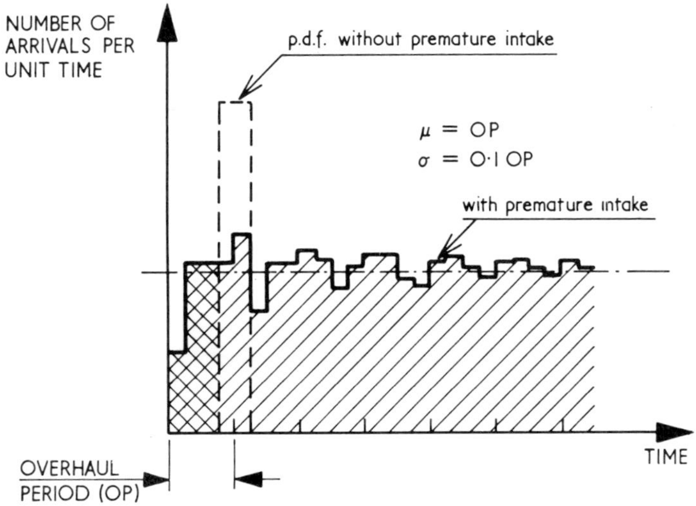

Figure 23 Numerical simulation of the arrivals for overhaul with premature intake

Figure 24 shows (for a real case [27]) how the decision to start premature maintenance can be made and followed up by means of the periodical sampling of the status. This approach has been investigated and tried out by the Royal Netherland Air Force for a practical application [27].

The maintenance of groups of identical, rather complex, units can be controlled in this way, including anticipating measures for their use in production and for their arrival in the maintenance process.

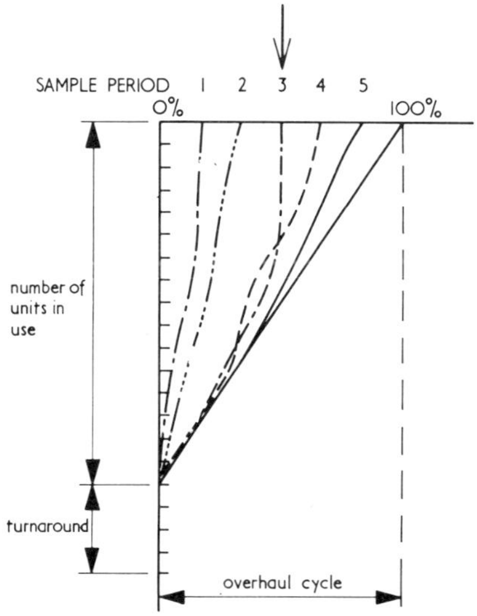

Figure 24 Five successive HSO status samples with premature overhaul start after sample at end of period 3

6.4 Conclusions

The failure rate, reliability and availability are of vital importance to maintenance because their required value can be derived from production and because maintenance results can be measured by means of these properties. The properties in real situations are different from the inherent values because of organizational influences. Within a limit to the achievable results, the maintenance conception and the control mechanisms for maintenance execution determine the actual results.

To arrive at a maintenance conception, the alternative maintenance policies have to be known to enable the selection of the optimal policy. This requires study of the production pattern.

In many cases the desire to apply preventive policies will require a study of the charactertistic properties that can serve as control values.

Singular objects, groups of simple objects and groups of complex objects require different approaches.

As yet, there is no theoretically developed methodology for the process of constructing the maintenance conception for objects.

7. INVENTORY CONTROL

7.1

Quite some effort has been devoted to the study of inventory control problems, in particular through the impact of operations research. The underlying philosophy is the recognition that having stocks involves costs of stock-holding, and that having too small or too large stocks in the end leads to higher costs than choosing the right amount. The models concentrate on finding minimum stock levels for a set minimum service rate. The service rate represents quantitatively the acceptability of having to wait.

Typical problems to be solved are: forecasting demand, determining the moment at which to order and the amount to be ordered [12] in echelonized inventories, developping distribution decision rules of a more complex nature because of inter-dependencies [13]. There is a vast amount of literature on inventory control that reflects the state of the art.

7.2 Typical aspects of inventories of spares

7.2.1. Multiple demand

The most striking aspect in comparing the inventory problems in a maintenance system with those in a production system is that the "raw material" available at the beginning of a production process is practically non-existent. The material needed is to a great extent of specific identity and is needed in some diversity at different points of time in the maintenance process. There is some resemblance with the production assembly processes. The delay penalty in case of stockout and the diversity of material that can cause it require a rather sophisticated inventory control system in order to prevent serious decrease of availability or high investments in stocks.

7.2.2. Diversity of frequency of demand

A second aspect that differs from the production process is the fact that the items to be carried are determined to a considerable degree by the mere fact of possessing a piece of equipment and having to maintain it. It is not possible to exclude an item from the inventory because it is unattractive from a point of view of inventory control itself. The same goes for the possibility of simplifying inventory control by increasing market share and hence demand. This rather rigidly defined assortment is more meaningful if the frequency of transactions is taken into account. Fig. 25 illustrates the diversity of an inventory running into over 200,000 different items and covering very low as well as very high frequencies of transaction. The most notable fact, typical

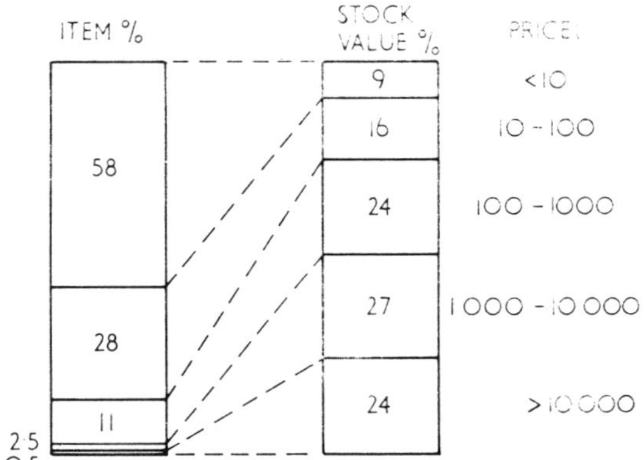

Figure 25 Value distribution of stocks

for maintenance stocks, is the large portion of items in the low frequency range (Fig. 26). Because a large number of these items are complicated, involving long and unreliable procurement times and high prices, they account for the greater part of the investment in stocks for maintenance. Owing to this diversity of the behavior of maintenance stocks, a few simple decision rules for the control of the inventories are not sufficient. The rules applied will have to cover the full range. The problem of finding workable decision rules is made worse by the limited effectiveness of statistical methods applied to small populations.

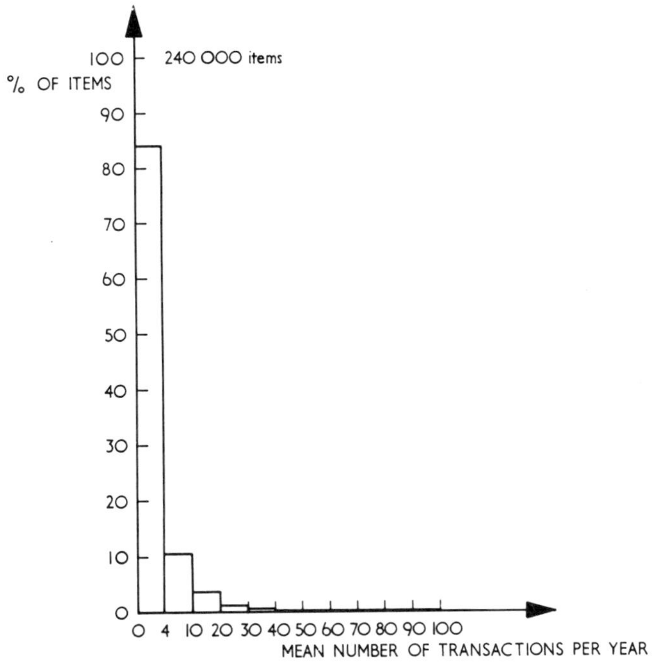

Figure 26 Item transaction frequency

7.2.3. Forecasting

A third problem is the limitation in the possibility of determining forecasting of demand. Research into the correlation between consumption and intensity of operations by the Department of Industrial Engineering of the Technological University of Eindhoven [28] confirmed the findings of similar research that application of this method–insofar as it is applicable at all–yields minor improvement as compared with statistical forcasting [29, 30]. However, the items investigated were in the range of over 20 transactions per year. Forecasting demand by a casual model for slow moving items will undoubtedly require rather complicated decision models with many variables. If the frequency of demand is high and if the demand is stable or has a reasonably well-known trend, several forecasting techniques can be applied.

One may be able to forecast via causality in those cases where the demand is mainly determined by maintenance activities of a repetitive nature. Short-term and mid-term demand can then be predicted by using a bill of material, in which the number of items needed per maintenance job is not rigid as in production, but is foreseen with a probability of expected demand.

7.2.4. *Service rate*

A variable of vital importance in inventory control theory is the service rate, usually defined as the fraction of all requisitions that could be honored by direct issue, measured over a representative period. This control variable is too simple to serve in inventory control for maintenance processes. As a maintenance job usually requires more than one item, the waiting times for material are of the touring car type. More relevant than the service rate per individual item is the probability that an item will cause a delay in a maintenance job. It is to be noticed that ordering and receiving material, together with the maintenance operations, are activities in a network structure. This means that material delivery delay will cause maintenance delay only when this activity is on the critical path. One step further would be to anticipate the actual demand and add to a requisition the latest time of arrival. The time element in material suggests that service rate should be viewed as a more complicated aspect than the usually mentioned ratio (fig. 27 and 28) [31]. What one would like to know

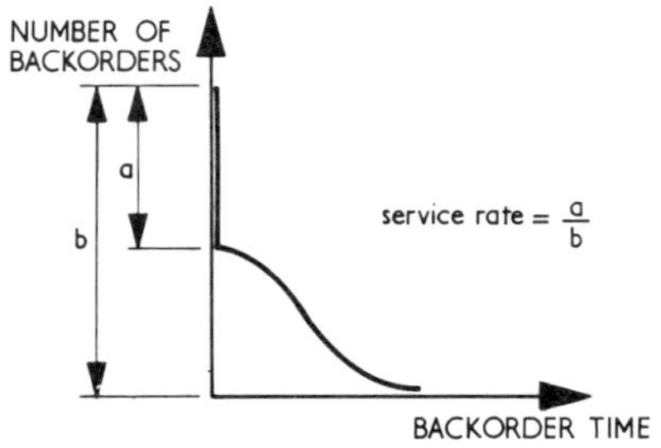

Figure 27 Backorder time characteristic

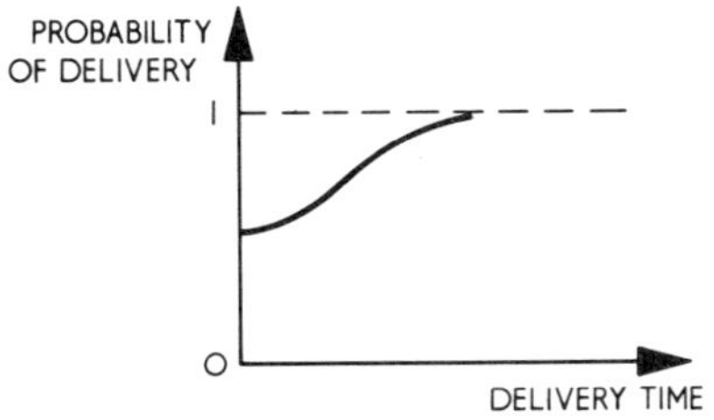

Figure 28 Delivery time probability characteristic

in the end is the quantitative effect on availability of different inventory decision rules. Trials in that direction have hardly been made, undoubtedly because of the complexity of the underlying models. KARR described an approach in which the spare-part essentially is qualified and serves as one of the decision parameters [32].

7.2.5. *Turnaround*

In this paper turnaround is defined as the number of units of an item (identity) held in addition to the number continually authorized for operational use, in order to be able to execute maintenance while affecting the number continually in use as little as possible. The decision to use a turnaround results from the maintenance conception chosen for the item concerned. If it concerns components, the maintenance of the component corresponds with delivery time as far as its behavior in the inventory is concerned; but as it is a closed loop system, the delivery time can be influenced by means of priority rules in the maintenance process. This puts more emphasis on the maintenance planning and control side than on the inventory control side. Behavioral aspects of systems of this type require fundamental research.

7.2.6. *Echelonzied structure*

Organizations having a maintenance task of some size usually show echelons in maintenance as well as in the inventories. Not much research has been done on the effects of interrelations between the different levels. The study of Forrester [13] has shown the dangers of multistage systems in which decision rules are developed independently for each stage. In the Royal Netherlands Air Force the introduction of third generation automatic data-processing equipment enabled us to cope with this danger by—apart from other reasons—centralizing the information of all echelons, instead of the usual hierarchical structure visualized in figure 29. A new problem,

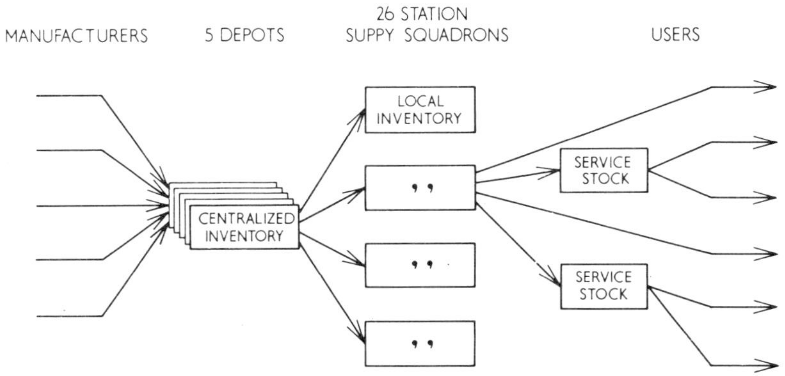

Figure 29 Two stage inventory chain

originating from this decision, is the manner in which the redistribution, vertical as well as lateral, should be controlled. Study of this problem by an analytical approach was started recently by the Technological University of Eindhoven.

7.2.8. *Initial demand and insurance type items*

There are two cases in which the forecast of demand cannot be based on historical figures. The first case is the forecast of initial demand at the time of introduction of a new piece of equipment. The second case is the forecasting of insurance type items. These are items that, for practical and economical reasons, can only be procured as long as the production line concerned is working. Though experience elsewhere and manufacturer's advice are helpful, the final decisions lie with the user. In these situations the only source available is the experience and know-how of personnel in the user's organization, assisted by historical data of equipment supposed to be similar. It is doubtful if analytical research into the behavior of technological systems will lead to the replacement of man by quantitative models for this type of forecasting. More reliable advice by the manufacturer can be expected as a by-product of the growing attention that is being paid to reliability and maintainability aspects in the design stage. Karr [32] mentions results of research into the reliability of estimation methods by mechanics that deserve more attention for standardized application.

7.2.8 Conclusions

The following conclusions can be summarized. Reviewing existing scientific knowledge in the area of inventory control, we find a vast amount of a more or less fragmentary nature. The many problems which strike one when reviewing inventory control of spares for maintenance seem to justify an independent theoretical approach fitting in fragments already developed.

To achieve that result it is necessary to find the underlying structure that enables the selection of policies and methods to be used in actual situations. This will lead to a better indication of the areas to be researched by analytical approach, by simulation and by empirical research in actual situations.

Research of this kind is being done by the Royal Netherlands Air Force in cooperation with the Department of Industrial Engineering of the Technological University of Eindhoven. At this stage the emphasis lies with analytical model-building and the use of real figures for testing, leaving out the maintenance aspect for the time being.

8. ADDITIONAL ASPECTS

8.1 Economics

The analysis of process behavior and the design of systems and decision rules to control them requires the selection from alternatives. It is evident that, in may cases, direct comparison is impossible and that measurement of the alternatives needs a common denominator. If possible, the most complete study is the cost-effectiveness approach, which includes the time span from design to the ultimate disposal. Studies of this kind have emphasized the importance of estimating the costs over the operational period at the time of design and procurement [25]. This aspect will have to be used whenever practical as a matter of course in the building of the quantitative models. As knowledge of the economics of the effectiveness of non-productive organizations is growing [33], its contribution can be expected to grow substantially.

8.2 Multidisciplinary approach

It is generally accepted that the classical dogma derived from a monodisciplinary organization theory can not hold out against critical scientific research and re-evaluation. Though these research activities are still going on and have not yet resulted in a new generally accepted theory, it is taken as evident that real situations in an organization have to be met with a multi-disciplinary systems approach. This goes also for maintenance, which has the nature of a sociotechnical system. The developments in this area of knowledge will undoubtedly lead to important contributions to the analysis and application of maintenance conceptions and of maintenance control systems.

8.3 Maintenance and design

Because maintenance as such is concerned with objects that enter the maintenance process as they have been procured, all inherent properties having relevance to maintenance are more or less fixed from the moment of arrival. Though design cannot be regarded as maintenance, the rapidly growing knowledge of the way in which to make allowance for maintenance aspects in the process of designing should be mentioned. Symptoms of it are modular constructions, accessibility, redundancy, built-in test equipment and such-like. Maintenance will not only benefit from this development, but will also have to supply empirical statistical information in order to indicate what aspects require special attention in new designs. The Netherlands' Society for Efficient Maintenance installed a working-group recently with the task of developping directives, designed for practical use, to be used in the formulation of requirements derived from maintenance, and to be incorporated in the procurement specification.

9. TOWARDS A THEORY OF MAINTENANCE

9.1 Conclusion

According to the point of view developed in the foregoing paragraphs, maintenance is to be regarded as a controlled process of activities. The need for better control of maintenance is evident. To arrive at a better control, there is a need for a theoretical, comprehensive structure that is unavailable today. Fragmentary knowledge in some areas appears to be applied scarcely, in spite of the depth covered scientifically. This indicates that further research concerning these areas should be directed not vertically but laterally, according to the systems view, and that capability of practical application should be a vital consideration. Knowledge in analogous areas requires little or no modification for application. On the other hand, other areas appear to have been investigated insufficiently and consequently require fundamental research.

Scientific fields that appear to have been, or are being, researched sufficiently in the present state of development of a maintenance theory are:

- production control theory
- OR theory in particular concerning failure rate behaviour, reliability and availability
- inventory theory as far as fundamental aspects are concerned
- modern organization theory
- economics.

Problem areas requiring fundamental research aimed at the maintenance process appear to be:

- maintenance policies including object behavior

- design of maintenance conceptions
- cybernetic mechanisms concerning the balance of the maintenance and production process
- inventory control of spare parts.

9.2 Recommendation

The development of general theories and models is one of the tasks of the universities, in particular in the departments of industrial engineering. There is a need for more attention to the maintenance area than in the past.

The systematic analysis of and approach to maintenance in real cases will primarily be a task of the using organizations. As the maintenance task is relatively small compared with other functions in an organization, activities of reasonable size in this field can be expected only from large organizations and from the military organizations in particular. In the first place they have to test the results of general models in real situations so as to correct the general models. In the second place, they have to report on the findings of their research in real situations so as to supply the experimental knowledge needed for inductive development of general models.

The first step to be taken now is the organization of more appropriate ways of exchanging information than are available now.

In more general terms, it could be stated that maintenance development requires that maintaining today's situation be prevented.

LITERATURE

1. Turban, E., The Pulse Beat of Maintenance Today. *Factory*, 124 (1966) nr. 6, pp. 98-110.

2. Rawson, K.J., The Navy's need for reliability. Paper 1, 1968-Conference on "The Reliability of Service Equipment", London 1969. The Institution of Mechanical Engineers, pp. 1-9.

3. Wharton, T., The need for Reliability in the Royal Air Force. 1968-Conference on "The Reliability of Service Equipment", London 1969. The Institution of Mechanical Engineers, pp. 10-15.

4. Geraerds, W. M. J. and Le Poole, L.H., Nomenclatuur 2. *Onderhoud,* 2 (1967) nr. 4, pp. 74-75.

5. Buffa, E.S., *Modern production Management*. John Wiley & Sons, Inc., New York, 1965.

6. Conway, R. W. and Maxwell, W.L., Network dispatching by the shortest operation discipline. *Oper.Research*, 10 (1962), nr. 1 pp. 51-73.

7. Conway, R.W., Priority dispatching and work-in-process inventory in a job shop. *Journal of Industrial Engineering*, 16 (1965) nr. 2, pp. 123-130

8. Conway, R.W., Priority dispatching and job lateness in a job shop. *Journal of Industrial Engineering*, 16 (1965) nr. 4, pp. 228-237.

9. Conway, R.W., Maxwell, W.L. and Miller, L.W., *Theory of Scheduling*. Addison-Wesley, Reading, Massachusetts, 1967.

10. Bakker, W., De levertijd in machinefabrieken. Doct. thesis, Waltman, Delft, 1965.

11. O'Brien, J.J., *Scheduling Handbook*. McGraw-Hill, New York, 1969.

12. Van Hees, R.N. and Monhemius, W., *Produktiebesturing and voorraadbeheer. Algemene beginselen*. Centrex, Eindhoven, 1964.

13. Forrester, J.W., *Industrial Dynamics*. M.I.T. Press, Cambridge, Massachusetts, 1964.

14. Hussey, P.A. and Thomas, S.G., Are scheduled component overhauls necessary? SAE–National Aeronautic and Space Engineering and Manufacturing Meeting, Los Angeles, Oct. 5-9, 1964, 8 pp.

15. Jorgenson, D.W., McCall, J.J. and Radner, R., *Optimal Replacement Policy*. North Holland Publishing Company, Amsterdam, 1967.

16. Morse, P.M. *Queues, inventories and maintenance*. Wiley, New York, 1958, 202 pp.

17. Goldman, A.S. and Slattery, T.B., *Maintainability: a major element of system effectiveness*. John Wiley & Sons, New York, 1964. 282 p.

18. Geraerds, W.M.J. and Le Poole, L.H., Nomenclatuur 4. *Onderhoud*, 2 (1967), nr. 10, pp. 216-218.

19. Newbrough, E.T., *Effective Maintenance Management*. McGraw-Hill, New York, 1967, p. 61.

20. Harris, D.H., The nature of industrial inspection. *Human Factors*, 11 (1969) nr. 2, pp. 139-148.

21. Carroll, J.M., Estimating errors in the inspection of complex products. *AIIE Transactions*, 1 (1969) nr. 3, pp. 229-235.

22. Barlow, R.E. and Proschan, F., *Mathematical theory of reliability*. Wiley, New York, 1965.

23. Thacker, R., Soap. *Flight Comment*, May-June 1967, pp. 2-6.

24. Scott, G., Aides Memoires. *Flight International*, 1968, 4 July, p. 17.

25. McFarland, J.W., Integrated logistic management. *The Journal of the Armed Forces Management Association*, Vol. 2, 1967, nr. 7, pp. 79-87.

26. R.L. Ackoff and Sasieni, M.W., *Fundamentals of operations research*. Wiley, New York, 1968, pp. 211-214.

27. Geraerds, W.M.J., LTP-beheersingssysteem voor motorisch aangedreven gronduitrusting bij de Koninklijke Luchtmacht. *Literatuuroverzicht TDCK*, 11, 1967, nr. OR 97, pp. 175-196 en 11, 1967, nr. OR 99, pp. 247-270.

28. Van 't Hullenaar, F.J.W.H. and Van Winkel, E.G.F., Forecasting spare parts demand for aircraft of the Royal Netherlands Air Force. Staff Section Management, R Neth AF, The Hague, Jan. 1970. Paper to be published.

29. Haber, S.E., A comparison of usage data among types of aircraft. *Naval Res. Logistics Quarterly*, 14 (1967) nr. 3, pp. 399-410.

30. Brown, B., Micro simulation in logistic research. A comparative study of prediction techniques. Rand report nr. P-1980.

31. Van't Hullenaar, F.J.W.H., Problematiek rondom servicegraad, Staff Section Management R Neth AF, 17 nov 1969. Unpublished paper.

32. Karr, H.W., A method of estimating spare-part essentiality. Rand paper P-1064. 17 April 1957.

33. Bogaert, K.A.M. and Kuypers, S.W.M., *Kosten en effectiviteit*. Universitaire Pers. Rotterdam, 1969.

DISCUSSION

O'Connor: In his discussion on the optimum type of preventive maintenance for various types of system, Col. Geraerds does not detail what in my opinion is a very important possibility, which has received very little study. I refer to the possibility of applying progressive maintenance to groups of identical complex objects. For these objects, e.g. aircraft, the ratio of unscheduled workload to scheduled workload is high, and will not probably decrease within the near future. The possibility of using unscheduled downtime, or other opportunity, e.g. weather, to perform scheduled maintenance therefore seems to commend itself. The RAF in fact uses this type of maintenance on some aircraft, but for various reasons the system is going out of favour, largely in my view because of a lack of comparative quantitative analysis. Have you considered carrying out such an analysis?

Geraerds: I mention progressive maintenance to distinguish it from other policies. When to apply these policies by means of a methodological selection procedure is what I mention the constructuion of the maintenance conception of the object concerned. I agree with you that this, as I state in my paper, is an underexplored area that requires research. Progressive maintenance is not restricted to large, singular objects. But we do not know what conditions and constraints have to be taken into account when deciding on application of it. I believe it is of benefit for the development of insight in the conditions and constraints of this policy, that you publish your findings in the case you mentioned. I did not consider to do analysis on this policy as such. Before starting research on this type of aspects there should be more insight in the methodology of construction of maintenance conceptions. I am studying that problem as it should, in my opinion, have priority over quantitative analysis of one of the policies we know now.

Buck: Has it been proved that in system analysis it is better to use someone else's model than to devise one for your own special needs.

Geraerds: The direct answer to this direct question is: no. I don't know of any research that led to an answer that can be validated as a proof. Neither do I have an opinion on the question as it is formulated. In my experience the applicability of models developed elsewhere differs from case to case. Therefore I think it is worthwhile to study methods and results of others as for their wider suitability than the originating problem.

An Application of Network Theory to the Determination of Minimum Cost Aircraft Pipeline Factors

S. Scott Sutton
Richmond Lloyd

1 INTRODUCTION

The purpose of this paper is to present a practical example of how operations research helped solve a logistic support problem. The emphasis is on application rather than on theory.

An important budgeting problem in the U.S. naval aviation program is to determine the number of aircraft to procure so that a desired level of operating inventory of aircraft is maintained, given that some aircraft will not be operational as a result of certain maintenance procedures. This problem became especially important during the Southeast Asia build-up when Navy procurement managers were urged to scrutinize more closely the trade-off's between buying new aircraft and reducing the number of aircraft tied up in maintenance.

This procurement decision was based on a planning factor called the "pipeline" factor. The term "pipeline" refers to aircraft unavailable to operating commands. Pipeline is defined to mean those aircraft tied up in Progressive Aircraft Rework (PAR), en route to and from PAR, and in Special Rework. PAR is a regularly scheduled maintenance action; Special Rework is depot maintenance involving unscheduled repairs and sometimes modifications. The operating inventory, then, is the number of aircraft not in the pipeline, and the pipeline planning factor is the ratio of pipeline aircraft (aircraft unavailable) to the desired operating inventory (aircraft available).

$$\text{Pipeline factor} = \frac{\text{Aircraft unavailable}}{\text{Aircraft available}}$$

Procurement managers applied this factor to the desired operating inventory to determine how many pipeline aircraft should be procured to get the desired level of operation. The problem was that Navy procurement managers lacked an analytical method for determining an aircraft pipeline factor based on an economic consideration of the trade-off's between investment in pipeline aircraft and in the PAR maintenance process. Consequently, the Center for Naval Analyses was requested to conduct an Aircraft Pipeline Study with the following objectives:

(i) Develop an applied analytical model which can be used by navy planners to predict the minimum cost pipeline planning factors for aircraft of a particular type, model, and series.

(ii) Demonstrate the capabilities of the Model with an actual analysis of the A-7 aircraft pipeline factor.

The A-7 aircraft reworked at the Naval Air Rework Facility (NARF) in Jacksonville, Florida was selected for demonstration of the model. The results desired were the determination of the optimal pipeline factor for the A-7 and its associated resource requirements and costs.

The model was to be developed under the following assumptions:

(i) The operating inventory of aircraft is the force level planned in the five year defense program.

(ii) The model is for peacetime planning.

(iii) The frequency of PAR, transit time, and special rework time are taken from guidance set by the chief of naval operations.

(iv) The model takes the existing NARF resources as a given base against which to apply additional resources.

(v) The arrival rate of aircraft for PAR is a constant.

(vi) A deterministic PAR process is assumed; that is, stochastic variations in job completion times in the PAR are not considered, and

(vii) The model is designed for a stable aircraft inventory.

2 DETERMINANTS OF THE PIPELINE FACTOR

2.1 Definition of the Pipeline Factor: the Pipeline Factor Equation

The pipeline factor was defined as the ratio of pipeline aircraft to the operating inventory. Expressed in terms of time, the number of pipeline aircraft depends on months in PAR, months in transit, and months in Special Rework, and the operational service life of the aircraft.

Thus, the determinants of the pipeline factor are:

(i) PAR Time
(ii) Special Rework Time
(iii) Transit Time
(iv) Operational Service Life

Expressed in terms of time, the pipeline factor equation is:

$$\text{Pipeline Factor} = \frac{\text{Months in PAR} + \text{months in transit} + \text{months in special rework}}{\text{Operational service life (months)} - \text{months in special rework}}$$

Transit time = 40 calendar days
Special rework = 11 months
Operational service life = 108 months
Time period = service life = 116 months

The assumptions of the study specified that all of these determinants, with the exception of months in PAR, were to be taken as given. The data for the A-7 is shown. Thus, the only variable in the above equation is PAR time. Pipeline here then always refers to PAR pipeline.

2.2 The Pipeline Factor Cost Equation

We can now restate the problem more precisely. For given values of special rework time, transit time, and aircraft operational service life, it is necessary to determine some PAR time for which the pipeline factor is optimal. An optimal pipeline factor depends on the costs involved in the PAR process. We want to identify these costs and develop an expression for pipeline factor costs. Derivation of the optimal pipeline factor can then be shown to be equivalent to a simple cost-minimization problem: finding the PAR time for which relevant pipeline costs are a minimum.

Since some desired level of operating inventory of aircraft is taken as given, additional aircraft must be purchased for pipeline to maintain such a level. Consequently, if the PAR time is reduced, the number of aircraft required for pipeline is also reduced. PAR time, however, cannot be reduced without incurring additional costs, because additional resources must be supplied to the NARF performing the PAR. Thus, there is a trade-off between investment in pipeline aircraft and investment in resources to reduce PAR time. This trade-off can be formulated as a cost-minimization problem, shown in figure 1.

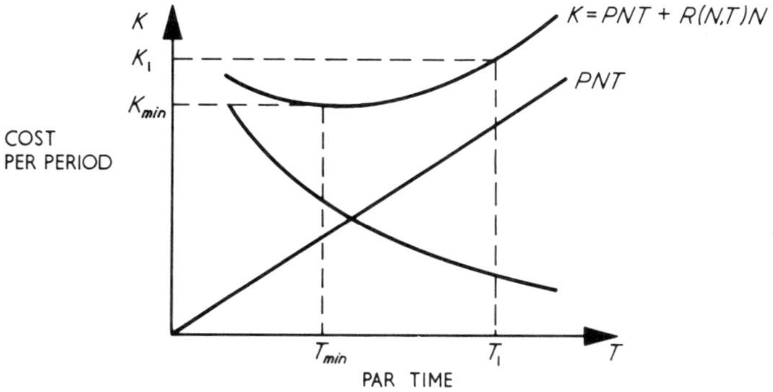

Figure 1 The pipeline factor cost equation

The pipeline factor cost equation is given by

$$K = PNT + R(N,T)N,$$

where

K = total cost per period
= (total pipeline aircraft investment costs per period) +
= (total PAR costs per period),

P = procurement costs per period per aircraft

N = number of aircraft arriving for PAR per period, the induction rate

T = PAR time and

$R(N, T,)$ = total PAR cost per aircraft as a function of T.

In this equation, PAR time, T, is the only variable under control of the decision-maker. The expression for total PAR cost per aircraft, $R(N, T,)$, is written to include N to stress that this cost may also depend on the induction rate, although in this study N is not a variable but a constant. The problem is to determine the PAR time, T, for which the total cost per period, K, is a minimum.

2.3 Determination of the Optimal Pipeline Factor

From the pipeline factor cost equation, it can be seen that determination of an optimal pipeline factor is reduced to the equivalent problem of determining the optimal PAR time. The procedure is straightforward, as shown in figure 1. First, determine the PAR time, T_{min}, which provides minimum total costs, K_{min}. Next, substitute T_{min} into the pipeline factor equation developed above. This yields the optimal pipeline factor. The pipeline factor determined in this manner is optimal, of course, only when the assumptions of the analysis hold.

The first-order condition for the total costs, K, to be a minimum are simple to state. Taking the first derivative of K and setting it equal to zero, we obtain:

$$PN = -dR(N, T)N/dT, \text{ or}$$

$$[\text{slope of } PNT] = -[\text{slope of } R(N, T)N]$$

In other words, comparison of these slopes is a means of determining the economic trade-off of extra savings in pipeline aircraft against extra investment in reducing PAR time. Thus, reductions in PAR time, T, are economically worthwhile so long as savings in pipeline aircraft procurement costs are greater than the additional PAR cost required to achieve reductions in PAR time.

2.4 Modification of the Pipeline Factor Cost Equation

This is all simple enough. But the discussion of pipeline factor costs has focused on total costs of the PAR process, and it is empirically difficult, if not impossible, to determine total costs from present Naval accounting procedures. The preceding discussion, however, emphasizes that an optimal pipeline factor can be determined from a study of the additional costs required to reduce PAR time. That is, it is not necessary to determine the total costs of reworking aircraft for various PAR times; it is only necessary to determine the additional costs incurred in reducing PAR time from some given current time. This concept allows a redefinition of $R(N, T)N$, the total PAR cost per period. This redefinition also greatly facilitates data collection on the costs of the PAR process which frustrated some earlier study efforts to determine an optimal pipeline factor.

This modification of $R(N, T)N$ is shown in figure 2. As previously defined, $R(N, T)N$ gives the total PAR cost per period for various PAR times. So, if the current PAR time is T_1 in figure 2, then the total PAR cost is given by an amount x, shown by the dashed line from T_1 to a point on the curve $R(N, T)N$. But since it is only the

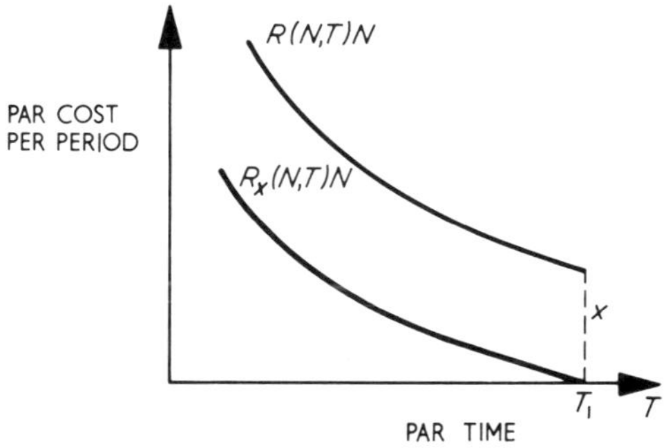

Figure 2 Modification of $R(N,T)N$

additional costs incurred when T is changed that are relevant, it is not necessary to calculate a value for x. Thus, the curve $R_x(N, T)N$ is drawn in figure 2 with a zero value at the PAR time, T_1. Note that the slope of $R_x(N, T)N$ is everywhere equal to the slope of $R(N, T)N$. The two curves differ only by the amount x, which measures the height of $R(N, T)N$ above $Rx(N, T)N$. The curve $Rx(N, T)N$ gives the additional PAR costs incurred per period as T is reduced. This is the cost curve of interest here, since it provides the same optimal value for PAR time, T_{min}, as the total cost curve, and since it requires less data collection.

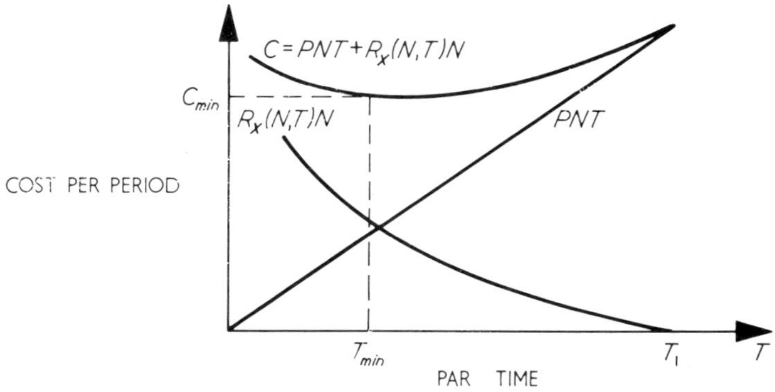

Figure 3 Modified pipeline factor cost equation

The modified pipeline factor cost equation to be minimized shown in figure 3 is now given by

$$C = PNT + R_x(N, T)N,$$

where

C = (total aircraft pipeline investment cost per period) + (additional PAR cost per period).

$R_x(N, T)$ = additional PAR cost per aircraft.

Again, it is emphasized that T_{min} is the same using either cost formulation. The optimum PAR time, T_{min}, occurs when

$$[\text{slope of } PNT] = -\ [\text{slope of } R_X(N, T)N]\,.$$

It should be kept in mind, of course, that C is nó longer the *total* cost per period K, since the amount x is not included. This is the basic pipeline factor cost equation used in the subsequent analysis.

3 AN APPLIED PIPELINE FACTOR MODEL: BASIC CONCEPTS

Determining an optimal pipeline factor has been shown to be a cost-minimization problem, one of finding that PAR time which minimizes the costs per period of the PAR process. The objective of this study was to develop an applied analytical model to perform this cost-minimization and to determine an optimal pipeline factor.

The choice of Network Analysis for this problem was inspired by the structure of the PAR process for the A-7 aircraft, and the network model was used to determine an optimal PAR time based on trade-off's between time and costs.

3.1 The A-7 PAR Process

Since the pipeline factor is a function of PAR time, the first step in development of the model was to structure the PAR as a production process. Figure 4 shows the A-7 PAR process as outlined by the Operations Analysis Division at the Naval Air Rework Facility in Jacksonville, Florida. Although it shows the PAR process as a sequence of

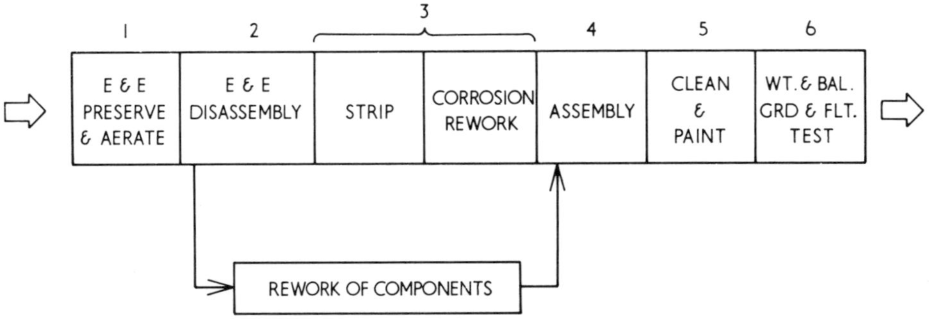

Figure 4 Flow chart of the A7 PAR process

ordered production stages, it is an over-simplification, since each stage is also composed of a sequence of jobs or activities. These activities consume time and resources. The total PAR time, therefore, depends upon that sequence of activities throughout the PAR process which takes the longest time to complete. Only a reduction in time of these activities can affect the total PAR time. The flow chart description of the production process does not explicitly relate the dependency of one activity to another and it does not reveal the critical path of activities which determine the PAR time. A model was required that can accomplish the following tasks:

(i) Identify the dependency of activities, their completion times, and associated resources and costs.

(ii) Permit determination of that sequence of activities which takes the longest time to complete, thus determining the PAR time.

(iii) Permit optimal allocation of resources to individual activities to produce the minimum possible PAR time for a given dollar budget, or, equivalently, determine the minimum cost allocation of resources for a given PAR time, and

(iv) Be implemented easily with the use of a computer.

3.2 Network of a Simplified PAR Process

Network Analysis clearly meets the above four requirements. This technique permits the generation of the previously defined additional PAR cost per aircraft, $R_X(N, T)$. The total cost function, C, is easily determined once $R_X(N, T)$ has been generated. The method used to generate $R_X(N, T)$ is sometimes referred to as the Project Cost Curve Algorithm, developed by Fulkerson in *Flows in Networks*. The technique is demonstrated by the device of a simplified PAR process, illustrated in figure 5. In other

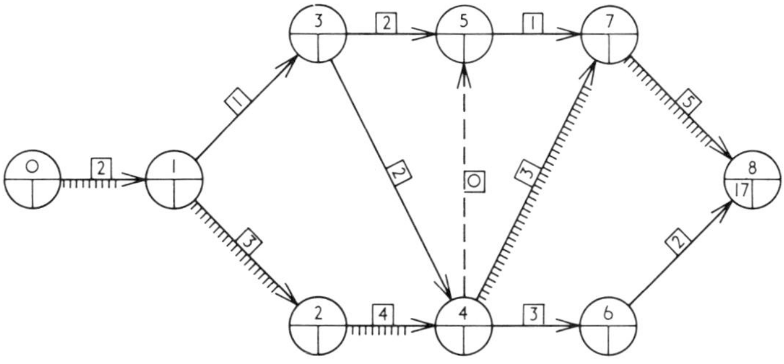

Figure 5 A hypothetical PAR process

words, the PAR process is represented as a directed graph, where the nodes and arrows have the conventional interpretation. The arrows represent activities in the PAR. The nodes represent the start and completion of an activity. The numbers assigned to the nodes are used to code the activities. The connection of the nodes by arrows displays the dependency among activities, with a broken arrow used to denote a "dummy" activity, that is, an activity which does not consume time or resources but is necessary to denote a precedence relationship among activities. The time to complete each activity is entered on the arrows. The critical path, also shown in figure 5 by the harsh-marked arrows, determines the minimum possible PAR time, in this case, 17 days.

The empirical application of this model to the A-7 required the following steps:

(i) Construction of the Network of the A-7 PAR based on existing production procedures at the naval air rework facility.

(ii) Determination of the additional resources required for the time reduction of critical activities in the network.

(iii) The pricing of the extra resources and the development of this data into time/cost trade-off curves or schedules for critical activities.

(iv) Development of a computer program to perform the time reduction analysis; that is, to reduce PAR time by buying time of these critical activities where the greatest reduction in time per dollar spent is obtained, from which followed

(v) Determination of the optimal A-7 pipeline factor, and

(vi) A comparison of alternative pipeline factors.

Perhaps the most difficult among these steps of translating methodology into practice are construction of the A-7 PAR network and development of the time/cost trade-off curves for the activities.

4. CONSTRUCTION OF THE A-7 NETWORK

The resultant network had the various characteristics. It contained 259 activities or arcs and 184 nodes. The initial critical path contained 63 activities with a total PAR time of 37.31 workdays, or approximately 52 calendar days. Forty of the activities on the critical path are in the cockpit area. These activities can be regarded as incompressible activities with respect to additional labor, although such activities could be reduced in time by use of overtime or a third-shift operation. Application of the time reduction algorithm to this network of the A-7 PAR permitted determination of the additional PAR cost per period, $R_X(N, T)N$.

4.1 The A-7 Pipeline Factor Cost Equation

The second most difficult problem in moving from theory to practice is the development of the pipeline cost equation for the A-7. It is necessary first to define the pipeline cost equation, C, for the A-7 PAR.

Recall that the cost equation to be minimized is:

$$C = PNT + R_x(N,T)N$$

In order to quantify these terms, it is necessary to specify the time period to which the analysis applies. The A-7 network applies to a period during which a given operating level of inventory has been achieved. This is taken to be equal to the service life of the A-7, a total of 116 months or 3,480 calendar days, as given by OPNAV planners. The procurement cost per aircraft for this time period is \$2.163 million. The arrival rate at NARF, Jacksonville of aircraft requiring PAR is 0.5 aircraft per work day. This arrival rate, N, must also be defined in terms of the period chosen for the analysis. Thus,

$$\begin{aligned} N &= (0.5 \text{ aircraft per workday})\,(22 \text{ workdays}/31 \text{ calendar days}) \\ &\quad (3{,}480 \text{ calendar days per period}) \\ &= (0.5)\,(0.71)\,(3{,}480) = 1{,}235.4 \text{ aircraft per period.} \end{aligned}$$

With this information, it is now possible to express each of the cost terms in the equation in terms of the period of the analysis.

4.*1.1. Total Aircraft Investment Cost Per Period, PNT.*

To use the term *PNT*, the PAR time, *T*, must be expressed in terms of the same period used in the definition of *P* and *N*.

However, it is convenient to define PAR time in terms of workdays, especially when using the network model to generate $R_x(N, T)$, the additional PAR cost per aircraft. In figure 9, the total aircraft investment cost per period is plotted as a function of PAR time, *T*, in workdays. To do this, the relationship PNT is modified as follows:

Total aircraft investment cost per period

$$= PNT \; \frac{31 \text{ calendar days}/22 \text{ workdays}}{3{,}840 \text{ calendar days}}$$

$$= PNT\,(4.05 \text{ x } 10^{-4}).$$

The PAR time, *T*, is multiplied by the factor (4.05 x 10^{-4}) to convert it to the same period used to define *P* and *N*. This permits expression of *T* on the horizontal axis in figure 6 in terms of workdays.

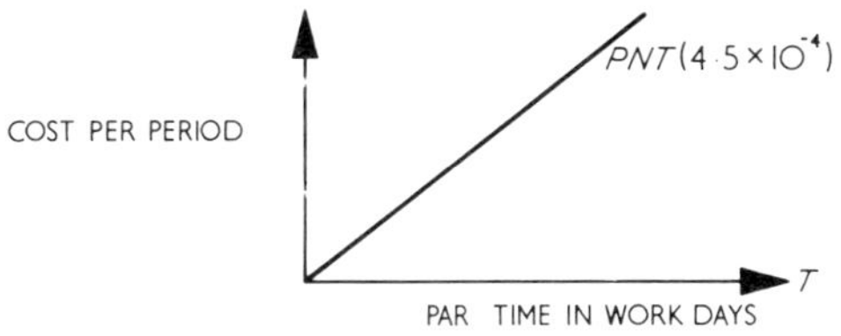

Figure 6 Aircraft investment cost per period

4.*1.2. The Additional PAR Cost per Period, $R_x(N,T)N$.*

The additional PAR cost period, $R_x(N,T)$, is generated using a computer program of the network algorithm. To use this program, it is necessary to transform the activity TCTO schedules into TCTO curves acceptable to the program. The program accepts TCTO curves only of the form shown in figure 7. That is, the cost must vary linearly between a normal and crash time, and the costs must increase as activity time is reduced. Also, the cost of completing an activity in the normal time is assumed equal to zero. Such TCTO curves are completely specified by their normal and crash times and the slope of the curve.

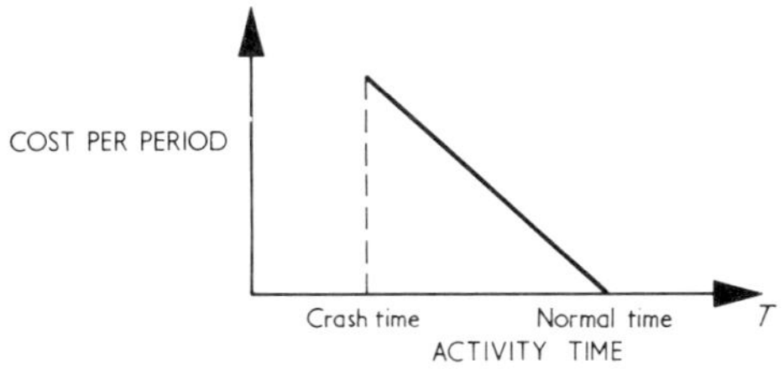

Figure 7 Acceptable TCTO curve for an activity

The types of resources used in developing the TCTO schedules for the A-7 network are labor, equipment, and spare components. We will not discuss the development of these schedules in detail, although it required a considerable data gathering effort through carefully designed questionnaires. We will give examples here of how TCTO curves can be developed for each type of resource used.

(i) *Labor.*

The completion time for activity (126,127) in the A-7 network is normally 0.56 days. The TCTO curve for this activity is shown in figure 8. If an additional mechanic is

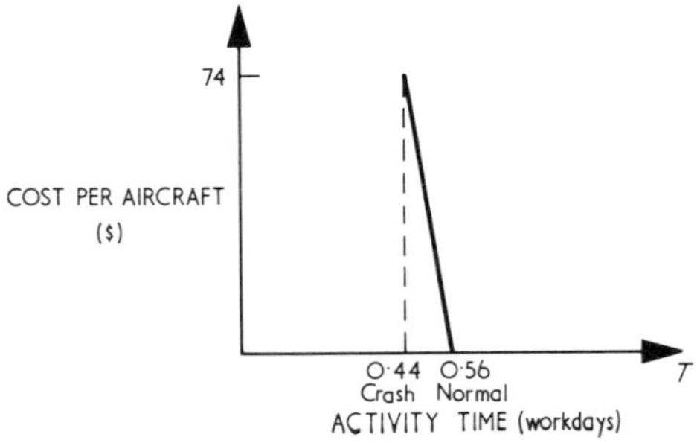

Figure 8 TCTO curve for activity (126, 127)

used, this completion time can be reduced to a crash time of 0.44 workdays, that is, the crash time can be achieved by adding a mechanic for 0.44 days. The additional cost per aircraft, the height of the TCTO curve, at the normal time is assumed equal to zero. The additional cost per aircraft to achieve the crash time is calculated as follows:

additional cost per aircraft

= (1 mechanic) (0.44 workdays/aircraft) ($168/workday)

= $74/aircraft.

The slope of this curve is

$$\text{slope} = \frac{(\$74 - 0)}{(0.56 - 0.44)} = -\ \$616.67.$$

As another example, consider network activity (50,59). The TCTO curve for this activity is shown in figure 9. This activity has two alternative applications of labor to achieve time reductions. The normal completion time for this activity is 0.56 workdays and, again, the additional costs per aircraft are assumed equal to zero for the normal time. The first alternative gives a completion time of 0.37 workdays at an

additional cost per aircraft
= (1 mechanic) (.37 workdays/aircraft) ($168/workday)
= $62/aircraft.

The second alternative gives a completion time of 0.31 workdays at an

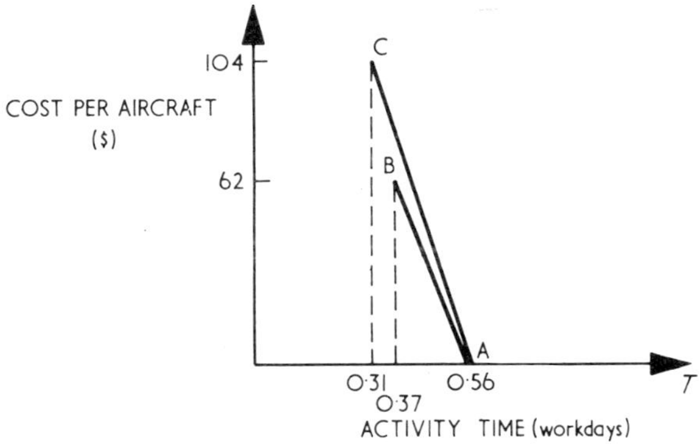

Figure 9 TCTO curve for activity (50, 59)

additional cost per aircraft
– (2 mechanics) (.31 workdays/aircraft) ($168/workday)
= $104/aircraft.

It should be noted that a piece-wise linear curve connecting points A, B, and C was not considered appropriate for this case. Instead, two TCTO curves are used, one for each alternative time reduction. The alternatives can be considered mutually exclusive. Times can be reduced by moving from point A to point B or from point A to point C, but not by moving from point A to point B and then from point B to point C.

This type of time reduction presents a special problem for the computer program, since it can accept only one TCTO curve for each activity. For activities involving more than one alternative TCTO curve, the program was run several times, each time using a different alternative TCTO curve for the given activity. The solution that gave the minimum additional PAR cost per aircraft for specific PAR times was used to formulate the final curve, $R_X(N,T)$.

(ii) *Equipment*

A reduction in time for activity (140,141) in the network can be achieved by use of additional equipment. The TCTO curve is shown in figure 10. Since each activity

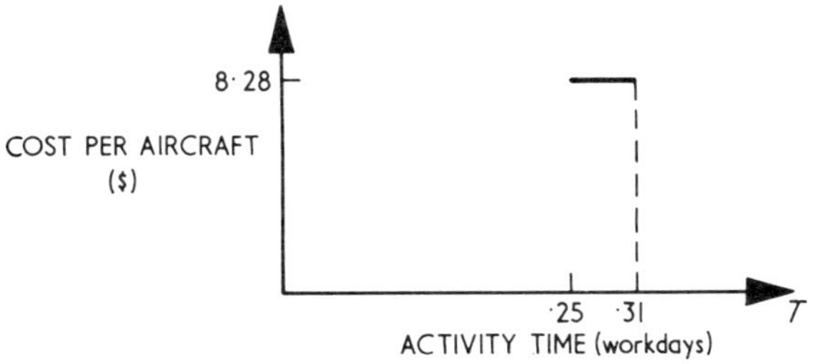

Figure 10 TCTO curve for activity (140, 141)

TCTO curve must be defined on a per-aircraft basis, the $R_X(N,T)$ generated by the activities must also be defined on a per-aircraft basis. In the case of labor, it is easy to calculate the cost of a specific PAR time on such a basis. Equipment presents a special problem, for to calculate cost on a per-aircraft basis requires consideration of the economic service life of the equipment.

In the present example, a line analyzer can be purchased for \$15,880 with an economic service life of 15 years. This piece of equipment will reduce activity time from a normal time of 0.31 workdays to 0.25 workdays. But the economic service life of the line analyzer is greater than the period chosen for the analysis. Therefore the cost of the line analyzer allocated to the A-7 PAR must be proportional to the difference between its economic service life and the service life of the A-7. Assuming straight-line depreciation (more sophisticated depreciation techniques could be used), the effective cost of the line analyzer for the period of 116 months is

$$\text{Effective cost per period} = (15{,}880)\,\frac{(116 \text{ months})}{(15 \text{ years})\,(12 \text{ months per year})} = \$10{,}223.$$

Thus the

$$\text{additional cost per aircraft} = \frac{\text{effective cost per period}}{N} = \frac{\$10{,}223}{1{,}235} = \$8.28 \text{ per aircraft}$$

The TCTO curve for this case should be represented as a step-function, as shown in figure 10. The TCTO curve takes this form because the entire piece of equipment must be purchased to reduce time, and thus for any time between the normal and crash times, the cost of the equipment is the same.

This possibility of TCTO curves as step-functions produces another problem for the computer program analysis. Since the program only accepts linear TCTO curves, the step-function is approximated by the straight-line shown in figure 11. If the activity is

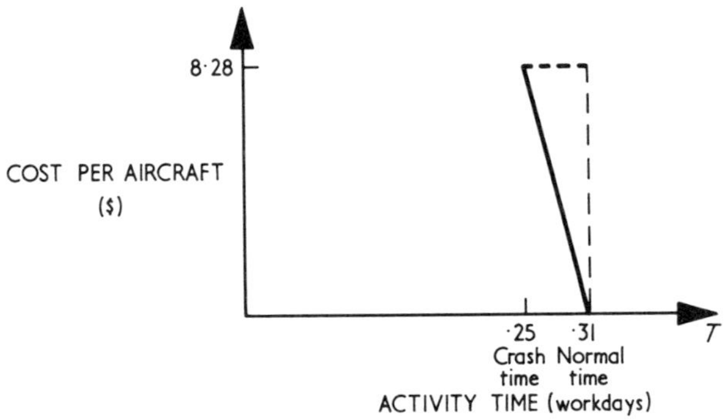

Figure 11 Straight line approximation of a step function TCTO curve

not reduced, or if it is fully reduced to its crash time, then $R_X(N, T)$ reflects the true cost of the equipment. This is because the straight-line approximation corresponds to the step-function only at the normal and crash times. If the activity were partially reduced to a time such as t, in figure 12, then the program would be re-run with a new TCTO curve, curve b in figure 12. If it is still economical to reduce the activity to t_1 then it would be properly costed.

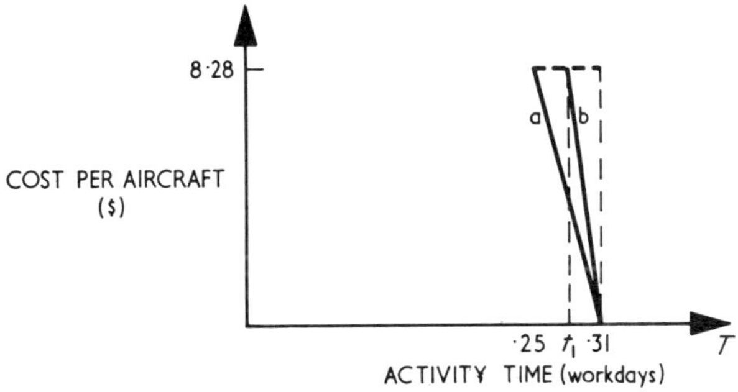

Figure 12 Partial reduction along a step function TCTO curve

(iii) *Components*

The completion times for several activities are reduced by stocking additional components. A method for determining the number of components required to achieve various completion times was developed. Space does not permit a detailed explanation of this method here. An example, however, of the use of components in reducing PAR time is activity (2,79), the repair of arresting gear. The total cost per PAR and the times achieved by the additional of components are as follows:

Activity Time	12.5	11.6	10.6	9.3
No. of Spares	0	2	3	4
Total Cost of Spares	\$0	5050	7560	8080
Cost per Aircraft	\$0	4.08	6.12	6.54

The activity times and number of spares required were calculated using the component model and the basic data for the arresting gear. We assume that the components last the full period of the analysis with zero salvage value at the end of that time. The time/cost points above were also graphed as a step-function in figure 13. Step-

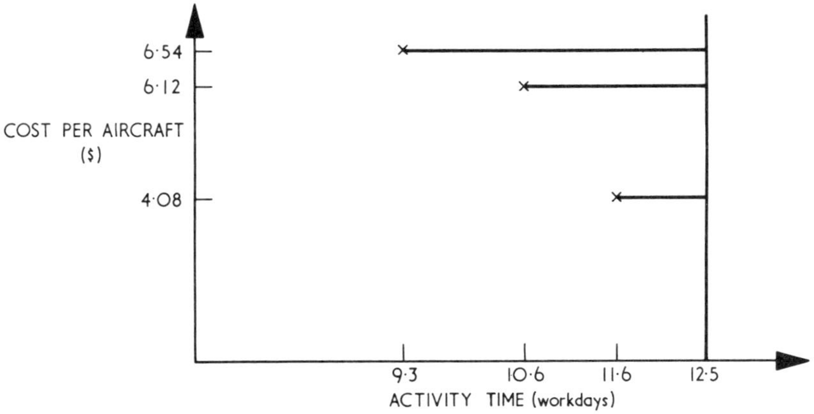

Figure 13 TCTO curve for activity (2, 79)

functions are produced because only integral numbers of components can be purchased. The computer program procedure was again to use straight-line TCTO curves to approximate the step-functions.

5. ANALYSIS OF ALTERNATIVE PIPELINE POLICIES

The purpose of this section is to demonstrate the usefulness of the network model by illustrating the results of its application to the A-7 aircraft. The network model was used to produce a set of alternative pipeline factor along with their associated costs and to identify an optimal pipeline factor for the A-7.

5.1 Alternative Pipeline Policies

A set of outputs were generated for various pipeline policies, shown in figure 14. A pipeline policy is described in terms of PAR time in workdays, T, and its associated pipeline factor expressed as a percentage.

The set of policies in this table include the pipeline factor at the start of the analysis, November, 1968 (Policy 1), the optimal pipeline factor (Policy 17), and the pipeline factor close to the one recommended by Department of Defense policy for procurement planning of most type, model, and series of aircraft in the Navy inventory (Policy 18). The purpose of this table is to demonstrate to the Navy decision-maker the capability of the model to compare the costs and benefits of alternative recommendations for planning pipeline aircraft procurement.

5.2 Least-Cost Analysis of Alternative Policies

These cost comparisons are shown in figure 14, where costs are determined over a

Alternative Policies	PAR Time (Work Days)	Pipeline Factor	Aircraft Investment Costs PNT	Additional PAR Costs $R_X(N,T)N$	Total Costs $C = PNT + R_X(N,T)N$
1–Present	34.31	19.36	$37.12	$0.0	$37.12
2	33.31	19.17	36.04	.01	36.05
3	32.31	18.97	34.96	.02	34.98
4	32.19	18.95	34.83	.02	34.85
5	31.16	18.75	33.71	.07	33.78
6	30.82	18.68	33.35	.08	33.43
7	29.52	18.43	31.94	.12	32.06
8	29.48	18.42	31.89	.13	32.02
9	29.42	18.41	31.83	.14	31.97
10	27.42	18.03	29.67	.64	30.31
11	27.39	18.02	29.64	.65	30.29
12	27.35	18.01	29.59	.67	30.26
13	27.11	17.97	29.33	.76	30.09
14	26.92	17.93	29.13	.83	29.96
15	26.83	17.91	29.03	.88	29.91
16	26.71	17.89	28.89	.99	29.88
17	26.65	17.88	28.84	1.03	29.87
18	19.99	16.59	21.63	34.41	56.04

Figure 14 Least cost analysis of alternative policies ($m)

116-month period and expressed in millions of dollars. For each policy, the table shows the total aircraft investment cost per period, *PNT*, the additional costs incurred in achieving a PAR time, $R_X(N,T)N$, and the total costs *C*.

In Policy 1, aircraft investment costs are \$37.12 million and additional PAR costs are zero, since no time reduction in the PAR process has yet occurred. The policy of principal interest is Policy 17. An optimal PAR time of 26.65 days was determined by reducing aircraft investment costs to \$28.84 million as the result of an additional investment of \$1.03 million to reduce PAR time. The result is a total cost policy of \$29.87 million, which is seen in the table to be the minimum total-cost policy.

Policy 18 is of interest because it approximates the DOD policy of a 15 percent pipeline factor for planning aircraft procurement. The time reduction costs for Policy 18 were approximated by using a third-shift operation in the Naval Air Rework Facility. We resorted to this method of time reduction because, coincidentally, Policy 17–the optimal policy–also was the technologically lower bound below which PAR time could not be reduced by adding additional resources of equipment, labor, and components. The result was a PAR time of 19.99 days and a pipeline factor of 16.59 percent.

5.3 Comparison of Optimal and Present Policies

The economic implications of these alternative policies were presented to the Navy decision-maker in a cost-benefit comparison of the present and optimal policies, shown in figure 15. Benefits here refer to savings in pipeline aircraft investment as a result of reducing PAR time. Benefits are calculated by considering the change in costs in moving from one policy to another policy. In figure 15, these benefits are shown in

C (Present Policy)	=	\$37.12
C (Optimal Policy)	=	29.87
Net Benefit		\$ 7.25

Total Benefits (Decrease in Pipeline Aircraft)	Cost (Additional Increase in PAR Costs)	Net Benefit	Cost-Benefit Ratio
\$8.28	\$1.03	\$7.25	8.0

Figure 15 Comparison of optimal and present policies (\$m)

two ways. One is the difference in total costs between the two policies. This shows a net benefit of \$7.25 million saved in a change from the present policy to the optimal policy. The second way is a comparison of the trade-off's between reduction in pipeline aircraft investment and the increase in PAR costs. Figure 15 shows that an investment of \$1.03 million in reducing PAR time saves \$8.28 million in pipeline aircraft investment, again producing a net benefit of \$7.25 million. The result is a cost-benefit ratio of \$8.0 (\$8.28/\$1.03). In other words, each dollar invested in reducing PAR time from 34.31 workdays to the optimal 26.65 days produces a savings of \$8.00 dollars in the procurement of pipeline aircraft.

The pipeline network model also identifies the type of resource investment required to attain the optimal PAR time. Figure 16 shows a break-down of resource allocations for the $1.03 million investment in the PAR process required to achieve a PAR time of 26.65 days. The largest expenditures required are for labor and reparable components: $0.36 million and $0.65 million respectively.

Method of Reduction	Additional PAR Costs $[R_x(N,T)N]$
Equipment:	
Line Analyzer	.010
Weight Balancing Jacks	.003
Labor:	
Mechanics	.220
Electricians	.092
Painters	.053
Components	.655
Total $[R_x(N,T)N]$	1.033

Figure 16 Additional resource requirements for optimal policy ($m)

5.4 Comparison of Optimal and Other Alternative Policies

To complete the analysis, figure 17 shows a comparison of the other alternative

Alternative Policies	PAR Time Work Days	Pipeline Factor Percent	Cost Difference (Alternative–Optimal)	Net Loss
1–Present (Nov. 1968)	34.31	19.36	(37.12–29.87)	7.25
2	33.31	19.17	(36.05–29.87) =	6.18
3	32.31	18.97	(34.98–29.87) =	5.11
4	32.19	18.95	(34.85–29.87) =	4.98
5	31.16	18.75	(33.78–29.87) =	3.91
6	30.82	18.68	(33.43–29.87) =	3.56
7	29.52	18.43	(32.06–29.87) =	2.19
8	29.48	18.42	(32.02–29.87) =	2.15
9	29.42	18.41	(31.97–29.87) =	2.1
10	27.42	18.03	(30.31–29.87) =	.44
11	27.39	18.02	(30.29–29.87) =	.42
12	27.35	18.01	(30.26–29.87) =	.39
13	27.11	17.97	(30.09–29.87) =	.22
14	26.92	17.93	(29.96–29.87) =	.09
15	26.83	17.91	(29.91–29.87) =	.04
16	26.71	17.89	(29.88–29.87) =	.01
17–Optimal	26.65	17.88	(29.87–29.87) =	0
18 (DOD)	19.99	16.59	(56.04–29.87) =	26.17

Figure 17 Comparison of optimal and other alternative policies ($m)

policies with the optimal policy. The table shows the net loss incurred in moving from the optimal policy to any alternative policy. A net benefit is realized by moving from any alternative to the optimal policy.

A major point of interest in this study was a comparison between the optimal policy and the approximated DOD policy of a 15 percent pipeline factor, Policy 18. If Policy 18 were imposed as a matter of policy, the net loss compared to a 17.88 percent pipeline factor would be $26.17 million. Furthermore, it should be noted that a 16.59 percent pipeline factor would require an investment of $34.41 million in the PAR process to assure a PAR time of 19.99 days. If this investment were not made, the policy would result in a decline in the operating inventory of aircraft. The $34.41 million investment in the A-7 PAR process should be compared to the $1.03 million PAR investment of the optimal policy, which gives the minimum total cost.

6. SOME IMPORTANT CONSIDERATIONS

Before concluding, it is important to discuss the assumptions on which the model was based and their implication for the proper use of the model:

- Recognition of a trade-off of resources between NARF resources and aircraft
- PAR process assumed to be deterministic:
    - *N*, arrival rate, assumed constant
    - *T*, PAR time, assumed constant for a given level of resources
- The relevant time period of the analysis
    - Aircraft inventory at a stable level
- The value of aircraft unavailable to the operator

An important consideration for the Navy decision-maker is the recognition of a trade-off of resources when planning the purchase of a new type, model, and series of aircraft. The procedure assumes that a desired level of operating inventory of aircraft is first determined. Planners then calculate the number of pipeline aircraft required to maintain this operating inventory. It is important, however, that planners recognize that they are trading NARF resources for aircraft when they make this decision: that is, they cannot realize savings by buying fewer aircraft in support of a lower pipeline factor unless they make available to the NARF the necessary resources for reducing PAR time. It is likely, however, that the planners who determine aircraft purchases are not the ones who determine the NARF budget. It is important, therefore, that the two groups communicate, for if pipeline aircraft corresponding to a given PAR time are purchased, and if the NARF is not given the resources required to achieve that PAR time, then the desired operating level of inventory cannot be maintained. Planners who determine pipeline aircraft procurement and planners who determine NARF budgets must coordinate plans if inconsistent results are to be avoided.

Another consideration is that the PAR process was assumed to be a deterministic process. The arrival of aircraft for PAR, *N*, was assumed constant and equally spaced through time. Furthermore, the PAR time, T, was assumed constant for a given level of resources. In practice, N and T will vary. The random character of N and T could be incorporated into the analysis using such techniques as Queuing Theory. The effect of such an extension of the analysis should be studied. In general, the NARF smooths out fluctuations in N by scheduling and liaison with a nearby master jet base. The PAR time, T, is a function of the time required to complete the individual jobs in a PAR. Some fluctuations will occur in these individual jobs, which, in turn, will cause T to vary. Although no empirical study was made, this effect was considered not significant except for the repair of components. This is because not all components will require repair. The analysis should also be extended to consider stochastic variations in job completion times.

Another question is the relevant time period to which the analysis is to be applied when determining the purchase of pipeline aircraft. This study is concerned with the operation of the system during the service life of the aircraft and it considers the aircraft inventory to be at a stable level during this time. Reduction of PAR time during the build-up and replacement phases of the inventory could be considered in order to achieve a higher operating inventory during these periods, but these alternatives were not considered here.

Still another consideration is how to value aircraft unavailable to the operator, or the value of aircraft tied-up in pipeline. This is given by the term P, the procurement cost per period per aircraft. The definition of P depends on alternatives available to the decision-maker. The present definition of P is valid when the analysis is applied to a new aircraft for which pipeline aircraft are yet to be purchased. Then, the savings attributed to a reduced number of pipeline aircraft is their procurement cost per period. The present analysis is intended to apply to such situations. If the analysis is applied to aircraft for which pipeline aircraft are already purchased, the P must be defined differently. For example, suppose the production of an aircraft type has ceased and tools and equipment to produce this type no longer exist. The only benefit from a reduction in pipeline aircraft of this type is an increase in the operating inventory. Then P must be defined as the value of an additional aircraft to the operating inventory and not as the cost of aircraft procurement.

7. SUMMARY

In summary, the determination of an optimal pipeline factor was reduced to a simple cost-minimization problem. Once the cost function, C, was specified, it was possible to determine the optimal PAR time, T_{min}. The optimal pipeline factor was then calculated by substitution of T_{min} into the pipeline factor equation. The problem was to develop an applied analytical model to make this determination. The major difficulty is that such a model must be capable of determining the additional PAR cost per aircraft, $R_x(N,T)$. This was accomplished by applying the technique of Network Analysis to the PAR production process.

Based on this application of network analysis to the A-7, the network model has been accepted by the Navy and is now being implemented by the Naval Air Systems Command.

DISCUSSION

Geraerds:

1. You state or assume that PAR time cannot be reduced without incurring additional costs. This is acceptable only if the activities constituting a PAR have been checked as for the presence of overmaintenance. Did you check this and if so what did you find?

2. Using networks requires that the activities constituting the network are known. In the case of maintenance however there is a certain probability of appearing in a network. For quite some activities that probability is not one. This means that one cannot have one network for all possible combinations. How did you solve this problem? Do you construct a network each time for each individual aircraft that arrived for maintenance? Did you consider network techniques containing probability, like GERT?

Sutton:

1. As stated in my paper, the PAR process was taken as a *given* production process, including any policies that would allow "over-maintenance", as you put it, and the specifications for performing each check, verification and rework activity. In other words, if there is over-maintenance, it is included in the cost curve of the PAR. Given this cost curve, PAR time cannot be reduced without incurring additional costs–that is, moving along this cost curve. My statement is not unacceptable if the PAR includes over-maintenance. I think you mean that the cost curve of a PAR can be *shifted* to some lower level by altering the set of production activities by eliminating "over-maintenance", and changing job specifications, and I agree. Having done that, one must still construct the resultant new cost curve of a PAR and again examine the reduction in PAR time associated with the additional costs of moving *along* this new curve. There are two effects at confusion here: shifting the entire cost curve and moving along any given cost curve. We looked at only the latter type of effect–no mean task in itself. We made no thorough examination of the type of effect you mention but are aware of its existence. In an earlier phase of our study, we did examine the efficiency of the NARF's and recommended investments in plant modifications which have the effect of shifting the cost curve. PAR policy as opposed to the PAR production process is also the question raised by Mr. Buck, and I elaborate on this point in my response to his question.

2. I noted in another answer that there are many types of maintenance. Your question is most relevant to unscheduled maintenance where probabilistic activities and events are a greater problem. A PAR is a regularly scheduled maintenance activity with a definite sequence of tasks to be performed, although some of these could be described by a probability distribution. Rework of components is one such case to which we did apply probability of occurrence. Since our model is for budget *planning* purposes and not for production management, this is not too severe an assumption: to treat the PAR deterministically; it certainly makes it a practical tool for use by intelligent but statistically untrained laymen. Recognizing the dynamic and stochastic elements in the PAR, we recommend that the static version be continually revised and up-dated to capture as many of these changes as possible. Practicality, not optimality, constrained our research.

Buck: Has any consideration been given to the trade-off between PAR frequency and operating maintenance?

Would this not also relate to DOD's 15% factor and be significant in achieving it?

Sutton: No explicit consideration was given to this question. As I understand the question, you are referring to the possibility of performing fewer PARs at the depot level and performing more maintenance at the squadron or organizational level. We did not consider the other echelons of maintenance as a variable in the pipeline factor equation, but took them as given as specified by our study directive. The two types of maintenance, of course, are fundamentally different and your question does raise important issues concerning the maintenance specifications that define the type of rework to be performed. Whether these considerations would or would not be economically significant in achieving a 15% pipeline factor cannot be answered without extension of the network technique to each type of maintenance. The question is worthy of affirmation or denial, however, and should be considered along with all the other factors that determine pipeline requirements. We did make some recommendations on PAR frequencies and tour lengths, as a result of another phase of the study which considered the economic service life of an aircraft. The economic service life was considered a function, among other things, of the cost of squadron maintenance and cost of rework. The economic service life is one input into determining the operational service life, and operational service life is one of the terms, taken as given, in our pipeline factor equation. Use of the economic service life model in determining the optimal operational service life value to use in the pipeline factor equation would then capture part of the trade-off you mention.

Baglow: What were the relative magnitudes of the linear and non-linear terms in the cost curve?

Sutton: The thrust of this question is: how do the additional costs of reducing PAR time (non-linear) compare to the additional costs of pipeline aircraft procurement (linear)? When compared with the procurement cost per aircraft per day over a ten-year period (as the value of an aircraft tied-up in PAR), not many investments in reducing PAR time were economically feasible. The catch is using procurement cost as the value of an aircraft tied-up in pipeline and not available to the fleet. This is proper under our assumption of a constant and given operating inventory. When this assumption is relaxed, how do we measure the value of an aircraft foregone to the operator? To commercial airlines, this is a clear loss of large revenue; and, investment in rework is more economically worthwhile–in fact, they are capital intensive, using replacement modules and parts, and the turn-around- time is considerably shorter than in the Navy.

Bax: Can you explain the different types or models of inspections you have checked before the PAR was introduced in US Navy?

Sutton: The question, as I understand it, refers to different types of maintenance procedures used in rework of aircraft prior to introduction of the PAR, and did our study investigate these? If so, explain them.

There is, indeed, a set of different maintenance actions performed on aircraft, some scheduled and some unscheduled. Even a PAR is divided into PAR-MOD (PAR-Modification) and a straight PAR. The PAR-MOD usually describes rework on a new type aircraft in the inventory during which alterations and modifications are made based on fleet experience with the new weapon system. The PAR is the, more or less, stable rework procedure that follows from this experience after all changes are made. Unscheduled Maintenance is also a rework category. We investigated only the PAR-MOD and the PAR, because the purpose of the study was to develop a pipeline factor model for aircraft given that the operating inventory had reached a stable level and new buys would be for pipeline aircraft. The technique, of course, could be used to determine the effects on pipeline for the other maintenance activities to the extent that they are production processes that can be structured as a network of activities. It would be useful to extend the network to all echelons of maintenance affecting pipeline.

Fischer: You mention in your paper, that the US Navy has accepted the presented network analysis approach for the A-7 system. I imagine, that it would be a logic consequence to apply the same technique to other aircraft systems, too. The question I'd like to ask is whether it has been considered to use the approach also for the improvement of other activities such as a ship overhaul, which is a very similar problem as concerns optimizing the tradeoff of maintenance resource requirements vs. downtime cost.

Sutton: I think this is an interesting area of application of network techniques given the great conceptual and analytical problems in efforts to measure the "material readiness" of ships and to use such a measure in planning maintenance and support budgets for ships. One could state the following problem.

Given the requirement to maintain a given number of ships on line per some time period and the overhaul arrival rate, determine that overhaul time that minimizes the sum of overhaul and downtime costs.

Morris: Whilst your paper demonstrates the sub-optimisation inherent in present policies when applied to known aircraft, policy makers are concerned to provision pipeline aircraft at an early stage in the development of the aircraft. In the present state of the art of maintainability prediction (i.e. prediction of MTTR) do you think that your model could be used?

Sutton: The answer is no. The reason is that the pipeline model was developed for application to an existing operating inventory of aircraft *after* it was built-up and procurement is for pipeline aircraft. In fact, it is *after* the build-up of the operating inventory that pipeline procurement is of greatest interest, not during the development stage when procurement is for the operating inventory. We did wonder about how one should phase pipeline procurement during the build-up phase of the operating inventory up to the point where it is stabilized. This is another dynamic consideration and our model is static. It would be risky to use a static model to answer dynamic problems. I agree that more consideration should be given to the relationship between aircraft design and development and maintenance policy and costs. Do you believe contractors should be charged with high maintainability costs of complex aircraft with low reliability during the development phase before issuing the first buy for the operating inventory? This is another question requiring another approach.

A New Approach to the Provisioning of Repairable Items

W.N. Hurst
C.S. Landau

1. INTRODUCTION

A repairable item is an item that can be repaired or overhauled (at least once and generally indefinitely) in distinction from a consumable item which typically is used once and then discarded. Hence the repairable item passes through a cycle which in its simplest form has two parts–in service and undergoing maintenance. (The term *maintenance* is used to include overhaul and repair.)

The usual practice with repairable items is to purchase them at the same time as the parent equipment (e.g. aircraft) of which it forms part, because very often production of the item ceases with the end of production of the aircraft. This entails making a forecast of the requirement for spare repairable items for the whole life of that particular aircraft. The total number of each type of repairable items is generally not large, especially when compared with the numbers of consumable items, but they are expensive and hence considerable effort is made to increase the reliability of the forecast and to keep down the size of the inventory of repairable items. Implicit in the latter aim is the hope that the investment in inventory is spread equitably across all the repairable items.

Any method of assessing the number of spares can be considered as essentially a two step process, the first of which is forecast of the average demand for spares. This first step requires estimates of such factors as mean time between removal and time spent in maintenance. The second step is to provide a buffer stock to cope with the variations in the demand. This note is concerned with the second step, i.e. with the factors that determine the size of the buffer stock.

Because a given parent equipment may go unserviceable because of an inability to meet a demand for any one of the many repairable items that comprise it, the buffer stock for any one item cannot be considered in isolation but must be considered in relation to all the other items. Now the broad aim of the logistics system is to minimize the cost of achieving a given effectiveness or to maximize the effectiveness for a given cost. Hence the broad aim of this paper is to determine the optimal allocation of the investment in the inventory of repairable items for a given parent equipment.

2. PRESENT METHODS OF ASSESSING THE BUFFER STOCK

2.1 Empirical Rules

The usual approach to spares assessing is to estimate the average number in the maintenance pipeline and then to add a buffer stock to allow for variations in demand, repair time, etc. There are three simple rules of thumb which are used frequently to estimate the buffer stock.

(a) The size of the buffer stock is made proportional to the number of installed repairable items. This is a common rule for aero-engines, the buffer stock for which is set equal to say one-fifth of the number of installed engines.

(b) The size of the buffer stock is proportional to the rate at which the items are removed for maintenance. Thus the buffer stock may be set equal to the average number of removals in either a three-months or a six-months period.

(c) The size of the buffer stock is proportional to the average number in the maintenance pipeline. Thus the buffer stock might equal the average number in the pipeline.

Cost is introduced into the assessing procedure by making the proportionality factor in the above rules depend on the unit cost of the item. Thus in rule (b) above, the buffer stock may equal an average six-months removals if the unit cost is less than \$5000 and three-months removals if the unit cost exceeds \$5000.

2.2 A Simple Probabilistic Model

The next level of sophistication is to assume that the demand follows a Poisson distribution and is independent of the maintenance part of the cycle. The average demand equals the number of removals that occur during a time equal to the mean time in maintenance, i.e. it equals the average number in the maintenance pipeline. The level of stock is adjusted to produce a selected probability of having a stockout. This method is recommended, for example, by the Lockheed Aircraft Co (J. K. Segar– Application of optimising techniques to critical spares provisioning, Lockheed California Co.)

Cost can be introduced in a manner similar to 2.1 above. The repairable items are separated into two or more categories according to cost and a different probability of stock assigned to each. Alternatively, Lockheed introduces a new factor defined by

$$D\,\delta P\,/\,C \qquad (1)$$

where D = average demand for that type of item,

δP = the decrease in the probability of a stockout caused by the acquisition of one more item, and

C = the cost of that item.

The inventory of spares is then built up in such a way that the factor for each type of item reaches approximately the same value.

2.3 A RAND Model

The RAND Corporation has developed a Base Stockage Model (H.S. Campbell & T.L. Jones, *A Systems Approach to Base Stockage*, RAND Paper P3345, May 1966.) in which the demand is assumed to follow a compound Poisson distribution, with a given mean time between removals. The maintenance facility has a given average turnaround time, but is assumed to be large so that the actual variation of turnaround times is not relevant. With these assumptions, the effectiveness of the system is determined as a function of the number of spares.

Cost is introduced by choosing the number of each type of item so as to make

$$\frac{\delta P_1}{C_1} = \frac{\delta P_2}{C_2} = \cdots \frac{\delta P_n}{C_n} \tag{2}$$

where δP_1 is the marginal decrease in the probability of a stockout in item "1" brought about by the last one of that type to be acquired, and C_1 etc. is the cost per item. In this way, additional items are acquired until the budget available for actual inventory is allocated completely; in so doing, a certain overall probability of a stockout will have been achieved. Alternatively, the information so derived can be presented as a relation between effectiveness (e.g. probability of a stockout) and total investment in inventory.

3. THIS APPROACH

There are two steps in this approach also. Firstly the behaviour of the sub-system or cycle for one type of item is determined (the term *system* will be used to describe the whole range of repairable items that comprise a given parent equipment). In the other step, the allocation of the budget among the whole set of sub-systems is considered. This latter step will include as a corollary the optimal allocation of budget within each sub-system.

This approach differs from other approaches in two respects. Firstly, the theory of cyclic queues is used to model the behaviour of each sub-system. Secondly, the model specifically allows for small numbers of parent equipments and a small maintenance facility.

4. REPRESENTATION OF THE REPAIR CYCLE

4.1 Description

The simplest representation of the repair cycle or sub-system, and the one used here, is shown schematically in Figure 1.

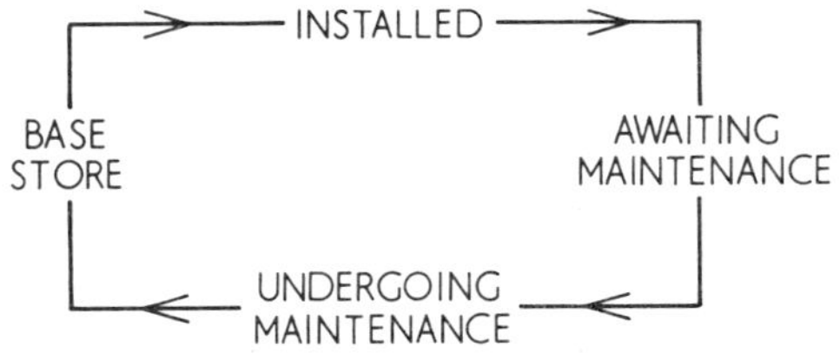

Figure 1 Representation of the rapir cycle

Each item is assumed to move around a cycle, starting with installation in a parent equipment, where it remains until removal is necessary. Thereupon, it is transported to a maintenance facility, where it is repaired and then transported back for reinstallation. If on its return there is no parent equipment requiring this type of item, it is placed in a store of serviceable items to be used when required. If there is no repair channel free when it reaches the repair facility, it is placed in a queue of unserviceable items until its turn for repair comes. The times spent in service or undergoing repair are described by probability distributions with specified means, thus taking account of the variability inherent in the system.

This process is much simplified compared with the real one, with the following principal assumptions:

a. There is only one repair loop.

b. Only one base is considered.

The factors required to describe the system are:

A = number installed

L = mean time between removal (MTBR)

K = mean time in maintenance (MTM)

M = size of maintenance facility

S = number of spare items

N = total number of items = $A + S$

4.2 Interpretation of the Factors

The number installed, A, is self-evident. It would change in magnitude only if the number of aircraft, say, were changed. In this model, it is assumed that sufficient aircraft will always be available to maintain the overall flying rate.

The mean time between removal, L, of the item is expressed in units of calendar time. It may change if the flying rate changes or if the item is modified. The variation of the times between removals is assumed to follow an exponential distribution.

The mean time in maintenance, K, is the time spent actually undergoing maintenance in the broadest sense. It includes transport time and administration, but does not include time spent waiting for some action to commence (unless this is part of the normal procedure). The time in maintenance is assumed to vary according to an exponential distribution.

The value of K can be changed by such measures as changing administrative techniques or by introducing overtime in the maintenance workshop.

It should be noted that the mean time in maintenance as defined here is not the same as the more common turnaround time in maintenance. In this model the two quantities differ by an amount equal to the mean time spent in the queue awaiting maintenance.

In the simple model assumed here, the items pass through the maintenance facility in parallel rather than in series. Hence the value of M equals the number of items that can be processed simultaneously. It can be changed by changing the allocation of personnel, tools and workshop space.

Unlike the other parameters, the practical value of M may depend on factors external to the particular item. In order to achieve almost full utlization of the maintenance facility, the value of M should be equal to AK/L, i.e. the average number of items in the maintenance facility. If the maintenance facility handles more than one type of item, say n types, the total size of the facility should be

$$\sum_{i=1}^{n} M_i = M_1 + M_2 + \ldots . M_n \tag{3}$$

where M_1 .... equals the average number of maintenance channels allotted to item Type 1 ....

If the trade skills and repair equipment etc. can be used with equal efficiency on any one of the n types of item, management has the power, in principle at least, to vary the day-to-day allocation of effort to any one type of item between

$$M = 0 \text{ and } M = \sum_{i=1}^{n} M_i$$

The number S of spares is the difference between the total number of items and the number of parent equipments. If there are no unserviceable equipments, S comprises spares in the maintenance pipeline in addition to those in the base store.

4.3 Criteria of Effectiveness

Several criteria have been suggested for measuring the performance of a repair cycle:

a. Probability of a stockout–this will show how often the system is working properly, but will not tell how serious any failure is.

b. Average store level–this measures the average size of the store of serviceable items, but, since variability is not considered, gives no idea of frequency of stockouts.

c. Availability of the parent equipment–this determines the average percentage of aircraft, say, which are *available* over a period of time, in the sense that they have the particular item considered; this does not show whether peak operating schedules can be met.

d. Fill rate–defined by RAND as the average number of demands that are met immediately or within a specified period; this does not give weight to demands that remain unfilled for long periods.

e. Back order date–defined by RAND as the average number of demands outstanding over a period; this takes account of the length of time a demand is outstanding, e.g. one demand for 10 days contributes the same to this criterion as 10 demands outstanding for one day.

Because this paper purposes to illustrate a particular approach, the use of one criterion will suffice. We have selected the first of the criteria listed above, i.e. the probability of a stockout, because its interpretation is straightforward and it does provide information that is of direct concern to the logistics manager. For added convenience in presentation of the results, however, we shall use its complement, i.e. the probability P of not having a stockout.

5. BEHAVIOUR OF THE REPAIR CYCLE

5.1 Method

The theory of cyclic queues has been used to determine the behaviour of the repair cycle. The mathematical equations that derive from this theory are not amenable to an exact general solution. Instead, specific values of the factors are taken and the corresponding probability P determined by numerical methods. With the aid of a computer, ranges of values can be fed in and an approximate solution obtained.

The simple theory as used here has the following principal assumptions:

a. There is no batching policy–items are transported from repair to unit store as soon as they are ready.

b. There is no interaction between items in that the time in service or repair of one item is independent of what is happening to other items.

c. The situation is *steady state*, in the sense that the means of the probability distributions do not change with time–this means that the introduction of overtime or a change in operating hours cannot be included directly.

d. Only exponential probability distributions can be treated; whilst this description is adequate for time in maintenance, it is strictly satisfactory for time between removal only if failure is random.

e. There is no distinction between scheduled and unscheduled repair.

5.2 Amalgamation of the Factors

It is a result of the theory of cyclic queues that the behaviour of the repair cycle does not depend on the absolute values of K the mean time in maintenance, and L the mean time between removal, but only their ratio K/L.

Further in Section 4.2 the assumption was made that sufficient of the A parent equipments would always be available to maintain the overall operating rate. A consequence of this assumption is that the overall average rate of removal of a given type of item will be independent of what happens to individual parent equipments and will remain equal to A/L. In these circumstances, the behaviour of the cycle will depend of the absolute values of A, K and L, but on the quantity D given by

$$D = AK/L.$$

With these assumptions, D equals the average demand for replacement items during a time equal to the mean time in maintenance and the average number of items in the maintenance facility. For ease of reference, D shall be called average demand.

Hence the behaviour of the sub-system for a given type of item can be described in terms of three factors: D, M, and S.

5.3 Results

5.3.1. The Probability Function

A relationship with three independent variables cannot be shown graphically, but since we are interested primarily in the interaction among D, M and S when achieving a given value of P, let us fix P, somewhat arbitrarily, at 0.9995. That this is the right order of magnitude of P can be shown by imagining an aircraft type that comprises 200 repairable items. In such a case there would be about 30 days per year on which a demand for at least one or another item remained unsatisfied. Thus the relationship can be transformed into one giving S as a function of D and M. This can be represented by a surface (in three dimensions); part of this surface is shown schematically in Figure 2 by projecting the contours of constant S onto the D-M plane. Remember that each point on this surface represents one combination of S, D and M that will give P equal to 0.9995.

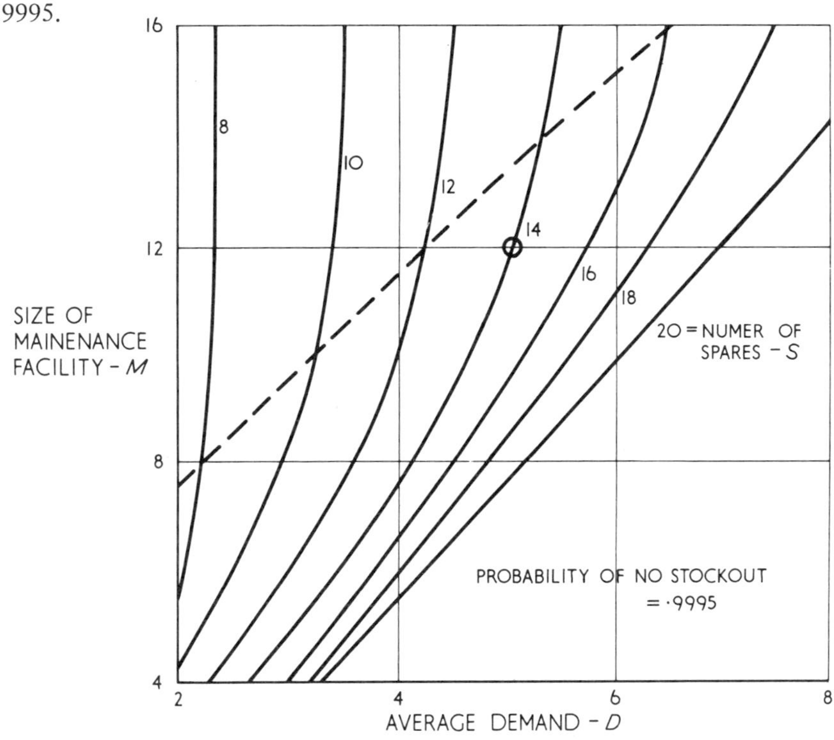

Figure 2 Contours of constant S on the surface $P = 0.9995$
(NOTE: along the dashed line, the probability that a queue froms is 0.0005)

Figure 3 shows the section through this surface at $M = 12$, while Figure 4 shows the section at $D = 5.0$. The circles indicate the point common to the three figures ($S = 14$, $D = 5.0, M = 12$).

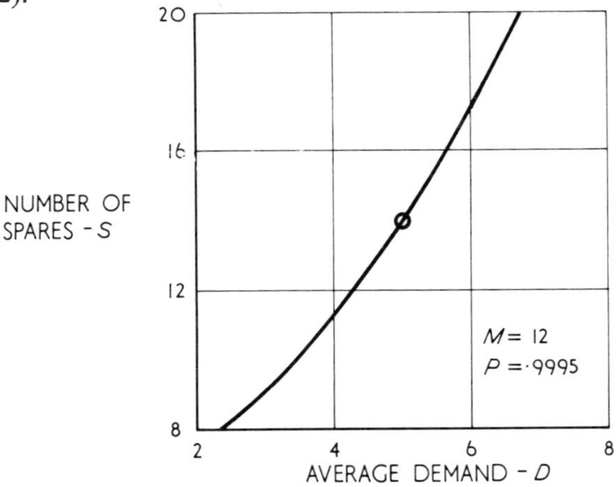

Figure 3 Variation of S with D for constant M and P

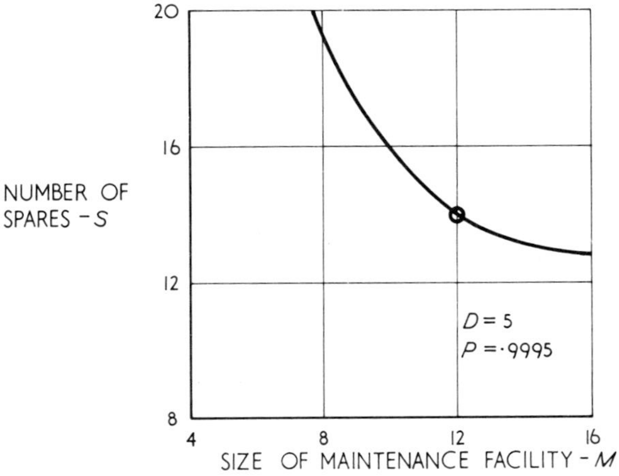

Figure 4 Variation of S with M for constant D and P

5.3.2. *Behaviour at the Maintenance Facility*

For high values of M with respect to D, i.e. the upper lefthand section of Figure 2 or the flat part of the curve in Figure 4, the probability of a queue forming outside the maintenance facility is small and approximately independent of the value of S. The values of M which give 5, 0.5 and 0.05 percent for this probability are shown in the table facing page, which also shows the corresponding utilization $U(M)$ of that size maintenance facility. It can be seen that as D decreases, a very much greater proportional increase in M is required to achieve the same decrease in the probability of a queue forming, and that the overall utilization is very much less at low values of D.

UTILIZATION OF THE MAINTENANCE FACILITY M
WHEN PROBABILITY OF NO STOCKOUT IS CLOSE TO 100%

Average Demand D	Probability of an Item Having to Wait for Maintenance					
	5%		.5%		.05%	
	M	$U(M)$	M	$U(M)$	M	$U(M)$
		%		%		%
0.5	2	25	3	17	4	12
1.0	3	33	4	25	5	20
2.0	5	40	6	33	7	29
5.0	9	56	11	45	13	38
10	16	62	18	56	21	48
20	28	71	32	62	35	57
50	62	81	68	74	72	69

NOTE: For approximately 100% utilization of M, the value of M should equal D.

5.3.3 Variation of P

The results presented thus far have had P fixed at 99.95%. Rather than attempt to illustrate the variation of P with each of the three factors, we show in Figure 5 the variation of P with S for particular values of D, while Figure 6 shows the variation of S with D for particular values of P. In each of these cases the value of M is large, so that the behaviour of the sub-system is approximately independent of the precise value of M.

6. Allocation of the Budget

6.1 Principle

The problem of determining the optimal combination of L K, M and S for one type of item must be considered in conjunction with all other types of items.

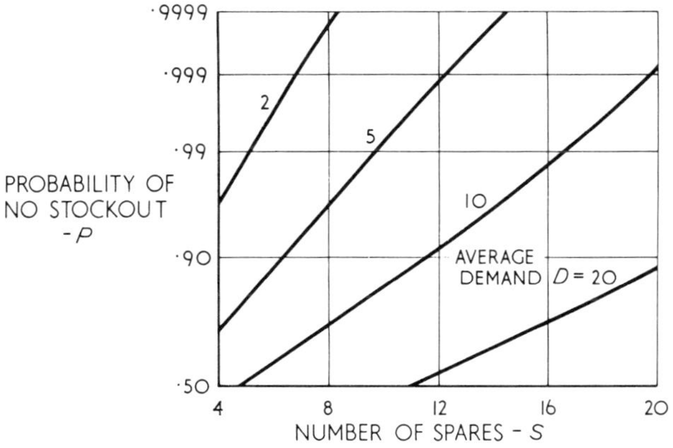

Figure 5 Variation of P with S for given values of D and with large M

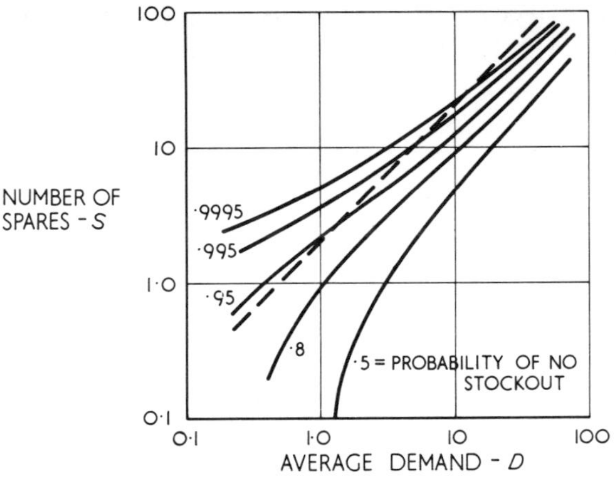

Figure 6 Variation of S with D for given values of P and with large M

Suppose that the parent equipment comprises n types of item and that the demand for each type is independent of the demands for other types. The probability P of not having a stockout of any one item is given by

$$P = P_1 \; P_2 \; \ldots . \tag{4}$$

where P_1 etc. is the probability of not having a stockout of item 1; P_1 is itself a function of L_1, K_1, M_1 and S_1.

Associated with L_1, K_1, M_1, and S_1, etc. are costs C_{L_1}, C_{K_1}, C_{M_1}, *and* C_{S_1} etc; assume that there is a functional relationship between L_1 and C_{L_1}, K_1 and C_{K_1} *etc.*

The problem is to maximize P for a given total cost C given by

$$C = C_{L_1} + C_{K_1} + C_{M_1} + C_{S_1} + C_{L_2} + C_{K_2} + \ldots . \tag{5}$$

It can be shown that the necessary conditions are given by the following equations:

$$\frac{1}{P_1} \frac{\partial P_1}{\partial C_{L_1}} = \frac{1}{P_1} \frac{\partial P_1}{\partial C_{K_1}} = \ldots . = \frac{1}{P_2} \frac{\partial P_2}{\partial C_{L_2}} = \ldots . \tag{6}$$

Generally the value of P is expected to be close to 1.0 (100%). In such a case Equation 4 implies that P_1, P_2 etc. must each be very close to 1.0. Equation 6 may therefore be written

$$\frac{\partial P_1}{\partial C_{L_1}} = \frac{\partial P_1}{\partial C_{K_1}} = \ldots . = \frac{\partial P_2}{\partial C_{L_2}} = \ldots . \tag{7}$$

6.2 Outline of Method

Each derivative, e.g. $\frac{\partial P_1}{\partial C_{L_1}}$, in Equation 7 can be considered as the marginal cost of the probability, in this example P_1, by a change in the relevant factor, in this case L_1. To assist in its evaluation, it can be written

$$\frac{\partial P_1}{\partial C_{L_1}} = \frac{\partial P_1}{\partial L_1} \frac{dL_1}{dC_{L_1}} \tag{8}$$

where $\frac{\partial P_1}{\partial L_1}$ can be determined by the method used in Section 4, and $\frac{dL_1}{dC_{L_1}}$ is the marginal cost of changing L_1 (which must be estimated separately).

A search for the optimum can proceed in steps, starting from any likely combination $L_1, \ldots L_2, \ldots$ etc. A method such as that used above in Section 5 can be used to determine P_1, P_2, etc. and $\frac{\partial P_1}{\partial L_1}, \frac{\partial P_2}{\partial L_2},$ etc. The other data required are estimates of the marginal costs of each factor, *i.e.* $\frac{dL_1}{dC_{L_1}}$ etc.

With these as input, the values of the terms $\frac{\partial P_1}{\partial L_1} \frac{dL_1}{dC_{L_1}}$ etc. in Equation 8 can be calculated. A comparison of these values should indicate the direction of the changes in L_1 etc. needed to bring them closer to equality. In this way, another and improved combination $L_1 \ldots L_2 \ldots$ is obtained. The procedure can be repeated until the approximation to the optimal condition is judged to be sufficiently close.

The big advantage that accrues from this approach is that the data requirement in the variation of the marginal costs of each factor is only a limited region rather than over the whole range of factors.

6.3 Special Cases

6.3.1. Attention Confined to the Factor S

If it can be shown (or assumed) that the only feasible way of changing P is by changing the number of spares, then attention can be confined to those terms in Equation 7 that contain S. In this case Equation 7 reduces to

$$\frac{dP_1}{dC_{S_1}} = \frac{dP_2}{dC_{S_2}} = \ldots\ldots \tag{9}$$

This can be approximated further to

$$\frac{\delta P_1}{\delta C_{S_1}} = \frac{\delta P_2}{\delta C_{S_2}} = \ldots\ldots \tag{10}$$

where δP_1 etc. is the change in P_1 brought about by a change of one in the number of spares of Item 1 and δC_{S_1} etc. is the cost of adding one more item to the inventory (generally proportional to the unit cost of the item and with the same constant of proportionality for all items).

The application of this particular result is straightforward (compared with the general case).

6.3.2. *Large Values of M*

For the reasons given in Section 4.2, the values of some (or all) of the factors M_1 etc. may be large. If this is so, the corresponding values of $\frac{\partial P_1}{\partial M_1}$ etc. will already be smaller than the likely optimal values. The particular terms in Equation 7 may therefore be neglected.

7. DISCUSSION

7.1 Effect of Numbers

The effect that the size of the fleet can have on the desirable level of spares is illustrated best by considering the simplest case where

(a) the maintenance facility is relatively large,

(b) the MTBR (Factor L) and MTM (Factor K) are more or less fixed.

In this case the results are embodied in Figure 6 which gives directly the spares requirement for selected values of P and any value of $D = AK/L$. For example, if a 99.95% probability of no stockout is selected, an item whose value of D is 20 requires 34 spares, i.e. an average pipeline of 20 and a buffer of 14; an item with a value of D of 4 requires 11 spares, i.e. an average pipeline of 4 and a buffer of 7. (In practice, the separate values of P are close to 1.0; this together with assumption (a) above means that D equals the average number in the maintenance pipeline.)

This result shows clearly the marked effect of varying the size of the fleet, remembering that with assumption (b) above, the value of D is proportional to the size of the fleet. For example, suppose that the ratio MTM/MTBR for a particular item is 0.25, and that one item is installed per aircraft say. Then a fleet of 16 aircraft (D = 4) requires 11 spare items of that type (i.e. 0.69 spare to each installed) while a fleet of 80 aircraft (D = 20) requires 34 spare (i.e. 0.43 spare to each installed).

7.2 Limitations of Empirical Rules for Buffer Stock

Figure 6 also points up the limitations of the empirical rules that have been devised for estimating buffer stock. One such rule is to make the buffer stock equal to the average pipeline, i.e. S equals $2D$. This is shown by the dashed line in Figure 6; note that if the turnaround time is also three months, the same line represents the other common rule of making the buffer stock equal to three months removals. It is seen that this rule is moderately satisfactory for values of D between 4 and 20, but under-estimates the spares requirement if D is smaller and over-estimates if D is larger.

7.3 Effect of Specialization

The problems of provisioning small numbers of aircraft are accentuated if the repairable items require specialized maintenance facilities. In this context, the facilities for one type of item are specialized if the trade skills and workshop equipment required for other items cannot be shared.

Consider the extreme case where a maintenance facility must be set up for one type of item only. Then Figure 4, for example, shows the trade-off between S and M for the given values of D and P. It is seen that if S is to be small, M must be large so as to reduce the probability of unserviceable items having to wait for maintenance. But as M increases, the average utilization of M decreases (see table in Section 5.3.2).

If M is made small so as to achieve a high average utilization, the value of S must increase very considerably with most of the items in the pool of unserviceable items awaiting maintenance.

The optimal solution of this conflict between low spares inventory and high utlization of maintenance facilities requires the application of the method outlined in Section 6.2. However it is clear that the level of spares will be higher and the maintenance utilization lower for the specialized item than for the non-specialized item.

This problem has not arisen in the past because the same maintenance facility has been shared by more than one type of item. Thus suppose that there are four types of items, each with $D = 5$ and each requiring the same kind of trade skills and workshop equipment. In this case a maintenance facility with a total capacity of

$$M = 4 \times 5 = 20$$

will achieve an average utilization close to 100%. (If P, the probability of no stockout, is close to 1.0, then D approximately equals the average number of items in the maintenance facility.)

In this situation, the maintenance manager has the option of switching effort from one type of item to another. Hence the effective value of M for each of the four types of item in this example is near 20. In this way a low probability of hold-up of unserviceable items (and therefore a low value of S) is attained whilst still utilizing fully the maintenance facility.

However if the number of different types of item that share the same facility decreases (e.g. with increasing specialization) or if D decreases (e.g. with smaller numbers of aircraft), the table in Section 5.3.2 shows that the maintenance manager's option may diminish to such an extent that the logistics manager is forced back to the general case discussed above.

DISCUSSION

Høgberg: You have chosen Probability of Stockout as the criterion for measuring performance for reasons stated. *Back order rate* is considered by some as a more relevant criterion. I suggest that if in fact the shop is structured with some degree of multiskill so that the channel structure $M = M_1, \ldots M_N$ may be changed, then you are in fact constructing a situation where to some degree it is true that 10 stockouts each lasting 1 day is equal to 1 stockout lasting 10 days. Your model results would then be valid, or close to it, for back order rate.

Landau: We have not given much attention to other criteria, but Major Høbgberg is probably correct about back-order rates. However, this work was directed mainly towards initial provisioning. At this stage, data are scarce and the uncertainties in MTBR, allocation of repair between echelons, etc. are so great that the choice of a specific criterion is of secondary importance. The main aim is to assist the technical assessor who estimates the initial provisioning of spares, and the principal value of this work is to indicate how the effect of number of installed items as well as their costs can be allowed for.

Fischer: You have defined the broad aim of your paper as to determine the optimal allocation of the investment in the inventory of repairable items for a given parent equipment. Did you ever try to determine also what a feasible size of the investment might be?

Landau: Some attention has been given to this problem, or rather to the problem of optimising and allocating the budget between total support and the "teeth" units. Assumptions were made about the variation of the marginal value of increased support and of additional operational aircraft. This led to results in accord with what data we had, but again the benefit of the study was the insight gained into what affects the allocation of budget. The study has been published as an Aeronautical Research Laboratories (Dept. of Supply) Note.

Asymptotic Performance of a System Subject to Cannibalization*

Warren M. Hirsch
Herman Henisch

INTRODUCTION

Among possible ways of maintaining a system* whose spares have been exhausted, cannibalization is one that has been practiced over a long period of time by the military of many nations, albeit not always with the consent of the commanding officer. By "cannibalization" we mean the removal of a failed part[2] from a system and its replacement by an operating part of the same type extracted from another part of the same system. In a system consisting of two aircraft, a simple example of cannibalization is the switching of a damaged tire on one aircraft with a good tire on the second. If, for instance, the radar on the second aircraft is damaged and a replacement is unavailable, the exchange of tires may at least restore one aircraft to operating condition.

The practice of cannibalization is highly controversial primarily because the very act of extracting an operating part from a complex system can cause extensive damage to other parts of the system. On the other hand, in situations where it is crucial to maintain the performance of a system at a high level, a viable alternative to cannibalization may be difficult to find. Because of the importance and polemical nature of cannibalization, one of the present authors was led in an earlier paper [1] to develop a mathematical model of a system subject to cannibalization. This model, which is applicable to a very wide class of systems and to parts subject to arbitary failure distributions, leads to closed form expressions for the probability distribution and expected value of the system performance level. In the present paper we carry further the analysis in [1] by studying in detail a special class of systems whose parts are subject to exponential failure rates. In this special case, as will be shown, it is possible to derive limit theorems that make it computationally feasible to assess the effects of cannibalization and to compare the performance levels and efficiencies of cannibalized and uncannibalized systems.

* This report represents results obtained at the Courant Institute of Mathematical Sciences, New York University, with the support of the Office of Naval Research, Contract N00014-67-A-0467-0004.

[1] The word "system" is not limited to a single physical entity such as an aircraft, ship, missile launcher, etc. It includes also any designated collection of objects whose maintenance and performance as a unit is of interest.

[2] The failed part can be either a single part or an entire module, i.e., a package of components of a system that can be removed and replaced as a whole. Throughout this paper the word "part" can be interpreted to mean "individual part" or "module".

I. DESCRIPTION OF THE SYSTEM

We consider a collection of n objects, $O_1, O_2, \ldots O_n$, to be used as a system to carry out a sequence of specified tasks, each task requiring τ units of time to complete. For simplicity we assume that the k-th task is initiated at time $(k\text{-}1)\tau$ and completed at time $k\tau$, $k = 1,2, \ldots$ [3] We shall refer to each object as a "machine" and to each task as a "mission". We define the size of the system to be the number, n, of objects it contains. A typical example of the kind of system we have in mind is a fleet of destroyers sent on voyages, or a squadron of aircraft sent on reconnaissance missions. We emphasize, however, that the nature of the machines and the kinds of missions performed may be quite arbitrary. Each machine O_j is composed of r distinct parts, $\gamma_{1j}, \gamma_{2j}, \ldots, \gamma_{rj}$, $j = 1,2, \ldots n$, where $r > 1$ and for each index i, $1 < i < r$, the part $\gamma_{i_1}, \gamma_{i_2}, \ldots, \gamma_{in}$ are assumed to be replicas of each other.

A machine is sent on a mission if and only if all of its parts are in good working order at the beginning of the mission. If one or more of its parts fails during the course of a mission, it is assumed that the machine returns from the mission, i.e., that the machine and its unfailed parts are available at the end of the mission.[4] We consider the case when no spare parts are available and damaged parts are irreparable, so that cannibalization is the only possible mode of maintenance. We assume that for each index i, $1 < i < r$, the parts $\gamma_{i1}, \gamma_{i2}, \ldots, \gamma_{in}$ are freely interchangeable, and that these are the only allowed exchanges. At the conclusion of a given mission k, the system is cannibalized in such a way as to maximize the number of machines that can be dispatched on mission k+1. The assumption that missions are dispatched at times $0, \tau 2\tau, \ldots$ and are of duration τ implies that all cannibalizations are instantaneously carried out.[5]

It is convenient to describe the state of each part at a given moment t by means of an ordered pair (x,y) whose first component x indicates the "employment status" of the part at time t and whose second component y indicates the "fitness' of the part at time t. For our analysis it suffices to distinguish two categories of employment: If, at time t, the part is installed in a machine that is then on a mission, we set $x = 1$ and say that the part is "engaged"; if it is in a machine that is not on a mission at time t, we set $x = 0$ and say that the part is "idle."[6] Concerning fitness, we assume that each part is either in good working order (operational) or has failed (unfit). We set $y = 1$ if the part is in good working order at time t; otherwise, $y = 0$. Thus, at each instant the state of a

[3] Only minor modifications are needed to treat the case when the tasks are carried out during arbitrarily specified non-overlapping intervals of time of length τ:

[4] This assumption is not as pernicious as it may seem, since the machines being considered may be subassemblies of larger pieces of equipment. For example, the machines may be periscopes in submarines; even if the periscope is damaged during the course of a mission, the submarine has the capacity to return to its base.

[5] This restriction can easily be removed without essential modification of our conclusions.

[6] We adopt the convention that the instants $0, \tau) 2\tau, \ldots$ when the missions are initiated, are engaged times for parts in the dispatched machines.

given part corresponds to one of the four vertices (0,0), (0.1), (1,0), (1,1) of the unit square. It is assumed that all parts are operational at time 0.

We shall analyze the performance of the system under two distinct assumptions concerning the periods when a part is exposed to risk of failure. The first of these, called the assumption of "synchronous clocks," postulates that at every instant t each part which is operational at time t, regardless of its employment status, is subject to risk of failure. This assumption (adopted in [1]) contrasts with the assumption of "asynchronous clocks" (adopted in [2]) according to which a given part is subject to risk of failure at time t if and only if it is in state (1,1) at time t, i.e., engaged and operational.[7] In both cases–synchronous and asynchronous clocks–it will be assumed that when parts are simultaneously exposed to risk, they fail independently of each other. Moreover, the failure rate of a part at any given instant when it is exposed to risk does not depend on the particular machine in which the part is installed nor on the sequence of machines through which it has passed.

2. THE CASE OF SYNCHRONOUS CLOCKS

In this section we restrict attention to the case of synchronous clocks. Suppose that at the time $k\tau$, when the $(k+1)$–st mission is dispatched, there are S_i operational parts of type i, $i = 1,2, \ldots, r$. Setting

$$\Phi = \min_{1 \leq i \leq r} S_i \ ,$$

it is clear that by appropriate cannibalizations exactly Φ machines can be assembled. all of whose parts are operational. It therefore seems reasonable to define Φ to be the "performance level" of the system at time $k\tau$.

We now associate to each part γ_i a random variable X_{ij}, called the "lifetime" of γ_{ij}, whose value denotes the time of failure of γ_{ij}. We shall assume that each randon variable X_{ij} is exponentially distributed with the distribution function

$$P \quad X_{ij} < t \quad = \begin{cases} 1 - e^{-\alpha_i t} & , \quad t > 0 \ , \\ 0 & , \quad t \leq 0 \ . \end{cases}$$

We emphasize that the failure rate α_i of γ_{ij} depends only on the part type i. Clearly the expected lifetime of a part of type i is

[7] The terms "synchronous clocks" and "asynchronous clocks" are motivated by the following interpretation: We imagine that each part is equipped with its own clock, all clocks being set to 0 and turned on at time 0. Under the assumption of clocks, no clock is ever turned off. Under the assumption of asynchronous clocks, the clock associated with a given part is turned off when the part is not engaged; hence, under this assumption two operational parts, one of which is in state (0,1) during the time interval $[(k\text{-}1)\tau, k\tau]$ and the other of which is in state (1,1) during this time interval, will, at time $k\tau$, have clocks showing different times if at time $(k\text{-}1)\tau$ they showed the same time.

$$E\,[X_{ij}] = \alpha_i^{-1} > 0 \, .$$

Under the assumption of independent failures (stated in Section 1) it follows that the random variables $(X_{ij})_{\substack{i=1,2,\ldots,r\\ j=1,2,\ldots,n}}$ are independent.

We obtain a convenient expression for the number of operational parts of each type at time $k\tau$ by introducing the indicators $Y_{ij}(k)$ of the events $[X_{ij} > k\tau]$, defined for $k > 0$ and the admissible values of i,j by the equations

$$Y_{ij}(k) = \begin{cases} 1\,, & \text{if } X_{ij} \geq k\tau\,, \\ 0\,, & \text{otherwise.} \end{cases}$$

Plainly, for each index k the random variables $(Y_{ij}(k))_{\substack{i=1,2,\ldots,r\\ j=1,2,\ldots,n}}$ are independent, and

$$Y_{ij}(k) = \begin{cases} 1\,, & \text{with probability } e^{-\alpha_i k\tau}\,, \\ 0\,, & \text{with probability } 1 - e^{-\alpha_i k\tau} \end{cases} \tag{1}$$

Let

$$S_{in}(k) = \sum_{j=1}^{n} Y_{ij}(k)\,. \tag{2}$$

Then $S_{in}(k)$ denotes the number of operational parts of type i at time $k\tau$, given that the size of the system is n. Setting

$$p_i(k) = e^{-\alpha_i k\tau}$$

and

$$q_i(k) = 1 - p_i(k)\,,$$

it follows from equations 1 and 2 and the independence of $(Y_{ij}(k))_{j=1,2,\ldots,n}$ that

$$P\left\{ S_{in}(k) = m \right\} = \binom{n}{m} p_i^{m}(k) q_i^{n-m}(k)\,, \quad m = 0,1,\ldots,n. \tag{3}$$

By definition, the system performance level at time $k\tau$ is given by the random variable

$$\Phi_n(k) = \min_{1 \leq i \leq r} S_{in}(k)$$

We observe now that for each pair of indices (n,k) the random variables $S_{1n}(k), S_{2n}(k), \ldots, S_{rn}(k)$ are independent, since the random vectors

$$(Y_{11}(k),Y_{12}(k),\ldots,Y_{1n}(k)),\ (Y_{21}(k),Y_{22}(k),\ldots,Y_{2n}(k)),\ \ldots,(Y_{r1}(k),Y_{r2}(k),\ldots,Y_{rn}(k))$$

are independent. Hence, the expected performance level at time $k\tau$ can be determined from the tail distribution by the well-known formula,

$$E[\Phi_n(k)] = \sum_{\lambda=0}^{n-1} P\left\{\Phi_n(k) > \lambda\right\} \tag{4}$$

$$= \sum_{\lambda=0}^{n-1} P\left\{S_{1n}(k) > \lambda,\ S_{2n}(k) > \lambda\ ,\ldots, S_{rn}(k) > \lambda\right\}$$

$$= \sum_{\lambda=0}^{n-1} \prod_{i=1}^{r} P\left\{S_{in}(k) > \lambda\right\} = \sum_{\lambda=0}^{n-1} \prod_{i=1}^{r} \sum_{m=\lambda+1}^{n} \binom{n}{m} p_i^m(k) q_i^{n-m}(k)$$

$$= \sum_{\lambda=1}^{n} \prod_{i=1}^{r} \sum_{m=\lambda}^{n} \binom{n}{m} p_i^m(k) q_i^{n-m}(k) \quad [8]$$

It follows immediately from the definition of $\Phi_n(k)$ that the expected system performance level is, for fixed k, an increasing function of n, i.e.,

$$E[\Phi_{n+1}(k)] \geq E[\Phi_n(k)]\ , \quad n = 1,2,\ldots\ .$$

Enlarging the size of the system, however, entails added cost, and it is therefore of interest to introduce some measure of the return per dollar of investment. To this end we define the "efficiency" $\rho_n(k)$ of a system after k missions, given that the system has size n, as the expected performance per machine; thus,

$$\rho_n(k) = \frac{1}{n} E[\Phi_n(k)] \tag{5}$$

We emphasize that, in general, the efficiency depends not only on the time $k\tau$ at which it is evaluated but also on the size, n, of the system. It is of interest to study the behavior of $\rho_n(k)$ as n varies and to compare the efficiencies of cannibalized and uncannabilized systems.

Let us consider the system described in Section 1, assuming that cannibalizations are not performed. Let $\Phi_n^o(k)$ and $\rho_n^o(k)$ denote, respectively, the performance level and efficiency after k missions, given that the system has size n. Since a given machine will be operational at time $k\tau$ if and only if all of its parts survive k missions, and this

[8] This formula is a special case of equations (4.3) and (4.4) in [1], which give the expected performance level of an arbitrary system to cannibalization.

occurs with probability $p_1(k)p_2(k) \ldots p_r(k)$, it is clear that

$$P\left\{\overset{o}{\Phi}_n(k)=m\right\} = \binom{n}{m}[p_1(k)p_2(k)\cdots p_r(k)]^m [1-p_1(k)p_2(k)\cdots p_r(k)]^{n-m}$$

Hence, by an elementary formula,

$$E[\overset{o}{\Phi}_n(k)] = np_1(k)\,p_2(k)\,\cdots\,p_r(k) ,$$

and

$$\overset{o}{\rho}_n(k) = p_1(k)p_2(k)\cdots p_r(k) . \tag{6}$$

From equations 4 and 5 we see that the efficiency of the system when cannibalization is practiced is given by the expression

$$\rho n(k) = \frac{1}{n} \sum_{\lambda=1}^{n} \prod_{i=1}^{r} \sum_{m=\lambda}^{n} \binom{n}{m} p_i^m(k)\; q_i^{n-m}(k) . \tag{7}$$

We note from equation 6 that the efficiency of the uncannibalized system is a constant depending only on the part failure rates and is independent of the system size, whereas from equation 7 it is clear that the efficiency of the cannibalized system varies with n, although it is by no means obvious precisely how the efficiency changes as the system is enlarged.

The tables in the Appendix show how the efficiencies of the cannibalized and uncannibalized systems compare in various special cases. These tables suggest that for each fixed time $k\tau$ the efficiency of the cannibalized system is a strictly increasing function of the system size. For simplicity we present here a proof of this result only in a very special case, leaving for another paper a discussion of the full theorem.

Theorem 1.

Let $\rho_n(k)$ denote the efficiency after k missions of the system described in Section 1, assuming that cannibalization is practiced, each machine is composed of two parts, the failure rates of the parts are equal, and clocks are synchronous. Then for each positive integer k,

$$\rho_n(k) < \rho_{n+1}(k) , \quad n = 1,2,\ldots .$$

Proof. Since k will be held fixed throughout our argument, for simplicity we surpress k in all our notation. From the definition of efficiency,

$$\rho_{n+1} - \rho_n = E\left[\frac{\Phi_{n+1}}{n+1} - \frac{\Phi_n}{n}\right] = \frac{1}{n(n+1)} E[n(\Phi_{n+1} - \Phi_n) - \Phi_n] . \tag{8}$$

Setting

$$\Phi_0 = 0$$

and

$$\Delta_n = \Phi_{n+1} - \Phi_n \ , \qquad n = 0,1,\ldots \ ,$$

from (8) we see that it suffices to prove

$$E[n\Delta_n - \Phi_n] > 0 \ , \qquad n = 1,2,\ldots \ . \tag{9}$$

Writing Φ_n in the form

$$\Phi_n = \sum_{j=0}^{n-1} \Delta_j \ , \qquad n = 1,2,\ldots \ ,$$

inequality 9 reduces to

$$E[n\Delta_n] - \sum_{j=0}^{n-1} E[\Delta_j] > 0 \ , \qquad n = 1,2,\ldots \ . \tag{10}$$

We define for $n \geqslant 1$,

$$u_n = P\{S_{1n} = S_{2n}\} \ . \tag{11}$$

It is convenient to introduce the additional definition,

$$u_0 = 1 \ .$$

Setting

$$p_1(k) = p_2(k) = p \ ,$$

it is easy to verify that for $n \geq 0$, Δ_n takes only the values 1, 0 with, respectively, the probabilities

$$P\{\Delta_n{=}1\} = u_n p^2 + (1-u_n)p \ ,$$

$$P\{\Delta_n{=}0\} = 1 - [u_n p^2 + (1-u_n)p] \ .$$

Since $E[\Delta_n] = P\{\Delta_n{=}1\}$, equation 10 is equivalent to the inequality

$$n[u_n p^2 + (1-u_n)p] - \sum_{j=0}^{n-1} [u_j p^2 + (1-u_j)p] > 0 \ ,$$

which in turn reduces to

$$p(1-p)\ (nu_n \ - \sum_{j=0}^{n-1} u_j) < 0\ . \tag{12}$$

By Abel's partial summation formula,

$$nu_n \ - \sum_{j=0}^{n-1} uj = \sum_{j=1}^{n} j(u_j \ - \ u_{j-1})\ .$$

Since $p(1-p) > 0$, to prove equation 12 it suffices to establish the inequality

$$\sum_{j=1}^{n} j(uj - u_{j-1}) < 0\ . \tag{13}$$

We shall in fact prove the stronger result that for all positive integers j,

$$u_j < u_{j-1}\ . \tag{14}$$

To this end we note from equations 3 and 11, and the independence of the random variables S_{1n}, S_{2n} that

$$u_n = \sum_{m=0}^{n} \binom{n}{m}^2 p^{2m}\ (1-p))^{2(n-m)}\ , \qquad n = 0,1,\ldots\ .$$

Let us recall that if f is the polynomial

$$f(z) = \sum_{m=0}^{j} a_m z^m\ ,$$

then

$$\frac{1}{2\pi}\int_0^{2\pi} f(e^{i\theta})\bar{f}\,(e^{i\theta})\ d\theta = \frac{1}{2\pi}\int_0^{2\pi} |f(e^{i\theta})\,|^2\ d\theta = \sum_{m=0}^{j} |a_m|^2\ . \tag{15}$$

Setting $q = 1-p$ and applying equation 15 to the particular function

$$f(z) = (p+qz)^j = \sum_{m=0}^{j} \binom{j}{m} p^m q^{j-m} z^{j-m}\ ,$$

we obtain

$$\frac{1}{2\pi}\int_0^{2\pi} |p+qe^{i\theta}|^{2j}\, d\theta = \sum_{m=0}^{j} \binom{j}{m}^2 p^{2m} q^{2(j-m)} = u_j \; . \tag{16}$$

Now noting that for $0 < \theta < 2\pi$,

$$|p+qe^{i\theta}|^2 < p+q = 1 \; ,$$

and for $\theta = 0$ or 2π,

$$|p+qe^{i\theta}|^2 = 1 \; ,$$

Equation 14 follows from equation 16. This completes the proof of the theorem.

A measure of the advantages of practicing cannibalization on a system of size n during a time period of duration $k\tau$ is given by the ratio of efficiencies

$$\mu_n(k) = \frac{\rho_n(k)}{\rho_n^{\circ}(k)}$$

The limit

$$\mu_\infty(k) = \lim_{n\to\infty} \mu_n(k) \; , \tag{17}$$

if it exists, can be interpreted as a measure of the advantage of cannibalizing a very large system in the time interval $[0,k\tau]$. We call $\mu_\infty(k)$ the "maintenance index for k missions." The limit

$$\mu^* = \lim_{k\to\infty} \mu_\infty(k) \; ,$$

if it exists, is called the "steady state maintenance index"; it provides an estimate of the long-run superiority of a very large cannibalized system over an uncannibalized one.

Theorem 2.

In the case of synchronous clocks, the maintenance indices of the system described in Section 1 satisfy the equations

$$\mu_\infty(k) = \frac{\min\limits_{1\le i\le r} p_i(k)}{p_1(k)p_2(k)\ldots p_r(k)} \tag{18}$$

and

$$\mu^* = \infty \; . \tag{19}$$

Proof. By the Strong Law of Large Numbers,

$$\lim_{n\to\infty} \frac{1}{n} S_{in}(k) = p_i(k) \qquad \text{a.s.,} \qquad i = 1,2,\dots,r.$$

Consequently,

$$\lim_{n\to\infty} \frac{\Phi_n(k)}{n} = \lim_{n\to\infty} \min_{1<i<r} \frac{1}{n} S_{in}(k) = \min_{1<i<r} p_i(k) \quad \text{a.s.}$$

Since $|\frac{1}{n}\Phi_n(K)| \leqslant 1$, by the Lebesgue bounded convergence theorem it follows that

$$\lim_{n\to\infty} E[\frac{1}{n} \Phi_n(k)] = \min_{1<i<r} p_i(k) \; .$$

Hence, using equation 6, we have

$$\mu_\infty(k) = \frac{\min_{1<i<r} p_i(k)}{p_1(k)p_2(k)\dots p_r(k)} ,$$

which, from the definition of $p_i(k)$, reduces to

$$\mu_\infty(k) = \left(\frac{\min_{1<i<r} e^{-\alpha_i \tau}}{e^{-\sum_{i=1}^{r} \alpha_i \tau}} \right)^k$$

Since

$$\frac{\min_{1<i<r} e^{-\alpha_i \tau}}{e^{-\sum_{i=1}^{r} \alpha_i \tau}} = \min_{1<i<r} e^{-\alpha_i \tau} e^{\sum_{i=1}^{r} \alpha_i \tau} > 1 ,$$

we have

$$\lim_{k \to \infty} \mu_\infty (k) = \infty$$

which completes the proof.

3. THE CASE OF ASYNCHRONOUS CLOCKS

The case of asynchronous clocks, although more realistic, is analytically more complicated because the lifetimes $(X_{ij})_{\substack{i=1,2,\ldots,r\\ j=1,2,\ldots,n}}$ are no longer independent random variables. We confine ourselves in this case to showing that the maintenance indices are precisely the same as they are in the case dealt with in the previous section.

Theorem 3.

In the case of asynchronous clocks, the maintenance indices of the system described in Section 1 satisfy the equations

$$\mu_\infty(\mathrm{k}) = \frac{\min\limits_{1 \leq i \leq r} p_i(k)}{p_1(k) p_2(k) \ldots p_r(k)}$$

and

$$\mu^* = \infty .$$

Proof. Let $(X_{ij}(k))_{\substack{i=1,2,\ldots,r\\ j=1,2,\ldots,n\\ k=1,2,\ldots}}$ be a family of random variables such that for each k the random variables $(X_{ij}(k))_{\substack{i=1,2,\ldots,r\\ j=1,2,\ldots,n}}$ are independent, and for all i, j, k,

$$X_{ij}(k) = \begin{cases} 1, & \text{with probability } p_i(1), \\ 0, & \text{with probability } 1-p_i(1) \end{cases}$$

The random variable $X_{ij}(1)$ corresponds to a part of type i used on the first mission; $X_{ij}(1) = 1$ if and only if the part to which it corresponds survives the first mission. Thus, the random variable $\sum_{j=1}^{n} X_{ij}(1)$ denotes the number of parts of type i operational at the outset of the second mission, and the random variable

$$N_1 = N_1(n) = \min_{1<i<r} \sum_{j=1}^{n} X_{ij}(1) \tag{20}$$

denotes the number of operational machines obtained by cannibalization after the first mission. For each fixed index i the random variables $X_{ij}(2)$, $j = 1,2,\ldots,N_1$, correspond to the N_1 parts of type i used on the second mission; $X_{ij}(2) = 1$ if and only if the part to which it corresponds survives the second mission. The assumed independence of the random variables $(X_{ij}(2))_{\substack{i=1,2,\ldots,r\\ j=1,2,\ldots,n}}$ reflects the fact that when parts are simultaneously exposed to risk, they fail independently of each other. Clearly the number of operational parts of type i available after the second mission consists of parts of type i dispatched on the second mission that survive the second mission, i.e., $\sum_{J=1}^{N_1} X_{ij}(2)$, plus parts of type i that survived the first mission but were not used on the second, i.e., $\sum_{j=1}^{n} X_{ij}(1) - N_1$. Thus, the number of operational machines obtained by cannibalization after the second mission is given by the random variable

$$N_2 = N_2(n) = \min_{1<i<r} \left[\sum_{j=1}^{n} X_{ij}(1) + \sum_{j=1}^{N_1} X_{ij}(2) - N_1\right]. \quad (21)$$

More generally, we define recursively a sequence of random variables

$$N_1, N_2, \ldots$$

by choosing N_1 according to equation 20; then, assuming $N_1, N_2, \ldots, N_{k\text{-}1}$ have been defined, we set

$$N_k = N_k(n) = \min_{1<i<r} \left[\sum_{j=1}^{n} X_{ij}(1) + \sum_{j=1}^{N_1} X_{ij}(2) + \ldots + \sum_{j=1}^{N_{k-1}} X_{ij}(k) - \sum_{\nu=1}^{k-1} N_\nu\right] \quad (22)$$

By an argument similar to the one used in deriving equation 21 it follows that N_k is the number of operational machines obtained by cannibalization after the k-th mission, i.e., it is the system performance level at time $k\tau$.

To obtain the maintenance index for k missions we shall prove inductively that for all integers k,

$$\lim_{n\to\infty} \frac{1}{n} N_k(n) = \min_{1\leq i\leq r} p_i(k) \quad (23)$$

To shorten notation we set

$$p_i = p_i(1) = e^{-\alpha_i \tau}$$

and

$$\delta = \min_{1 \leq i \leq r} p_i > 0.$$

By the Strong Law of Large Numbers,

$$\lim_{n\to\infty} \frac{1}{n} \sum_{j=1}^{n} X_{ij}(1) = p_i \text{ a.s.}$$

Hence,

$$\lim_{n\to\infty} \frac{1}{n} N_1(n) = \delta \quad \text{a.s.,} \tag{24}$$

i.e.,

$$N_1(n) \sim n\delta \quad \text{a.s.}$$

Assume that for $\nu \leq k-1$,

$$N_\nu(n) \sim n\delta^\nu \quad , \quad \text{a.s.}$$

Since this implies that $N_\nu(n) \to \infty$ as $n \to \infty$, we obtain from the Strong Law of Large Numbers the relationship

$$\lim_{n\to\infty} \frac{1}{n} \sum_{j=1}^{N_\nu} X_{ij}(\nu+1) = \lim_{n\to\infty} \frac{N_\nu}{n} \frac{1}{N_\nu} \sum_{j=1}^{N_\nu} X_{ij}(\nu+1) = \delta^\nu p_i \ , \quad \text{a.s.}$$

Hence, we conclude from (22) that as

$$\lim_{n\to\infty} \frac{1}{n} N_k(n) = \min_{1 \leq i \leq r} (p_i + \delta p_i + \ldots + \delta^{k-1} p_i - \sum_{\nu=1}^{k-1} \delta^\nu)$$

$$= \min_{1 \leq i \leq r} (p_i \sum_{j=0}^{k-1} \delta^j - \sum_{\nu=1}^{k-1} \delta^\nu) \ \cdot$$

$$= (\sum_{j=0}^{k-1} \delta^j)(\min_{1 \leq i \leq r} p_i) - \sum_{\nu=1}^{k-1} \delta^\nu$$

$$= \sum_{j=1}^{k} \delta^j - \sum_{\nu=1}^{k-1} \delta^\nu = \delta^k$$

This completes the induction. Since

$$\delta^k = (\min_{1 \leq i \leq r} p_i(1))^k = (\min_{1 \leq i \leq r} e^{-\alpha_i \tau})^k = \min_{1 \leq i \leq r} e^{-\alpha_i k \tau} = \min_{1 \leq i \leq r} p_i(k),$$

equation 23 is proved. From the definition of N_k it is obvious that $|\frac{N_k}{n}| \leq 1$, so that by the Lebesgue bounded convergence theorem,

$$\lim_{n \to \infty} \frac{1}{N} E\,[N_k(n)] = \lim_{n \to \infty} \rho_n(k) = \min_{1 < i < r} p_i(k)$$

Plainly, the efficiency of the uncannibalized system after k missions does not depend on whether or not clocks are synchronous. It follows that the maintenance indices are the same as in Theorem 2.

ACKNOWLEDGMENT

We should like to acknowledge that our interest in the subject matter of this paper was stimulated by results of A. J. Rolfe [2], whose numerical computations suggested the validity of the theorems we have proved. We are indebted also to Professor Agnes Berger for many stimulating discussions in connection with Theorem 1.

APPENDIX

TABLE 1

Efficiency After k Missions of a System with 2 Part Types

System Sizes, 2–25

$p_1(1) = .9, p_2(1) = .95$

No. of Mis. k	Cannibalized System of Size																								Uncan. System of Any Size
	2	3	4	5	6	7	8	9	10	11	12	13	14	15	16	17	18	19	20	21	22	23	24	25	
1	.859	.863	.866	.869	.871	.873	.875	.877	.878	.880	.881	.882	.883	.884	.885	.886	.886	.887	.888	.888	.889	.889	.890	.890	.855
2	.745	.755	.763	.769	.774	.778	.781	.784	.786	.788	.790	.791	.793	.794	.795	.796	.797	.798	.798	.799	.800	.800	.801	.801	.731
3	.649	.665	.676	.685	.691	.696	.700	.703	.706	.708	.710	.712	.713	.715	.716	.717	.718	.718	.719	.720	.720	.721	.722	.722	.625
4	.568	.589	.602	.612	.619	.624	.628	.632	.634	.637	.639	.640	.642	.643	.644	.645	.646	.647	.648	.648	.649	.650	.650	.651	.534
5	.499	.523	.537	.547	.554	.560	.564	.568	.570	.573	.575	.576	.578	.579	.580	.581	.582	.583	.584	.584	.585	.585	.586	.586	.457
6	.439	.464	.479	.490	.497	.502	.507	.510	.513	.515	.517	.519	.520	.521	.523	.523	.524	.525	.526	.526	.527	.527	.527	.528	.391
7	.387	.412	.428	.438	.446	.451	.455	.459	.461	.464	.465	.467	.468	.469	.470	.471	.472	.473	.473	.474	.474	.475	.475	.475	.334
8	.340	.366	.382	.392	.399	.405	.409	.412	.415	.417	.419	.420	.422	.423	.423	.424	.425	.426	.426	.427	.427	.427	.428	.428	.286
9	.299	.326	.341	.351	.358	.363	.367	.370	.373	.375	.377	.378	.379	.380	.381	.382	.382	.383	.384	.384	.384	.385	.385	.385	.245
10	.263	.289	.304	.314	.321	.326	.330	.333	.335	.337	.339	.340	.341	.342	.343	.344	.344	.345	.345	.346	.346	.346	.347	.347	.209
11	.231	.256	.271	.281	.287	.292	.296	.299	.301	.303	.305	.306	.307	.308	.309	.309	.310	.310	.311	.311	.311	.312	.312	.312	.179
12	.203	.227	.241	.251	.257	.262	.266	.269	.271	.273	.274	.275	.276	.277	.278	.278	.279	.279	.280	.280	.280	.281	.281	.281	.153
13	.178	.201	.215	.224	.230	.235	.238	.241	.243	.245	.246	.247	.248	.249	.250	.250	.251	.251	.252	.252	.252	.252	.253	.253	.131
14	.156	.178	.191	.200	.206	.210	.214	.216	.218	.220	.221	.222	.223	.224	.225	.225	.226	.226	.226	.227	.227	.227	.227	.228	.112
15	.136	.157	.170	.178	.184	.188	.192	.194	.196	.198	.199	.200	.201	.202	.202	.203	.203	.203	.204	.204	.204	.204	.205	.205	.096

TABLE 2

Efficiency after k Missions of a System with 3 Part Types

System Sizes, 2–25

$p_1(1) = .9, p_2(1) = .92, p_3(1) = .96$

No. of Mis. k	Cannibalized System of Size																								Uncan. System of Any Size
	2	3	4	5	6	7	8	9	10	11	12	13	14	15	16	17	18	19	20	21	22	23	24	25	
1	.807	.817	.825	.832	.837	.842	.846	.850	.853	.856	.858	.860	.862	.864	.865	.867	.868	.869	.870	.871	.872	.873	.874	.875	.795
2	.667	.691	.709	.721	.731	.738	.744	.749	.754	.757	.761	.763	.766	.768	.770	.772	.773	.775	.776	.778	.779	.780	.781	.782	.632
3	.561	.595	.617	.632	.643	.652	.659	.665	.670	.674	.667	.680	.683	.685	.688	.690	.691	.693	.695	.696	.697	.698	.699	.700	.502
4	.476	.515	.540	.556	.568	.577	.585	.591	.596	.600	.604	.607	.610	.613	.615	.617	.619	.620	.622	.623	.625	.626	.627	.628	.399
5	.406	.448	.473	.490	.502	.512	.519	.525	.531	.535	.539	.542	.545	.548	.550	.552	.554	.556	.557	.559	.560	.561	.562	.563	.317
6	.346	.389	.414	.431	.444	.454	.461	.467	.473	.477	.481	.484	.487	.490	.492	.494	.496	.498	.499	.500	.502	.503	.504	.505	.252
7	.295	.337	.363	.380	.392	.402	.410	.416	.421	.425	.429	.432	.435	.438	.440	.442	.444	.446	.447	.448	.450	.451	.452	.453	.200
8	.250	.292	.317	.334	.347	.356	.364	.370	.375	.379	.383	.386	.389	.391	.393	.395	.397	.399	.400	.402	.403	.404	.405	.406	.159
9	.212	253	.277	.294	.306	.315	.323	.329	.334	.338	.341	.344	.347	.350	.352	.354	.355	.357	.358	.360	.361	.362	.363	.364	.126
10	.179	.218	.242	.258	.270	.279	.286	.292	.297	.301	.304	.307	.310	.312	.314	.316	.318	.319	.321	.322	.323	.324	.325	.326	.100
11	.150	.188	.211	.226	.238	.247	.253	.259	.264	.268	.271	.274	.276	.279	.281	.283	.284	.286	.287	.288	.289	.291	.292	.292	.080
12	.126	.161	.183	.198	.209	.218	.224	.230	.234	.238	.241	.244	.246	.249	.251	.252	.254	.255	.257	.258	.259	.260	.261	.262	.064
13	.105	.138	.159	.173	.184	.192	.198	.203	.208	.211	.215	.217	.220	.222	.224	.225	.227	.228	.229	.231	.232	.233	.234	.234	.051
14	.087	.118	.137	.151	.161	.169	.175	.180	.184	.188	.191	.193	.196	.198	.199	.201	.202	.204	.205	.206	.207	.208	.209	.210	.041
15	.072	.100	.118	.131	.141	.148	.154	.159	.163	.166	.169	.172	.174	.176	.178	.179	.181	.182	.183	.184	.185	.186	.187	.188	.033

TABLE 3

Efficiency After k Missions of a System with 4 Part Types

System Sizes, 2–25

$p_1(1) = .9, p_2(1) = .92, p_3(1) = .95, p_4(1) = .97$

No. of Mis. k	Cannibalized System of Size																								Uncan. System of Any Size
	2	3	4	5	6	7	8	9	10	11	12	13	14	15	16	17	18	19	20	21	22	23	24	25	
1	.781	.796	.808	.817	.825	.831	.837	.841	.845	.849	.852	.854	.856	.859	.860	.862	.864	.865	.866	.868	.869	.870	.871	.872	.763
2	.634	.668	.690	.706	.718	.728	.735	.741	.746	.751	.754	.758	.761	.763	.766	.768	.770	.771	.773	.774	.776	.777	.778	.779	.582
3	.527	.572	.600	.618	.632	.642	.650	.657	.663	.668	.672	.675	.678	.681	.684	.686	.688	.690	.692	.693	.695	.696	.697	.698	.444
4	.444	.494	.523	.543	.557	.568	.577	.584	.590	.595	.599	.603	.606	.609	.611	.614	.616	.618	.620	.621	.623	.624	.625	.626	.339
5	.375	.427	.457	.477	.492	.503	.512	.519	.525	.530	.534	.538	.541	.544	.547	.549	.551	.553	.555	.557	.558	.559	.561	.562	.259
6	.318	.369	.399	.419	.434	.445	.454	.461	.467	.473	.477	.481	.484	.487	.489	.492	.494	.496	.497	.499	.500	.502	.503	.504	.198
7	.268	.318	.348	.368	.383	.394	.403	.410	.416	.421	.425	.429	.432	.435	.438	.440	.442	.444	.446	.447	.448	.450	.451	.452	.151
8	.224	.274	.304	.324	.338	.349	.358	.365	.370	.375	.379	.383	.386	.389	.391	.394	.396	.397	.399	.401	.402	.403	.404	405	.115
9	.187	.235	.264	.284	.298	.309	.317	.324	.329	.334	.338	.342	.345	.347	.350	.352	.354	.356	.357	.359	.360	.361	.362	.363	.008
10	.115	.202	.230	.249	.262	.273	.281	.287	.293	.297	.301	.305	.308	.310	.313	.315	.317	.318	.320	.321	.322	.324	.325	.326	.067
11	.128	.172	.199	.217	.230	.241	.248	.255	.260	.265	.268	.272	.274	.277	.279	.281	.283	.285	.286	.287	.289	.290	.291	.292	.051
12	.105	.146	.172	.190	.202	.212	.220	.226	.231	.235	.239	.242	.245	.247	.249	.251	.253	.254	.256	.257	.258	.259	.260	.261	.039
13	.085	.124	.148	.165	.177	.186	.194	.200	.205	.209	.212	.215	.218	.220	.222	.224	.226	.227	.229	.230	.231	.232	.233	.234	.030
14	.069	.104	.127	.143	.155	.164	.171	.176	.181	.185	.188	.191	.194	.196	.198	.200	.201	.203	.204	.205	.207	.208	.209	.209	.023
15	.056	.087	.109	.124	.135	.143	.150	.156	.160	.164	.167	.170	.172	.174	.176	.178	.180	.181	.182	.184	.185	.186	.186	.187	.018

TABLE 4

Efficiency After k Missions of a System with 5 Part Types

System Sizes, 2–25

$p_1(1) = .9, p_2(1) = .92, p_3(1) = .95, p_4(1) = .97, p_5(1) = .99$

No. of Mis. k	Cannibalized System of Size																								Uncan. System of Any Size
	2	3	4	5	6	7	8	9	10	11	12	13	14	15	16	17	18	19	20	21	22	23	24	25	
1	.775	.791	.804	.814	.823	.830	.835	.840	.844	.848	.851	.854	.856	.858	.860	.862	.863	.865	.866	.867	.869	.870	.871	.871	.755
2	.627	.663	.688	.705	.717	.727	.735	.741	.746	.750	.754	.758	.761	.763	.766	.768	.770	.771	.773	.774	.776	.777	.778	.779	.570
3	.521	.569	.598	.617	.631	.641	.650	.657	.663	.667	.672	.675	.678	.681	.684	.686	.688	.690	.692	.693	.695	.696	.697	.698	.430
4	.439	.491	.522	.542	.556	.568	.576	.584	.589	.595	.599	.603	.606	.609	.611	.614	.616	.618	.620	.621	.623	.624	.625	.626	.325
5	.371	.425	.456	.476	.491	.503	.512	.519	.525	.530	.534	.538	.541	.544	.547	.549	.551	.553	.555	.557	.558	.559	.561	.562	.245
6	.314	.367	.398	419	.434	.445	.454	.461	.467	.472	.477	.481	.484	.487	.489	.492	.494	.496	.497	.499	.500	.502	.503	.504	.185
7	.265	.317	.348	.368	.383	.394	.403	.410	.416	.421	.425	.429	.432	.435	.438	.440	.442	.444	.446	.447	.448	.450	.451	.452	.140
8	.222	.273	.303	.323	.338	.349	.358	.365	.370	.375	.379	.383	.386	.389	.391	.394	.396	.397	.399	.401	.402	.403	.404	.405	.106
9	.185	.235	.264	.284	.298	.309	.317	.324	.329	.334	.338	.342	.345	.347	.350	.352	.354	.356	.357	.359	.360	.361	.362	.363	.080
10	.153	.201	.229	.248	.262	.273	.281	.287	.293	.297	.301	.305	.308	.310	.313	.315	.317	.318	.320	.321	.322	.324	.325	.326	.060
11	.126	.172	.199	.217	.230	.240	.248	.255	.260	.265	.268	.272	.274	.277	.279	.281	.283	.285	.286	.287	.289	.290	.291	.292	.045
12	.103	.146	.172	.189	.202	.212	.219	.226	.231	.235	.239	.242	.245	.247	.249	.251	.253	.254	.256	.257	.258	.259	.260	.261	.034
13	.084	.123	.148	.165	.177	.186	.194	.200	.205	.209	.212	.215	.218	.220	.222	.224	.226	.227	.229	.230	.231	.232	.233	.234	.026
14	.068	.104	.127	.143	.155	.164	.171	.176	.181	.185	.188	.191	.194	.196	.198	.200	.201	.203	.204	.205	.207	.208	.209	.209	.020
15	.055	.087	.109	.124	.135	.143	.150	.156	.160	.164	.167	.170	.172	.174	.176	.178	.180	.181	.182	.184	.185	.186	.186	.187	.015

REFERENCES

1. Hirsch, Warren M., Meisner, Morris, and Boll, Charles, 1968. Cannibalization in multicomponent systems and the theory of reliability. *Nav. Res. Log. Qu.* Vol. 15, No. 3.

2. Rolfe, Alan, J., 1970. Markov chain analysis of a situation where cannibalization is the only repair activity. *Nav. Res. Log. Qu.* 17, Vol. 17, No. 2.

DISCUSSION

Benyon-Tinker: Has the Distribution of like components been studied, and what difference does this make to the results of cannibalization versus non-cannibalization?

Hirsch: My answer to your question is similar to the one that I gave to Mr. Durbin. The main purpose of limiting our study to a system of identical machines, each machine being a series system, was to make the problem mathematically tractable. Certainly the model would be more wisely applicable if the assumptions were relaxed by allowing the system to be composed of possibly different kinds of machines and by permitting each machine to be more complicated in structure, for example, some kind of series-parallel structure.

Rosenman: Wouldn't it be more natural to consider a measure of relative efficiency: the improvement available thru cannibalization versus improvement available thru increasing the number of spares?

Hirsch: There are many ways of measuring the advantages of cannibalization. As you saw, we compared the efficiency of cannibalized to non-cannibalized systems. Another possibility, which I understand you to be suggesting, is to compute the cost of the minimum spares inventory necessary to achieve without cannibalization the efficiency obtained by cannibalizing the system during k-missions. I don't see why this latter method is any more natural than the former, although I do think it is a reasonable method and would shed additional light on the value of practising cannibalization.

Baker: My limited experience of a system subject to cannibalization leads me to suppose that the analysis presented in this paper may be unnecessarily complex. Components which have to be cannibalized are normally subject to very low failure rates indeed with the result that increasing time gives an increase in the number of different failed components but very rarely does more than one component of a particular type fail. Even when this happens the system can often be restored by cannibalization at a lower level. As a result the effective level of n systems can be maintained at n-l for most periods of interest without resupply, if facilities to permit cannibalization are provided. The corresponding level for provisioning of spares is rather difficult to determine, but the aim should be to avoid more than one failure of each cannibalizable part at the lowest level of equipment breakdown which can be achieved in the field.

Durbin: If there exists a danger of damage to a part during cannibalization, why not include probability of degradation even in cannibalization? Then your chosen measure would be testable against your hypothesis of probability of damage.

Hirsch: Certainly it would be a good idea to include in the cannibalization model the probability that a working part is damaged by translocation. We chose to ignore this real contingency in order to simplify the mathematical problem. Many refinements of our model are possible and I agree that the one you suggest would be interesting to pursue.

Ferrier: Le papier présenté à Athènes, il y a 3 ans, par le Professeur Hirsch était d'un niveau mathématique très élevé. Personne ne lui a posé de questions. Personne n'a contesté ses hypothèses ni ses résultats.

Son papier d'aujourd'hui, très simplifié sur le plan mathématique, mais impressionnant par ses résultats pratiques lui a valu au contraire quelques questions embarrassantes, certaines hypothèses ayant été jugées un peu sommaires. C'est la preuve que le dialogue a été bien engagé, et à la bonne hauteur, celle si l'on se trouve encore à portée de flèche. Je souhaite que le prochain papier de Mr. Hirsch satisfasse, à partir d'hypothèses un peu plus sophistiquées, mais sans développements mathématiques trop éprouvants, tous ceux qu'intéressent les remarquables possibilités encore mal explorées de la cannibalisation.

Information System Design in Large Scale Logistic Systems

E. P. Durbin*

1. INTRODUCTION AND SUMMARY

Operations researchers and systems analysts are increasingly concerned with information system design. Operations research has traditionally been concerned with physical production, distribution, and stockage problems. In these areas operations researchers using techniques such as simulation, Delphi methods, and operational gaming, generally aim at finding strategies–decisions that take account of important variables at the time the decision is made. The Operations Research approach has not been noticeably successful in improving the Information Systems Design process or the resulting product.

This paper develops the following points:

(i) Various factors cause transition to new information systems.

(ii) The traditional systems analysis approach is a "top-down" design–specifying goals, objectives, decision variables, policies, and finally information system specifications.

(iii) Organizations typically follow a "bottom up" or "inside out" design approach.

(iv) "Bottom up" design occurs because of institutional incentives and because of the complexity of modern systems.

(v) The "bottom up" design process typically yields degraded performance.

(vi) We must consider incremental or phased development that preserves design options.

2. LARGE INFORMATION SYSTEMS

An information system is an evolving organization of people, computers, and other equipment (associated communication and support systems) and their integrated operation to regulate and control selected environmental events to achieve systems objectives.

In a complex organization, an information system performs the same function as the

* Any views expressed in this paper are those of the author. They should not be interpreted as reflecting the views of The Rand Corporation or the official opinion or policy of any of its governmental or private research sponsors. Papers are reproduced by The Rand Corporation as a courtesy to members of its staff.

nervous system in the human body. This paper deals with information systems used by managers and planners in very large organizations. Such systems may be as simple as item stock level reports in a chain of warehouses or as complex as systems that come into play when an expensive spare part is required by an out-of-commission aircraft at a remote airfield. Stockage, airlift, and procurement information as well as repair computations may be required to determine the point of origin for resupply of the required part. A typical *logistics* information system consists of several complexes of computers tied together with owned or leased communication facilities. Logistics managers may interact with the information system in making daily decisions, and may enter their decisions back into the information system.

Information system concepts have developed more slowly than hardware. Most attention in large systems has been directed at the transition between second and third generation computer equipment. Second generation systems are characterized by serial memory (tape units), sequential batch processing, and only one user at a time on the central processor. Third generation systems are characterized by direct access memory, various terminal options, multi-programming, and multi-user time sharing.

The complexity of an information system depends on many factors. A few are listed below.

- Nature of Use: An information system used for interactive analysis is more complex than one simply used for data processing and report generation.

- Number of Installations: Complexity increases as the number of interconnected installations increases.

- Diversity of Equipment: Problems of standardization and conversion are compounded with diverse equipment.

- Number of Simultaneous Users: The executive routines to handle many simultaneous users are complex and costly.

- Size of Data Base: Data management systems for very large files and very large records are still being developed.

- Frequency of Data Changes: The efficiency of file management systems is increasingly important as the frequency of data updates increases.

- Interaction of Transactions (Cascading): Performance of systems in which one event triggers several others is less predictable.

- Extent of Imbedded Decisionmaking: Complexity increase as algorithms for decisions are imbedded in the information system and triggered automatically.

3. SYSTEM DESIGN OR CONVERSION

Several factors may lead to initiation of system change—either design anew or conversion. Workloads increase over time. Facilities begin to wear out and require increased maintenance. Fashions in computing change. Increased operating flexibility is desired. Speed of hardware (not software) improves and arguments concerning decreased dollar costs of each computing operation are difficult to ignore.

Transition may involve shifting transactions from one system to another, perhaps only through software or hardware changes, but perhaps also by policy changes, and perhaps by changing from batch to on-line processing.

Even "simple" hardware or software conversion can involve several considerations. Standardization can be a significant problem in multifacility system. In converting hardware in such systems we find that physical separation leads to operating differences. This is compounded if generations and manufacturers of equipment are different. File conversion and program conversion are more difficult in a multi-facility system. Staff training in a new system is always required, and planning for additional physical space and power is too often overlooked.

4. TOP-DOWN DESIGN

Systems analysts tend to recommend top-down design. Top-down design proceeds from the ideas of constrained optimization. It attempts to look at an overall organization, its policies, and their interactions.

Changes in decision and operating procedures appear to be the source of the major dollar and effectiveness gains in organizations. New technology may be required to implement desired decision and operating procedures, and introduction of new technology may be an essential step. However, resources available for system development are generally limited, often severely. When an initial decision is to take a very large step in introducing new technology, policy improvement will inevitably suffer. The proble associated with simply making new technology run absorb most of the staff. Top-down planning stresses policy and upgrades technology only as necessary. Only a policy base exists, the full range of new technology can be introduced. Processing requirements generate costs. The comparison of policy benefits with processing costs dictates the choice of both policy and information processing schemes. Processing parameters and available technology then lead to hardware selection.

Top-down design typically relies on analytic decision procedures. Such planning emphasizes decisions procedures to avoid trouble rather than ad hoc procedures to get out of trouble. Analytic models, simulation, and cost-effectiveness analyses are used to evaluate the worth of policy improvements.

Expending resources on modeling and experimentation requires a tradeoff between time and uncertainty. The more effort put into experimentation and analysis, the greater the reduction in uncertainty about the performance of the ultimate system and reduction in the consequent risk that it will be inadequate. The less effort put into experimentation and analysis, the faster a system gets designed and implemented; but with more attendant uncertainty and risk about ultimate performance. Obviously, when time is available, simulation can be of great benefit.

Top-down design can be likened to a complex decision tree with successive branches in policy, information processing concepts, and hardware configuration.

Policies might include decision procedures for Stock Distribution, Industrial Repair

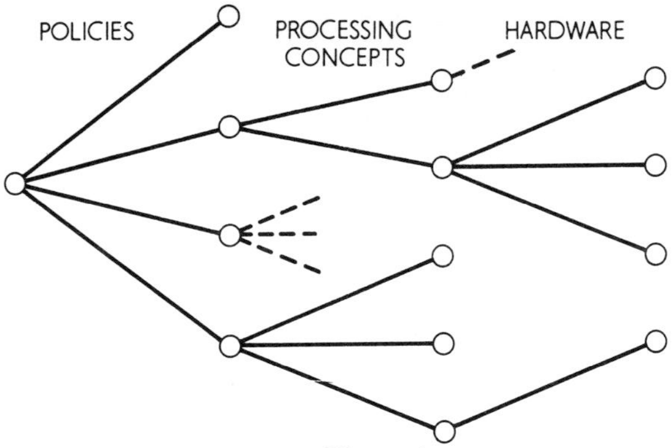

Figure 1

Scheduling, Procurement Policy, or EOQ Purchases. Processing concepts consider the degree of man-machine interaction, mix of batch and on-line processing, types of data base, file management systems, and the degree of system autonomy or manual override. Hardware considerations include the tremendous range of processors, terminals, and storage units available.

Very little is known about where to stop analyzing and experimenting and start implementing. While sound analysis and experimentation is necessary, we can only base our decisions on subjective estimates of the utility of additional research.

Rand's general approach has been to:

(1) Develop formal simulation models of the entire system under consideration to use in evaluating alternative management policies;

(2) Develop detailed simulation or analytical models of the proposed software (data management and multiprogramming systems for example) and experiment with these. For example see how systems behave against different input distributions and study the tradeoffs between data redundancy and information retrieval times.

(3) Develop both gross and detailed simulation models for use in cost-effectiveness and decision-rule studies. These models are elaborations of the system models first developed. They incorporate more of the functional details developed during the system requirements determination and system integration phases and the computer processing details developed during the software requirements development phase. Thus, they are able to attach costs to specific procedures and processing methods and evaluate the benefits achieved through their use.

(4) Survey similar industrial and military systems and collect statistics on software performance. Determine what overhead factors are being incurred and how existing multiprogramming monitors work.

(5) Finally select software structure.

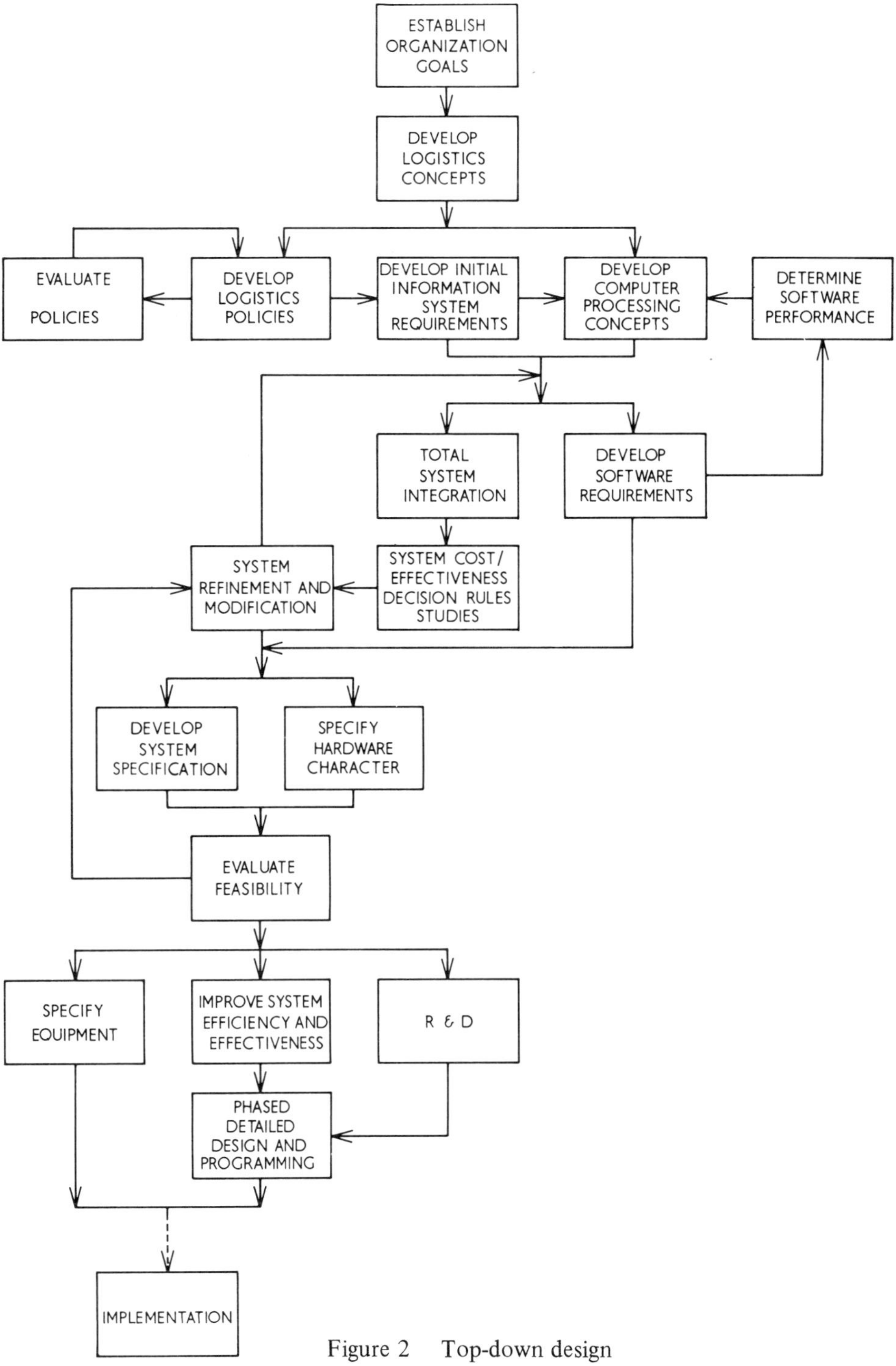

Figure 2 Top-down design

In summary, top-down design of information systems is characterized by a strong degree of sequential decisionmaking based on improved information. It is illustrated in Figure 2 and summarized below.

- Establish organizational missions and goals.
- Develop evaluations of policies through analysis and simulation.
- Explore alternative information processing concepts:
  - Degree of on-line vs. batch processing
  - Degree of man-machine interaction
  - Frequency of update
  - Degree of data base redundancy
  - Test a variety of information system options for feasibility
  - Compare costs, performances of information system options
  - Develop and follow an implementation plan:
  - Delivery, checkout tests
  - Plans for tuning the new system to attain high performance
  - Support, training plans

5. BOTTOM-UP DESIGN

The opposite of the top-down approach is "bottom-up" or "inside-out" analysis. It starts with arbitrary decisions at some detailed level of design, or decisions about specific management policies of the system. It may contain a detailed model of one part of the system, and by a process of addition, build to an overall system picture. As an example, equipment modernization in most corporations has led from 407 punch card equipment to 704 computers to 7090 computers to 360/65 computers with no change in processing rules or frequency or extent of interaction between transactions.

Bottom-up design forces low-level decisions in restricted contexts. It is further characterized by arbitrary selection of hardware. Organization policies are set before any overall planning or cost/effectiveness studies are undertaken, and important policy decisions are made without evaluation of their consequences. Bottom-up decisions frequently reflect a desire to utilize new and perhaps attractive computer hardware rather than a commitment to improvement in the organization's overall performance. Policy innovations receive marginal attention, and enthusiasm is directed at modernizing the processing equipment. Given this initial conceptual set, system design effort tends toward feasibility rather than system performance or cost.

Bottom-up design is frequently also characterized by strong parallel or concurrent structure. Simultaneous choice of management policy and hardware configuration occurs, and software must bridge a possibly unbridgeable gap.

Most design efforts observed in practice appear to initially emphasize estimate system hardware requirements, and later emphasis is on modification to achieve feasibility rather than on design exploration and experimentation to improve organizational performance.

If we characterize the effects of bottom-up design on the policy, information concepts, and hardware tree, the tree gets bare rapidly.

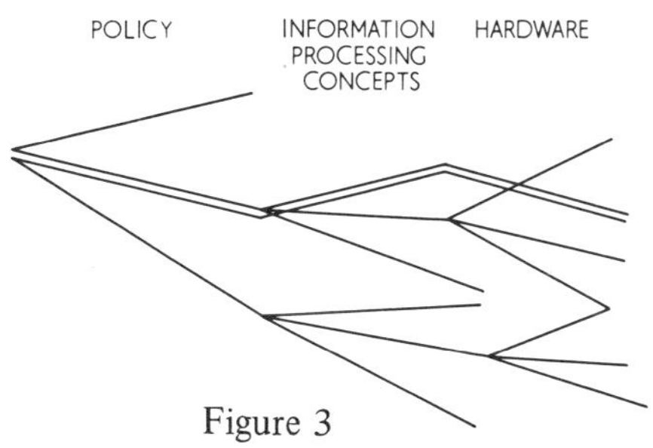

Figure 3

6. DANGER OF BOTTOM-UP DESIGN

Bottom-up design leads to system deficiencies. Arbitrary schedules, policies, development paths, and system boundaries lead to independent inclusion of different policies at different time points without knowledge of their interactions and implications. This can stunt the creativity of designers since there is no overview from which to predict interactions or effects. Thus there is no formal way to introduce new policies, and no valid means of predicting or evaluating policy and system performance.

The rapid and frequently simultaneous pruning of the decision tree in policy and hardware configurations leaves software to fill the gap. This may lead to:

(i) *Infeasibility*: A management evaluation system may provide no data. Or there may be no interfaces between transactions. Or a communication system may break down under heavy volume.

(ii) *Performance degradation and system rigidity*: After implementation the *entire* information system cannot be altered. Therefore fewer applications may be run, or less frequent updating than originally planned may be permitted.

(iii) *Schedule slippage* may occur in constructing software or hardware specifications, or in equipment deliveries, or in development of feasible policy applications.

(iv) *Increased equipment cost* may be incurred when extra equipment is procured to permit minimally acceptable system performance.

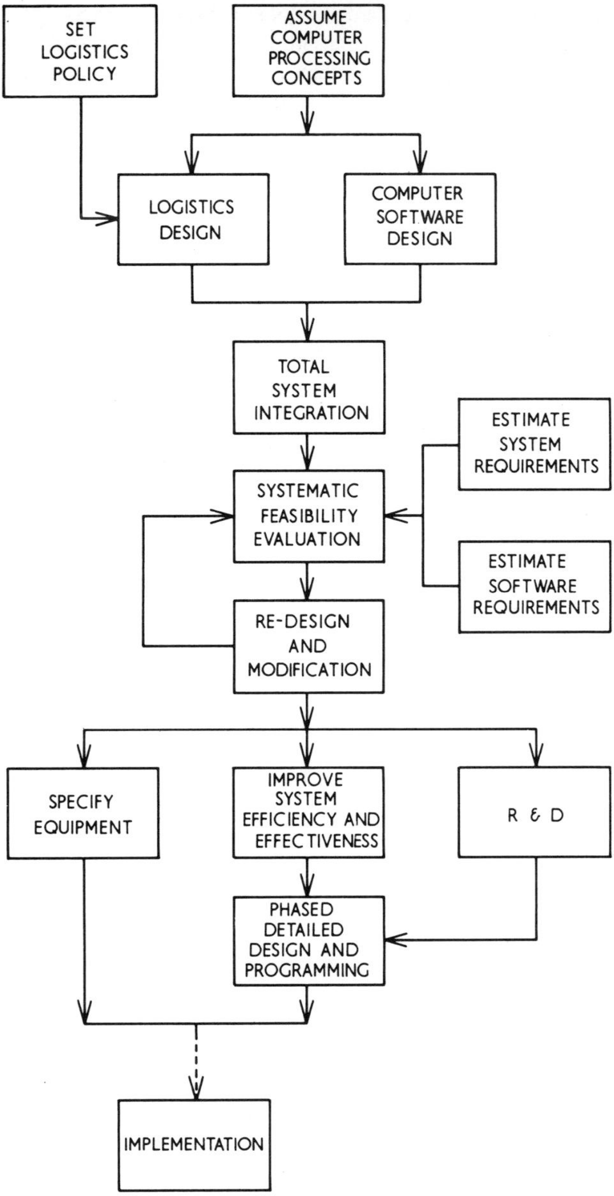

Figure 4 Bottom-up design philosophy

7. WHY DOES BOTTOM-UP RATHER THAN TOP-DOWN DESIGN OCCUR?

Bottom-up design is simpler. Top-down design requires determination of organizational and policy goals. These are difficult to obtain. Policy effects and interactions are difficult to model. Moreover, bottom-up design rapidly eliminates uncertainty and yields—on paper—a straightforward implementation plan which is easy to monitor. The arbitrariness is unnoticed until too late.

Management is generally not involved in system design. Technicians are typically in actual charge, and it cannot be assumed that the organization's data processing professionals understand corporate managers' responsibilities. It is easier for the design organization to prepare equipment specifications than to assess management needs of diverse organizations.

Procurement mechanisms in large organizations are tedious. System specifications are sometimes desired rapidly. "Buying in" ahead of other capital expenditures in the organization may be desirable. Speed drives the designer to concurrency in policy and hardware selection which requires development of general purpose software independent of any machines. Such software may itself be difficult to develop and probably can contain only a small subset of standard languages.

Bottom-up design flourishes because costs and performance evaluation of system specification is difficult. The performance of a computer is not determined by either the hardware or the software alone. The performance of an installation (hardware, software, and procedures) depends strongly and markedly on the hardware configuration. Computing standards do not yet exist for many areas. Metrics have not been identified or established. Costing, especially as it relates to procurement, does not reflect true consumption of resources. Because of all these factors the technical evaluation process is sometimes weak, and lags behind the complexity of systems. Thus, top-down design is not usually used. It is expensive in dollars, time consuming, and may lead to loss of momentum. Short time schedules for system implementation frequently preclude appropriate planning efforts.

8. ACHIEVING FEASIBILITY IN THE BOTTOM-UP WORLD

In the presence of such realities how may we assure feasibility for large systems?

(i) "Buy it and try it." A common approach, but not good for very large systems since the expense is infeasible.

(ii) Tune the *existing* installation and add equipment as necessary. This is especially attractive since cost-benefit arguments for new systems are generally not borne out.

(iii) In designing a new system use experts—they are much less expensive than in-house personnel. Also modify organizational procedures to fit canned routines and systems (accounting, payroll).

(iv) Use a phased, stated, prototype approach to implementation of large systems.

(v) Preserve flexibility and back up by keeping the old system as long as you can, and buying program-compatible equipment.

(vi) Rent equipment with an option to return.

(vii) If you must buy equipment, buy modular equipment with excess core capacity.

(viii) Weigh procurements in favor of vendors with families of equipment.

(ix) Following installation, conduct performance audits, tune configurations for additional performance, and return unneeded equipment.

9. SUMMARY

Top-down design will *not* in general be accomplished unless a relatively long time for development exists, and an excellent consultant group is on hand. In cases where the developing organization has had information system experience, the design team will probably use an incremental approach rather than a top-down design approach.

Since modern information systems are complex beyond intuition, consultants must realize that simple historical examples and homilies presented to the acquiring organization will not work. Criticism will be ignored. Specific implementable suggestions and specific citations and demonstrations of infeasibility are required.

Consultants must emphasize design for flexibility and change, as well as continuing to advocate modeling and analysis. Rand's empirical studies and observations of past development projects lead us to believe that a highly phased development strategy is preferred. Rather than allocate available resources across many subsystems, focusing resources on one or several critical subsystems has several advantages. First, subsystems are available in the shortest possible time. This strategy permits a staff of modest size and thus a higher quality of level can be maintained. Subsystems that appear later in the effort can profit from learning, further policy development, field tests, and simulation exercises. Management and control of phased development are easier since managers do not have to make decisions and follow progress in as many concurrent efforts. Phasing also allows management to recognize that areas differ in terms of (a) payoff, (b) amount of prior work, and (c) ease of development. Phasing does present some difficulties. Some parts of the system must be redesigned and reprogrammed but evidence suggests that the total cost of the phased approach is lower. The real danger of the phased approach is cancellation or loss of momentum prior to completion. Resolution of this problem depends on the organization's procurement policy and on the role taken by top management.

Backup and flexibility must be pressed as key factors. Development is difficult and uncertain. Since management systems always take longer, cost more, and work less well than planned, backup is crucial. Existing systems should be maintained so that they can operate longer if necessary. Buying new equipment that is program compatible with existing equipment provides extensive backup but may be an unavailable or undesirable option for other reasons. Adequate backup gives the development manager important flexibility. If he encounters a need for modification or additional testing that will delay his program, he can make his decision on the costs and gains involved rather than being forced to meet the schedule. Modularity in design can be achieved by selecting equipment to allow rental or purchase add-ons, to change terminal equipment, and number of peripherals, and to change data transmission volumes. Rental flexibility is especially important since they permit return of parts of the system on short notice.

Prototyping portions of the installation should be encouraged. "System" utlization is a misleading phrase if it is not known which part is critical—memory, communications, or the CPU. One can install, and measure the utlization rate, with actual loads and then add equipment where necessary. Moreover "system" performance in the abstract generally ignores software and staff skills which are observable in the installation.

Management planning is required to produce the system plan and to build the system. Organization is not a final answer to any problem, but it is important that (1) a strong management role be present throughout development to maintain the policymaking or management function, (2) the project be reviewed at top management level to achieve a cross-function view, and (3) the project group include both functional and computer personnel to allow the close interaction needed in modern systems.

Modern systems analysis is an effort to apply structured rationality to problems of choice. To be of use in Information System design in large organizations the analyst must be aware that his techniques of analysis requre time and data. Neither may be available. The analyst must also understand that institutional factors cause real design to proceed from simultaneous policy and hardware selection through software to the final system. The analyst must therefore supply advice on policy phasing, equipment phasing, flexibility, and backup. Our emphasis must be on creativity-preserving options, and we must plan for the freedom of the next planner of the system.

DISCUSSION

Ferrier: L'orateur a parfaitement mis en évidence les inconvénients du "bottom-up". A-t-il analysé les inconvénients du "top-down"? En particulier celui de ne pas faire participer suffisamment à la conception du système ceux qui auront la charge et la responsabilité de le faire fonctionner. Il semble que le "top-down" fasse la part trop belle au technocrate. Ne doit-on pas rechercher une double approche, à la fois "top-down" et "bottom-up"?

Durbin: Arguments against "top-down" design:

Basically the problem of complexity; the multiplicity of options to be explored; the multiplicity of interactions to keep account of. In this context (Logistics Information Systems).

- Examine many logistics policies
  - Material management
  - Repair location concepts
  - Etc.
- Examine many information concepts
  - Network versus free standing
  - Batch, on line

- Etc.

- And then the actual *design* and development are difficult. One man or one team finds it almost impossible to insure the accuracy of all the lower level work.

The evaluative and analytic tools are not yet available.

Tyler: On the need to take account of historical experience, how would you advice getting the right balance between experience and enthusiasm?

Durbin: On your Aero Engine question, Arinc of Annapolis, Md., did a study for the Navy on engine management. I believe it included information system considerations. On the balance, I believe through *partial* implementation early—perhaps as a field test—of a limited segment.

(Reference is made to the question introduced by Tyler during the discussion following the next presentation: Royal Airforce Aero Engine Logistics Model by J. Hough and P. D. T. O'Connor.)

Høgberg: Design will follow both Top-down and Bottom-up. The first step should be Top-down to establish modules and their interaction. Then follows Bottom-up under the constraint of the criteria established in Top-down.

Taking into account hardware and software development, could you comment on future Information Systems Architecture; specifically the seemingly possible future of computer Network Systems rather than the past belief in one large central computer?

Durbin: This is certainly the route we seem to be following. The University of Colorado even has laser communication over a 10 or 20 mile path between computers. So long as the communication technology continues to advance this portion seems technologically feasible and economical.

The drawback seems to be in the software needed to control the diverse programmes and data resident in core at any one time. High "overhead" which reduces useable capacity.

Morris (L.J.): Can you comment on evaluating the cost-effectiveness of alternative systems solutions?

Durbin: I have a tremendous amount of material in Santa Monica on this. It's really a very young field. If you drop me a note I'll start some materiel in the mail to you. The simple answer is that we *can't* do it well; and we
1) overestimate manpower and cost savings and
2) seriously underestimate manpower and cost requirements.

Young: While total top-down may not be practical, have you any *techniques* in mind which might redress the balance to some extent in favour of top-down design? (e.g. design disclosure documentation systems).

Durbin: I would be happy to send you any relevant Rand material on this topic if you drop me a note. Basically it (your question) would involve (1) knowing and using the array of simulation and analytic models already constructed to deal with various logistics policy alternatives, (2) knowing the capabilities of existing software and hardware simulators, and (3) being aware of the experience of other system developers.

Solstrand: Isn't it a considerable danger involved in implementing such an enormous system–in that it may be very vulnerable *in a conflict or war situation?*

What I mean; isn't it very important in building up such a large system to build in an amount of redundancy sufficient to prevent systems of this size from getting into some sort of a dinosaur status?

Durbin: The question of data redundancy is very significant–not just from war or physical catastrophe, but because of system blackout–i.e. loss of part of the data base through

Power failure

Human error

An errant programme

Etc.

A comprehensive system of journal tapes or disc must be kept.

Hubeaux: My question concerns the "Management Systems Design and Development" approach, i.e. "top-down" method.

In this context Mr. Durbin stated that it was difficult, even impossible, to obtain from the "top" the definition of the policy to be followed and objectives to be attained, thus the framework for the system to be designed.

My question is twofold:

1. Has the speaker experience of the setting up of a Management System for which the policies were not clearly defined in advance? If the answer is yes,

2. What were the consequences of setting up a system on these bases, by consequences, I mean a) effectiveness and validity of the system in relation to undefined objectives, b) final acceptances of the system by the authorities concerned.

Comment: Personally, I am of the same opinion as Major Høgberg that both methods are necessary, "top-down" and "bottom-up", it being certainly necessary to define the objectives to be attained.

Durbin: I agree with the thrust of your question. Both design orientations are necessary. But many system designs are *simply* automation on the next generation of equipment. The problem of involving the functional user in design is crucial whether the designer and developer is civilian or military. We tend to not remember that. Even the commercial corporations tend to allow separation between the ultimate user and the designer, with poor results.

Morris(L.J.): How do you suggest that higher management should state their requirements for a logistics system by, for example, specifying standards of service, or by saying which conditions are mandatory?

Fiorello: (A written reply was provided by Mr. Marco Fiorello, Rand Management Sciences Department, on Mr. Durbin's behalf.)

The writer had, for a year, been investigating the problems and characteristics of Large Scale Management Information Systems (ISMIS), with specific application to material control and inventory management environments (logistics). As a partial caveat, he noted that there did not yet exist a formal set of guidelines, let alone a methodology, for the successful design and development of LSMIS. There were, however, a number of "rules of thumb" and a proliferation of visceral utterances. Rand were then currently attempting to prepare a set of guidelines with the eventual goal of defining a methodology for the design and analysis of LSMIS.

Regarding Mr. Morris' question, the writer, speaking frankly, found it rather broad and, consequently, found it difficult to compose a short reply that did not border on the trivial. If the question was interpreted as being directed at information system designs for large scale logistics systems, then the writer's reply was that large scale systems tended not to work well as opposed to not working at all. As such, experience had indicated that specific provisions with penalty clauses (mandatory conditions?) tended to cost as much or more to enforce, as the accruable benefits derived from their realization. Many systems analysts preferred instead a statement of general requirements (standards of service?); such as, the system would successfully process a certain workload within a certain range of response times, etc., or the system was unacceptable.

The question, in effect, was really addressing the issue of performance measures. It had to be recognized that the idiosyncracies of LSMIS were sufficiently different from their smaller scale predecessors, that many new performance metrics were necessary for their successful management. For example, in the development of a LSMIS, it was entirely appropriate to view the initial design as a precedent, and not an immutable specification.

The writer enclosed the following short list of some Rand publications in logistics that he thought might be of interest:

RM-2871-PR *The Costs and Benefits of Responsive Support Operations*, by J.W. Petersen, et al., October 1962.

RM-2901-PR *Information System Design in a Complex Organization: RAND's LP-II Manned Simulation*, by Robert W. Johnson, May 1962.

RM-4563-PR *Computer-Assisted Maintenance Planning*, by P.J. Kiviat, July 1965.

RM-4778-PR *Unifying Resource Allocation, Control, and Data Collection*, by I. Cohen, et al., November 1965.

RM-4852-PR *Logistics Systems Capacity*, by Stoller.

RM-5078-PR *METRIC : A Multi-Echelon Technique for Recoverable Item Control*, by C. Sherbrooke, November 1966.

RM-5678-PR *An Aggregate Stockage Policy for EOQ Items at Base Level*, by J. Lu, et al., June 1968.

RM-5687-R *A Real Time Metric for the Distribution of Serviceable Assets*, by B.L. Miller, October 1968.

RM-6094-PR *Alternative Measures of Supply Performance: Fills, Backorders, Operational Rate, and NORS*, by R.B. Brooks, et al., August 1969.

D-18847-PR *SCAM, A Source-Coders Cost Analysis Model,* by R. Kaplan, et al., May 20, 1969.

Royal Air Force Aero Engine Logistics Model*

J. Hough
P.D.T. O'Connor

1. INTRODUCTION

This paper describes the development of a computer simulation model of the logistics of the Avon engine in some of the marks of Lightning twin-engined fighter aircraft. The model was developed by Rolls-Royce Ltd under contract to MOD(Air). The paper describes the structure of the model, gives the results achieved during validation tests, and explains how such models could be used to aid decision making. It also explains the advantages which such models have over the previously used models for reserve engines provisioning and for analysis of the effects of decisions concerning engine logistics, reliability, overhaul life, and repair and overhaul policy.

The optimum provisioning of reserve engines is a very difficult exercise, as it is based on predictions of overhaul life and reliability progressions as well as on predictions of transit and delay times, and aircraft utilisation. Similarly, the problem of optimising expenditure on modifications for overhaul life improvement or reliability growth, repair and overhaul policy, and logistics policy are very difficult owing to the many interactions involved, and the time lags experienced before a decision actually begins to affect the logistics situation. Because of the present very high and rising costs of procurement and maintenance of aero engines, and the problems of optimising logistic policies, it was considered that more refined methods than have been used in the past for logistics planning should be investigated.

2. CURRENT METHODS AND POLICY

2.1 Reserve Engine Buy

The RAF at present use a simple formula to decide on the number of reserve engines required. The formula is as follows:

$$N = \frac{A.\ h.\ n.\ P}{MTBR}$$

where N	=	number of reserve engines
A	=	number of aircraft
h	=	flying hours per aircraft per month
n	=	number of engines installed per aircraft
$MTBR$	=	mean time (engine flying hours) between engine rejection (scheduled and unscheduled) involving return to contractors

P = average pipeline time (months) for rejected engines, plus a 'provision-factor' to allow for variability, errors in assumptions, and contingency stocks. The size of the factor is determined on the basis of past experience.

The formula is evaluated repeatedly to take account of predicted changes in the parameters during the period when aircraft are being introduced into service, and flying rate, overhaul life, and reliability are increasing. The decision on the number of engines to buy is made on the basis of the largest requirement thus determined.

The main weakness of this 'provisioning formula' is that the provisioning factor has to include a number of independent parameters. Consequently the value to apply has been derived empirically, based on experience, and no attempt has been made to analyse the effects of the separate constituents. The formula does not allow for the fact that not all repaired engines start a new overhaul life (in the case of the Avon, engines exceeding 50% of overhaul life on return to the contractor were overhauled, and then commenced a new life). It also does not take account of the engines repaired in service.

The use of more complex mathematical models has been investigated, but it was apparent that any such model which took account of the many factors involved, and of variability and time lags, would be too complex and unwieldy to be practicable.

2.2 Management of In-Service Engines.

Until recently all overhaul work and most repairs were carried out by Rolls-Royce Ltd, the RAF carrying out only minor repairs involving little or no dismantling. The RAF plans in future to carry out deeper in-service repair and modification work. The movement of engines between bases and the contractor, and to and from the single central depot (which provided buffer and contingency stocks) is controlled by supply staffs, who attempt to maintain spare engine stocks at the bases up to the recommended levels. They do this by allocating repaired or overhauled engines as they leave the contractor's works, and by setting production targets for the contractor. The description of the model illustrates the processes used, and the various policies in force.

3. POTENTIAL ADVANTAGES OF MODELLING TECHNIQUE

A simulation model can be used to assess the lifely effects of any decision which affects the outputs from the model. These outputs can be selected to represent the factors that are important to the staffs concerned, such as numbers of AOGs (AOCPs), overhaul and repair arisings, stock levels, pipeline times, and trends in such factors. Decisions that can influence these factors are the number of reserve engines purchased, policy for repair and overhaul at bases, depots and contractor's works, movement policy, and stockholding policy. Modifications or other action which affects reliability or overhaul life will also affect the model outputs, as will aircraft utilisation.

As a model can assess these effects, it can be used by management to forecast requirements and to compare policies and assumptions. During use early in the life of a project, such as for reserve engine forecasting, much of the input data, such as reliability and overhaul life progression, would be assumptions. Instead of single point assumptions, a range of figures can be used, and a model can show the effects on reserve engine requirements.

Simulation models could not be used to analyse decisions which have only short-term effects, such as the movement of one engine. However, they could be used in conjunction with an ADP system, which would provide day-to-day control, and which could automatically record data suitable for use with the model.

A computer simulation model is able to take account of very many parameters simultaneously, and of their variations with time. In the case of an engine logistics model the parameters of interest are overhaul life, premature rejection rate, flying rate, and transit, delay, and repair and overhaul times. A simulation model can also take account of different forms of distribution of such parameters as variation of rejection rate with time since overhaul, and flying rate. The model can be re-run as often as necessary to analyse the effects of changes in policy, or of engineering action taken to improve reliability or overhaul life, and to test the sensitivity of the outputs to any of the assumptions used.

4. DESCRIPTION OF MODEL

4.1 General

The logic of the model concerns the flow of engines around the system, which is shown in flow chart form in Figure 1. An engine begins its movement when it is rejected from an aircraft. It then passes, either to a base work-shop, or in the case of major repairs or overhaul, to the contractors. Having been processed the engine is then dispatched to a depot where it is held until needed at a base, or dispatched directly to a base in urgent cases. This then, is the cycle taken by an engine from its removal from an aircraft to its arriving back at a base. The logic which dictates how and when an engine moves is contained in certain modules in the simulation model and a detailed description of these is given in the following paragraphs.

4.2 Structure

The model is written in FORTRAN IV programming language and can be conveniently divided into three levels:

(a) *Level 1*—Level 1 is the main programme and contains the basic structure of the model. It is from this level that all other levels are controlled.

(b) *Level 2*—Level 2 comprises several large subroutines, each of which has a specific function. These are:

(i) *SETUP*. This routine reads in all the input data to the model and sets up the initial conditions. For example, allocating aircraft and spare engines to bases and depots, adding engines to the Rolls-Royce system, etc.

(ii) *UPDATE*. This routine is called every seven days. It increases the flying hours on installed engines, and samples rejections. It initiates the movements of engines from the bases to the contractor.

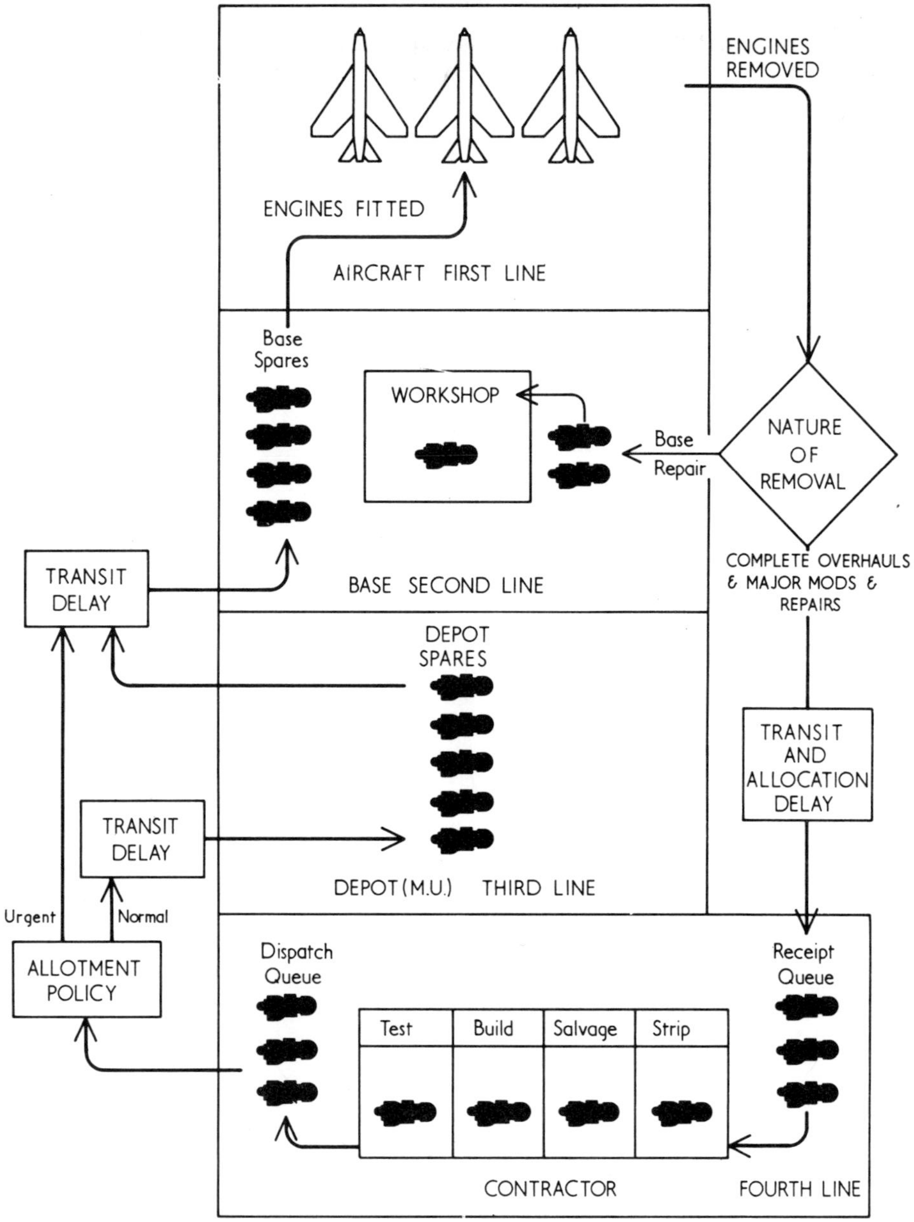

Figure 1 Model flow chart outline

(iii) *ALLOC.* The entire object of the model is to maintain as many aircraft as possible in a flyable condition (ie with two serviceable engines installed). This routine allocates engines to aircraft in need of them, either from base spares, depot spares, or the contractor's dispatch bay. This routine is called whenever an aircraft requires an engine.

(iv) *CHECK*. This routine is called at thirty-day intervals and has a variety of tasks to perform. It allocates spare engines from the depot to bases, where the complement of spare engines is below the desired level. It also monitors the performance of the contractor in supplying engines to programme and, should this be unsatisfactory, makes changes in the priorities and work in progress at the contractor's works to restore the performance to the necessary level.

(v) *ROLLS*. This routine is a simulation model in its own right and is designed to represent the processing of engines at the contractor's works. Engines arrive at the works, are checked in to a receipt queue and then, as capacity allows, pass through the works via any or all of the strip, salvage, build, test and fitting facilities. They eventually reach the dispatch queue from where they are allocated either to bases or to the depot. This routine is called whenever an engine within Rolls-Royce is scheduled to move from (or to) any of the facilities.

(c) *Level 3*. Each of the two previous levels is exclusive to the logistics model. Level 3 contains many small subroutines which are required for standard simulation procedures. This level is common to all models written in FORTRAN IV, although the details vary. It includes set-up, time scale, queueing, sampling, and other general subroutines. A separate report exists which gives details of this level, and of operating procedures for the programme.

4.3 Engine Rejection Logic

Each week the flying hours are increased, by sampling from a distribution, for all engines fitted to operational aircraft. If, in the course of this weekly update, the overhaul life for an engine is exceeded it is removed from the aircraft, flagged as 'time-expired–return to works (TX–RTW) and added to a queue of engines awaiting allotment. The aircraft is added to another queue to await a replacement engine. Should the engine(s) not have exceeded the overhaul life, they are then checked for premature rejection.

The probability of an engine being prematurely rejected is derived from the convolution of two separate distributions.

a. The overall premature rejection rate for all engines which changes with calendar time as service use of the engine accumulates and as modifications are introduced.

b. The premature rejection rate for individual engines based upon time since overhaul.

These distributions were obtained from data collected on Avon rejections during 1966/68. Therefore, when an engine is tested for premature rejection the probability is of the form

$$\text{Probability of rejection} = p(j) \,.\, F$$

where $p(j)$ = overall probability of premature rejection in month j

F = factor based upon achieved life

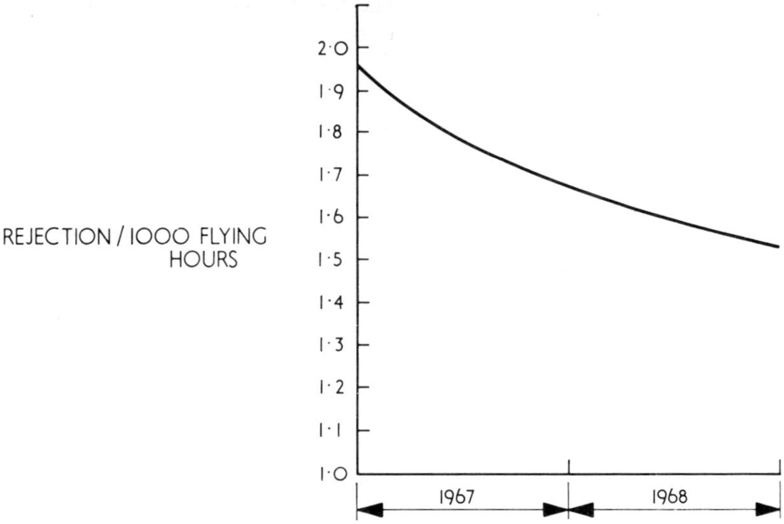

Figure 2 Premature rejection rate

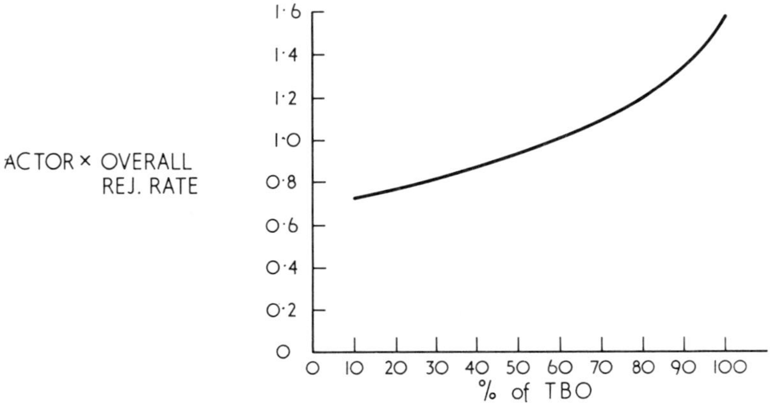

Figure 3 TSO Modifier to premature rejection rate

Engines prematurely rejected are added to a queue to await allotment and flagged as 'Repair' engines. The aircraft are added to another queue to await replacement engines. A further check on prematurely rejected engines is carried out at this point to determine which engines can be repaired at base. Those which can be repaired are flagged as 'repair–continue in service (CIS)' otherwise they are flagged as 'repair–return to works (RTW)'. Should an engine still be installed after both life expiry and premature rejection checks are complete the engine is not referred to further until the next weekly update. At the end of each weekly update, two major queues exist: the queue of rejected engines awaiting allotment and the queue of aircraft on the ground awaiting replacement engines.

Rejected engines are allotted as follows. All 'TX–RTW' and 'Repair–RTW' engines are dispatched to Rolls-Royce from the bases. The transit time has a delay included as an allowance for time taken to receive allocation instructions. Engines flagged as 'Repair–CIS' are added, without delay, to the queue of engines awaiting repair at the relevant base. When the base work-shop is available, the engine enters, spends time being repaired and then returns to the spares queue at the base in question, in accordance with the relevant time distributions.

Spare engines arriving at a base are always added to the base spares queue on arrival, irrespective of whether they were intended for an aircraft on the ground or merely for stock adjustment. The process of fitting an engine to an aircraft is carried out whenever the requisite conditions are met. These are:

(a) An aircraft is present with engine(s) removed at Base (I) (Condition (A)).

(b) A spare engine is available at the relevant base (Condition (B)).

The check for these conditions is continuous throughout the running of the model and is independent of all other operations. Should, during such a check, Condition (A) prove positive and Condition (B) be negative, immediate action is taken to satisfy Condition (B). This is done by dispatching a spare engine from the depot stocks to the base which is in need.

When the engine is dispatched the aircraft on the ground is flagged to show that the AOG has been recognised and actioned. The process of allocation of spare engines to aircraft, therefore, will only take place on aircraft on the ground which have not already been actioned. When the transit time has elapsed the engine emerges from transit and is added to the base spares queue. This has now made Condition (B) positive and the fitting process can take place.

Should the depot be unable to produce an engine for an AOG, the next resort is to allocate an engine, where possible, from the dispatch queue at Rolls-Royce. The procedure regarding the transit phase and the flagging of AOGs as actioned is identical to the previous case of allocation from a depot. Should an aircraft on the ground still be unflagged after both procedures have been concluded it remains unflagged until an unallocated spare engine appears at either the depot or Rolls-Royce, whereupon the allocation procedure commences as before. When both Conditions (A) and (B) are positive the engine(s) are fitted to the aircraft which is then returned to full operation. During the fitting process, priority is given to aircraft with only one engine removed since this is the quickest way to return aircraft to service.

4.4 Allocation of Spare Engines to Bases

So far, the only cases of allocation of engines considered are those where an aircraft is waiting for an engine. However, the most common allocations are made simply to re-distribute the stocks of spare engines. At the commencement of the simulation run each base has a recommended spares stock level, which is derived from the expected monthly arisings. Should allocation be made only in the type of situation mentioned before, the stock levels at the bases would rapidly decrease until all engines fitted would have to come directly from either Rolls-Royce or the depot. There is a mechanism in the model, as in real-life, for periodically reviewing actual stock-levels and maintaining them at the recommended level.

Every 30 days, the stock levels at the bases, including engines in transit, are monitored and compared with recommended levels. Whenever the actual stock is below the recommended stock sufficient engines are dispatched from the depot to remove the deficit. These engines, upon dispatch, are included in any future stock level calculation for the base until such time as they are fitted to an aircraft. This mechanism operates on a 30-day cycle and does not attemtp to deal with any shortage within the 30 days. A further procedure is designed for this purpose.

Whenever an engine enters the dispatch queue at Rolls-Royce it may be allocated either to a base, or, more normally, to a depot. Should there be no aircraft on the ground, a scan is made to locate any base where the stock level is zero. This situation is unacceptable because it is potentially dangerous; any engine removal would result in a delay before fitting could take place. Thus, in this zero-stock situation the engine is always sent from Rolls-Royce direct the base.

When all the foregoing procedures have been completed any engines remaining for dispatch at Rolls-Royce are sent directly to the depot.

4.5 Rolls-Royce Facilities

When an engine arrives at Rolls-Royce it is already flagged as 'TX–RTW' or 'Repair–RTW'. It is now examined further to comply with certain overhaul policies. All engines which are time-expired are scheduled for overhaul and are flagged as such. If the achieved life of any repair engine is in excess of a certain percentage (the policy overhaul level), of its overhaul life (TBO), it is re-scheduled as a repair/overhaul engine and subsequently follows the same route as TX engines. Engines which do not exceed this policy level are scheduled as repair. A summary of this is as follows:

CLASSIFICATION	REASON FOR REMOVAL	ACHIEVED LIFE
Overhaul	TX	Life $\geqslant$ TBO
Repair/Overhaul	Premature	TBO$>$ Life $\geqslant$ TBO x f
Repair	Premature	Life $<$ TBO x f

('f' is the 'policy overhaul level' and is set at 50% in the case of the Avon)

The route followed by an engine passing through Rolls-Royce is shown in Figure 2. All engines enter the strip/inspection facility and upon completion of this operation, repair/overhaul and overhaul engines enter the salvage phase and repair engines enter the repair facility. The two paths converge again at the unit/engine build stage and the final phases, via test beds and dispatch fitting are the same for all engines. Not all engines go directly from test-beds to dispatch; a proportion, which is input as data, is returned to the build stage after failing on the final test, and these pass through the final stages again. The time spent by an engine in any of the facilities mentioned is found by sampling from the relevant distribution.

The difference between the processing of repair engines and overhaul engines has been partly described with regard to the salvage and repair phases. The difference in overall throughout time between the two types of engine is explained by this. There are, however, other differences between overhauled and repaired engines in the fields of modification and engine life. A repaired engine leaves Rolls-Royce having the same expired life with which it entered and is also unchanged as far as modifications are concerned. The overhauled engine, on the other hand, is treated as a new engine when it leaves and also has incorporated into it the latest level of modifications. Should the modification level be a necessary condition for an overhaul life increase, this increase is also recorded for the engine concerned.

The logic of the processing of engines is simplified for the benefit of the model. Certain real-life decisions are based upon experience and prior knowledge and possess a flexibility which cannot be built easily into a computer model. In these cases simplifying assumptions are made when they are necessary to the efficiency of the model. A particular area which is difficult to simulate is the progressing and control within the workshops. For the sake of simplicity the decision about when to add an engine to the shop floor is based upon two major conditions:

(a) How many engines are currently in the shop and where are they located?

(b) How is the supply of engines to the RAF meeting the agreed monthly programme?

These two points are closely connected, since the rate of engine supply depends upon the work-in-progress level on the shop floor.

At any point in time, the number of engines which are required in work-in-progress is never less than the programme for the next three months. The actual level may fall below this desired level only when the arrivals from the bases are lower than expected.

The performance of Rolls-Royce in terms of supply of engines to the RAF is monitored against the programme on a monthly basis. Should the supply be in arrears the decision is taken to increase the work-in-progress sufficiently by taking engines from the receipt queue. This should increase the rate of supply at some point in the future. In times when the pressure for engines is high, the first in, first our priority system, (FIFO) is suspended in favour of a system which chooses repair engines first. The lower throughput time for these engines enables the rate of supply to respond more quickly.

This solution to the problem of arrears is not a good one because the response is so slow–the average throughput time is 90 days. To overcome this problem a further monitoring system is used which attempts to predict arrears situations before they occur and thus reduce the response time. The prediction is quite simple and merely totals next month's expected output, (ie all engines which have completed the salvage phase), and compares this with the program for the next month. Should a potential arrears situation occur, the engines currently undergoing salvage are examined and certain of them, dependent on their current state of completion, are expedited within the next few days into the build stage. This increases the expected output and should reduce the arrears at the end of the month. This mechanism is an attempt to mirror the operation of progressmen in the shops, who are able to isolate potentially damaging siatuion and accelerate work to meet them.

5. VALIDATION OF MODEL

Since major decisions may have to be taken on the strength of recommendation derived from the model, it is vital to make certain that the output is reliable. To validate the model the following approach was adopted:

(i) The model was set up to mirror the situation existing at the beginning of 1967 (eg aircraft and engines, flying rate, etc).

(ii) The model was run to simulate the two year period from January 1967 to December 1968.

(iii) Actual data was used for 1967, but predicted data extrapolated from 1967 information was used for 1968.

(iv) The data which the model generated for 1967/68 was compared with what actually happened.

5.1 Through-put time at Rolls-Royce

Through-put time at Rolls-Royce was chosen as a validation check because of the type

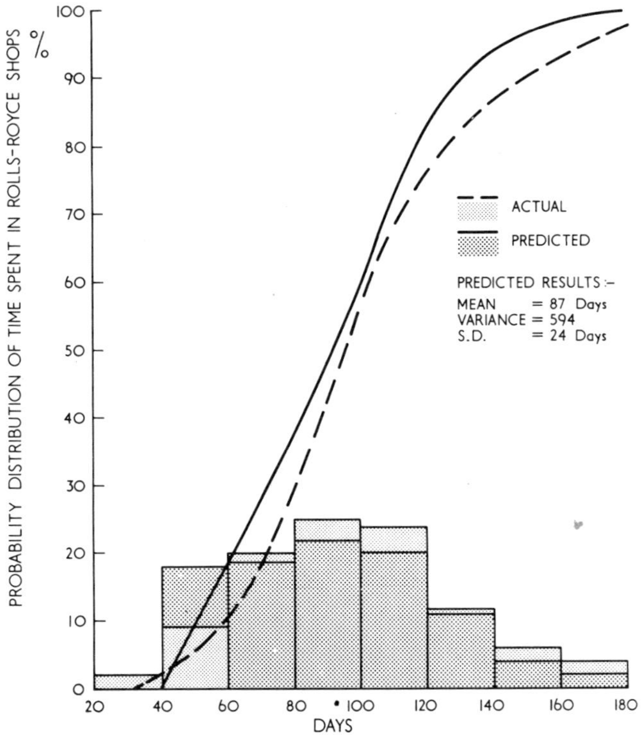

Figure 4 Time spent on shop floor at Rolls-Royce

of data used for the Rolls-Royce module of the programme. The results relating to time spent in Rolls-Royce shops are shown in Figure 4. Both the distribution of actual time and the distribution of predicted time are plotted. There is good alignment of curves although, owing to the large variance of the actual data available, statistical significance tests are inconclusive. It is apparent that the predicted curve is truncated, at the upper and lower extremities, compared to the actual curve. These truncations, however, represent very few engines and would have a marginal effect on the performance of the model. The actual throughput times were obtained by examining all data on engines passing through the Rolls-Royce workshops in 1966-68. The only times available, however, were the time an engine entered the shops and the time that it was dispatched to the RAF. For the model, however, the time that an engine spends in the shop is obtained from the separate time distributions for all operations undergone during repair/overhaul plus the effect of test-bed rejections. The type of work required on an engine also influences the time spent in the shops (eg overhaul is a longer process than repair).

5.2 Queueing time at Rolls-Royce

As well as the time spent on the shop floor, engines at Rolls-Royce also queue in the receipt bay before entering the shops. This queueing time depends on the following:

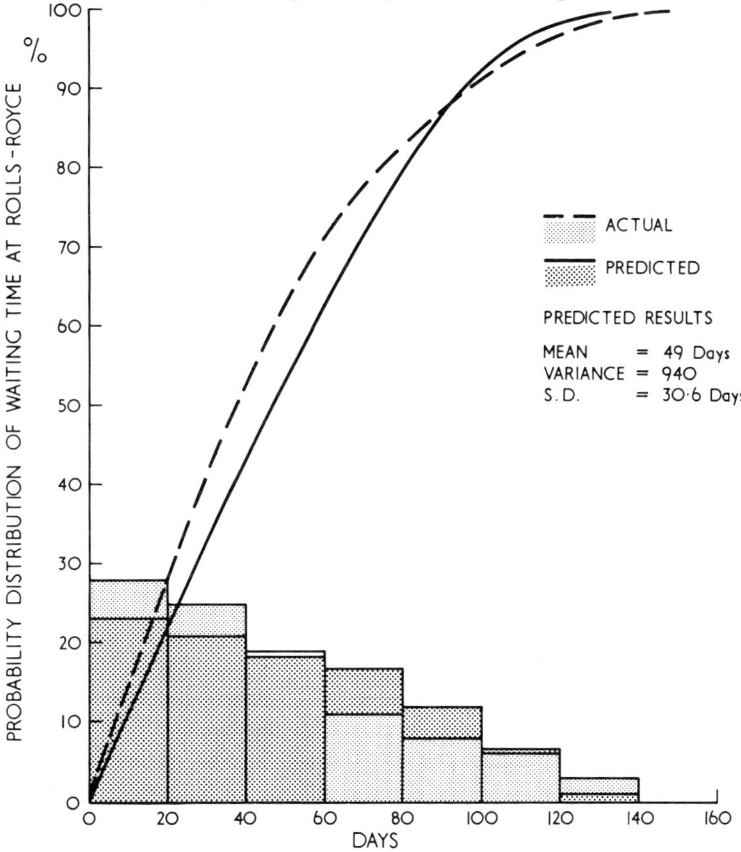

Figure 5 Waiting time at Rolls-Royce

(i) Length of queue when engine arrives

(ii) Priority system

(iii) Level of work-in-progress

These factors all influence the queueing time, and, by examining the same data as in para 5.1, a distribution of actual queueing time was derived. The queueing time distribution which emerged from the simulation was also plotted and the results shown in normal and cumulative form in Figure 5. There is reasonable agreement between actual and predicted results, although, for the reasons explained in para 8.1, significance tests are inconclusive.

5.3 Depot Stock Movement (1967/68)

The results of this stage of the validation are shown in Figure 6. The solid line represents the actual movement of depot stocks and the points represent the depot stock levels, on a monthly basis, which emerged from the separate simulation runs.

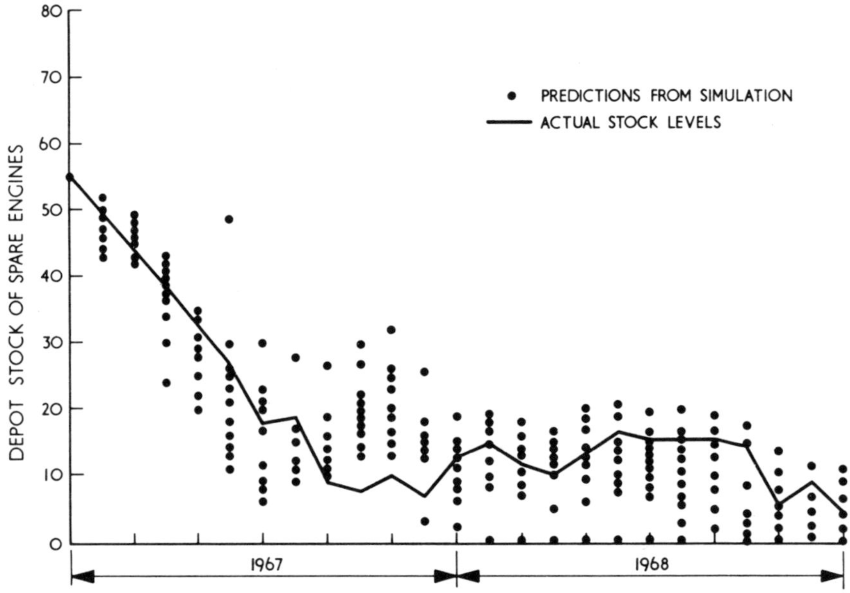

Figure 6 Depot stock levels

The results agree well with the recorded data. The variance between runs is due largely to the fact that the stock level in a particular month is closely correlated with that in the previous period. Thus any unusual deviation from the mean value will be reflected in the stock values in the subsequent months. However, in any model used for management purposes for in-service engines the data will be updated to take account of deviations which actually occur. Thus events occurring in the past would not affect the accuracy of the predictions produced by the model. Variability of predictions would still depend in part on the length of time concerned. However, it is possible to programme the model to correct for fluctuations which occur during a run. In the real situation such corrective action is taken and this can be simulated. The model does this to a limited extent in the case of arrears or expected arrears situations in the supply of serviceable engines, as described in para 4.5, and similar logic can be applied to other aspects, such as fluctuations in rejection rate.

5.4 Total Stock Movement (1967-68)

The total stock, in this case, is the depot stock plus the base stock, plus any serviceable engines in transit. The actual and predicted movement is shown in Figure 7. The agreement here is good. The points made in para 5.3 regarding variability also apply to this section.

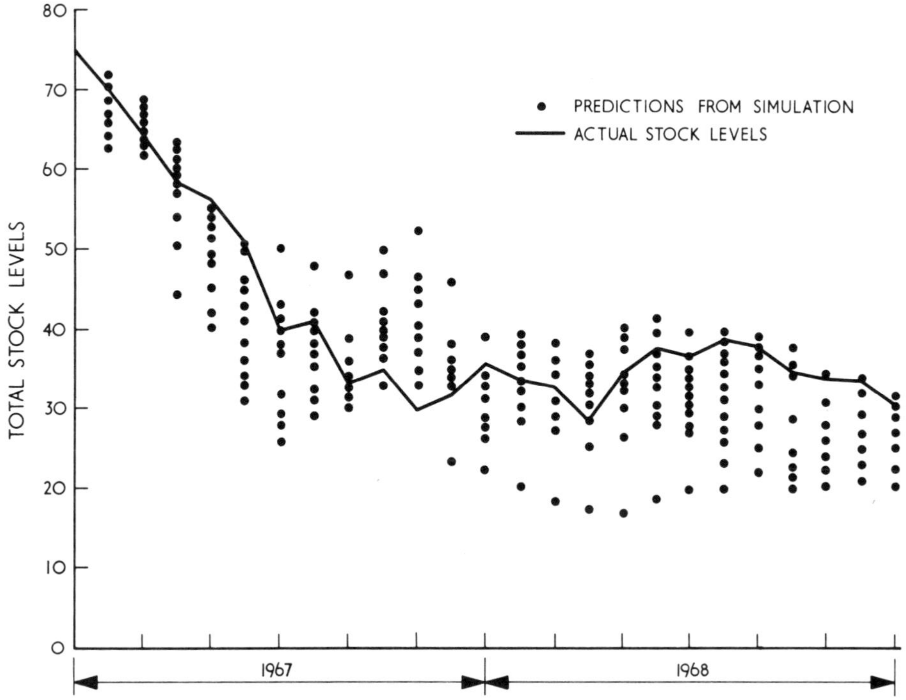

Figure 7 Total stock levels

5.5 Prediction of Arisings 1968

The final validation involved the prediction of engine repair and overhaul arisings for 1968. The arisings were classified into three types:

a. Life-expired engines, returned to works for overhaul (TX)

b. Rejected engines, returned to works for repair (RTW)

c. Rejected engines, continued in service after base level repair (CIS).

The results of the validation are shown opposite.

	Actual	Predicted	90% Limits (1)
TX	23	24	± 3
RTW	42	45	± 8
CIS	25	24	± 6

(1) This assumes that the predicted results, from a series of simulation runs, are distributed normally.

TABLE 1

The results show that the logic used for generating arisings, based on overhaul life progression, rejection rate progressions, and flying rate distributions, accurately simulates the real situation, with a fairly small deviation from the mean values.

5.6 Validation Conclusions

The validation tests carried out show that the model simulated the real situation accurately, within the limitations of the data input. It should be noted, however, that the confidence with which the predictions can be used by management depends on the confidence which can be placed on the input data. Thus, if the input is in the form of assumptions of long-term behaviour, such as reliability progression, the input will reflect the uncertainty of the forecast. This would be characteristic of the model used, for example, for reserve requirement forecasting. However, if the input is in the form of actual data and relatively short-term extrapolation from it, the model would accurately predict outcomes. The variability of the predictions would depend in part on the length of time over which they were made. It is likely that outputs of models using continually updated inputs would have less variability than those of the Avon model, which uses data averaged over a period (thus not all trends are taken into account), and which is not automatically updated. As a result, any fairly large deviation from a mean value (eg depot stock movement in a period) affects all subsequent periods. It is possible to programme a model to compensate for such fluctuations, as these would occur and be compensated for in the real situation.

6. EXAMPLES OF APPLICATION

The following paragraphs present examples of use of the model to analyse the effects of assumed changes of certain parameters. The parameters changed were transit and delay times, and the decision level at which an engine is overhauled instead of being repaired. In each case five runs of the model were made, each covering a two-year period. The same type of analysis would be used to carry out tests on the sensitivity of the output to variations in assumptions.

6.1 Transit and Delay Times

The transit and delay times derived from service records (Figures 8, 9, 10, 11) were replaced with a common distribution of times (Figure 12). The results of this change are shown in Tables 2–4.

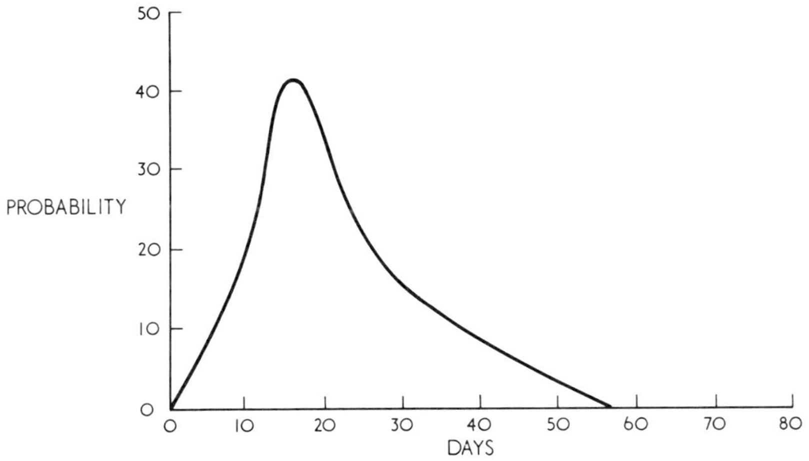

Figure 8 Base delay time

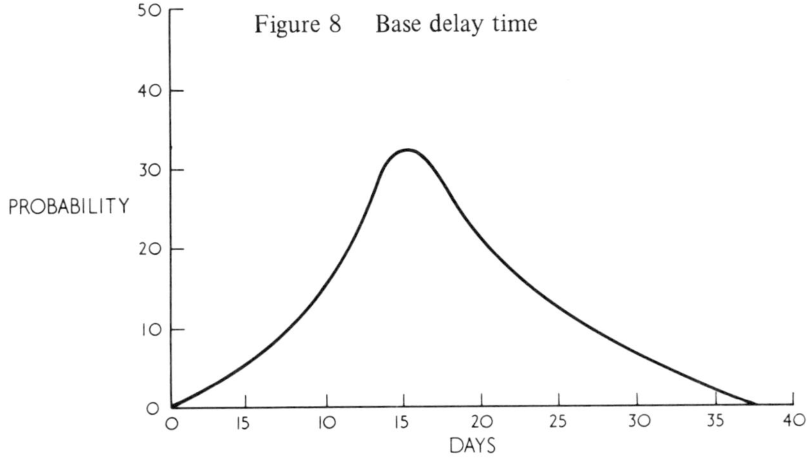

Figure 9 Transit times to contractor (abroad)

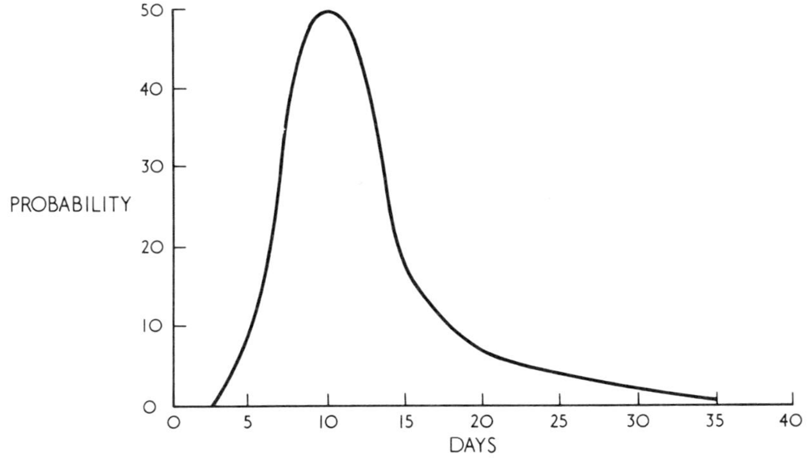

Figure 10 Transit times to contractor (home)

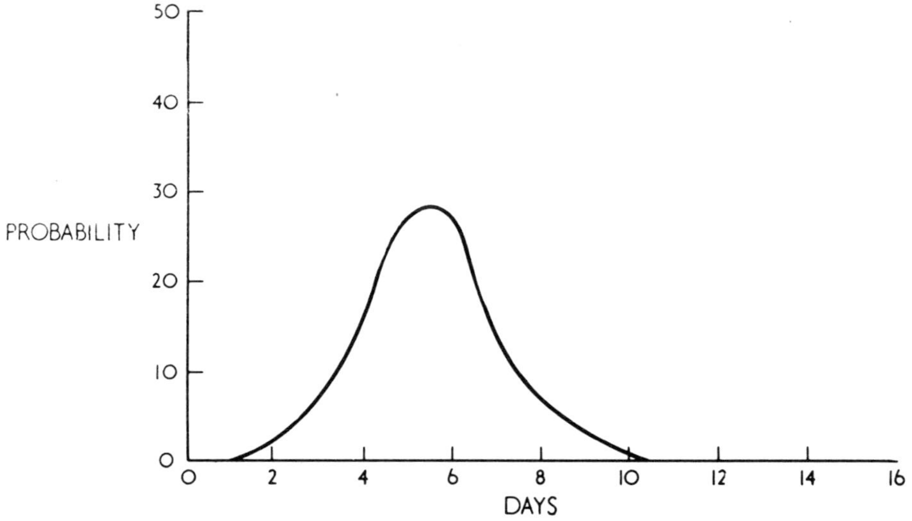

Figure 11 Reduced transit and delay times

TABLE 2 – *Number of AOGs*

	Actual Times	Figure 11 Times
No of rejections unsatisfied by base stocks (AOGs)	6.6%	2.4%
99% of AOGs satisfied in:	6 days	4 days
100% ,, ,, ,, ,,:	12 days	10 days

TABLE 3 – *Base Stock Levels*

	Actual Times	Figure 11 Times
Recommended levels	20 engines	20 engines
Average Stock throughout runs	15 engines	16 engines

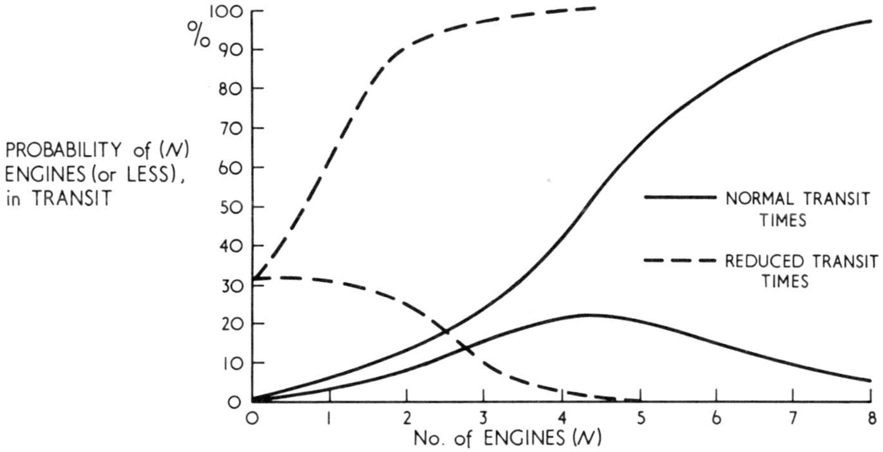

Figure 12 Rejected engines at base or in transit

TABLE 4

Number of Rejected Engines still at Base or not yet at Contractor (See Figure 12)

Actual Times

No. of Engines	0	1	2	3	4	5	6	7	8
Probability of this number	.01	0.1	.12	.09	.21	.21	.15	.11	.1

Mean Number = 5 engines

Figure 11 Times

No. of Engines	0	1	2	3	4	5^+
Probability	.31	.31	.28	.06	.02	.02

Mean Number = 1 engine

The tables show that a reduction in average transit and delay times of rejected engines from the recorded levels to the substituted time distribution would slightly decrease the time by which most AOGs are satisfied, but would have marginal effects on AOG arisings and base stock levels. This is because the stock position of the Avon was never critical in the period concerned, but if it had been, the result would have been more marked. However, the reduced transit time does result in the average number of unserviceable engines in the pipeline (excluding those in the queue at the contractor's works) being reduced from 5 to 1.

6.2 Policy Overhaul Level

A further example demonstrates the effect of varying the decision point at which an engine is overhauled and then commences a new life, instead of being repaired and being returned to complete its life (for the Avon, this point was 50% of overhaul life in the period covered, ie engines were repaired and overhauled if they had exceed 50% of overhaul life on rejection.) The results of the analysis are shown in Tables 5–8.

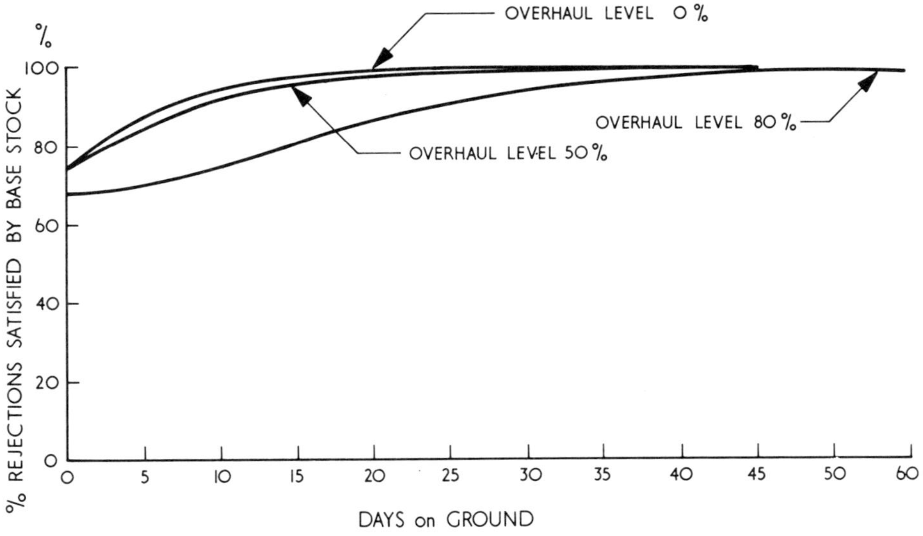

Figure 13 Time between rejection of engine and replacement

TABLE 5 – *Times for Replacement of Engine following a Rejection (See Figure 13)*

Policy Level	% Rejections Satisfied by Base Stock	90% of Rejections Satisfied in	% of Rejections Not Satisfied Within 20 days
0%	76%	6 days	1%
50%	77%	6 days	2%
80%	69%	22 days	14%

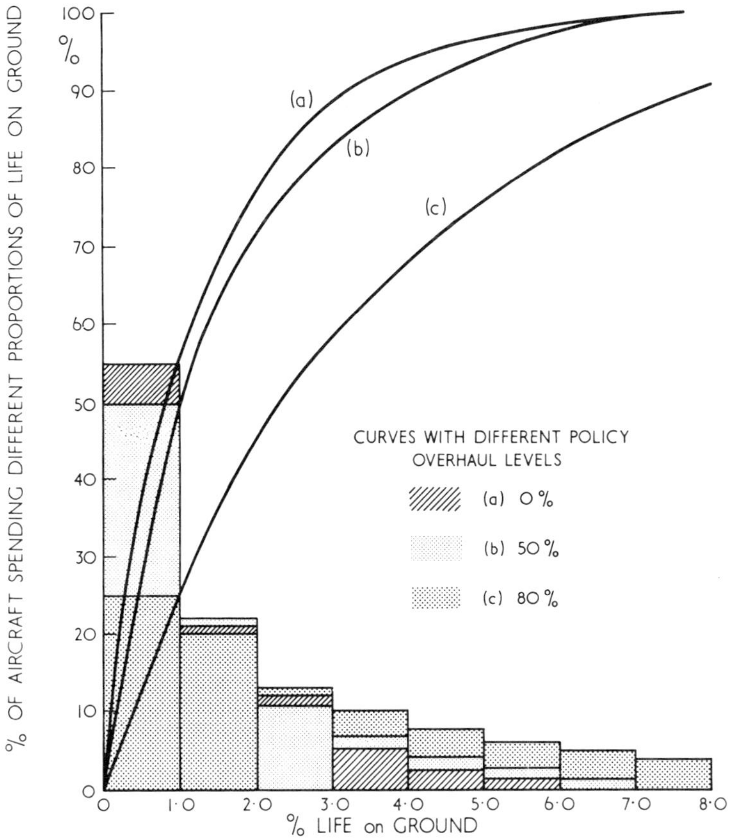

Figure 14 Percentage of life that aircraft spends on ground through engine shortage

TABLE 6

Time AOG Owing to Engine Shortage (See Figure 14)

Policy Level	No of A/c Spending less than 1% of Life on Ground	No of A/c Spending less than 5% of Life on Ground	90% of A/c Spend Less Than
0%	55%	97%	3.1% of life on ground
50%	50%	95%	4% of life on ground
80%	25%	76%	7.7% of life on ground

TABLE 7

Time Spent by Engine at Rolls-Royce (Days)

Policy Level	Waiting Time		Shop Floor Time		Total Time
	Mean	SD	Mean	SD	Mean
0%	150	51	98	15	248
50%	160	65	87	25	247
80%	178	59	79	26	257

TABLE 8

Causes of Rejections and Effect on Work-in-Progress at Rolls-Royce

Policy Level	Total Rejections	Base Repairs	Premature Rejections (RTW)	Time Ex-Rejections (RTW)	No of Engines Requiring Overhaul	Average Achieved Life (AAL) % of Overhaul Life (All Rejections)
0%	184	45	86	53	139	64.4%
50%	193	45	88	60	110	69%
80%	204	43	86	75	95	74%

The preceding results are not directly comparable with the validation results since the runs upon which the latter were based had a conversion programme included. This resulted in more engines being overhauled at the contractors than in the validation runs.

The tables show that if all engines returned to works are overhauled (0% level), the effect on the AOG situation would be negligible, but that the situation would be worsened if an 80% level were selected. The average time spent by engines at the contractors would be slightly increased, the increase in waiting time more than compensating for the reduced average working time. The total number of rejections would rise, but the average cost of a rejection would fall, since a smaller proportion would require overhaul. Also the average achieved life of all rejected engines would increase from 64.4% of overhaul life to 74% of overhaul life at the 80% overhaul level. The effects on total repair and overhaul costs of these changes could be analysed to enable an optimum overhaul level to be chosen for the particular situation, in relation to stock levels (and hence risk of AOGs).

7. LIMITATIONS OF MODEL

The model has certain limiations as currently programmed. These are:

(i) Though it was developed as far as practicable as a general purpose model, there are certain parts of the logic which refer particularly to the Avon. Also, the data is unique to the Avon. Thus the model could not be used for other engines without slight modification and a considerable data collection effort.

(ii) It is not programmed to include the build-up phase of a fleet of new aircraft and concurrent introduction of engines into service, though slight modification would allow these aspects to be fed in as data.

(iii) It cannot simulate a modularly constructed engine.

(iv) It cannot compensate for the fact that large deviations from mean values affect subsequent periods.

8. FUTURE DEVELOPMENT OF MODELS

It is planned to develop the model for management of engines already in service and for engines due to enter service. The logic of the model is suitable, with slight alterations, for most engines, but each engine will require a separate data collection exercise. Future models will include logic to compensate for larger fluctuations from mean values, in accordance with actual practice.

The model is now operating with the facility to introduce new aircraft and engines into service, so that the fleet build-up phase is simulated. Such a facility will be useful for determining the optimum timing of spare engine production, and the effects of varying the production rate or introduction to service rate, as well as for predicting the likely number of spare engines required.

Modular engines present difficult problems in logistic forecasting, owing to the fact that modules may have separate reliabilities and lives, and as different policies may apply to individual modules. Moreover, there will be a complex interaction between modules so that they cannot be considered entirely separately. Rolls-Royce have, since this paper was first presented, written a programme for a model which will simulate all these factors. In view of the current uncertainty on provisioning and logistics policy for modular engines, such a model is required to assist decision making on the range of these engines which will enter service in the near future.

The possible interfaces between logistic models and an ADP control system for engines is being investigated, and future models will be developed with these in mind. Similarly, ADP systems development should take account of the data requirements of models.

9. CONCLUSIONS

It has been demonstrated that the computer model which has been developed to simulate the logistics of the Avon provides an effective method of forecasting the long-term effects of logistics decision and of reliability, overhaul life, and other predictions. It possesses advantages over curently used ad hoc methods of predicting such effects in being able to take account of a large number of assumptions, and of time lags, trends and statistical probabilities and distributions. It can also be used to test the sensitivity of the outputs to changes in the assumptions made, as it can be re-cycled often in a short time.

With small changes to the structure of the original model, it has been developed to simulate the build-up period of a fleet of new aircraft and predict the effects of different levels of reserve provisioning. Thus, it would have the same advantages over the current method of reserve engine provisioning as it has over other logistics decision-making methods in use. The uncertainty inherent in reliability and overhaul life forecasts for new engines will be reflected in the outputs, but predictions of requirements will be more accurate, as many other factors can be considered quantitatively. Furthermore, it could analyse the effects of different rates of engine and aircraft production, so that the engine production rate could be optimised.

A logistics model is now being used at Rolls-Royce which can simulate the operational characteristics of a modular engine. The model, which is a general purpose one, will eventually supplement the model described, which was designed specifically for non-modular engines.

The combined use of an ADP system for engine logistic control and of simulation models will result in advantages both to the ADP system and to the models. Economies could be made in data collection and analysis, and the system considered as a whole would be more efficient than the two systems used separately.

DISCUSSION

Buck: The System you have modelled considers a very convenient peacetime situation. Has consideration been given to a wartime upheaval of the system? i.e. movement of bases, greatly increased utilization, damage to engines from enemy action.

Hough: The model was designed around a peacetime situation. However, if the wartime situation simply involved redeployment of aircraft, increased flying rate and enemy damage, the simulation of it would be trivial.

An actual wartime situation is likely to differ so much from what is expected, the the modelling would have to be carried out after the war.–Not particularly useful. In conclusion, wartime presents too many imponderables to be realistically considered in a model such as this.

Iving: This is a comment on Dr. Buck's question. We do not envisage fighting a war which will last longer than the average engine repair time. Therefore the peacetime solution is still valid, as we shall not need engines out of repair to fight the next war.

Landau: I think that Mr. Hough is being unduly modest about his simulation. A few years ago, Dept. of Supply Australia built a rather similar engines logistic simulation which was used, among other things, to determine what would happen if aircraft flew at a wartime rate while the engine contractor continued to operate at a peace-time rate i.e. how long could the increased flying rate be maintained before the squadron ground to a halt. I feel sure that Mr. Hough's simulation could do the same.

Geraerds: What are the elements out of which the given figures for the waiting times with the contractor have been built up?

Did you consider a way to influence the production control quality of the contractor as the waiting times shown are relatively large and as the user pays for this in having to buy an additional number of engines?

Hough: Waiting times quoted in the text of the report refer only to time spent by an engine before work commencing on the engine in the overhaul shops.

The waiting times referred to by the questioner (i.e. those in Table 7) are derived from a hypothetical situation–the actual waiting times (see Fig.6) are far smaller. As far as using the model to develop production control techniques–this is now being closely examined by the Rolls-Royce O.R. group.

Tyler: This is a request rather than a question. We are about to start a feasibility study into an aero-engine control system which might, as you have heard, be linked to the Model described.

We have had access to USAF experience in this field. We would be most grateful for information on anyone else's practical experience and would be happy to exchange ideas with anyone working in this same area i.e. control of engines by serial number from the cradle to the grave. My address is in the List of Participants.

A Data Collection and Planned Maintenance System for Warships

F. Hormann

1. THE NEW TECHNICAL-LOGISTIC SYSTEMS OF THE FEDERAL GERMAN NAVY

The purchase of three Guided Missile Destroyers DDG 28, 29 and 30 by the Federal Germany Navy caused the first time as well a necessity as an opportunity to develop an up-to-date and comprehensive servicing and maintenance system.

1.1 Reasons for introduction

Technical-logistic procedures developed especially for that new system shall be applied initially on DDG type ships and later on will be introduced for the entire fleet. There are a lot of reasons for the development of a modern up-to-date servicing and maintenance system.

Caused by proceeding technical development both the number of equipment and parts of a weapons system and the dependence of each other among themselves is increasing permanently. If the complexity of a system reaches a certain limit it is impossible to maintain this system with success by customary methods. Furthermore the training and drill of military personnel and the duration of their detached service is not as long as formerly, thus the training hardly can keep up with the growing progress in the technical area. In addition there are difficulties in personnel affairs, as well in strength as in higher fluctuation of personnel. In the meaning of "cost efficiency" furthermore it is necessary to form the servicing and maintenance procedures by assistance of modern methods to an outmost economy.

1.2 Tasks of "Materialinformationszentrum der Marine GmbH"

The responsibility of development, introduction and handling of the various procedures is assigned to the Federal German Navy.

It is the task of the "Material Information Centre of the Navy" (Materialinformationszentrum der Marine GmbH–called "MIZ")–a private managed business corporation–to cooperate in developing technical-logistic procedures and methods.

After completion of these procedures and methods the "MIZ-Corporation" has to maintain the equipment, installations and weapon systems of the Federal German Navy using the aforementioned procedures and methods.

1.3 General task of new systems

It is the main purpose of the new procedures, to keep and to increase the readiness for operation of the fleet. In detail it means the attainment of the following objectives: As many as possible steaming hours, short shipyard refit periods, few failures, low costs and short periods in time for repair and technical correction of installations and equipment.

Especially in the area of maintenance it is the declared aim to assure an effective, fast and economic utilization of existing maintenance capacity, i.e. personnel, material and repair facilities. For this reason all maintenance and material consumption data must be collected, stored and processed with the intention to reach a higher efficiency of equipment at a lower maintenance expense. Improvements apply to both technical improvements of equipment and improvements of procedures. It is a further aim to unburden the activity for material maintenance from management work as much as possible by using the new procedures. Accordingly a modern management for maintenance has the task to plan carefully all necessary maintenance actions concerning the a.m. procedures, to support during execution, to supervise and to evaluate after that.

1.4 Presuppositions for operation of new systems

The main presupposition for developing the new procedures appropriately is the fact to consider them as part of a higher servicing system. That means that the system-idea must be realized really also on this field. Contrary to this at present a number of procedures exist within the Federal German Navy, these procedures are relatively independent and therefore don't consider too much the "Interface-Problems".

For solving these problems by a "system approach" above all unifying points of view must be considered. These points of view apply to performance of systems, design of administrative means–especially preprinted forms–and also material identification and general coding.

A standardized coding is particularly necessary because all modern systems are not conceivable without electronic data processing. In consequence of growing complexity of to-days weapon systems the flow of information is raising permanently. Only an electronic data processing system of latest design provides properties of storing and processing very fast a great deal of information with relatively low costs–compared with manual handling of the same matter. With that initially the possibility exists to learn from the reservoir of past knowledge and to take corrective measures.

Furthermore a presupposition for a smoothly working of the systems is the fact, that they are flexible in performance and not rigid. The systems must be applicable for all weapon systems and ship types. defects must be corrected easily and fast, and the systems must be easy adaptable to varying conditions. A means for that are the "Feedback-Reports" built-in in a number of systems.

Very important is also a psychological factor. Because on every system participates as well the electronic data processing system as also a great number of persons, the system does not operate automatically. The entire mental readiness of all participating persons to cooperate with a corresponding sense of responsibility is the presupposition for the faultless performance of the system in question. This may be facilitated by intensive and repeated training and also feedback-reports, thus informing the cooperator aboard that his cooperation was successful. Furthermore it is necessary that the work and activity of the individual cooperator is simplified as much as possible.

2. U.S. SERVICING SYSTEMS

2.1 General

The U.S. Navy operates with a great lot of systems existing relatively independent, which are moreover different for various weapon systems. The system-idea is considered with a relative small sector only. However, for some weapon systems, e.g. for the TARTAR-System, the servicing system is developed to a remarkable high level. For this example this applies to "Configuration Control" and also to the system-tailored procedure "PMS/SMS" (Planned Maintenance System/Surface Missile System).

2.2 3-M System (Maintenance and Material Management System)

In the field of maintenance the U.S. Navy started in 1961 with the development of the "Maintenance and Material Management System", the so-called "3-M System". Model and prototype was an already existing U.S. Air Force system. The 3-M System was initiated into the U.S. Navy in 1964; since that time several improvements have been made.

The "3-M System" is an integrated management-system, providing for that the maintenance on organizational level will be planned in due time, the performance will be made in proper manner and important maintenance data will be stored and processed. The system consists of two subsystems, the "Planned Maintenance Sub-System" and the "Maintenance Data Collection Sub-System".

2.2.1 MDCS (Maintenance Data Collection System)

The MDCS-System (Maintenance Data Collection System) covers capacities, material consumption, periods of time and costs, required for maintenance on organizational level. This data is processed for fulfilment of various tasks. They furnish informations for accomplishment of reliability tests concerning weapon systems, the data evaluation serves the improvement of maintenance generally, and especially the technical improvement of equipment and material.

2.2.2 PMS (Planned Maintenance System)

The PMS-System (Planned Maintenance System) serves for the preventive maintenance on organizational level and its assignment is to plan periodically and reasonably personnel, material and time for future utilization. Moreover the system states maintenance instructions, simple and as uniform as possible, for accomplishing systematic works–daily, weekly, monthly, quarterly or yearly–as required. This system shall assure the continuous operation of installations and equipment; it has been proved highly up-to-date as a very good system.

3. GERMAN SERVICING SYSTEMS

3.1 The modern maintenance system of the Federal German Navy

The following systems are the most important:

"PME" (Verfahren der Planmässigen Materialerhaltung), a planned maintenance system, comparable to PMS (Planned Maintenance System),

"MDS" (Materialerhaltungs-Daten-System), a system for maintenance data collection, comparable to MDCS (Maintenance Data Collection System),

"Formänderungs-Verfahren", a system covering alterations,

"Verfahren der Bauzustandsverfolgung" a configuration control system and

"Materialbewirtshcaftungsverfahren", a logistic support system.

Moreover a number of systems are in development, some of them will closely resemble the following U.S. systems:

SOAP (Supply Operations Assistance Program)

LOGSAT (Logistics Support Assistance Team)

PUBSAT (Publications Support Assistance Team),

and others.

Basic foundation of every other activity within the field of technical-logistic servicing is moreover the existence of a documentation, permanently maintained on the latest up-to-date level and at every time ready for disposal.

As in the following will be explained in detail both the systems PME and MDS correspond largely to the systems of the U.S. Navy, that is PMS and MDCS.

3.2 Maintenance

There are two kinds of trouble, impeding the readiness for action of weapon systems. Troubles caused by continuous wear can be foreseen and threfore avoided. However, if extraordinary strains occur, if the equipment has been maintained not satisfactorily or if hidden damages in design or of the material exists so these kinds of trouble cannot be foreseen; such kinds of trouble must be removed by a repair action. Considering both kinds of trouble, i.e. as well conservation as restoration for full operability the noun phrase "Maintenance" is used in such cases. The systems of maintenance shall assure the following:

In due time capacities, funds and materials must be prepared and actions must be started.

During accomplishment technical informations must be collected and service inspection must be effectively guaranteed. Moreover the execution of the maintenance action in question must be reported. After that collected data has to be processed in order to make improvements.

Troubles which can be foreseen shall be avoided by the system of PME (planned maintenance system); for repair actions the system of MDS (a system for maintenance data collection) is competent.

3.2.1 MDS (Materialerhaltungs-Daten System), comparable to MDCS (Maintenance Data Collection System)

The MDS (Materialerhaltungs-Daten-System) is a comprehensive system for reporting and information, including the entire field of maintenance, except PME-I.

On organizational level the accomplishment of a repair action will be reported by MDS, on field and depot level both the requisition and the accomplishment will be reported through that channel.

The report for accomplishment of approved alterations is also made on MDS-forms. Start and report of accomplishment for works of planned maintenance on field and depot level is made also by MDS procedures. By assistance of stored MDS-data a reasonable processing is possible for multiple purposes. Trouble analysis gives information about present reliability of equipment and material and about the readiness for action of ships. Furthermore the analysis is a basic document for making decisions concerning measures to be taken on the base of repeated troubles. This kind of measures refer to planned maintenance, to the logistic support system or to the training of maintenance personnel. In addition alterations may be proposed.

3.2.2 PME (Planmässige Materialerhaltung), comparable to PMS (Planned Maintenance System)

The system of planned maintenance, called PME, makes sure that the required capacities, funds and materials for the planned maintenance will be prepared in due time, the performance is made properly, and later evaluations for improvements can be made. Besides that PME may be used as a training aid for equipment and for combat practice.

Especially by the PME-system—on organizational level—it is assured that all work necessary for a quarter of a year can be planned in advance. The quarterly schedule, prepared by section chief ("Hauptabschnittsleiter"), shows the weekly work. The station master (Stationsleiter) enters every week the daily work again into the weekly schedule; the weekly schedule is an order. The PME instructions itself states which equipment is meant, which work especially has to be done, which kind of tools and test equipment are necessary, how the cycle of operation looks, which kind of spare parts and expendable supply has to be prepared, which kind of operating safety precautions are necessary and also states the time needed for executing corresponding actions.

After accomplishing of corresponding maintenance actions these actions must be marked in the individual schedules, that means "order carried out". On this way it is easy to control, and find out any work not performed.

Systems not yet prepared, pertaining to the PME on field and depot level, will cover in future in a similar manner the required preventive maintenance actions for these maintenance categories.

4. COMPARISON OF U.S. AND GERMAN SYSTEMS

4.1 General comparison of maintenance systems

A comparison of U.S. and German systems shows as well the respective advantages and disadvantages as possible difficulties too, occuring during the conversion of U.S. systems into German ones.

The German MDS-System shall be used on all three maintenance categories–organizational, intermediate and depot level. Moreover the a.m. system shall cover the planned maintenance, corrective maintenance of material and also report accomplishment of approved alterations. However, the U.S. 3-M System refers to maintenance on organizational level only. Moreover the MDCS does not cover alterations.

Within the German systems a large deal of attention has been paid to a standardized coding system. The so called "Materialnummer"–created for equipment identification –is used in all systems. The "Materialnummer" must identify the U.S. equipment of the DDG's–for which equipment only U.S. documents and data exist–as well as the German equipment. The "Materialnummer" consists of 21 digits, gives information about operation, manufacturer, and type-No. of the equipment in question, and shows the construction of the equipment up to the finest detail. Consequently the "Materialnummer" covers both codes used within the U.S. Navy, the "U.S. Equipment Identification Code (EIC-Code)" and the "U.S. Component Identification Code (CID-Code)". There exists a cross reference to the NATO Stock Number, required for maintenance and in MILSTRIP and MILSTRAP.

The preprinted forms of PMS and MDCS are very like to that of PME and MDS. However, the flow of information is very different–caused by the different organization–and also the arrangement of data files within the data bank.

4.2 Comparison MDCS–MDS

As aforementioned the newly created MDS-System is more comprehensive than the MDCS-System, because .besides other reasons it also reports the accomplishment of approved alterations. Besides this the MDS-System is not only an information system–as the MDCS-System is–but the form 2 of the set of preprinted forms capital "A" is simultaneously the work order to the repair activity. The external arrangement of MDS-forms corresponds largely to the MDCS-forms, valid until January 1, 1970. However, a comparison of the number of digits provided for, shows that MDS cares more about technical evaluation and identification of equipment. However, for the U.S. Navy the control of the engaged maintenance organizations is more important. The storage and processing of data by a data processing system is completely different within the U.S. Navy compared to the German one.

During input of data into the data bank the MDS-data will be tested and compared against two basic data files–a "ship orientated" and an "equipment orientated". The data reduction by special programs is very similar to that of the "Störmeldeverfahren" of the German Air Force, a trouble and failure reporting system.

4.3 Comparison PMS–PME

The biggest difference between the systems PMS and PME is the fact that in future the system of planned maintenance will be extended onto the field and depot level. Additionally PME can unlike PMS be processed completely by ADP; that applies to the handling of changes as well as to the recording of all expendable supplies and material required for each equipment.

Concluding it may be considered that the main problems during preparation of the new systems PME and MDS arose with the preparation of the required material basis and with the establishment of new data files. However, preprinted forms including coding and technical contents could be took over in a large extent.

DISCUSSION

Cruon: Pourriez-vous préciser ce que vous appelez "organizational level"?

Hormann: Corresponding to the grade of difficulties and to the extent of work the "Deutsche Bundeswehr" distinguishes between three categories of maintenance.

The category I (Truppen Instandhaltung) and the category II (Feld Instandsetzung) refer as much as possible to mobile maintenance capacities of the "Deutsche Bundeswehr" itself. On the category III (Depot Instandsetzung) stationary installations of both the "Bundeswehr" and the private industry are used.

The U.S. Navy also has three levels of maintenance comparable roughly to the aforementioned categories.

The organizational level corresponds to category I, the intermediate or field level to category II, and the depot level to category III.

Solstrand: My question relates directly to a statement made by the author in his presentation–but also has a strong connection to many of the other presentations we have heard on this conference. I quote: "In consequence of growing complexity of todays weapon systems, the flow of information is raising permanently"

The question now is–arising more from some sort of a feeling, rather than a conviction: Do we force this flow of information to grow more than justified by the increased complexity of modern weapon systems, just because the electronic data processing systems of latest design offer us a capability and capacity which we more or less feel forced to utlize?

Or to put it another way: Is it right that there is a danger–maybe a serious one–that we use the powerful tool, which a modern data handling system really is, more to increase the amount of information circulating in our system–which might be great enough as it is–rather than to help providing a more thorough and deep understanding of what the real need for and value of such information is?

Hormann: I agree with the questioner to the statement, that at present the powerful tool, which a modern electronic data processing system really is, is not always utilized in the right way. It should be the main task of the ADP (Automatic Data Processing) to accomplish a reasonable data reduction of the big amount of incoming information by assistance of suitable programmes. Among other things this output also should be utilized to reduce the input. Especially in the area of maintenance the trouble and failure analysis has among other things the assignment to reduce the number of troubles.

But on the other hand it was only possible by the assistance of the modern ADP to plan complex systems, to realize these systems in a relative short period of time and to control such systems.

Briot: Entre le moment où un défaut apparaît dans le système d'armes–par exemple, une erreur ou un défaut de la procédure de test–et le moment où les mesures correctives sont effectivement appliquées, il se passe un certain temps. Pouvez-vous donner une indication quant au temps nécessaire pour aboutir au redressement de défaut?

Hormann: In case of maintenance on organizational the downtime of an equipment–i.e. the time period from the moment of breakdown up to the moment of report for operational readiness–will be recorded in full hours and in tenth of an hour. This statement covers the time period used to disassemble and to install the equipment, troubleshooting, real repairwork and final checkout.

In case of field and maintenance the downtime–measured in days–will result as the difference from the date of completion and date of trouble reporting.

Ballow: How do you foresee the development of systems for utilization of the data you are collecting?

At what levels of management will it be used?

How are you insuring enthusiastic participation of your data inputs personnel?

Hormann: As defined in paragraph 3.2.1. of my presentation the MDS (Materialerhaltungs-Daten-System)–data-analysis serves many purposes. There exists the possibility to give information about the present readiness for action of the ships and on the other side material for improvement of future readiness for action can be prepared and furnished.

In paragraph 3.2.2. a brief statement is made pertaining to utilization of collected PME (Planmässige Materialerhaltung)–data.

Significant data stored by PME and MDS are selected and utilized at all management levels.

At the end of paragraph 1.4, I mentioned the two possibilities of positive influence to data input personnel: Repeated training and feedback-reports back to the originator. During the intensive instruction the task of the individual should be explained within the frame of the entire objective.

Geraerds: In section 2.2.2 you state that the PMS System has been proven highly up to date as a very good system. That is much more than satisfactory, and is proven. What was the absolute criterion you used for this evaluation of the quality of the system?

Hormann: Within the U.S. Navy inquiries have been made before and after introduction of the PMS System.

The result showed that the extent of repair is smaller than before.

To make more exact statements it is necessary to consult the corresponding documents and data.

Fischer: I am glad to hear, that you managed to extend the Data Collection System over all maintenance levels. I feel that is a serious lack of most presently existing systems, that they don't cover all levels and therefore don't close the loop.

The question I have is to which extent you included manhour reporting into the MDS system?

Hormannn: Manhours required for one maintenance action on organizational level will be recorded on the MDS-form filled in aboard the ship.

In the case of repair on field depot level in general no manhours will be stated.

However, placing orders to the industry the production hours may be entered on an extra form if the account shows these statements.

Combes: Comment est effectuée la collecte des informations au niveau de l'utilisation?

En d'autres termes comment vous assurez-vous de l'impartialité des infos-collectives?

Hormann: All information will be recorded on forms, while an extensive code-collection takes care of the fact, that only thoroughly selected data will be collected.

On the way from the recording place to the data collecting activity–the "Material-informationszentrum der Marine GmbH–MIZ–"–consequently it is impossible to change data.

A part of entered data will be checked for correctness manually on shipboard and in the collecting activity (MIZ) mechanically after having fed into the computer.

Geluykens: Speaker says that "PME will in future cover field and depot level". On that particular point I would like to know:

1. how far it is intended to go for both of these levels and if the system is going to cover *corrective* as well as *preventive* maintenance?

2. how is it planned to implement the full system and when will it be operational?

Hormann: The already developed system PME-I applies at present to preventive maintenance only. However, considerations are made to extend this system to corrective maintenance also.

The system PME-I, that is the planned preventive maintenance on organizational level, will be introduced on DDG 28 probably on October 1st, 1970.

The realization of the order, to introduce this system on all ships of the Federal German Navy, is depending upon the available working capacity.

Mathmatical Model for Budgetting for the Replenishment of Repair Parts

J.P. Thu

1. INTRODUCTION

In the beginning of the 60's US Grant Aid expired. National funds had to be made available each year to cover the purchase of replenishment repair parts. The term "replenishment repair parts" is used here to distinguish this class of inventory from "war reserve of repair parts" and also from "provisioning of repair parts" for new equipment being introduced into the army; both are other classes of inventory.

The prediction of required funds for yearly procurement of replenishment repair parts as expressed in the annual defence budget, was causing a great deal of concern at top level. It fell upon my staff and myself to propose the size of these funds for the entire army under the appropriate budget account in the defence budget.

Theories on modern scientific supply management was being introduced in Norway in conjunction with the entry and increasing use of ADP. There was, however, very little aid to be had, except what could be found in sparce foreign literature, and in the application program descriptions provided by the ADP equipment manufacturers.

At this time an IBM 1410 was being installed in our Defence Data-processing Centre. We were also studying IBM's "Inventory Management Program and Control Techniques". Chapter 6 of this publication E20-0174 and R.G. Brown's *Statistical Forecasting for Inventory Control* lead to this case study.

We were fortunate in having an inventory control system on PC equipment, and some historic data on requirements. The first study was based on statistical material from 1964, the next from 1965. Both showed great similarity.

In 1968 a major Norwegian consultant firm "Asbjørn Habberstad A/S" was approached to verify the correctness of the study, and this was performed by Mr. K. Mjelde. A Technical note was presented to the Norwegian Army Materiel Command by this firm. It confirmed the results of the study. This document is the "backbone" of this presentation.

2. MODEL RELATIONSHIP TO OTHER ROUTINES IN THE ADP SUPPLY MANAGEMENT SYSTEM

In the supply and maintenance of technical equipment in a large logistic system one will find many models to choose from. The environment causing item demand and the item demand pattern is considered when choosing the appropriate order model. Now each item demand is run through the chosen formula every time at operational level. This is how it should be. The great advantage of ADP is this rapidity of computation, giving this ability to treat each item demand individually extremely fast.

When moving up to next higher management level above the operational, one will normally find managers selecting and supervising the order models mainly by deciding parameter values, or giving directives.

Above these two stands top level management where policies are decided upon. This is extremely difficult in a large logistic system. At this level one wants a birds eye view of the whole large logistic system, preferably as one entire unit. This wish is hardly possible to fulfil. It is, however, necessary to find ways in summarizing chosen parts of the large logistic system, and then presenting discriptive results in such a manner that top level management may in it, find alternative policy-values to choose from.

The budget model for top level management differs, therefore, from the various order models by looking at the inventory class concerned as a whole package. The order models, on the other hand, treat each item singly or in small groups of related items.

Order models also differ from the budget model as they are not restricted by charge number accounting. The order model is chosen to minimize total costs like, cost to hold, cost to buy, cost to manufacture etc. The budget model is naturally budget

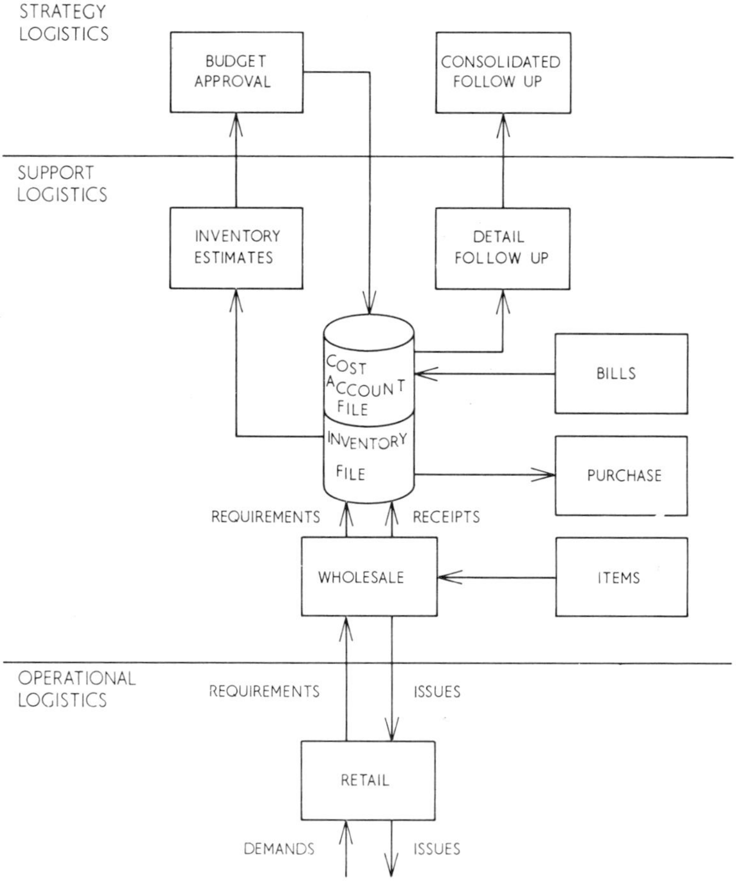

Figure 1 Simplified information flow

account orientated. This has the advantage in this case, that one is only concerned about consumption value charged against the account number concerned, and need not be detailedly worried about the order models used. One is actually only treating this single variable of the various variables in any order model formula. However, this budget model does reflect certain guides which may indicate the appropriate order model for the various segments of inventory.

Fig. 1 shows how the budget model is placed.

The budget accounting numbers, where this model is intended used, follow a major equipment category division and comprise funds for replenishment repair parts. The study includes 5 of these categories which in this text are termed as category 1-5.

Uncertainty in the future is a common problem. Statistically this is taken care of by allowing fluctuations or spread from the average usually measured by standard deviations. This is often termed as "service level" in the various order models, and concerns the fluctuations of the items average demand per period. The budget model also uses an average. In this case it is the value of all the item demands for 1 year divided by the number of the various items. The spread about this average is similarly expressed, but in this case it is more a question of safety of not exceeding the budget proposed.

If there is any way of relating these two types of spreads, then one could use the usual performance standard for depots, namely the value of issues and against the value of demands, not only when comparing performance results with order models, but also for budgetting. This has not been studied by us as yet.

3. DESCRIPTION

3.1 Distribution by Value

The study was restricted to national procurement of replenishment repair parts. Each of the 5 major equipment categories were treated separately. Historical data was not available for another 2 categories. In all, these 7 categories contained approximately 20 000 items including common hardware.

Each item was coded as to which category of equipment it belonged. A data processing routine produced a list as shown in Fig. 2.–one list for each of the 5 categories.

Results from the list were plotted on log normal graph paper. The curve for each category corresponded to a log normal distribution extremely well. A manual computation gave the same result. Fig. 3 shows the results obtained for category 1 and Fig. 4 shows the results for the 5 categories.

The sample size n for each category was nearly as large as the corresponding population size N.

Two characteristic statistics describe the log normal distribution:

$\bar{x}$ = the average "annual consumption value per item type"

ITEM NUMBER	%	QTY	UNIT PRICE	ANNUAL CONSUMPTION	CUM.ANN.CONS.	%
2	0.0	276.00	00101.30	50038.80	108902.80	9.64
3	0.0	1272.00	00026.30	33453.60	142356.40	12.60
4	0.0	1320.00	00022.70	29964.00	172320.40	15.25
5	0.1	8.00	02300.00	18400.00	190720.40	16.88
6	0.1	6.00	03018.00	18108.00	208828.40	18.49
7	0.1	69.00	00256.00	17664.00	226492.40	20.05
8	0.1	302.00	00056.90	17183.80	243676.20	21.57
9	0.2	85.00	00181.30	15410.50	259086.70	22.94
10	0.2	40.00	00376.00	15040.00	274126.70	24.27
11	0.2	310.00	00043.10	13361.00	287487.70	25.45
12	0.2	43.00	00310.00	13330.00	300817.70	26.63
13	0.3	20.00	00649.00	12980.00	313797.70	27.78
14	0.3	1383.00	00009.35	12931.05	326728.75	28.93
15	0.3	23.00	00552.00	12696.00	339424.75	30.05
16	0.3	946.00	00012.80	12108.00	351533.55	31.12
17	0.4	66.00	00183.00	12078.00	363611.55	32.19
18	0.4	7.00	01660.00	11620.00	375231.55	33.22
19	0.4	185.00	00062.60	11581.00	386812.55	34.25
20	0.4	154.00	00070.50	10857.00	397669.55	35.21
21	0.5	48.00	00222.00	10656.00	408325.55	36.15
22	0.5	250.00	00042.50	10625.00	418950.55	37.09

Figure 2 Distribution by value listing

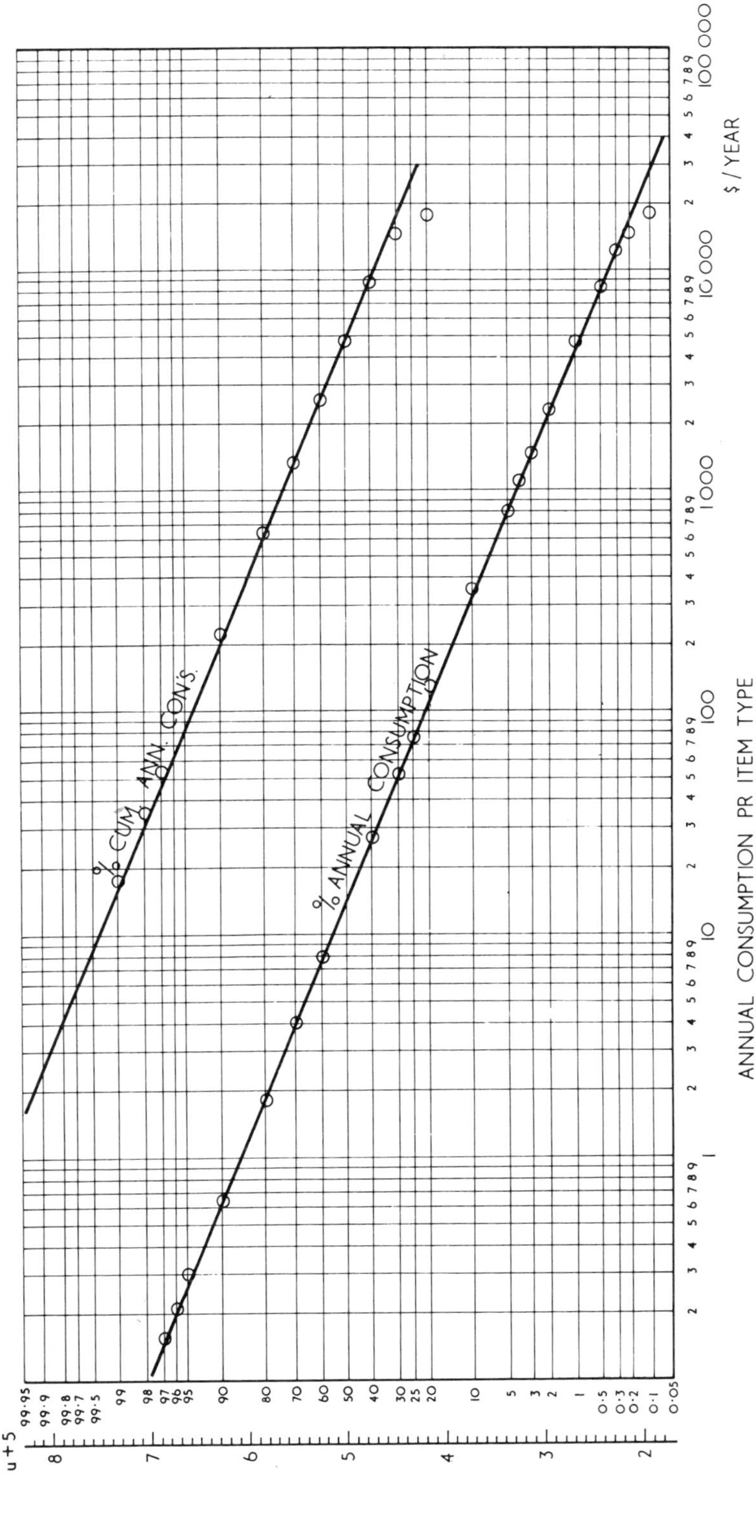

Figure 3a & 3b Distribution by value plot

Major Equipment Category	Assortment Population Size n	Assortment Sample Size N	Standard Ratio p	Mean x	Median x_m
1	4171	5039	11.3	271	14.2
2	926	1024	10.2	482	10.5
3	580	676	8.4	119	12.15
4	99	128	7.4	66	8.8
5	4080	4080	12.7	165	3.7

Figure 4 Assortment study results

p = standard ratio describing the variation of the annual consumption value per item type, and also the slope of the lines on the graph.

3.2 Connection between standard ratio and quantity of item types

Figure 5 shows a plot between N qty of item types and standard ratio p for each of the five categories.

It shows that when N increases the value of the annual consumption for the various items distribute over a larger span and this means that p increases. A correlation test was performed and the following formula resulted

$$p = p_0 N^a$$

where $p_0 = 3{\cdot}92$

and $a = 0{\cdot}1315$

From this follows that, when the quantity of item types N in a category is known, one can compute the standard ratio p of the assortment, granted of course that the environment causing consumption are similar to that from which historic data has been taken.

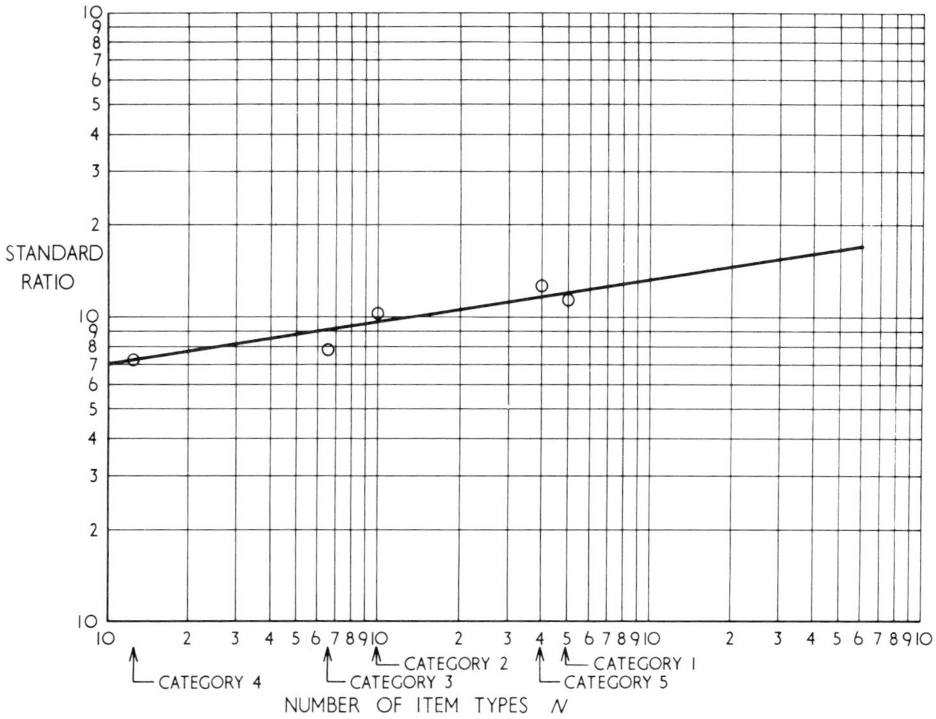

Figure 5

3.3 Ranges of medians and means

The 5 categories gave following results:

– median between 8.8 and 14.2
– mean ,, 66 ,, 482

3.4 Comments on the findings of the study

As mentioned previously, a budget model should reflect certain guides concerning order models.

In this case figure 3 shows that approximatley 20% of the item types influence 95% of the total annual consumption value. It is also worth while to note that 35% of the item types singly have less than \$ 10,–annual consumption. These and other similar information from the graph can be used as guides like deciding where to stratify the inventory in "control intensity layers" A, B and C. Such directed measures might of course influence later historical data and give a different plot. This would have to be taken into consideration when it occurs.

The log nǒrmal distribution is in itself a tool in predicting inventory estimates and actually gives the birds eye view one needs in order not to be tangled up in order model details. I would go so far as to say it gives "inventory rules of thumbs". See appendix 1, fig. A1.

Presuming that standard ratio, the quantity of item types and the average is known, one can construct the corresponding log normal distribution and produce a tentative distribution by value list.

3.5 Mathematical Model Recommendation

The following is a translation of the recommendation.

The annual consumption value per item type in kroners/year is denoted by x. The assortment studies show that $y = \ln x$ is normally distributed. The expectance value and the standard deviation are denoted respectively y_m and d

$$E(y) = y_m$$
$$Var(Y) = d^2$$

If x_m is median of x, then

$$y_m = \ln x_m$$

The expectance value of x is given by

$$E(x) = u = x_m \; exp(d^2/2)$$

and the variance by

$$Var(x) = d_x^2 = u^2(e^{d^2} - 1)$$

One calculates a prediction F for the annual consumption by

$$F = \bar{x} \cdot N$$

where the average $\bar{x}$ is based on a sample size n.

Then:

$$var(F) = \frac{N^2}{n} d_x^2$$

and the standard deviation for the prediction is given by the formula

$$d_F = \frac{F\sqrt{e^{d^2} - 1}}{\sqrt{n}} \approx \frac{F.exp(d^2/2)}{\sqrt{n}}$$

for $d \geq 2$

If the sample contains a fraction of the item types ($f = \frac{n}{N}$) and d_F is based on N instead of n, then the relative error will be ½(1–f).
The probability that the consumption will not exceed the budget is denoted by q and is termed "safety level." This is a policy parameter and must be decided upon in advance. The budget B is given by

$$B = F + t_q \cdot d_f$$

The relation between q and t_q is given in fig. 6. The value t_q indicates the difference between the budget and expected consumption or prediction, measured with the prediction standard deviation as unit.

Safety Level q	t_q
0.50	0
0.60	0.26
0.70	0.53
0.75	0.68
0.80	0.84
0.85	1.04
0.90	1.28
0.925	1.44
0.95	1.64
0.975	1.96
0.9875	2.24
0.995	2.57

Figure 6

3.6 Budgetted Consumption

We consider a specific group of item types and shall decide

. Prediction of consumption F

. The budget B corresponding to safety level q

We get two cases, depending on whether the average x of the annual consumption per item type is known or not.

3.6.1 Method 1:

We start with x which is found by historical data and the number of variables N.

1. Predicted consumption is calculated by

   $$F = \bar{x} \cdot N$$

2. The group (of item types) is related to a specific major equipment category, where at least some have been assortment analyzed, so that the standard ratio can be computed by the formula:

   $$p = p_o \cdot n^a$$

   p can be found directly from figure 5 for the categories shown. In this manner one takes into consideration that the numbers of item types may vary from year to year.

3. The standard deviation d is calculated

   $$d = \ln p$$

4. The standard deviation d_F for the prediction is found

   $$d_F = \frac{F \cdot exp(d^2/2)}{\sqrt{N}}$$

5. The probability q for exceeding the budget is decided upon. See figure 6. The budget is given by

   $$B = F + t_q \ d_F$$

3.6.2 Method 2:

The average annual consumption for item types in a category is not known. This will be the case for new major equipment categories. One can, however, often relate the category to a set of categories with historical data. The median for the various categories (in this set) vary very much less than the average, and this is used to find a budget based on the number of variables N in the category.

1. The median x_m for the category is equated with the average median for the category set.

2. The standard ratio is decided upon by the relationship

$$p = p_o \cdot N^a$$

and $d = \ln p$

3. The prediction F is computed by

$$F = N.x_m \cdot exp(d^2/2)$$

and the standard deviation by

$$d_F = \frac{F.exp(d^2/2)}{\sqrt{N}}$$

4. The budget corresponding to the safety level is decided upon.

A good result (for a new category) by this method depends on the right classification of the category.

3.7 Choosing Confidence Level

The confidence level of a category is fixed by the value of t_q, see figure 6. The choice depends on if the category must by predicted isolated, or if it is possible to predict a class of categories from the fact that money may be transferred from one category to another within the class when required.

If the class consists of n categories, and we choose the confidence level (t_k) for the total class, then the confidence level t for each category is given by the relationship

$$t \approx \frac{t_k}{\sqrt{n}}$$

The budget then becomes less with the same total safety.

3.8 The use of Model on an Example

The calculations are made for major equipment category 1. See fig 8.

3.8.1 Method 1.

We start with the historical average annual consumption per item type $x = 271$ and the number of item types $N = 5039$. The standard ratio is found from figure 5.

The budget is calculated in fig 7, and one sees how it grows in size with the safety level. When the category is budgetted as a class of 8 categories, the budget is from 30% to 10% less than if the category had been predicted alone, and the reduction is greatest for the higher safety levels.

	Category 1 singly	Set of 8 Categories
Safety Level *q*	Budget B $ 10000	Budget B $ 10000
0.995	246	175
0.9875	232	170
0.975	220	166
0.95	207	161
0.925	198	158
0.90	191	156
0.85	181	152
0.80	172	149
0.75	165	146

Figure 7 Category 1: expected consumption £1 360 000
standard deviation $ 430 000

Annual Consumption pr Item Type	Commulative % Item Type Qty	Commulative % Total Consumption
18.000	0.1	18.5
15.000	0.2	24.3
12.000	0.4	32.2
5.000	1.0	49.1
2.000	2.2	63.0
1.000	4.2	73.0
400	9.1	84.5
100	22.7	94.8
40	35.2	97.8
10	56.3	99.5
4	70.0	99.8
1	86.5	99.97

Figure 8 Annual consumption, category 1

3.8.2, Method 2.

The number of item types $N = 5039$ is the basis of the computation. The median is calculated from the median of the category set 2, 3 and 4, while the median of group 5 is disregarded as it is very low. The result is

$$x_m = 10.6; \quad p = 12; \quad F = 120.10^4, \quad d_f = 38.$$

The prediction F and the standard deviation d_F is nearly 10% less than by method 1. The budget may now be calculated for the various safety levels and the result can be produced as shown in fig. 7.

3.9 Practical Suggestion is Using the Model

The computation of the budget for various safety levels is programmed and a listing is produced for each category.

The category is defined by a category code which also indicates the budget account the category belongs to. Each inventory item is category coded. This work is simplified when the budget account number is structured according to major equipment categories, as proposed by NATO, and repair parts are grouped according to NATO Stock Number classification.

In order to use the model it is only required and sufficient for the user to report the number of item types in each group to the ADP centre. One naturally considers any trend in the quantity of item types. If a category of the annual consumption value is known and stored then method 1 is used for computing the budget, if not, then method 2.

A suggested users instruction of the model is:

1 For each category to be budgetted the following is reported to the ADP centre:

- category code
- number of item types.

2 A listing is produced which shows the budget figure for corresponding safety level.

3 Safety level is chosen, and the budget is thus decided upon.

4. LEAD TIME AND EXPECTED PAYMENTS

4.1 Lead Time Study

Lead time study was performed by the major depot and can be accepted as normally distributed. See Fig. 9.

Here the mean is 210 days and the standard deviation is 96 days. 95% of the lead times are shorter than 360 days.

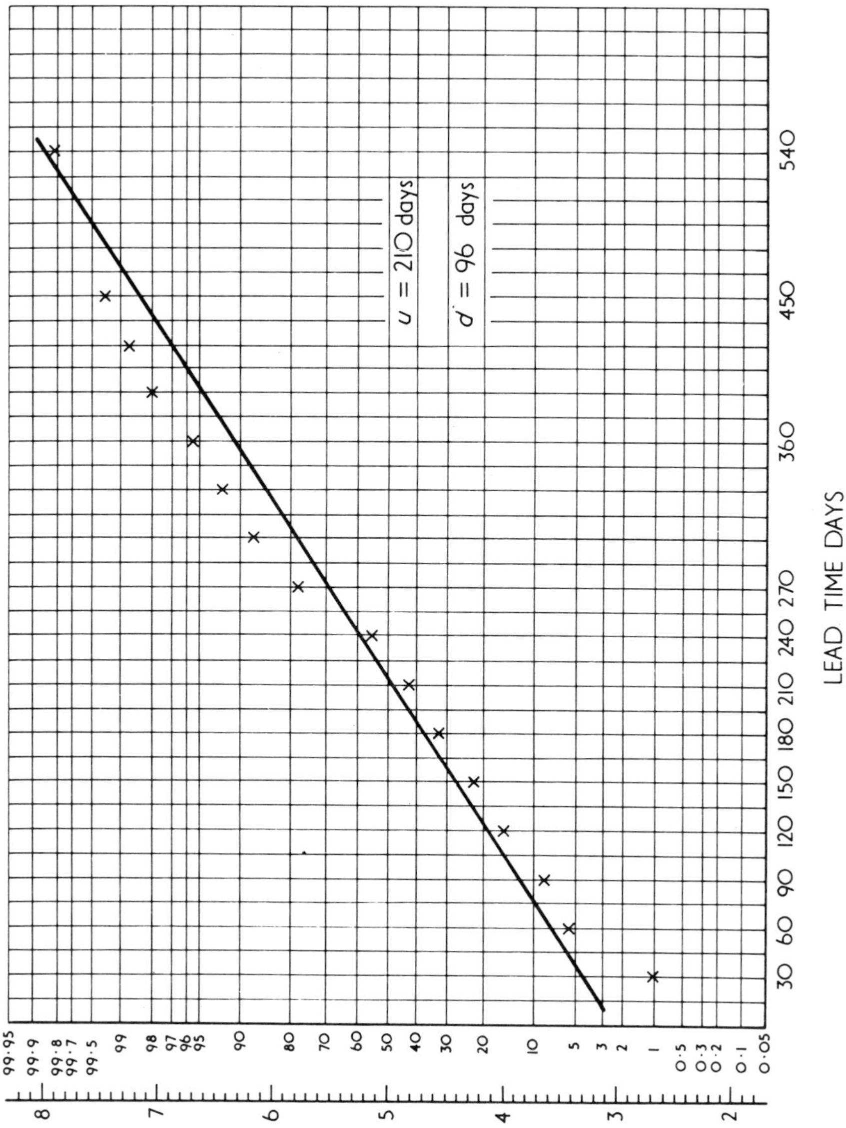

Figure 9 National procurement lead time

Month Number	Proportion Payment %
1	2.5
2	2.5
3	4.5
4	7
5	10
6	11
7	12
8	13
9	12
10	9
11	6
12	5
13	2.4
14	1.4
15 ⟶	1.2

Figure 10 Expected payments

Expected payments falling on succeeding months after ordering can be taken directly from Fig. 10.

4.2 Expected payments

The recommendation here is also a translation.

The time t in months from the time the requirement is registered till receipt of stock payment, is normally distributed:

$$g(t) = \frac{1}{d\sqrt{2\pi}} exp\ (-½(\frac{x-u}{a})^2)$$

The probability that the lead time lies between a and b is

$$P(a \leq t \leq b) = \int_a^b f(t)dt = G(b) - G(a)$$

where $G(t)$ is the cummulative distribution.

If the requirement B occurs in a particular month, then the fraction $R(k)$ expected paid k months later is given by

$$R(k) = G(k) - G(k-1)$$

The expected payments in the succeeding months can then be recorded based on the month's registered requirement B.

The lead time distribution for national procurement is given in fig. 10. As examples

$$P(6 < t < 9) = 75 - 38 = 37\%$$

and

$$P(t < 12) = 95\%$$

In practice the purchase requirements will with sufficient approximation distribute evenly and with (1/12) of the annual requirement each month. Payments will then too occur in this proportion.

5. USE OF TABLES

The study so far has resulted in suggesting the use of table listings which do not require great ADP programming. The tables would be according to the two methods described.

There is a third area where a table set using method 2 and producing a table for each ascending quantity of 200 item types may perhaps be used, that is in "the conceptual phase" of a new major equipment, before actual "bill of material" is available for more correct computation.

This study has indicated that it is necessary to look more closely into two fields,

- The relation between safety factor used in order models and safety level in the budget model,
- the lead time complex.

Both the above mentioned influence the investment in safety stock which, after all, also has to be taken care of by the annual budget if there are great changes in item types. It is felt, however, that in the suggested budget model this is taken care of by the "safety level".

In order to extend the studies a "follow up" routine based on requisition and purchase history file is now being introduced this summer. This type of feed back is also required to indicate the reliability of the budget model.

As indicated earlier, top level management must be given a birds eye view level and meaningful information from which to choose alternative policy values from. It is hoped that tables with accompanying log normal graphs can be of value in inventory management of replenishment repair parts.

APPENDIX: LOG NORMAL DISTRIBUTION

Definition: If $y = \ln x$ is normally distributed then we define x as log normally distributed.

The normal distribution is defined by

$$g(y) = \frac{1}{d\sqrt{2\pi}} \; exp\; (-½ (\frac{y-y_m}{d})^2)$$

where

d is the standard deviation of y, and y_m is the median and the mean of y.

The distribution equation of x is

$$f(x) = \frac{1}{d\sqrt{2\pi}} \; exp\; (-½(\frac{\ln x - \ln x_m}{d})^2)$$

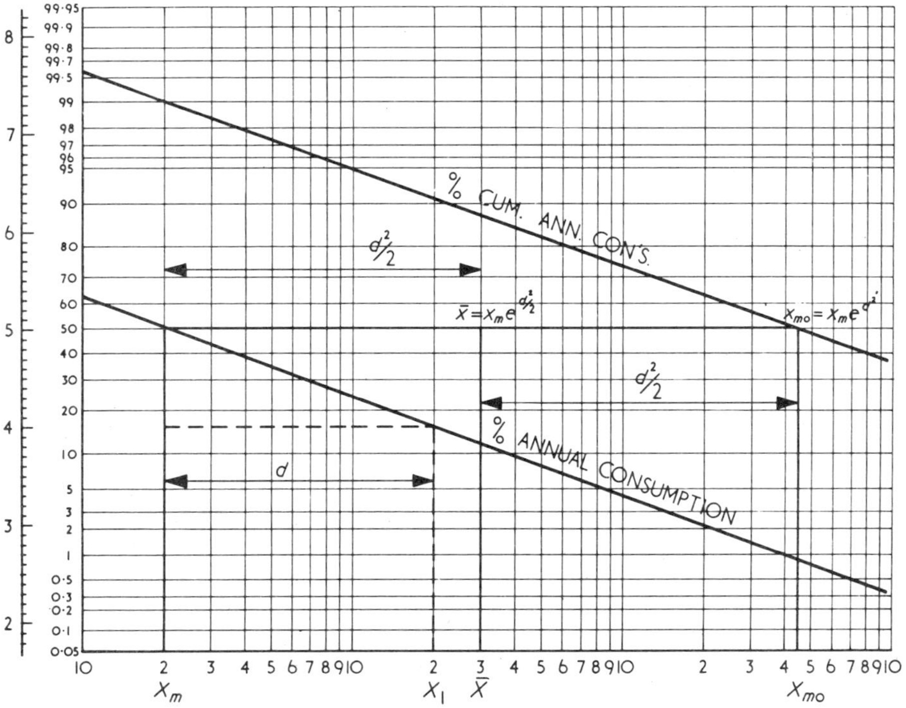

Figure A1 Log normal distribution plot

where

x_m is the median of x,

$y_m = \ln x_m$

The kth moment about the origin is given by

$$E(x^k) = \int_0^\infty x^k f(x)dx = u^k \; exp \; (½k(-1)d^2)$$

where $E(x) = u$ is the expectance valus of x

$$E(x) = u = x_m \; exp \, (d^2/2)$$

The variance of x is

$$d_x{}^2 = E(x^2) - E(x)^2 = u^2(e^{d^2} - 1)$$

The cummulative consumption median x_{mo} is given by

$$x_{mo} = x_m \cdot e^{d^2} = u.e^{d^2/2}$$

For the categories within a set of related categories, the point x_m will be approximately equal, and the slope of the 2 parallel lines will decrease when p and d increase.

DISCUSSION

Olivier: This is a comment on low consumption items. Numerous airlines are carrying inventory where 40 to 50% of the item are used less than 1 time per year–Such items make 25 to 30% of the inventory amount.

Bureau:

– Ceci n'est pas une question, mais une suggestion.

– Vous avez fait apparaître que dans l'inventaire figure un percentage important de pièces de faible valeur unitaire et de faible valeur d'achat.

– Vous avez signalé que ceci pose un problème d'achat (courtage)

– Il est suggéré que cette question fasse l'objet d'échanges de vue entre NAMSA et ses clients.

Tomorrow's Provisioning Tool—Today

R.A. Barnard
J.K. Seger

1. BACKGROUND

The provisioning of spare parts has always been more an art than a science. The results, good or bad, have been exclusively dependent upon the judgment of, and application of that judgment by, a relatively small number of "experts"

With the growth in size, complexity, and cost of modern weapons systems and even commercial airplanes, there has been significant escalation in the cost of spares investments. Thus, an evolving need has been arising for better provisioning techniques and better tools for decision-makers to use in managing the provisioning process.

In 1965, Lockheed initiated a company-funded research project to investigate and, if feasible, to develop a new provisioning concept which would have the following general characteristics:

(i) It should be comprehensive, covering all types of spare parts, all maintenance levels, and adaptable to all types of aeronautic systems—both military and commercial.

(ii) It should provide alternate decision opportunities, weighing cost and effectiveness for each potential decision choice.

(iii) It should aid the decision-maker, but it should not diminish in any way the opportunities for analysis and judgment by customer experts.

(iv) It should be capable of rapid operation so that the provisioning process can be performed in less time and using less manpower.

We have developed such a system. It is computerized and operational. It has been tested and validated. It is being employed by several airlines in a large commercial transport program. It is equally applicable to military programs.

2. MAJOR FEATURES

2.1 Scope

The new provisioning technique employs a series of mathematical models to determine the optimum quantity of spare parts required for initial support of aircraft fleets. It calculates spares quantities providing the highest level of overall spares availability* within customer specified limitations of funding, weight, or volume.

The technique is applicable to both military and commercial programs.

It computes requirements for both consumable and repairable spare parts. But, for practical purposes, the model does not consider bulk materials (such as grease and oil) or standard parts (such as rivets, nuts and bolts) that are normally provisioned in minimum unit packs.

The model can accommodate supply requirements for up to 200 stocking locations.

And it recognizes the special supply demand characteristics of each of the 4 levels of supply.

. Depot

. Intermediate

. Organizational

. Deployed Site

2.2 Application

The provisioning model is currently programmed for operation on an IBM 360-65 computer. It can easily be adapted for operation on other smaller computers if there is a requirement to do so.

Special attention has been given to consideration of the customer's operating plan–the environment in which the aircraft will be utilized, the deployment plan, the maintenance plan, and other operational characteristics that can have an effect on provisioning policy.

The output report displays all of the relevant data influencing quantity recommendations as well as the recommended provisioning quantities. Those recommendations can be selectively reviewed for reasonableness.

The decision-maker is freed from the need of performing manual calculations. He can concentrate upon those areas of concern requiring the exercise of judgement.

Although the model effectively makes use of the most recent data available, it does not perform the data estimation. Rather, it is limited to testing the input data to determine whether those data meet certain criteria of admissability.

The model is in no sense a substitute for human judgment in the quality of data that are providing for processing, although it does test the data that are provided for logical consistency.

* Availability is defined as the ratio of filled requests for a spare part to the total number of requests for that spare part. A request is considered to be filled if a spare part is immediately obtainable when requested. As an example, if there is a total of ten requests for a spare part and nine parts are immediately available, then there has been a 90% spare parts availability. This same definition is applied to all spares taken as a group, so that the overall availability of spare parts is the ratio of total filled demands to total demands for all spares.

2.3 Benefits

Among the important characteristics of the new provisioning technique is the uniform and consistent application of all input data. The model eliminates variations that are inherent in repetitive manual processes, thus providing consistent output results.

Because the model is computerized, very rapid response can be assured. Great savings in manhours at provisioning conferences can be realized as there is no longer a need to manually calculate requirements for each line item. For example, a provisioning conference conducted last year according to conventional techniques required the participation of 20 customer personnel during a 4 month period. The new technique could have produced better results using approximately 10% of the manpower.

The model can be used as a tool for analyzing the cost-effectiveness of various levels of spare expenditures on aircraft supportability. The decision-maker can then select the best level of availability considering the corresponding spares cost.

2.4 Cost/Effectiveness

The provisioning model is designed to show the relationship between spares availability and spare parts cost. The essence of that relationship is depicted in Figure 1, in which the principal of diminishing returns is apparent.

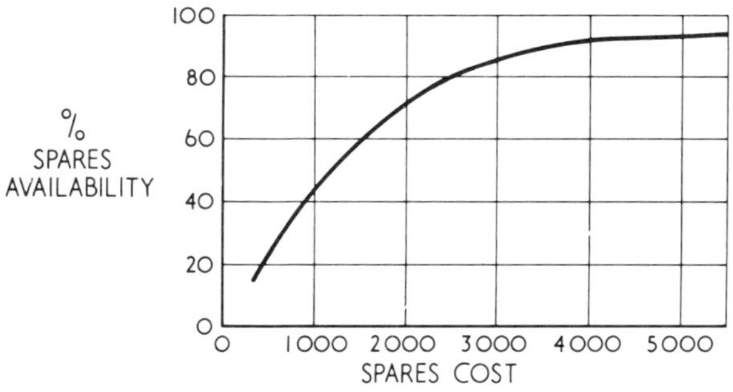

Figure 1 Spares availability versus cost

Obviously as additional units of an item are provisioned, both availability and cost are increased. However, the rate of 'availability' increase is not the same as the rate of cost increase.

Whereas total cost rises generally at a constant rate as more units of a given item are provisioned, availability increases at a diminishing rate. This means that as more money is spent for spares support, less additional availability is obtained per dollar expended.

What the optimum level of availability should be depends upon the particular characteristics of the customer's mission and the consequences of failing to achieve that mission. Whereas 90% availability might be adequate for some missions, 95% availability might be required for others. The choice is available to the decision-maker to determine the level that is optimum for his program. Regardless of what level of availability is optimum, the provisioning model indicates the total spares cost to obtain that availability.

2.5 Support System Trade-offs

The decision maker can use the provisioning model to perform a wide variety of trade-off studies for optimizing the design of logistics support systems.

2.5.1 Depot vs. Base Repair

Concentrating all repair capability at the depot will minimise the cost of support equipment, repair facilities, and personnel training. However, the pipeline time between base and depot tends to increase spare costs. In order to shorten the pipeline, repair capability may be established at one or more bases, thus raising the costs for repair capability and lowering the costs of spares. The provisioning model can identify the costs of spares for various pipeline lengths, which can be matched against the costs of establishing repair capability at bases. The overall minimum support cost can be identified as the smallest sum of spares costs and costs for establishling repair capability. This trade-off can be applied to individual components, such as engines, electronics systems, etc., as well as to the entire aircraft. The model can be used in conjunction with an optimum repair level analysis.

2.5.2 Spares Cost vs. Depot Repair Time

The provisioning model identifies the cost of spares support for different repair time. The shorter the repair time, the smaller is the spares requirement, yet the more costly it becomes to perform repairs due to tighter scheduling, additional repair equipment, more repair personnel, etc. The optimum repair time is the one having the lowest total cost.

2.5.3 Spares Cost vs. Number of Bases

The provisioning model can be used to identify the cost of spares support for aircraft deployed to various numbers of bases.

2.5.4 Spares Cost vs. Aircraft Utilization

Spares support costs increase with higher levels of aircraft utilization. The provisioning model can show spares cost for various rates of aircraft utlization.

3. STRUCTURAL ELEMENTS

The provisioning model requires two different kinds of input data as shown in Figure 2.

(i) Spares technical data	–describing the characteristics of each individual spare part. These data are developed by the manufacturer as the aircraft design evolves.
(ii) Program environment data	–describing the program for maintenance and operation of the aircraft. These data are furnished by the customer.

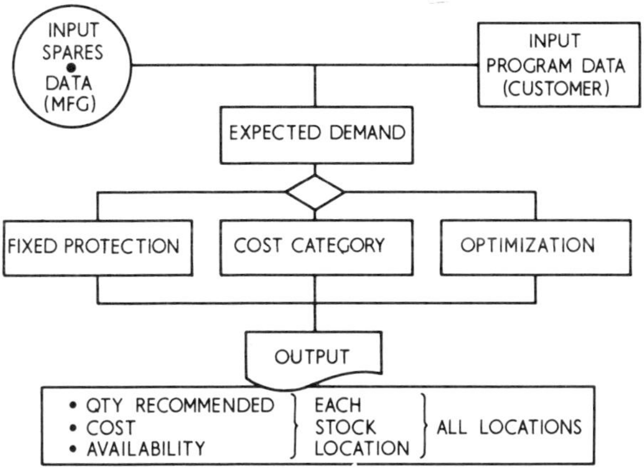

Figure 2 Structural elements, part 3

Once the data are obtained and introduced into the computer, an EXPECTED DEMAND is computed for each spare part at each proposed stocking location.

Next, one of three alternate models is applied to the EXPECTED DEMAND in order to determine the recommended stock quantity.

The alternate models are:

. FIXED PROTECTION

. COST CATEGORY

. OPTIMIZATION

After stock quantities have been calculated, the computer prints the results for each stock location and summarizes cost and availability statistics.

3.1 Spares Data

As stated earlier, the provisioning model requires two different kinds of input data. The first kind of data, SPARES technical data, describes the characteristics of each individual spare part, and is developed by the manufacturer as the aircraft design evolves. The second kind of data, PROGRAM environmental data describes the maintenance and operations program, and is provided by the customer shortly before the date of the provisioning conference.

There are many kinds of SPARES technical data which are input into the computer. Some of the most important kinds are:

(i) the quantity per aircraft of the specific part, i.e., the total number of applications of the spare part on each aircraft;

(ii) the unit cost of the spare part; this is a most important data element since it affects the total cost of provisioning;

(iii) the predicted scheduled and unscheduled removal rates for a spare part;

(iv) the condemnation rate for repairable items;

(v) the maintenance and overhaul factors for consumable items, which relate the frequency with which the spare is used to the frequency of repair and overhaul of its next higher assembly;

(vi) the repair capability of each spare at each level of maintenance;

(vii) the criticality to flight of each spare part.

3.2 Program Data

PROGRAM environmental data describe the manner in which the customer plans to operate his aircraft. The parameters required by the provisioning model include:

(i) the desired overall level of spares availability,

(ii) either a budgetary limitation or other constraint which may be in terms of spares weight or spares volume,

(iii) the length of time for which the aircraft is to be supported by the provisioning exercise;

(iv) the amount of flying that is programmed to take place at each base;

(v) the resupply shipment time between the depot and each base;

(vi) the source and frequency of resupply for each base;

(vii) the maintenance level classification for each base;

(viii) the programmed repair cycle time at each maintenance level.

These PROGRAM environmental data represent general policy constraints established by the customer and can be provided by responding to a relatively simple questionnaire.

3.3 Expected Demand

The EXPECTED DEMAND for a given spare part is the fundamental factor for computing recommended quantities and associated spares availability. It is the average requirement for a spare part at a given location during a specified period of time.

There are 36 formulas for computing expected demand. The following factors are used to select the proper formula for each spare's computation.

(i) the character of the spare part–consumable or repairable

(ii) the maintenance level for repair–i.e., base or depot

(iii) the level where overhauled

(iv) the lowest level where authorized for stockage

(v) the stock replacement policy

Each formula may be modified in a wide variety of ways depending upon the customer's support structure. The computer is programmed to automatically select and modify the appropriate formulas in accordance with the input data.

The formulae compute the demand at each using base and at each higher level base or depot which supports the using base. The model automatically summarizes all contributions to EXPECTED DEMAND for the same spare part at each depot, intermediate base, organizational level base, and deployed site.

3.4 Demand Variability

The expected demand formulas indicate the *average* requirement for a spare part. However, the *actual* demand for spares can be expected to fluctuate about the average from one time period to the next. Demand variability is expected to follow the Poisson distribution. The actual stock quantity recommended is based upon meeting a certain percentage of total demands. This can be visualized as the area under the probability curve, where the percent of the total area (i.e., availability) increases the higher the stock level is. No stock at all would provide zero percent availability, and a stock equal to the average demand would provide very roughly 50% availability. An infinite stock would of course provide 100% availability. The example in Figure 3 indicates that a stock level of 15 parts provides an availability of 91.6% when the expected demand is 10.

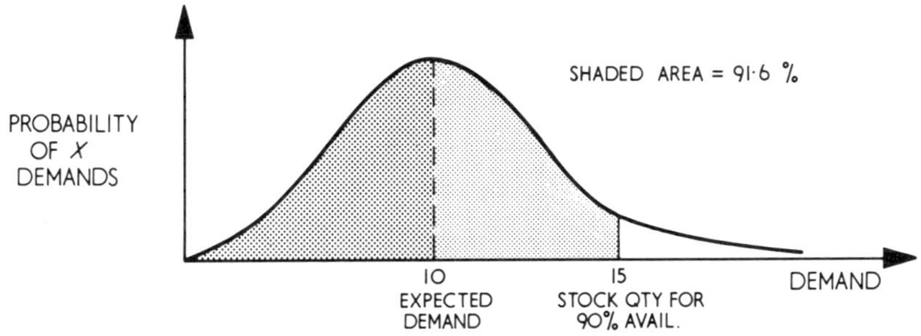

Figure 3 Variation of actual about expected demand

3.5 Fixed Protection

Now that the expected demand has been computed, one of the three models is applied to compute a stock quantity.

The simples of the provisioning models is the FIXED PROTECTION Model. It provisions all spares to the same level of availability without considering any individual characteristics of the spares. For example, the same level of availability, such as 90%, would be applied to a 50¢ washer as well as a $3,000 radio. The overall level of availability for all spares is the same as the individual availability of each spare. The same availability is required of each spare part "across the board".

EXAMPLE

Part No.	*Noun*	*Unit Cost*	*Availability*
ABC	Washer	$ 0.50	90%
BCD	Radio	3000.00	90%
CDE	Bracket	15.00	90%

Generally, this FIXED PROTECTION Model is selected when precise spares cost information is not known or is available only as estimates.

3.6 Cost Category

The second model is the COST CATEGORY Model. It is similar to the FIXED PROTECTION Model, with the difference being that each spare is provisioned to an availability level which depends upon the spare part's unit cost. For example, a 50¢ washer might be provisioned to 98% availability and a $3,000 radio to only 85% availability. The total cost to achieve an overall average availability level for all spares is always less with the COST CATEGORY Model than the FIXED PROTECTION Model. This model makes use of the high value concept, where the expensive items are bought in smaller quantities than the inexpensive items but are managed very closely.

EXAMPLE

Part No.	*Noun*	*Unit Cost*	*Availability*
ABC	Washer	$ 0.50	98%
BCD	Radio	3000.00	85%
CDE	Bracket	15.00	97%

Recognizing that an inexpensive part can cause an aircraft to be grounded just as effectively as an expensive part, and that a limited budget must somehow be allocated among all spares, it follows that buying more of the inexpensive spares and fewer of the expensive spares can result in a higher availability with no change in budget. The expensive spares represent a small percent of total line items, yet comprise a high percent of total cost. It is feasible to monitor and manage carefully these few items in compensation for buying them in relatively small quantities, thus achieving in practice a higher availability than forecast by the model. The provisioning model permits up to six categories to be specified.

This model is normally selected when spares cost information is reasonably accurate, and when the customer desires to specify the cost of total initial provisioning investment by varying the level of planned availability for individual items according to pre-established cost categories.

3.7 Optimization

The third and most complex model is the OPTIMIZATION Model. It incorporates a technique which selects a mix of spares that will maximize overall spares availability for any given amount of money. Each unit of each spare part at each base and depot is evaluated according to how much availability it provides per dollar. Those spares providing the highest availability per dollar are selected first. The selection of spare parts continues until either the desired overall availability is achieved or the budget limit is reached. In this manner each additional spare selected provides a small availability per dollar than all preceding units selected. This means that all units yet to be selected would provide less availability per dollar than those already selected, thus assuring the maximum availability for any level of expenditure.

If a plot of availability versus cost were made, it would trace out a curve of constantly diminishing slope such as that in Figure 1. The first unit selected would provide the highest availability per dollar, and each successive unit selected would provide constantly decreasing additions to the overall availability per dollar. The technique used in the Model is a straightforward application of marginal analysis.

If spares weight or volume is the critical factor in a support situation, the OPTIMIZATION Model can be used by supstituting weight or volume for spares unit cost, and spares are then selected according to how much availability they provide per pound or per cubic foot rather than per dollar. Spares weight is a critical factor in desigining flyaway kit of spares. Similarly, spares volume is sometimes a critical factor on board submarines or aircraft carriers.

Consider the following example of the use of the OPTIMIZATION Model: a system consists of three spare parts, Part A, Part B, Part C. It is desired to find the best mix of these spares to stay within a budget of $3,500. That is, how many units of A, of B, and of C should be bought to maximize overall availability. Their expected demands are, respectively, 3.0, 2.0, and 1.3, and their unit costs are $400, $200, and $300.

EXAMPLE

Part No.	*Expected Demand*	*Unit Cost*
A	3.0	$ 400
B	2,0	200
C	1.3	300

Figure 4 indicates the ranking of each unit of each spare part. The column headed "Increase in Protection per Dollar" shows how much protection against unfilled demands per dollar is obtained for the spare part under consideration. For example, the first unit of spare part "A" provides a protection per dollar ratio of .237. The first unit of spare part "B" provides the highest ratio, .432, and it is therefore assigned the rank of 1. This is shown under the column headed "Spares Rank".

The computer performs the ranking computation, selecting quantities of each part according to protection provided per dollar of expenditure. When the budget limit is

PART NO.	UNIT OF PART NO.	INCREASE IN PROTECTION PER DOLLAR	SPARES RANK
A	1	.237	4
A	2	.200	5
A	3	.144	7
A	4	.088	9
A	5	.046	12
A	6	.021	14
A	7	.0085	16
B	1	.432	1
B	2	.297	2
B	3	.166	6
B	4	.071	10
B	5	.0265	13
C	1	.242	3
C	2	.124	8
C	3	.048	11
C	4	.014	15

Figure 4 Optimization ranking process (Rank $= \frac{\Delta \text{ Protection}}{\text{unit cost}}$)

reached, the computer then summarizes the quantity of each item, the availability afforded, and the investment cost. Overall availability for the combined group of items, together with total cost, is also summarized.

The solution to this problem is to buy four units of A, four units of B, and three units of C. The overall availability of these spares is 93.0% and the total cost is $3,300. The next spare that would have been selected is a fifth unit of Part A; however, this would have added $400 to cumulative cost, causing the budget constraint to be exceeded. Thus, there is no other mix of these spares which could provide a higher availability for $3,300.

SOLUTION

Part No.	*Expected Demand*	*Optimum Buy Quantity*	*Unit Cost*	*Part Availability*	*Total Cost*
A	3.0	4	$ 400	89.4	$ 1600
B	2.0	4	200	96.2	800
C	1.3	3	300	95.7	900

OVERALL AVAILABILITY = 93.0%, COST $3300

As a practical matter, it is not feasible to optimize an entire aircraft's range of spares, which typically would contain more than 10,000 unique line items. The computer run time would involve several hours, depending on the number of bases in the support structure. The OPTIMIZATION Model is best applied to the high-value spares or to certain critical systems, such as the radar system or power plant. The remaining spares can be provisioned by the FIXED PROTECTION Model or the COST CATEGORY Model.

4. VALIDATION

A large number of validation studies have been performed to test the accuracy of the provisioning model. The tests have involved both commercial and military aircraft programs. Some of the results are summarized in the following paragraphs.

4.1 Validation–727

This validation study involves spares used on the 727 aircraft. It represents 150,000 flight hours, accruing over a 10-month period, with aircraft flying between 24 different stations. Actual initial provisioning for the range of spares in the study amounted to $1,940,000 and achieved 92.0% availability. Given the same data that was used in the actual provisioning, selections computed by the Lockheed model would have achieved 3.8% greater availability for the same cost. Alternatively, for the same level of availability, computations by the Lockheed provisioning model would have saved $755,000.

Actual Provisioning		*Model's Provisioning*		*Availability Improvement*	*Cost Savings @ Same*
Cost	*Avail.*	*Cost*	*Avail.*	*@ Same Cost*	*Availability*
$1,940,000	92.0%	$1,940,000	95.8%	3.8%	
$1,940,000	92.0%	$1,185,000	92.0%		$755,000

4.2 Validation–DC-8

The second validation study involves spares used on DC-8 aircraft, representing 17,700 flight hours over a 5-month period. The aircraft flew from 14 different stations. The cost of actual provisioning amounted to $637,000 and achieved 97.2% availability. The model would have achieved 1.8% higher availability for the same cost, or would have reduced cost by $222,000 for the same availability.

Actual Provisioning		*Model's Provisioning*		*Availability Improvement*	*Cost Savings @ Same*
Cost	*Avail.*	*Cost*	*Avail.*	*@ Same Cost*	*Availability*
$637,000	97.2%	$637,000	99.0%	1.8%	
$637,000	97.2%	$415,000	97.2%		$222,000

4.3 Validation–F-104G

This validation study involves spares on F-104G aircraft, representing 18,800 flight hours accruing over an 18-month period. The aircraft flew from Luke AFB in Arizona. The cost of provisioning for the range of spares in the study amounted to $412,000 and the availability achieved was 58.2%. Given the same data that was used in the actual provisioning, the model would have achieved 6.3% higher availability, for the same cost, or would have reduced cost by $269,000 for the same availability.

Actual Provisioning		*Model's Provisioning*		*Availability Improvement @ Same Cost*	*Cost Savings @ Same Availability*
Cost	*Avail.*	*Cost*	*Avail.*		
$412,000	58.2%	$412,000	64.5%	6.3%	
$412,000	58.2%	$143,000	58.2%		$269,000

5. CONCLUSION

With the growth in complexity and cost of modern weapon systems there has been significant escalation in the cost of spares investment. Thus a special need has been evolving for better provisioning techniques and more effective tools for the decision-maker to use in managing the provisioning process and in keeping pace with ever changing military plans.

We believe that application of a properly designed optimizing technique will provide important benefits in the initial selection and allocation of spare parts.

(i) *Uniform Application of Data*

The model I have described will eliminate variations that are inherent in repetitive manual processes, thus assuring consistent output results.

(ii) *Cost/Effectiveness*

The optimizing technique can be used for analyzing the relative cost-effectiveness of many alternative possibilities–providing quantitative data to assist the decision maker in selecting the optimum level of availability considering the cost of providing that availability. It can be used to perform a wide variety of trade-off studies.

(iii) *Time and Cost Savings*

Because the new provisioning technique is computerized, very rapid response can be assured. Great savings in manhours at provisioning conferences can be realized as there is no longer a need to manually calculate requirements for each line item. Changes in input data can be rapidly processed to reveal the impact on cost or availability.

The system is now in use by several major airlines in connection with Lockheed's L-1011 program, and, with U.S. Navy concurrence, it will be used in the new S-3A program.

DISCUSSION

Bureau: The standard inventory includes a small percentage of (F.M.) high consumption items and a big percentage of low consumption (S.M.) items. Some of the latter have a low value and others are of high value (so-called "security"): these absorb a large share of the investment.

1. What a criterion do you use to distinguish between F.M. and S.M. items?

2. Does the model provide a reply to the problem of "security" items?

3. If the reply is affirmative, what is the principle used?

Barnard:

1. Our model does not designate the category of items as to fast moving or slow moving. It does, however, establish an EXPECTED DEMAND for each item so there is a general relationship among all items in terms of expected usage.

2. The model does not treat "security" items differently than other items. The fact that security items have low demands and high unit costs does have a decisive impact on the quantity recommendations. The model does possess a criterion for determining when a quantity of zero should be recommended for an item; the criterion is a minimum demand threshold, such as one demand per three years, in which case all items having smaller expected demands would not be recommended for stockage.

Moreoever, there is another category of spare parts which we call "insurance" items—for which no realistic estimate of usage can be predicted. We prepare listings of such items but do not recommend the quantity that should be provisioned. Whether or not to procure such items is a matter of policy to be determined by the decision-maker.

3. The minimum demand criterion is the only feature in the model which is specifically directed to the problem of slow moving "security" items.

Benyon-Tinker: I note that your model assumes the Poisson distribution is a good description of spares demands. Even if the underlying failure mechanism of the piece parts gives rise to a negative exponential life distribution, it is not obvious that this will cause a Poisson distribution of spares demands, when additional factors such as the variability of operating hours are taken into account.

1. Have you validated use of Poisson; if not, have you carried out a sensitivity study to determine the effects of using different distributions? (I should mention that for some spares usage data that we have analyzed, Poisson was definitely not a good description).

2. How does your model cope with problems which arise when field usage data shows initial reliability assessment to have been wrong? How non-optimum does the solution become after the ensuing adjustment to stock levels?

Barnard:

1. We have not validated the standard Poisson as a measure of actual spares demand variability. However, our tests of the overall model upon actual case studies show that the standard Poisson, in combination with all of the other elements of the model, produces reasonable results. The only valid test of any model, being necessarily a simplification of the real-world, is the comparison of its predictions with experience. For a detailed discussion of this concept, see Milton Friedmann, "The Methodology of Positive Economics", *Essays in Positive Economics*, University of Chicago Press, 1953.

2. The model does not cope with data inaccuracy.

Young: Have you observed the results in the field of spares outfit as predicted by the model?

Barnard: No. We have thus far not had an opportunity to evaluate the results of an initial provisioning using this model. However, we are now doing initial provisioning for a number of commercial customers, and within about 2½ years we will have extensive information.

Monhemius

1. In your report you assume that demand per period follows a Poisson distribution.

   a. Are there technical, statistical or computational reasons for not using the more general compound Poisson distribution, as some others do?

   b. Do you think that the results, as you describe them would be very sensitive for the form of the distribution one uses?

2. This is a rather trivial remark, yet I like to make it. It strikes me each time again that the improvement in availability (or fill rate) does not sound impressive; it is trivial of course that *if* one uses here availability or non-availability as a criterion (and after all that describes your troubles better), *then* the improvement sounds much better: it would be a decrease of some 50%.

Barnard:

1. a. There are four reasons why we did not use the *compound* Poisson distribution:

   (i) It requires determination of a variance-to-mean ratio, thus increasing the complexity of computing quantitative provisioning requirements.

   (ii) There is no dependable basis for determining variance-to-mean ratio.

   (iii) Application of the *compound* Poisson distribution would tend to increase the quantities of parts (and costs) to achieve the same level of availability.

   (iv) Results from tests and validation studies that we and the airlines have performed, using the *simple* Poisson distribution, appear to be satisfactory.

b. We have not tested the sensitivity of overall results to various demand distributions. However, use of the *compound* Poisson distribution would have increased the number of parts required to achieve the stipulated AVAILABILITY levels–with corresponding increase in provisioning costs.

2. We agree with you that the results of our tests would appear more favourable if they had been expressed in term of non-availability improvement.

Geraerds:

1. I understand that in your terminology availability does not stand for availability as we know it but as fill rate?

2. You assumed that there exists a linear relation between demand and flying hours. Research in this field has shown that this cannot be found yet.

3. I understand that "initial" in your model does not stand for the problem of estimating or forecasting at the time that no historical data are available yet.

Barnard:

1. Yes, your understanding is correct.

2. Yes, the model does show a linear relationship between expected demand and flying hours. Expected demand is directly proportional to several factors, such as repair time and quantity per aircraft of an item. We realize that the linearity between flying hours and expected demand may depend upon the manner in which flying hours and expected demand may depend upon the manner in which flying hours change, just as with numerous other factors. However, we have not found a practical alternative to assuming linearity.

   The actual quantity recommended is definitely *not* linear with respect to flying hours. In the following example, note that doubling the flying hours doubles the expected demand and increases less than proportionately the recommended stockage quantity to provide 90% availability (effect of Poisson distribution).

Flying Hours	*Expected Demand*	*Recom. Qty. for 90% Availability*
50,000	20.0	27
100,000	40.0	49

3. "Initial" in the model means simply the first provisioning. It does not imply necessarily a new aircraft or new stocking locations, although these are typical conditions encountered in using this model. In the event the model is used for provisioning a new location for operating an old aircraft for which good historical data are available, then the "inputs" should be more factual and the "outputs" therefore should be superior.

Hennico:

1. Que pensez-vous de la prise en considération de la différence de coût entre l'approvisionnement initial et un ré-approvisionnement ultérieur en vue de déterminer le niveau de disponibilité désirable?

2. Lors des vérifications du modèle, avez-vous examiné ce qu'aurait coûté en argent et en disponibilité l'application du modèle?

Barnard:

1. The costs of initial supply and resupply are not comparable. The optimum initial supply covers a period of time which is selected on the basis of such factors as the time schedule for aircraft introduction, stability of design, quality of input data, etc. In the case of resupply, the optimum provisioning is attained by minimizing the sum of spares holding costs (e.g. insurance, cost of capital, risk of obsolescence) and spares reorder costs, considering, of course, the desired level of stock protection.

   Resupply, by definition, takes place after some operating experience has been gathered, and therefore faces a different set of circumstances than initial supply. The longer the timespan of initial supply, the greater is the risk of overprovisioning. The shorter the timespan of initial supply, the smaller is the amount of experience gathered upon which to base subsequent resupply decisions. The influence of initial supply on resupply diminishes as time goes on. However, the magnitude of initial supply costs bears no simple or fixed relationship to the cost of resupply but varies rather by the circumstances of support.

2. Yes, the objective is always to obtain the highest availability for a given cost, or, conversely, the lowest cost for a given availability.

Kay: The model does not seem to take into account the cost of stock-outs explicitly. As it optimizes on the basis of cost, which in itself is at least partially dependent on lead time, this surely is an important factor that militates against the optimization by ranking?

Barnard: Yes the model does not explicitly consider the cost of stock-outs. However, the model does show the least cost combinations of spares to attain any specified level of availability, i.e., protection against stock-outs. It is the decision-maker who must stipulate that desired level of availability.

We have provided opportunity in the model for the decision-maker to express judgement. He can do so by specifying desired availability for spares to support the overall weapon system, discrete subsystems, or even individual components. Moreover, he can specify different levels of desired availability at different stocking locations—so as to provide higher availability (fewer stock-outs) at remote locations than at locations nearby the main depot if he desires to do so.

Conrath: Under the optimization scheme, why not go up to the budget constraint, e.g. higher availability is likely to be had with a 5, 4, 2 breakdown of spares for your example? (Total expenditure $3,400).

Barnard: Yes, it is sometimes possible to obtain more availability by spending more of the unexpended budget. It must be recognized that doing this will produce a non-optimal result, i.e. a cost effectiveness which, if plotted on a chart showing the optimal cuve, would fall below the curve. However, in any practical problem, the optimization solution will come very close to exhausting the budget constraint. For instance, if the example had been to optimize a mix of 300 spares within a budget limit of \$350,000, and if an overall availability of 93.00 % were achieved at a cost of \$349,600, then it would be trivial to show that the overall availability could be improved to, say, 93.02 % by using part of the unexpended \$400.

Fischer: Could you indicate for some of your case studies the cost of operating your model (on which computer facility)?

Barnard: The cost of operating the provisioning model is approximately proportional to the computer run time, which in turn is proportional to the product of the number of different spares and the number of stocking locations. The optimization model consumes very roughly five times the computer run time as the other two models, other things being equal. For example, the Cost Category model required 75 minutes of IBM 360-65 run time for 6,300 spares at 3 stocking locations. In addition, the run time for operating the output printer is about 4 seconds per page, and typically 3 parts are printed per page.

Plumier: Dans le cas analysé, nous nous trouvons dans une optique d'approvisionnement initial.

Vous nous avez signalé que malgré cette situation d'approvisionnement initial le programme effectue des révisions des informations de départ lorsque nécessaire. Une variation d'information importante peut provenir d'un écart dans les délais de livraison prévus des fabricants de sous-ensembles. Ce même problème d'écart entre les délais de livraison prévus et les délais réels se pose, peut-être même avec plus d'intensité dans la phase suivante qui est celle du réapprovisionnement.

Pouvez-vous nous donner une idée de la technique que vous utilizez pour balancer les écarts de livraisons des fournisseurs?

Barnard: The model we have presented is strictly an initial provisioning model, and it does not take lead time in to account. The model recommends stock quantities, but it does not address the problem of *when* orders should be placed.

Rougerie: J'ai été heureux de voir réaliser un outil analytique à fonction coût-efficacité. Les exemples que vous avez donnés portent sur des pièces à coût unitaire allant de 1 à 2, ce qui rendrait efficace le choix de l'ordinateur $\frac{\Delta \text{ protection}}{\text{coût unitaire}}$.

Pouvez-vous m'indiquer si des coûts très différents, par exemple de 1 à 100, ne favorisent pas exagérémment les articles à coût unitaire faible? Si oui, comment avez-vous opéré pour éliminer dans votre modèle une surdotation en pièces de coût unitaire faible?

Barnard: Yes, we found that use of the optimizing technique for provisioning parts of widely different unit costs resulted in inordinately high quantities of the lower cost items. Therefore, we have incorporated a method for limiting the extent of provisioning for each part. For example, if an overall spares availability of 90% is desired, it would be reasonable to limit the availability of each spare to no more than 95% or 96%.

As stated in the presentation, it is not practical to use the OPTIMIZING MODEL over a wide range of spare parts. It is more suitable to use it in limited applications involving relatively small numbers of high value items. In such applications a further safeguard against overemphasis on the lower cost units can be established by programming in the optimizing technique to take place only for quantities above the EXPECTED DEMAND. In this manner, the quantity recommended for any items is never less than the EXPECTED DEMAND.

Buck: Is it possible in your model to give any weight to the essentiality or criticality of the part considered? Essentiality relates to the part contribution to mission performance.

Barnard: Yes the model can account for "essentiality" in two ways:

1. The decision-maker can specify the desired level of availability of spare parts, on the basis of essentiality, for the overall weapon system, for discrete sub-systems, or for individual components.

2. It also includes a "criticality" factor which can be used to establish "super stock" for selected items at the discretion of the decision-maker.

We have not yet discovered a meaningful or practical manner in which to apply the essentiality theory in the provisioning of spare parts–although the model can accommodate such an application.

Shepherd: In developing your model you have assumed that the design is frozen. This is an ideal situation which in practice is rarely met, as contractors have a habit of introducing or manipulating the introduction of modifications during the production phase. How does your model cope with the introduction of modifications and the risk of provisioning parts subject to quick obsolescence?

Barnard: In cases where design is relatively fluid and there exists a high degree of risk of obsolescence, it will be advantageous to compute initial provisioning for as short a period of time as is consistent with other support requirements. The model does not solve the problem of design changes. Protection against the impact of design changes must be negotiated with the supplier as a separate issue.

O'Connor:

1. In the validation of the model, did the data benefit from hindsight, or did you use the same data and/or assumptions that the original provisioners used?

2. To what extent does the model take account of the fact that MTBF and scheduled overhaul periods often increase during the life of our equipment? Do you consider that simulation is a better technique for initial provisioning of Hi-value components of this type?

Barnard:

1. It is, of course, impossible to reconstruct exactly the situation under which the

actual provisioning was accomplished. The anticipated removal rates, aircraft utlization rates, and other such data were used wherever documented. Undoubtedly there are places where the model operated upon superior as well as inferior information as compared to that available to the actual provisioners. For this reason the results of the validation studies provide only an indication, rather than a proof, of the model's accuracy. Even if it were possible, however, to recall perfectly the knowledge available to the original provisioners, the fundamental problem remains that is is impossible to separate errors stemming from data inaccuracy and those stemming from the model in any real-life test.

In summary, the tests conducted provided results that the airlines considered sufficiently persuasive for them to use the model in their initial provisioning of spare parts for the L-1011 program.

2. The model can not explicitly handle changing MTBF and scheduled overhaul periods. In those infrequent cases where we can reasonably estimate the growth of removal rates, the model can be used by approximating the growth with a "step-function", and operating the model against each step. If such an approximation is not acceptable, then resort to simulation may be necessary. However, for any sizable number of spares, simulation is a time-consuming exercise due to the large number of iterations required to obtain a reasonable degree of statistical stability. Thus, simulation should be used to the extent that analytical techniques, such as our model, are not adequate.

Investment in Repairables Spares Support for a Complex Weapon System—A Case Study

R.L. Baglow

This paper describes some studies which have been carried out while the operational research scientists have been working with the technical staff in the Canadian Forces. We have called this a case study, because the main object of the work was to arrive at a determination of spares support for a new weapon system. In the course of working with the staff, however, we have also examined the possibility of developing new approaches which on theoretical grounds are expected to be better.

The determination of spares support for a weapon system comes rather late in a program, just before the system is introduced into use. Not many options remain open at this stage. Typically, for example, the proportion of the overall budget for spares support will have been allotted. Commonly this is guessed to lie in the range of 25% to 30% of the cost of the capital equipment. This is a rough rule which has been used in the past in the Canadian Forces.

Also of course the equipment will have been selected so that its inherent reliability and maintainability have been more or less fixed.

The plan for deployment of the equipment has probably been determined as well, so that the problem of determining spares support can be viewed as one of optimizing over a limited range of resources. Nevertheless the sums of money involved are large enough to be interesting.

We come then to the provisioning conferences. For a major weapon system a considerable number of these will be required, say twenty or thirty, each devoted to a major or minor equipment or set of equipments. At least three groups of personnel participate in these conferences, each group with a distinct point of view. Technical personnel provide inputs concerning the reliability and maintainability of the equipments, supply personnel are interested in the procurement, distribution and stock levels of spares; and manufacturer's personnel, who provide much of the data on equipments and components, especially, of course, the cost. A question which immediately comes to mind is this: are these different groups making their decisions on the same basis? Are their judgments reasonably consistent with one another?

The first observation we wish to make is that at the provisioning conference the attempt is made to collect all the basic information *and* make the determination of the quantity of spares which will be bought. There is no question that to do this precludes considering a range of options and selecting a favourable one. The task of the provisioning conference is both tedious and onerous. In a relatively short time several thousand items have to be screened. In our experience about one in five of the total list will finally be selected. The items are considered one by one, and only relatively brief consideration can be given to each. Of course when the total cost is totalled up at the conclusion of the conference it is most unlikely that it will be matched by the funds which have been earmarked, and something will have to give–either the budget or the buy.

It is also very noticeable that a fairly small fraction–5% or 10% of the items–account for a fairly large fraction–80% or more–of the total cost. These are the expensive items- repairable assemblies, instruments, test equipment, tools, special handling equipment and the like. Yet the effort devoted to these items in the provisioning conference is by no means proportioned to the investment they entail.

Finally, a fundamental difficulty which faces the provisioners is that the conferences are spread out over a considerable period of time. Typically, therefore, determinations are made and contracts let on many sub-systems before the complete weapons system has been looked at. It would be extremely difficult, if not impossible, to delay raising contracts until the complete account had been drawn up. It can easily happen, and of course often does, that the initial allotment of moeny is spent before all the sub-systems of the program have been considered. One can easily imagine that some sub-systems get short-changed. This is decision making under difficulties, and our object is to find ways of rationalizing the process.

We have concentrated our attention on repairable assemblies because we have a good inventory model for these items developed by the RAND Corporation–the so-called METRIC model. It would certainly be desirable to have inventory models for

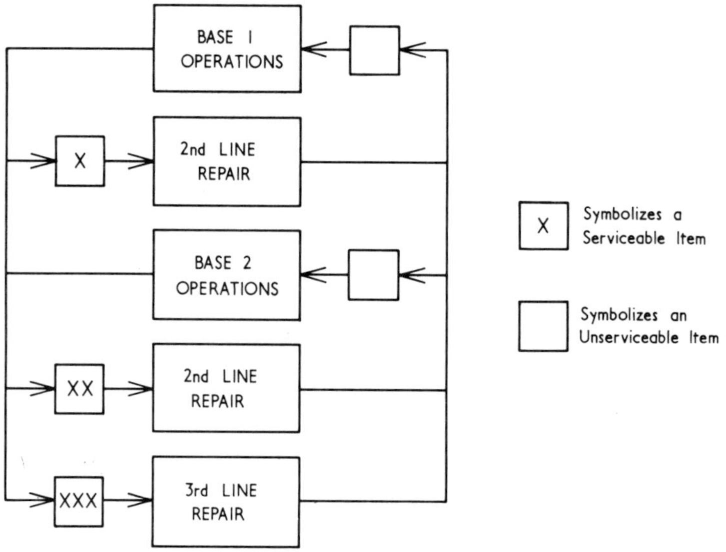

Figure 1 RAND inventory model (METRIC)

equipments like instruments, test gear, tools and special handling equipment. At present we have to rely on the judgment of technical officers for the determination of the requirement for such items. The basic idea we have is a very simply one. The RAND model enables us to make a prediction of the average number of backorders of a repairable item, or a set of repairable items, given a number of logistics factors which are fixed by management. As a useful approximation, which is sufficiently good for the rather low backorder rates we are considering, we can take this as a measure of the percentage of time a backorder condition exists. This enables us to translate the

THE AVERAGE NUMBER OF BACKORDERS OF A REPAIRABLE ITEM AT A RANDOM POINT IN TIME GIVEN THE FOLLOWING PARAMETERS:

MEAN FAILURE RATES (ASSUMED STATIONARY AND COMPOUND POISSON)

PROPORTION OF REPAIRS ACCOMPLISHED AT EACH OF TWO LEVELS

MEAN REPAIR TIMES

MEAN SHIPPING AND ORDERING TIMES

RELATIVE WEIGHT OF BACKORDERS FOR DIFFERENT ITEMS

UNIT PRICE

ACTIVITY FACTORS AT EACH POINT WHERE THE ITEMS ARE INSTALLED

VARIOUS CONSTRAINTS

Figure 2 RAND inventory model (METRIC) backorder rate

SYSTEM BACKORDER RATE: B

NUMBER OF INDIVIDUAL WEAPONS SYSTEM: N

REDUNDANCY OF EQUIPMENTS IN EACH WEAPONS SYSTEM = r

THEN THE PERCENTAGE OF TIME EQUIPMENT IS DOWN BECAUSE OF LACK OF SPARES IS

$$(B/N)^r$$

AVAILABILITY OF EQUIPMENT = $1 - (B/N)^r$

Figure 3 Availability of equipment

measure to the percentage of time a weapon is "up" *insofar as it depends on the supply of repairable assemblies*. In this paper I will call this the "availability" of the system. If this use of the term is non-standard I apologize.

The values of the parameters have been supplied by the logistics staff. In some cases it turned out they had not explicitly considered some of these parameters in arriving at a determination of spares support at the provisioning conference, and a good deal of "digging" was necessary.

Among these was the parameter describing the "failure rate" of the repairable assembly. In new equipments, particularly, this is a difficult parameter to estimate, when there is no experience to go on. For electronic equipments a lot of data is available for components, but the exact calculation of the failure rate of complex assemblies is not usually possible. For mechanical equipments the technical personnel argue from experience with like equipments. Their estimates would certainly be improved if a systematic effort were made to record initial estimates and subsequently compare them with experience. When such comparisons have been made it has been found that the predictions are extremely pessimistic, typically the predicted failure rates are too high by *factors* of 2 or 3. [Refs. 1 and 2], though the correlation between predicted and observed failure rates is high. That the technical people tend to err on the side of over-estimating failure rates is made abundantly clear as they explain the basis of their estimates. The *relative* failure rates of assemblies are known much more closely than their absolute values. Thus these estimates can be used to *rank* inventory policies with confidence. If one is interested in attaching quantitative significance to the measures, however, it is necessary to have estimates of the absolute values of the failure rates, because the measure we are using (availability of weapons systems) is a quite non-linear function of the failure rates. We have followed the practice of asking the technical personnel to make their best estimates of the failure rates, and we have then halved them. The resulting estimates are probably still pessimistic, nevertheless, they permit appreciably sharper comparisons of our selected measure of utility.

An example of the kind of data we are examining is shown in Figure 4. The RAND

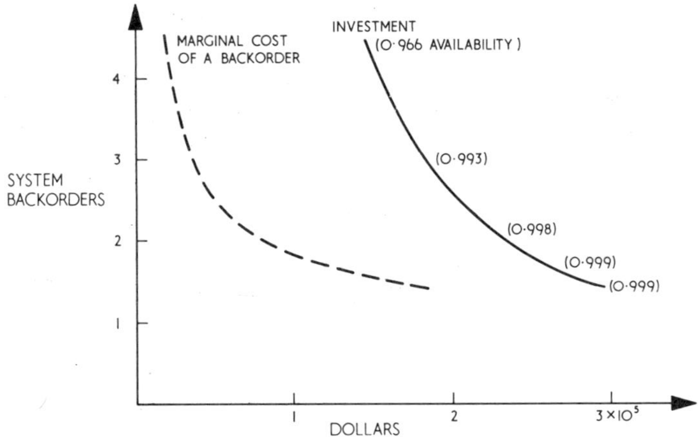

Figure 4 Main generator

algorithm has been used to allocate a fixed total investment among a set of repairable spares for a particular sub-system–in this case the main electrical generating sets–in such a way as to minimize the average number of backorders for the system. This can be done in a straightforward manner because this utility measure is separable. By a well known result–which has been referred to as the fundamental theorem of operational research [Ref. 3] –this optimal allocation is achieved when the ratios of the unit cost of each item to the change in the backorder rate resulting from adding one of this item to the stock of spares are all approximately equal. This function–the Lagrange multiplier–is also shown–it is simply the graph of the tangent to the first curve. Of course, the curve of optimal investment is not smooth because the number of items is necessarily integral. However to a good approximation we have such a representation. The calculation shows that you cannot buy your way into as high a level of service as you please simply by investing more and more money in spares. There is a limit where diminishing returns set in so sharply that it is not profitable to go further. This limit is most easily perceived by examining the marginal cost of a backorder–the Lagrange multiplier curve. The great utility of the RAND model is that it enables one to approach a natural limit on the allocation of money to spares. The closeness to which we can estimate this limit depends on how well we can correct the bias in the estimates of failure rates. To the extent that failure rates are over-estimated we will tend to over-invest in spares. Hence it is really important to try to get unbiassed estimates of failure rates. The model gives markedly better support than the methods previously in use, as shown by point A on the diagram. As we have explained before we can be confident that the result is better, regardless, of whether the failure rates are biassed or not. We also show on the diagram the "availability" of the equipment–the percentage of time the equipment is "up" *insofar as it depends on the availability of spares*. In the case shown the availability is relatively high because of the triple redundancy in the power generating equipment–a design which is justified by the importance of electrical power to practically everything aboard a ship.

Similar results are shown (Figure 5) for another important sub-system, in this case the central computer for action information. We see here that the investment is much

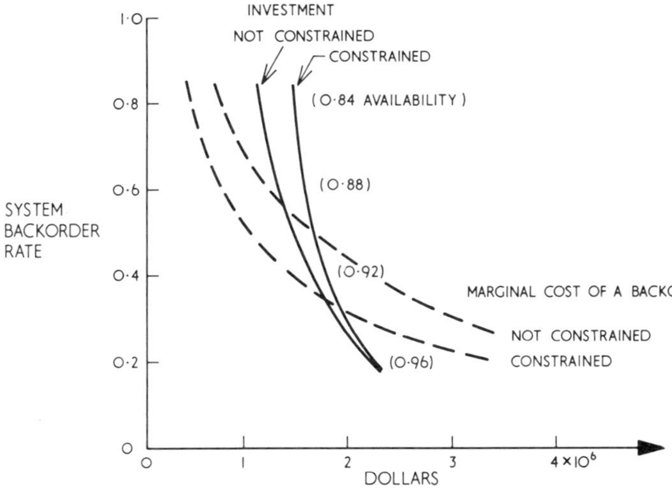

Figure 5 Computer system

greater than in the previous system, reflecting a more complex equipment. The availability is lower, because of no redundancy. The diagram shows two investment curves. This came about because at the time the calculations were made certain commitments had already been made with respect to a few items for which advance orders had been placed. The model was constrained to accept these items. The result of the "constrained" calculation can be compared with an "unconstrained" calculation, and it can be seen that the effect of the constraint was a small increase in the total cost of support, of the order of 5%. In this case the methods previously in use gave a result which was much poorer, so much so that it could not be plotted on the slide.

So far, then, our findings suggest that different provisioning conferences, that is, different groups of technical and supply personnel, are not likely to make investment decisions which are consistent, from the point of view of the overall weapon system program. Indeed we have noted individual cases where the technical experts were clearly carried away into some economic never-never land.

We now wish to consider whether the RAND model can be used to tie together, in some sense, decisions made at separate provisoning conferences for separate sub-systems of the complete weapon system. Our first naive idea was to suppose that we could extend the fundamental theorem to tie investment decisions in two separate sub-systems. For example, that we should allocate investment between the main generator sub-system and the computer sub-system with the same value of the Lagrange multiplier. Now the result obtained this way is simply wrong.

This can easily be seen by an example. The following table gives the results of an allocation of investment arrived at in this way, and compares it with a different allocation.

ALLOCATION OF INVESTMENT

SYSTEM	*INVESTMENT*	*LAGRANGE MULTIPLIER*	*AVAILABILITY*
Main Generator	.463	\$774,000	.999
Computer	1.592	\$774,000	.894
Combined	2.055 x 10^6		.999 x 894 = .893
Main Generator	.182	\$ 36,400	.992
Computer	1.875	\$1865,000	.936
	2.057		.992 x .936 =.928

The second allocation is clearly better, in fact in this simple case approximately optimum. Clearly this arises because the measure of utility for these two systems are not additive but multiplicative. A necessary condition for the computer to be operative is that power be available from the generator system.

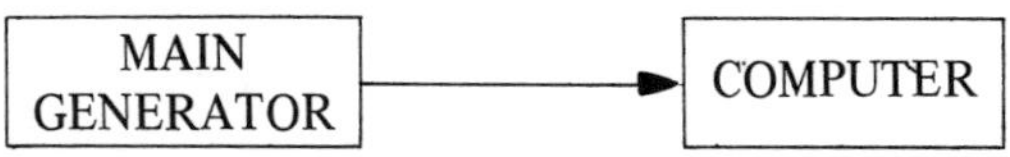

In short the measure of utility is not separable. The example is rather trivial. To extend the idea it would be necessary to arrange the principal sub-systems of the complete ship in a hierarchy of elements connected in parallel and in series, and attempt to find a measure of utility for the entire structure which could be related to the investment in spares in each of the sub-systems. The problem then would be to find an allocation of a fixed budget which to all the various sub-systems which would maximize the measure of utility. Even if we could overcome the conceptual problems it is possible that the computational difficulties would be considerable. However as the

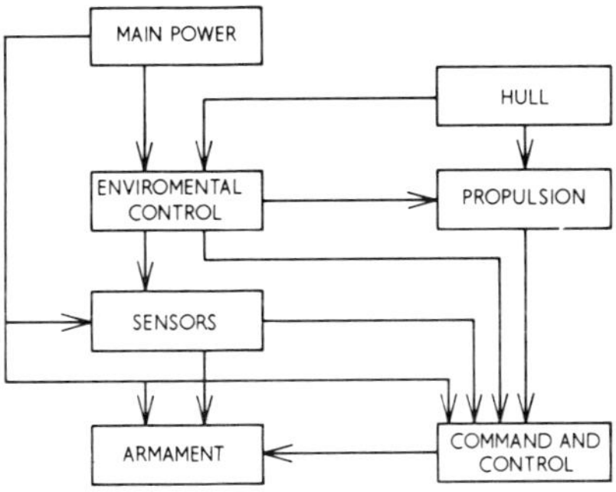

Figure 6 Hierarchy of sub-systems

provisioning of the current program proceeds we hope to acquire the data to study such an approach.

This means, however, that we have not overcome the difficulty that the series of provisioning conferences dealing with various sub-systems of the program are stretched out over a considerable period of time. One of our principal objectives is to find means of controlling the allocation of funds to provisioning various sub-systems of a weapons program in some way reasonably consistent with maximizing the effectiveness of the overall system, while still allowing the decisions to be made in sequence. At present we see no easy way to do this.

We offer the following suggestions which, although they do not lead to the optimal solution, would lead to some significant improvements.

The first is to use a suitable inventory model such as the RAND, in order to make the best determination of the quantities required of different repairable items, and in order to at least try not to invest beyond the point where diminishing returns set in sharply. For this purpose it is extremely desirable that initial estimates of failure rates be not too severely biassed on the high side.

Secondly, it would be desirable to take account of the hierarchy of sub-systems in the overall program and to *carry out the provisioning in this hierarchical order* beginning with the most fundamental systems. For a ship these might be hull, propulsion, and electrical power. Perhaps next would be sensors, with command and control and armament last of all. As the systems lower in the hierarchy are provisioned it might be possible to take into account their dependency on the higher systems and to a degree sub-optimize the investments.

Thirdly, a serious and systematic effort should be made to document thoroughly the initial determination, and to follow up with careful comparison of the predicted with the actual after the system has become operational. This may seem elementary but we don't do it. Only in this way can we learn to make better predictions.

Finally, we believe it would be worth while attempting preliminary calculations of spares requirements earlier in the program, even before the design is frozen. Many uncertainties would exist, and data would be rather imprecise. Nevertheless, we think estimates could be made, which while not very precise, would be useful in improving the allocation of funds to the different systems in advance of the detailed provisioning calculations.

REFERENCES

1. Feeney, G.J., *A System Approach to Base Stockage of Recoverable Items.* RM-4720-PR, December 1965, RAND Corporation.

2. Baglow, R.L., Ewashko, T.E., and Richard, E., *The RAND Metric Model–A Case Study of Its Application to Provisioning the Hercules Aircraft in the Canadian Forces.* DRAE Memorandum M11, October 1969.

3. Danskin, J.M., *The Theory of Max-Min*; *Application to Weapons Allocation Problems*. Springer-Verlag, 1967.

DISCUSSION

Volten: When the engineers knew that you were going to halve their (failure) estimates, would you think they would start to double their predictions? This might be the real underlying difficulty in trying to get better predictions by post-mortem analysis.

Baglow: I think not. In the provisioning process we did not attach much significance to the absolute values of the objective function. Rather we used the function to *rank* provisioning policies, and for this they are much more reliable.

Monhemius: I think Mr. Baglow treated one of the most important and critical points in this field of stock control when we discussed the human capability to estimate demand.

In your report you describe how you halved the estimates made by the technical personnel.

Did you consider having reduction factors per estimator or per group of products?

I ask this question because I have seen this done in industry by a producer of electronic valves who kept records of reduction factors for each one of his customers (set makers) individually, updating those factors with the aid of single exponential smoothing.

Baglow: This would be difficult to do in our organization because military personnel are posted so frequently.

Oliver: We have been recommended to provision spares in order of sub-systems importance. However taking the example of aircraft structure is defined before avionics which are more important. Do you investigate any solution in this area?

Baglow: Clearly, many administrative constraints may prevent us from doing what we wish to do. Still we should try. In any case, is it true that avionics are more important than the aircraft structure? It seems to me that the structure is more fundamental than the weapons system which is attached to it.

Buck: I should like to mention that Mr. Baglow was provided with certain information concerning a method of analysis developed by General Electric under the acronym NERAM, that is "Network Reliability Assessment Model".

Høgberg: You have most correctly pointed at the importance of a data structure. I suggest that whether you use Top down or Bottom-up methods in systems development, the problem-oriented data structure must be defined at a very early stage of development.

Do you have any advice on arriving at a good initial setting of the Lagrange multiplier?

Baglow: The form of the algorithm we are now using does not require an initial setting of the Lagrange multiplier.

1. In the Evaluation calculation the LM is meaningless.

2. In the Optimization calculation we may set an investment constraint *or* a constraint on the value of the objective function.

In either case the algorithm computes the LM which characterizes the contrained solution.

On Minimizing Non Separable Functions Defined on the Integers, with an Inventory Application*

Bruce L. Miller

1. INTRODUCTION

This paper was motivated by the study of the reparable inventory system discussed in Section 4. Sections 2 and 3 contain results which are all used in Section 4. The results of these two sections are presented in the general formulation described in the following paragraph since it is felt that they could be useful in other problems where the object is to minimize nonseparable funtions defined on the integers.

The general formulation of the problem we consider is:

Find the minimum of $f : X \to R$, where (1)

X is a discrete rectangle, that is to say, $X = \{x : x \in I^n,$

$a_i \leq x_i \leq b_i, i = 1, ..., n\}$,

I is the set of integers $\{..., -2, -1, 0, +1, ...\}$,

a_i, b_i are integers of infinite, and

R is the real line $(-\infty, +\infty)$.

Naturally, some conditions will have to be placed on the function f. The main contribution here is that we are able to treat cases where f is nonseparable.

This problem is of interest in its own right. It can also be important in the problem of maximizing a function f of n variables over a finite discrete set $X = \{\xi_1, \ldots, \xi_N\}$ subject to the nonlinear constraints $g_i(\xi) < b_i$, $i = 1, \ldots, m$. This problem is considered by Brooks and Geoggrion [1], who come up with a Lagrangean approach that requires the solution of the subproblem [1, Eq. (5)],

$$\text{Maximize}_{\xi \in X}, f(\xi) - \sum_{i=1}^{m} \mu_i g_i(\xi) \tag{2}$$

The subproblem (2) is an example of (1) (after the trivial change from max to min) if the set X is a discrete rectangle.

The study of Problem (1) dichotomizes as follows. First, we want to know for which functions f a local optimum is a global optimum; this is the topic of Sec. II. Second, we

* This research was supported by the United States Air Force under Project RAND Contract No.F44620-67-0-0045.

want a solution strategy for the hopeful case where a local optimum is a global optimum; this is the topic of Sec. III.

Perhaps the most important part of the paper, even with respect to the general formulation, is the inventory application mentioned in the first paragraph because it demonstrates the possibility of solving problems of many hundreds of variables. In the application we seek to minimize a nonseparable Lagrangean function arising from an inventory model subject to the nonnegativity of the variables. We prove that for this function a local optimum is a global optimum. Subsequently, we describe an algorithm that finds the optimal solution. The author programmed this algorithm in SIMSCRIPT II on the IBM 360-65. The average computation time (of five runs) as a function of the number of variables was:

Number of variables	25	50	100	500
Average time in seconds	5.5	14	27	170

2. DISCRETE CONVEXITY

In the nondiscrete case we say that a set Y is convex if $y^1\ y^2 \in Y$ implies $ay^1 + (1-a)y^2 \in Y$, where $0 \leq a \leq 1$. A function $f: Y \to R$ is convex if Y is convex, and for $y^1, y^2 \in Y$,

$$f(\alpha y^1 + (1-\alpha)y^2) \leq \alpha f(y^1) + (1-\alpha)\ f(y^2),$$

where $0 \leq \alpha \leq 1$. In the nondiscrete case, convexity is a useful property of functions in the study of minimizing a function of n variables, because it implies that a local optimum is a global optimum. In the case of differentiable functions, we have the nice result that a point y° in the interior of Y minimizes f if and only if the n partial derivatives of f at y° are zero.

In the discrete problem (1) the analog of the n partial derivatives of f being zero is

$$f(x_1, \ldots, x_i, \ldots, x_n) \leq f(x_1, \ldots, x_i + s_i, \ldots, x_n) \tag{3}$$

for $s_i = +1, -1$, and $i = 1, \ldots, n$. Unfortunately, in the case where f is nonseparable, (3) is too weak to imply optimality in cases of any generality. Instead we shall look for a definition of "discrete" convexity for f which says that if f is discretely convex and $f(x^\circ) \leq f(x)$ for all x such that $\|x-x^\circ\| = 1$, then x° is optimal. Here we are using the max norm, $\|x\| = \max_i |x_i|$.

We begin with a definition.

Definition. The *discrete neighborhood* of a point $y \in R^n$ is the set $N(y) = \{x : x \in I^n, \|x-y\| < 1\}$.

Note that if y itself is in I^n, then $N(y) = y$. The set $N(y)$ has 2^m elements, where m is the number of coordinates of y that are not integers.

Lemma 1. Suppose $X \subset I^n$ has the property that if $x^1, x^2 \in X$, then $N(\alpha x^1 + (1-\alpha)x^2)$

$\subset X$ for any α, $0 \leq \alpha \leq 1$. Then $x^1, x^2 \in X$ implies.

$V = \{v : v \in I^n, \min(x_i^1, x_i^2) \leq v_i \leq \max(x_i^1, x_i^2), i = 1, ..., n\} \subset X.$

Proof: Let $v \subset V$ and $\|x^1 - v\| = k$. For any sufficiently small $\alpha > 0$,

$$v^1 = x^1 + (\delta_1, ..., \delta n) \in N((1-\alpha)x^1 + \alpha x^2),$$

where $\delta_i = 0$ if $v_i = x_i^1$, $\delta_i = 1$ if $v_i > x_i^1$, and $\delta_i = -1$ if $v_i < x_i^1$. Here we use the property of V that says $v_i > x_i^1$ implies $x_i^2 \geq v_i$, and $v_i < x_i^1$ implies $x_i^2 \leq v_i$. The

point $v^1 \in X$ since it is in the neighborhood of a weighted average of two points in X. Also, $\|v^1 - v\| = k - 1$.

We reapply the above method $k - 1$ times with the successive v^i replacing x^1 to obtain $v^k \in X$ and $\|v^k - v\| = 0$, which implies $v^k = v$. Since v is arbitrary, the proof is complete.

Theorem 1.
The set $X \subset I^n$ has the property that $x^1, x^2 \in X$ implies $N(\alpha x^1 + (1-\alpha)x^2) \subset X$ for any α, $0 \leq \alpha \leq 1$, if and only if X is a discrete rectangle.

Proof: Let $X = \{x : a_i \leq x_i \leq b_i\}$ be a discrete rectangle. If $x^1, x^2 \in X$, then $a_i \leq (\alpha x_i^1 + (1-\alpha)x_i^2) \leq b_i$ for $i = 1, ..., n$, and $0 \leq \alpha \leq 1$. This implies $N(\alpha x^1 + (1-\alpha)x^2) \subset X$ and completes the first part of the proof.

Suppose that $X \subset I^n$ has the property that $x^1, x^2 \in X$ implies $N(\alpha x^1 + (1-\alpha)x^2) \subset X$ for any α, $0 \leq \alpha \leq 1$. For $i = 1, ..., n$, let $a_i = \inf x_i$, $b_i = \sup x_i, x \in X$. Clearly, $X \subset V = \{v : v \in I^n, a_i \leq v_i \leq b_i, i = 1, ..., n\}$. To show $V \subset X$ we enumerate the countable set X as $\{x^1, x^2, ...\}$. By applying Lemma 1 repeatedly we get at stage n, $V^n = \{v : v \in I^n, a_i^n \leq v_i \leq b_i^n, i = 1, ..., n\} \subset X$, where $a_i^n = \min\{x_i^1, ..., x_i^n\}$, $b_i = \max\{x_i^1, ..., x_i^n\}$. Since $V^n \to V$, $V \subset X$, which completes the proof.

Definition. The function $f : X \to R$ is *discretely convex* if X is a discrete rectangle and given $x^1, x^2 \in X$, $0 \leq \alpha \leq 1$,

$$\min_{X \in N(\alpha x^1 + (1-\alpha)x^2)} f(x) \leq \alpha f(x^1) + (1-\alpha) f(x^2).$$

Theorem 2.
Let the function $f : X \to R$ be discretely convex and $x^\circ \in X$. If $f(x^\circ) \leq f(x)$ for all $x \in X$ such that $\|x^\circ - x\| = 1$, then $f(x^\circ) \leq f(x)$ for all $x \in X$, i.e., x° is optimal.

Proof. Let $x^1 \neq x^\circ \in X$. For any sufficiently small $\alpha > 0$, $\|((1-\alpha)x^\circ + ax^1) - x^\circ\| < 1$. Each point $x \in N((1-a)x^\circ + ax^1)$ must satisfy $\|x - x^\circ\| < 2$ by the triangle inequality and must lie in X by Theorem 1. This result and our hypothesis imply that

$$\min_{x \in N((1-\alpha)x^\circ + \alpha x^1)} f(x) \geq f(x^\circ).$$

Since f is discretely convex,

$$\min_{x \in N((1-\alpha)x^\circ + \alpha x^1)} f(x) \leq (1-\alpha) f(x) + \alpha f(x^1) .$$

These two inequalities imply $f(x^1) \geq f(x^\circ)$, which completes the proof since x^1 is arbitrary. I thank the referee for indicating that Theorem 2 and later Theorem 3 can be strengthened by requiring that the set X only satisfy the property that if $x^1, x^2 \in X$ and $0 \leq \alpha \leq 1$, then $N(\alpha x^1 + (1-\alpha)x^2) \cap X$ is not empty (as opposed to $N(\alpha x^1 + (1-\alpha)x^2) \subset X$). This weaker requirement on X also gives us results analogous to the convex case such as that X satisfies the weaker requirement if and only if the function f defined by

$$f(x) = 0 \quad \text{if } x \in X$$

$$f(x) = 1 \quad \text{if } x \notin X, x \in I^n$$

is discretely convex. However, from the computational aspect it makes a great difference that X satisfy the stronger definition and be a discrete rectangle. For instance, Theorem 4 requires that X be a discrete rectangle.

The referee also suggested the following example which shows the restriction of a convex function $f: Y \to R$ to a discrete rectangle $X \subset Y$ does not necessarily yeild a discretely convex function. This example is simpler than the author's original one.

Example. Let $f(y_1, y_2) = (2y_1 - y_2)^2 : Y = R^2 \to R$ and $X = I^2$.

The function f is convex since the matrix of second partial derives is

$$\begin{bmatrix} 8 & 4 \\ -4 & 2 \end{bmatrix}$$

for all $y \in Y$. Consider the following points in X:

$f=2$, (1,0) .	. (1,1) ($f=1$)	. (1,2), $f=0$
$f=0$, (0,0) .	. (0,1) ($f=1$)	

For $y = (1/2, 1), N(y) = \{(1,1), (0,1)\}$ yet

$$\min_{x \in N(y)} f(x) = 1 > 1/2\, f(0,0) + 1/2\, f(1,2) = 0$$

The difficulty this example reveals is that the value of f at a point y may have little relation to the value of f at the points $x \in N(y)$. Although $f(1,½) = 2.01$ less than $f(0,0)$, $f(1,0)$ and $f(1,1)$ are greater than $f(0,0)$. This lead us to

Definition. The function $f: Y \to R$ is *weighted* if Y is the convex hull of a discrete rectangle X and for $y \in Y$,

$$f(y) = \sum_{x \in N(y)} w_x(y)\; f(x),$$

where the coefficients $w_x(y), x \in N(y)$, are weights.

The *convex hull* of any set E is the intersection of all convex sets containing E and is itself convex. The convex hull Y of a discrete rectangle X is of the form $Y = \{ y : y \in R^n, a_i \leq y_i \leq b_i, i = 1, \ldots, n\}$.
The first part of the proof of Theorem 1 can easily be changed to show that if $y \in Y$, $N(y) \subset Y$. Thus, all the terms in the previous sum are defined.

The term *weights* applied to a set of numbers $w_1, \ldots, w_n$ means they are nonnegative and sum to 1.

THEOREM 3.

If a weighted function $f: Y \to R$ is convex, its restriction to the discrete rectangle $X = Y \cap I^n$ is discretely convex.

Proof: Since $f: Y \to R$ is convex, for arbitrary $x^1, x^2 \in X \subset Y$, and $0 \leq \alpha \leq 1$,

$$f(\alpha x^1 + (1-\alpha)x^2) < af(x^1) + (1-\alpha) f(x^2).$$

Since f is a weighted function,

$$f(\alpha x^1 + (1-\alpha)x^2) = \sum_{x \in N(\alpha x^1 + (1-\alpha)x^2)} w_x(\alpha x^1 + (1-\alpha)x^2) f(x) \geq$$

$$\min_{x \in N(\alpha x^1 + (1-\alpha)x^2)} f(x).$$

These two inequalities complete the proof.

Corollary 1. Consider the function $f: X \to R$, where X is a discrete rectangle, and let Y be the convex hull of X. Suppose that there exist a constant K and weights $w_x(y)$, $X \in N(y)$, for every $y \in Y$ such that given any $\epsilon > 0$, there is a convex function $f_\epsilon : Y \to R$ satisfying

$$\sup_{y \in Y} \left| f_\epsilon(y) - \sum_{x \in N(y)} w_x(y) f(x) \right| < K\epsilon .$$

Then f is discretely convex.

Proof: Consider the function $F(y) = \sum_{x \in N(y)} w_x(y) f(x) : Y \to R$. F is convex, since for any positive sequence ϵ_i, $\epsilon_i \to 0$, $f_{\epsilon_i} \to F$ (uniformly) and the f_{ϵ_i} are convex. For $x \in X$, $N(x) = x$, so that $w_x(x) = 1$ and $F(x) = f(x)$. Therefore $F(y) = \sum_{x \in N(y)} w_x(y) F(x)$ and F is a weighted function. By Theorem 3 its restriction to X, f, is discretely convex, which completes the proof.

I thank Robin Brooks for suggesting this line of proof to me.

An example of a nonseparable discretely convex function is given in Sec. IV. There we prove (Theorem 5) that

$$g(x) = \sum_{j=0}^{m} \left(1 - \prod_{i=1}^{n} F_i(x_i + j)\right) + \sum_{i=1}^{n} \sum_{j=m+1}^{\infty} (1 - F_i(x_i + j))$$

$$+ \lambda \sum_{i=1}^{n} C_i x_i \; : \; X = \{ x : x \in I^n, x_i \geq 0, i = 1, \ldots, n \} \to R,$$

where $F_i(\cdot)$ is the cumulative distribution function of a Poisson random variable with mean λ_i, is discretely convex using Corollary 1.

3. A SOLUTION STRATEGY

This section outlines a general computational approach to Problem (1) when $f : X \to R$ is a discretely convex function. If X is infinite we also assume that, given any $x^\circ \in X$, there is a constant $M(x^\circ)$ such that if $\|x\| > M(x^0)$ then $f(x) > f(x^0)$. This insures that there can only be a finite number of improvements from any point in X.

The term "solution strategy" is borrowed from Geoffrion [2]; it refers to the process of reducing a problem into a sequence of simpler optimization problems. This contrasts with the concept of an algorithm, where all operations are specified. Of the four (2,3,5,6) principal steps in the solution strategy, two (2,5) are specified in detail. The other two steps will vary according to the particular form of f. The application described in Sec. IV provides an example of fully implemented versions of steps 3 and 6. Along the same line of thinking, we concede that this "general" approach has been greatly influenced by experience with the application described in Sec. IV.

Before describing the solution strategy, we restate some results of Section 2 and develop some new ones.

Definition. A point $x \in X$ is *stationary* if

$$f(x_1, ..., x_i, ..., x_n) \leq f(x_1, ..., x_i + s_i, ..., x_n)$$

for s_i = +1,-1 and $i = 1, \ldots, n$. Of course, the above inequality is only considered when $x + s_i\epsilon_i \in X$, where ϵ_i is the vector with a 1 in the ith position and zeros elsewhere.

Definition. A set of *potentially improving displacements* of a vector x, $S(x)$, is any subset of $\{ s : x + s \in X, \|s\| < 1 \}$ that satisfies the condition that if x is not optimal, then there is an $s \in S(x)$ such that $f(x + s) < f(x)$.

Theorem 2 says that $\{s : x + s \in X \; \|s\| < 1\}$ is a set of potentially improving displacements of x if f is discretely convex. Also, $S(x) = \{0\}$, the zero vector, implies x is optimal vacuously. Because X is a discrete rectangle, $\{ s : x + s \in X, \|s\| < 1 \}$ can be written $\underset{i=1}{\overset{n}{X}} M^i$, where $\{0\} \subset M^i \subset \{-1, 0, +1\}$. We have $M^i = \{-1, 0, +1\}$ when $a_i < x_i < b_i$. Throughout the solution strategy $S(x)$ will be of the form $S(x) = \underset{i=1}{\overset{n}{X}} M^i$.

Toward reducing the set $S(x)$ we will use the following observation, which is stated as a theorem.

THEOREM 4.

Let $S(x)$ be of the form $S = \underset{i=1}{\overset{n}{X}} M^i$, where $\{0\} \subset M^i \subset \{-1, 0, +1\}$. Also suppose that for some $w \in M^j \sim \{0\}$,

$$\min_{\{s : s_j = 0, s \in S(x)\}} (f(x + s + w\epsilon_j) - f(x + s)) \geq 0. \tag{4}$$

Then $S'(x)$ is also a set of improving displacements of x, where $S'(x) = \underset{i\ell}{\overset{n}{X}} M'^i$,

$$M'^i = M^i, \; i \neq j, \; M'^j = M^j \sim \{w\}.$$

Proof: It suffices to show that the minimum of $f(x + s)$ for $s \in S(x)$ is attained in $S'(x)$. Let s^* minimize $f(x + s)$ for $\{ s: s_j = w, s \in S(x)\}$. By (4), $f(x + s^*) \geq f(x + s^* - w\epsilon_j)$. Since $\{0\} \in M'^j$, $s^* - w\epsilon_j \in S'(x)$. This completes the proof, since $S(x) = S'(x) \cup \{ s : s_j = w, s \in S(x)\}$.

When the conditions of Theorem 4 are met, we reduce the number of elements in $S(x)$ by a third if M^j had three elements and by a half if M^j had two elements.

As a corollary to Theorem 4, we have the well-known result that for a separable discretely convex function f, stationarity implies optimality. The argument is that stationarity implies (4) is nonnegative for $s = 0$, the zero vector, and separability implies (4) is independent of the s which satisfy $s_j = 0$, $s \in S(x)$. Thus $S(x)$ can be immediately reduced to $\{0\}$, which implies that x is optimal.

The ease with which (4) can be verified depends almost exclusively on the mixed second differences $\Delta_{kj}f(x+s), k \neq j$, defined by $(x+s) = (f(x+s+ek+\epsilon_j) - f(x+s+\epsilon_k)) - (f(x+s+\epsilon_j) - f(x+s))$. For any coordinate k, if $\Delta kjf\ (x+s)$ is ≥ 0 for all $s \in S\ (x)$, then the coordinate s_k^* of the minimizing vector s^* in (4) is the value of $s_k^* \subset M^k$ which minimizes (maximizes) $s_k^* w$. To see this we fix w in (4) at, say, -1 and $\Delta_{kj}f(x+s)$ as ≥ 0. The condition $\Delta_{kj}f(x+s) \geq 0$ for all $s \in S(x)$ implies that $(f(x+s+\epsilon_k) - f(x+s+\epsilon_k-\epsilon_j))$ is greater than $(f(x+s) - f(x+s-\epsilon_j))$ for all $s - \epsilon_j \in S(x)$. Since we want to find the s^* that minimizes $(f(x+s-\epsilon_j) - f(x+s))$, we want s_k^* to be equal to the largest element in M^k, i.e., we want s_k^* to minimize $s_k w$.

We now describe the solution strategy. In the description, x and $S(x)$ refer to the *current* value of the solution vector and set of potentially improving points, respectively.

Step 1. Initialization. Begin with a pount $x \in X$.

Step 2. Finding a Stationary Vector. For each index i,

$1 < i < n$, find the value of x_i given $(x_1, \ldots, x_{i-1}, x_{i+1}, \ldots, x_n)$ such that either $\Delta_i f(x - \epsilon_i) \leq 0$ and $\Delta_i f(x_i) > 0$, or $x_i = a_i$ and $\Delta_i f(x) > 0$, or $\Delta_i f(x - \epsilon_j) \leq 0$ and $x_i = b_i$, where $\Delta_i f(x) = f(x + \epsilon_i)$. If the current value of x_i satisfies these conditions, then do not change it.

If x does not change value during these n iterations, go to Step 3 as x is stationary. If not, repeat the preceding paragaph.

Comment. Step 2 consists of the simple procedure of optimizing one variable at a time. By our assumption that there can be only a finite number of improvements, Step 2 is repeated only a finite number of times, since a change in vector x is necessarily accompanied by a strict decrease in the value of the objective function.

Step 3. A Heuristic Approach to Obtain an Improvement. Based on the characteristics of f, try different vectors of the form $x' = x + s$, where $s \in \{s : x + s \in X, \|s\| < 1\}$. If we find an x' such that $f(x') < f(x)$, we set $x = x'$ and return to Step 2. If no such x' is found, go to Step 4.

Eventually (again because of our assumed finite number of improvements), we come to Step 4, where we begin our efforts to prove that x is optimal using Theorem 2. If Step 3 were not heuristic, then x would be optimal when no improvement is obtained.

Step 4. A Second Initialization. Let $S(x)$, the set of potentially improving displacements of a point x, be $\{s : x + s \in X, \|s\| < 1 = \overset{n}{\underset{i=1}{X}} M^i, \{0\} \subset M^i \subset \{-1, 0, +1\}$

Step 5. Reducing S(x) via Theorem 4. For $i = 1, \ldots, n$, and $w \in M^i \sim \{0\}$, check the hypothesis of Theorem 4. Each time it is satisfied, the set $S(x)$ is reduced by taking w out of M^i.

If $S(x) = \{0\}$, the zero vector, then x is optimal by Theorem 2. If not, we check to see if $S(x)$ was reduced. If it was, we repeat the preceding paragraph since the smaller the set $S(x)$, the easier it is to satisfy Eq. (4). If $S(x)$ was not reduced we go to Step 6.

Step 6. Eliminating a Specific $w \in M^j \sim 0$ *by Contradiction.* In this step we pick a $w \in M^j \sim \{0\}$ according to some heuristic rule and try to prove $S'(x) = \overset{m}{\underset{i=1}{X}} M'^i$, $M'^i = M^i$, $i \neq j, M'^j = M^j \sim \{w\}$ is an improving set. We assume the contrary, which implies that the vector s^* that minimizes $f(x + s)$ for $s \in S(x)$ has $s_j^* = w$ and that $f(x + s^*) < f(x)$. (Hence we will want to have picked a $w \in M^j \sim 0$ such that $s_j^* = w$ seems very unlikely.) From these implications we try to make assertions about the other components of s^* with the hope of eventually obtaining a contradiction.

If Step 6 succeeds, we reduce $S(x)$ to $S'(x)$ and return to Step 5. If it does not succeed, we abandon efforts to prove x is optimal and go to Step 7.

Step 7. A Further Heuristic Approach to Obtain an Improvement. Using the same heuristics as in Step 2, we try different points of the form $x' = x + s$, where $s \in S(x)$. This may succeed where Step 2 failed since $S(x)$ will be a proper subset of, $\{s : x + s \in X, \|s\| < 1\}$ if Step 5 or Step 6 was partially successful. If we find an $x' \neq x$ such that $f(x') \leq f(x)$, we set $x = x'$ and go to Step 2. If no such x' is found we stop with an unverified x. This concludes the solution strategy.

Our solution strategy gives us an x that is optimal, or an unverified stationary point. The possibility of an unverified solution is of our choosing since there is only a finite number of improving points. This possibility could be eliminated by expanding Step 6 to produce either an improving vector or a contradiction. All test cases have produced an optimal x, and in fact Step 7 has never been reached. We now proceed to the application from which these test cases arose.

4. APPLICATION

We consider the initial procurement problem for a reparable inventory system. This system is a single-echelon version (we assume a given depot stock level) of that of Sherbrooke [3]. It is of continuing interest at RAND since the annual Air Force purchases of reparable spare parts exceed half a billion dollars.

An Air Force base operates in continuous time with independent Poisson demand (failure) rates λ_i^D, λ_i^B, for each item i, $i = 1, \ldots, n$. The superscripts D and B indicate whether the item is to be repaired at the depot or the base. When item i fails at the base at epoch t, and j units of i are currently on hand (j is negative if a backorder condition exists), the number on hand goes down by one and a one-for-one replenishment cycle is begun. A serviceable item will return to the base and raise the total on hand by one, at some time in the future given by the distribution function $F_{ij}^D | X_s$, $s \leq t$, or $F_{ij}^B | X_s$, $s \leq t$, according to whether the item is repaired at the depot or the base. Here X_s, $0 \leq s \leq t$, is the sample path associated with item i at the base from time 0 to time t. $X_s = 2$ means there are 2 items at the base at epoch s.

Under the assumption that the $F_{ij}^D | X_s$, $s \leq t$, and the $F_{ij}^B | X_s$, $s \leq t$, for $-\infty < j < +\infty$, do not depend on X_s, $s \leq t$, or j, Palm's theorem holds and the steady-state number of items in the replenishment cycle has a Poisson distribution with mean of $\lambda_i = \lambda_i^D \mu_i^D + \lambda_i^B \mu_i^B$, where μ_i^D and μ_i^B are the means of F_i^D and F_i^B, respectively. For a recent generalization of Palm's theorem see [4]. In fact, the functions $F_{ji}^{D}|X_s, s<t$, do depend on X_s, $s<t$, and j, and the above is only an approximation. Simon [5] gets exact results for the steady-state number of items in repair when $F_{ij}^{D}|X_s, s<t$, is made up of deterministic components. We choose to accept the Poisson approximation.

Spare parts are purchased to take the place of items in the replenishment cycle. If x_i spare parts are purchased, the probability of j or fewer backorders is $F_i(x_i+j)$, where $F_i(\cdot)$ is the c.d.f. of a Poisson distribution with mean λ_i. These items are used for the common purpose of supporting aircraft at the base. Operations are constrained by the inventory item with the most number of backorders under the assumption of "cannibalization" of nonoperational aircraft for spare parts, and cannibalization is, in fact. employed. The probability that the item in worst shape is backordered j or less is the probability that all n items are backordered j or less and equals $\prod_{i=1}^{n} F_i(x_i+j)$, since the distribution functions F_i (.) are independent.

We therefore would like to minimize $\sum_{j=0}^{\infty}(1-\prod_{i=1}^{n} F_i(x_i+j))+\lambda\sum_{i=1}^{n} C_i x_i$. The first term is the steady-state expected number of backorders of the (random) item with the most backorders or, under the assumption of cannibalization, the steady-state number of aircraft not operationally ready because of the lack of a spare part. The second term is the cost of spares purchases, with C_i the cost of a unit of item i. The two terms are connected by a λ whose interpretation is 1/cost of a backorder.

For computational purposes we approximate the previous function by

$$\sum_{j=0}^{m}(1-\prod_{i=1}^{n} F_i(x_i+j))+\sum_{i=1}^{n}\sum_{j=m+1}^{\infty}(1-F_i(x_i+j))+\lambda\sum_{i=1}^{n} C_i x_i, \tag{5}$$

and use (5) as our objective function in what follows. Since $\prod_{i=1}^{n} F_i(x_i+j)$ is very close to 1 for $j>m$ (this will determine our choice of m), the approximation (5) is far superior to truncation (not including the middle term of (5)). To see this, note that for arbitary y.

$$(1-\prod_{i=1}^{n} y_i)=\sum_{i=1}^{n}(1-y_i)\prod_{j=1}^{i-1} y_j.$$

4.1 Discrete Convexity

We want to show that

$$g(x)=\sum_{j=0}^{m}(1-\prod_{i=1}^{n} F_i(x_i+j))+\sum_{i=1}^{n}\sum_{j=m+1}^{\infty}(1-F_i(x_i+j))$$

$$+\lambda\sum_{i=1}^{n} C_i x_i : X = \{ x : x \subset I^n, x_i \geq 0, i = 1, ..., n \} \to R$$

is discretely convex, where $F_i(k) = \sum_{m=0}^{k} e^{\infty\lambda_i} \lambda_i^m/m$. We will do this by showing that there is a class of continuous convex functions $g_\epsilon(\cdot)$ satisfying the hypotheses of Corollary 1 of Theorem 3. The approximations $g_\epsilon(\cdot)$ use continuous approximations to the discrete functions $F_i(\cdot)$ which have "spikes" right before the integer values. Also, the sums in $g(\cdot)$ are replaced by integrals.

The functions $g_\epsilon(\cdot)$ are defined by

$$g_\epsilon(y) = \int_0^{m+1} (1 - \pi_{i=1}^{n} F_{i\epsilon}(y_i + w))dw + \sum_{i=1}^{n} \int_{m+1}^{\infty} (1 - F_{i\epsilon}(y_i + w))dw$$

$$+\lambda\sum_{i=1}^{n} C_i y_i : Y = \{ y : y \subset R^n, y_i \geq 0, i = 1, ..., n \} \to R.$$

For fixed $\epsilon > 0$, the function $F_{i\epsilon}(\cdot)$ may be any fixed function that satisfies the following conditions:

$F_{i\epsilon}(\cdot)$ has a differentiable density function $f_{i\epsilon}(\cdot)$

$f_{i\epsilon}(y) = 0$, except for $y \in (j - \epsilon, j)$ for $j = 1, 2, ...$

$F_{i\epsilon}(j) = F_i(j)$ for $j = 0, 1, ...$

These conditions imply that $F_{i\epsilon}(y) = F_i(j)$ for $y \in (j, j + 1 - \epsilon)$.

We will prove that g is discretely convex by showing that there exist a constant K and weights $w_x(y)$, $x \in N(y)$, for every $y \in Y$ such that, given any $\epsilon > 0$, there is a convex function $g_\epsilon : Y \to R$ satisfying

$$\sup_{y \in Y} |g_\epsilon(y) - \sum_{x \in N(y)} w_x(y)\, g(x)| < K\epsilon$$

Lemma 2. Let $h_\epsilon : Y \to R$ and $h : X \to R$, where X is a discrete rectangle and Y is its convex hull, be functions such that $h_\epsilon(y) = \sum_{j=0}^{N-1} h^j_\epsilon(y)$ and $h(x) = \sum_{j=0}^{N-1} h^j(x)$. We also assume that h is strictly positive and the h^j are nonnegative. If there exist weights $w^j_x(y)$, $x \in N(y)$, for every $y \in Y$, $j = 0, ..., N - 1$ such that $\sup_{y \in Y, j=0,...,N-1} |h^j_\epsilon(y)$

$\sum_{x\in N(y)} w_x^j(y)\, h^j(x)\, | < \epsilon$, then there exist weights $w_x(y)$, $x \in N(y)$, for every $y \in Y$ such that $\sup_{y\in Y} |h_\epsilon(y) - \sum_{x\in N(y)} w_x(y)h(x)\, | < N\epsilon$.

Proof: The proof is immediate upon letting

$$w_x(y) = \sum_{j=0}^{N-1} w_x^i(y) h^j(x)/h(x).$$

Lemma 3. There exist a constant K and weights $w_x(y)$, $x \in N(y)$, for every $y \in Y$ such that given any $\epsilon > 0$

$$\sup_{y \in Y} |g\epsilon(y) - \sum_{x \in N(y)} w_x(Y) g(x)| < K\epsilon.$$

Proof: The infinate integral in $g_\epsilon(\cdot)$ and the infinate sum in $g(\cdot)$ converge so that they can be truncated at some large N with an error less than ϵ. We can apply Lemma 2 if we can show there exist weights $w_x^j(y)$, $x \in N(y)$, for every $y \in Y, j = 0, \ldots, N + n - 1$, such that

$$\sup_{y\in Y, j=0,\ldots,N+n-1} |g_\epsilon^j(y) - \sum_{x \in N(y)} w_x^j(y) g^j(x)| \leq n\epsilon \quad , \tag{6}$$

where for $j \leq m$,

$$g_\epsilon^j(y) = \int_j^{j+1} (1 - \prod_{i=1}^n F_{i\epsilon}(y_i + s))ds$$

for $m + 1 \leq j \leq N - 1$,

$$g_\epsilon^j(y) = \sum_{i=1}^n \int_j^{j+1} (1 - F_{i\epsilon}(y_i + s))ds$$

and for $N \leq j \leq N + n - 1$,

$$g_\epsilon^j(y) = \lambda\, C_{j-N+1} y_{j-N+1}$$

We will carry out the details only for the difficult case, $0 \leq j \leq m$, For arbitrary $y \in Y$ and $0 < j < m$, the set $s_{jy\epsilon} = \{s : s \in (j, j + 1), y_i + s \in (k - \epsilon, k)$ for some integer k and and variable $i\}$ has length (measure) no greater than $n\epsilon$. For $s \in (j, j + 1)$ and not in $s_{jy\epsilon}$,

$$(1 - \prod_{i=1}^n F_{i\epsilon}(y_i + s)) = (1 - \prod_{i=1}^n F_i(x_i + j)) = g^j(x), \tag{7}$$

where x_i is the greatest integer less than $y_i + s - j$ (hence $x \in N(y)$). Here we use strongly the fact that $F_{i\epsilon}(z) = F_i(j)$ for $z \in (j, j + 1 - \epsilon)$.

Let $w_x^j(y)$ be the measure of $\{s : s \in (j, j+1), x_i$ is the greatest integer less than $y_i + s - j\}$ and independent of j. The $w_x^j(y)$, $x \in N(y)$, are weights since they are nonnegative and sum to 1. The integral defining $g_\epsilon^j(\cdot)$ and (7) imply that

$$|g_\epsilon^j(y) - \sum_{x \in N(y)} w_x^j(y)\; g^j(x)\;| < n\epsilon$$

Since y was arbitrary, (6) holds and the conclusion of the lemma follows from Lemma 2, letting $K = n(N + n - 1)$.

We now show that $g_\epsilon(\cdot)$ is convex. The proof is very close to that of [6, Theorem 1], where it is shown that the function $g(y) = \int_0^\infty (1 - \pi_{i=1}^{n} F_i(y_i + w))dw$ is convex if the c.d.f. $F_i(\cdot)$ have finite means and density functions $f_i(\cdot)$ that are differentiable and satisfy $\lim_{y \to \infty} f_i(y) = 0$.

Lemma 4. For $\epsilon < 1$, the function $g_\epsilon(\cdot)$ is convex.

Proof: The proof consists of showing that the Hessian matrix of second partial derivatives is positive and semidefinite.

For any i, $1 \leq i \leq n$,

$$\frac{\partial g_\epsilon(y)}{\partial y_i} = -\int_0^{m+1} f_{i\epsilon}(y_i + w) \;\pi_{k \neq i}\; F_{k\epsilon}(y_k + w)\,dw$$

$$-\int_{m+1}^{\infty} f_{i\epsilon}(y_i + w)\,dw + \lambda C_i$$

$$\frac{\partial^2 g_\epsilon(y)}{\partial y_i^2} = -\int_0^{m+1} \frac{df_{i\epsilon}}{dy_i}(y_i + w) \;\pi_{k \neq i}\; F_{k\epsilon}(y_k + w)dw + f_{i\epsilon}(y_i + m + 1)$$

$$= - f_{i\epsilon}(y_i + w) \;\pi_{k \neq i}\; F_{k\epsilon}(y_k + w)\;\Big]_0^{m+1}$$

$$+ \sum_{j \neq i} \int_0^{m+1} f_{i\epsilon}(y_i + w) f_{j\epsilon}(y_j + w) \;\pi_{k \neq i,j}\; F_{k\epsilon}(y_k + w)\,dw$$

$$+ f_{i\epsilon}(y_i + m + 1)$$

using integration by parts. Also,

$$\frac{\partial^2 g_\epsilon(y)}{\partial y_i \partial y_j} = -\int_0^{m+1} f_{i\epsilon}(y_i + w) f_{j\epsilon}(y_j + w) \prod_{k \neq i,j} F_{k\epsilon}(y_k + w)\, dw.$$

Therefore, the Hessian matrix of $g_\epsilon(\cdot)$ can be written as

$$A = \begin{bmatrix} \sum_{k=1}^{n} a_{ik} & -a_{12} & \cdots & -a_{1n} \\ -a_{21} & \sum_{k=1} a_{2k} & \cdots & -a_{2n} \\ \vdots & & & \\ -a_{n1} & -a_{n2} & \cdots & \sum_{k=1}^{n} a_{nk} \end{bmatrix}$$

where

$$a_{ii} = -f_{i\epsilon}(y_i + w) \prod_{k \neq i} F_{k\epsilon}(y_k + w) \Big]_0^{m+1} + f_{i\epsilon}(y_i + m + 1)$$

$$= f_{i\epsilon}(y_i) \prod_{k \neq i} F_k(y_k) + f_{i\epsilon}(y_i + m + 1) - f_{i\epsilon}(y_i + m + 1) \prod_{k \neq i} F_{k\epsilon}(y_k + m - 1),$$

and

$$a_{ij} = \int_0^{m+1} f_{i\epsilon}(y_i + w) f_{j\epsilon}(y_j + w) \prod_{k \neq i,j} F_{k\epsilon}(y_k + w)\, dw.$$

Clearly, a_{ij}, $j \neq i$, is nonnegative, and a_{ii} is nonnegative since each function $F_{k\epsilon}(\cdot)$ is bounded by 1.

We will have shown that $g_\epsilon(\cdot)$ is convex if we can show that $rAr \geq 0$ for any r.

$$rAr = \sum_{i=1}^{n} \sum_{j=1}^{n} a_{ij} r_i^2 - \sum_{i=1}^{n} \sum_{j \neq i} a_{ij} r_i r_j$$

$$= \sum_{i=1}^{n} \Big(a_{ii} r_i^2 + \sum_{j>i} a_{ij}(r_i^2 + r_j^2 - 2r_i r_j)\Big)$$

using the symmetry of A. This last sum is nonnegative since the a_{ij} are nonnegative, which completes the proof.

I thank Steve Jacobsen for suggesting this approach of direct expansion of rAr.

THEOREM 5.

The function $g : X \rightarrow R$ is discretely convex.

Proof: Lemmas 3 and 4 show that $g(\cdot)$ satisfies the conditions of Corollary 1.

4.2 The Algorithm

The function $g : X \rightarrow R$ is discretely convex by Theorem 5 and clearly satisfies the finite number of improvements assumption. Our objective is to go from the solution strategy of Sec. 3 to an algorithm that will minimize g. Like the solution strategy, our algorithm will have seven steps.

The algorithm makes important use of a property of $g(\cdot)$, which we now display. As in Sec. 3 we let $\Delta_{i\ell}g(x)$, $i \neq \ell$, denote the mixed second difference of $g(\cdot)$.

$$\Delta_{i\ell}g(\cdot) = -\sum_{j=0}^{m} (F_i(x_i + j + 1) - F_i(x_i + j)).$$

$$(F_\ell(x_\ell + j + 1) - F_\ell(x_\ell + j)) \ \pi_{k \neq \ell, i} F_k(x_k + j),$$

which is nonpositive for all feasible x. As explained in Sec. 3, this condition is sufficient to make Theorem 4 expedient.

We gain another benefit from this result. Les $s \subset I^n$, and consider the vectors s^+ defined by $s_i^+ = \max(0, s_i)$ and s_i^- defined by $s_i^- = \max(0, -s_i)$. Thus $s = s^+ - s^-$.

Theorem 6.
The function g satisfies the condition

$$g(x + s) + g(x) \geq g(x + s^+) + g(x - s^-)$$

where $x, x + s \in X$.

Proof: We begin with the identity

$$g(x + s^+) + g(x) = g(x + s^+) + g(x)$$

and add the vector $-s^-$ to the appropriate arguments, unit vector by unit vector.

The vector $-s^-$ decomposes into the sum of a finite number of negative vectors as by $-s^- = \sum_{i=11}^{k} n^i$. We prove the theorem by showing that for any j, $0 \leq j \leq k - 1$,

$$g(x + s^+ + \sum_{i=1}^{j+1} n^i) - g(x + s^+ + \sum_{i=1}^{j} n^i) \geq g(x + \sum_{i=1}^{j+1} n^i) - g(x + \sum_{i=1}^{j} n^i)$$

or

$$-\Delta_{m(n^{j+1})} g(x + s^+ + \sum_{i=1}^{j+1} n^i) \geq -\Delta_{m(n^{j+1})} g(x + \sum_{i=1}^{j+1} n^i) ,$$

where $m(x)$ is the nonzero coordinates of the unit vector x.

The vector $+s^+$ likewise decomposes, in this case into the sum of a finite number of positive unit vectors by

$+s^+ = \sum_{i=1}^{r} p^i$. We have $-\Delta_{m(n^{j+1})} g(x + s^+ + \sum_{i=1}^{j+1} n^i) = -\Delta_{m(n^{j+1})} g(x + \sum_{i=1}^{j+1} n^i)$

$$-\Delta_{m(n^{j+1})m(p^1)} g(x + \sum_{i=1}^{j+1} n^i)$$

$$\ldots -\Delta_{m(n^{j+1})m(p^r)} g(x + \sum_{i=1}^{j+1} n^i + \sum_{i=1}^{r-1} p^i) .$$

The second differences will be mixed since no coordinate is nonzero for both $-s^-$ and $+s^+$. The nonpositivity of the mixed second differences completes the proof.

A corollary of Theorem 6 is that if $g(x + s) < g(x)$, then either $g(x - s^-) < g(x)$ or $g(x + s^+) < g(x)$. But this means the set $\{s : x + s \in X, s_i \geq 0 . i = 1, ..., n, \|s\| \leq 1\} \cup \{s : x + s \in X, s_i \leq 0, i = 1, ..., n, \|s\| \leq 1\}$ is a set of potentially improving displacements. Because of the form of X, this set can be written as

$$S(x) = \mathop{X}_{i=1}^{n} M^{+i} \cup \mathop{X}_{i=1}^{n} M^{-i} ,$$

where $\{0\} \subset M^{+i} \subset \{0,+1\}$ and $\{0\} \subset M^{-i} \subset \{-1,0\}$. For example, we have $M^{-i} = \{-1,0\}$ when $x_i > a_i$.

We now describe the algorithm. As with the description of the general solution strategy, we let x and $S(x)$ refer to the current value of the solution vector and the set of potentially improving displacements, respectively.

Step 1. Initialization. A range in which the optimal value for each x_i must lie can be obtained by looking at the first differences $\Delta_i g(x) = g(x + \in_i) - g(x)$. We have

$$\Delta_i g(x) = -\sum_{j=0}^{m} ((F_i(x_i + j + 1) - F_i(x_i + j)) \cdot \pi_{k=i} F_k(x_k + j)) - 1 + F_i(x_i + m + 1) + \lambda C_i.$$

Since the $\pi_{k \neq i} F_k(x_k + j)$, $0 \leq j \leq m$, are bounded by 0 and 1, upper and lower bounds for $\Delta_i g(x)$ are $-1 + F_i(x_i + m + 1) + \lambda C_i$ and $-1 + F_i(x_i) + \lambda C_i$, respectively. We therefore find the w_i such that $-1 + F_i(w_i) + \lambda C_i \leq 0$ and $-1 + F_i(w_i + 1) + \lambda C_i > 0$, and set x_i equal to some value between w_i and max $(0, w_i - m - 1)$. Our choice of an initial point is

$$x_i = \max(0, w_i - k), \; i = 1, ..., n,$$

where $0 \leq k \leq m + 1$, and k is left as an input parameter to the algorithm.

Step 2. Finding a Stationary Vector. We use Step 2 of the general solution strategy.

Step 3. A Heuristic Approach to Obtain an Improvement. We are looking for a vector x' such that $g(x') < g(x)$ and $\|x' - x\| = 1$. The result that $S(c) \; \mathop{X}_{i=1}^{n} M^{-i} \cup \mathop{X}_{i=1}^{n} M^{-i}$ is an improving set of x suggests that we try an "up" sequence of vectors $x + s$, where $s \in \mathop{X}_{i=1}^{n} M^{+i}$; and a "down" sequence of vectors x + s, where $s \in \mathop{X}_{i=1}^{n} M^{-i}$.

For the "up" sequence of vectors we rank the n variables in ascending order according to the value of $\Delta_i g(x + s^{+i})/\Delta_i g(x)$, where s^{+i} is a vector of 1's except for the ith position, which has a zero. Since x is stationary, $\Delta_i g(x) > 0 (x_i < \infty)$ while $\Delta_i g(x + s^{+i})$ may be < 0.

Then for $i = 1, \ldots, n$, we evaluate $g(x + s^i)$, where s^i is a vector with components equal to +1 for variables ranked 1 to i and 0 to those ranked $i + 1$ to n. If we find a point $x' = x + s^i$ such that $g(x') < g(x)$, we replace x by x' and return to Step 2. If not, we test a "down" sequence of vectors obtained in an analogous fashion. If this yields an improving point we go to Step 2; otherwise we go to Step 4.

Step 4. A Second Initialization. We use the simplification obtained as a corollary to Theorem 6 and let $S(x)$, the set of potentially improving displacements of

$$x, = \mathop{X}_{i=1}^{n} M^{+i} \cup \mathop{X}_{i=1}^{n} M^{-i}, \text{ where } \{0\} \subset M^{+i} \subset \{0,1\} \text{ and } \{0\} \subset M^{-i} \subset \{-1, 0\}.$$

It turns out to be computationally expedient to reduce the two subsets of $S(x)$ one at a time rather than concurrently. The alogrithm therefore goes through the "up" phase, which tries to reduce $\mathop{X}_{i=1}^{n} M^{+i}$ to the zero vector, and a "down" phase, which tries to reduce $\overset{n}{\underset{i=1}{X}} M^{-i}$ to the zero vector. If both are successful x is optimal. If either of them fails we go to Step 7. As one would guess, they are analogous and only the "up" phase is described. Step 7 also is not displayed as it is the same as Step 7 of the general solution strategy. The "up" phase comprises Step 5 "up" and Step 6 "up", and they interact exactly the same way as do Step 5 and Step 6 of the general solution strategy.

Step 5 "up". Reducing S(x) via Theorem 4. We use Step 5 of the general solution strategy except that only $w \in M^{+i} \sim \{0\}$ are considered. Neither the statement of Theorem 4 nor its proof is altered by the change in form of $S(x)$. As noted earlier, we are making use of the nonpositivity of the mixed second differences in applying Theorem 4.

Step 6 "up". Eliminating a Specific $w = 1 \in M^{+j}$ by Contradiction. In this step we pick a $1 \in M^{+j}$ according to the rule given in the subsequent paragraph and try to prove that $S'(x) = \overset{n}{\underset{i=1}{X}} M'^{+i} \cup \overset{n}{\underset{i=1}{X}} M'^{-i}$, where $M'^{-i} = M^{-i}$, $M'^{+i} = M^{+i}$, $i \neq j$, and $M'^{+j} = M^{+j} \sim [w]$, is an improving set. We assume the contrary, which implies that the vector s^* that minimizes $g(x+s)$ for $s\ \overset{n}{\underset{i=1}{X}} M^{+i}$ has $s_j^* = w$ and that $g(x+s^*) < g(x)$.

We determine the desired j by the same ratio test as Step 3. Our rule is to pick the i, i satisfying $1 \in M^{+i}$, which maximizes the ratio $\Delta_i g(x+s^i)/\Delta_i g(x)$, where s^i is defined by $s_k^i = 1$, if both $k \neq i$ and $1 \in M^{+k}$, and $s_k^i = 0$ otherwise. We note that since x is stationary, $\Delta_i g(x) > 0 (x_i < \infty)$, while the fact that Step 5 "up" failed implies $\Delta_i g(x+s^i) < 0$. We let j be the maximized variable.

Before going to Part 1 we summarize where we stand. By hypothesis $s_j^* = 1$ and for all i such that $1 \notin M^{+i}$, $s_i^* = 0$. The other coordinates of s^* as of now are undetermined.

Part 1
In Part 1 we try to prove sequentially by contradiction that each undetermined coordinate of s^* equals 1.

The optimality of $s_j^* = 1$ implies that $\Delta_j g(x + s^* - \epsilon j) \leq 0$, or $-\sum_{i=0}^{m} ((F_j(x_j+i+1) - F_j(x_j+i) \prod_{k=j} F_k(x_k+s_k^*+i) - 1 + F(x_j+m+1)) + \lambda C_j \leq 0$. We will attempt to contradict the hypothesis that $s_\ell^* = 0$ given $s_j^* = 1$, where $+1 \in M^{+\ell}$, by showing $\Delta_\ell g(x+s^*) = -\sum_{i=0}^{m} ((F_\ell(x_\ell+i+1) - F_\ell(x_\ell+i)) \prod_{k \neq l} F_k(x_k + s_k^* + i) - 1 + F_\ell(x_\ell+m+1) + \lambda C_\ell < 0.$

An upper bound on $\Delta_\ell g(x + s^*)$ given $\Delta_j g(x + s^* - \in_j) \leq 0$ can be obtained from the following linear programming problem:

$$\Delta_\ell g\,(x + s^*) + b_0 \leq \phi = \max_w c_0 w_0 + \ldots + c_m w_m$$

$$\text{subject to } a_0 w_0 + \ldots + a_m w_m \leq b_1, \; -w_i \leq 0, \text{ and } w_i \leq U_i,$$

where

$$c_i = -(F_\ell(x_\ell + i + 1) - F_\ell(x_\ell + i))/F\ell(x_\ell + i)$$

$$a_i = -(F_j(x_j + i + 1) - F_j(x_j + i))/F_j(x_j + i + 1)$$

$$U_i = \prod_{[k\,:\,k\neq\ell,+1M^{+k}]} F_k(x_k + i) \prod_{[\text{all other } k]} F_k(x_k + i + 1)$$

$$- F_j(x_j + i + 1) \prod_{k\neq j} F_k(x_k + i)$$

$$b_0 = \sum_{i=0}^{m} c_i F_j(x_j + i + 1) \prod_{k\neq j} F_k(x_k + i) + 1 - F_\ell(x_\ell + m + 1) - \lambda C_\ell$$

$$b_0 = \sum_{i=0}^{m} a_i F_j(x_j + i + 1) \prod_{k\neq j} F_k(x_k + i) + 1 - F_j(x_j + m + 1) - \lambda C_j.$$

The unknowns w_i are interpreted as $\pi_k F_k(x_k + s_k^* + i) - F_j(x_j + i + 1) \prod_{k=j} F_k(x_k + i)$.

We have $b_0 = \Delta_\ell g(x + \in j)$ and $b_1 = -\Delta_j g(x) < 0$.

Rather than solve the linear program, we apply the weak duality theorem of linear programming to obtain an upper bound for $\Delta_\ell g(x + s^*)$, which though weaker (higher) is much easier to calculate. If $b_0 \geq 0$, a contradiction is immediate since $\phi \leq 0$, and we assume $b_0 < 0$.

Let $\lambda_1 = 1.1 b_0 / b_1 (\geq 0)$, and for $i = 0, 1, \ldots, m$,

$$\lambda_i^+ = 0, \qquad \lambda_i^- = c_i - a_i\lambda_1, \text{ if } c_i - a_i\lambda_1 \geq 0,$$

and

$$\lambda_i^+ = a_i\lambda_1 - c_i, \; \lambda_i^- = 0, \quad \text{if } c_i - a_i\lambda_1 < 0.$$

Therefore for $i = 0, \ldots, m$, $c_i - a_i\lambda_1 + \lambda_i^+ - \lambda_i^- = 0$, and our nonnegative dual solution

λ is feasible. By the weak duality theorem of linear programming, $\lambda_1 b_1 + \sum_{i=0}^{n} U_i \lambda_i^-$ $\geq \phi$.

For example, if $\lambda_i = 0$ for all i, hten $\lambda_1\ b_1 = 1.1\ b_0 \geq \phi \geq b_0 + \Delta_\ell g(x + s^*)$ and we have shown that $\Delta_\ell g(x + s^*) < 0$ since $b_0 < 0$.

If the weak duality test shows that $s_\ell^* = 1$ implies $s_\ell^* = 1$, we take this into account in our linear programming formulation in a straightforward way when we consider a new l' such that $1 \in M^\ell$. Taking $s_\ell^* = 1$ into account causes the upper bounds U_i to be lower and the b_0, b_1 to be higher than they would have been if s_ℓ^* were considered undetermined. This strengthens our test, and there is now a possibility that either b_0 or b_1 or both may become positive. If b_0 is positive, a contradiction is immediate, since $\phi \leq 0$ by vitrue of the fact that the w_j are nonnegative and the c_j are nonpositive. If b_0 is negative and b_1 is positive, a contradiction is impossible, since w equal to the zero vector is a feasible and necessarily optimal solution.

When we are finished with a pass (testing all i, $i \neq j$, such that $1 \in M^{+i}$, once) we do not repeat as in Step 5 but go directly to Part 2.

Before beginning Part 2, let us summarize again where we stand. By hypothesis $s_j^* = 1$ and for all i such that $1 \in M^{+i}$, $s_i^* = 0$. None, some, or all of the other coordinates of s^* have been shown to equal 1 in Part 1. The remaining coordinates of s^* are still undetermined.

Part 2.
Let s be the "smallest" vector s^* could be. Therefore, $s_i = 1$ if s_i^* is necessarily 1, and $s_i = 0$ otherwise. Let $\{i_1, ..., i_k\}$ be the set of undetermined coordinates of s^* and we relabel this set $\{1, ..., k\}$ to avoid the subscript. This will be appreciated by the reader in what follows.

Let $\{i_1, ..., i_r\}$ be an arbitrary ordered subset of $\{1, ..., k\}$. Then for any p, $1 \leq p \leq r$, $\Delta_{i_p} g(x + \epsilon_1 + \epsilon_2 + ... + \epsilon_{i_{p-1}}) \leq \Delta_{i_p} g(x + \epsilon_{i_1} + \epsilon_{i_2} + ... + \epsilon_{i_{p-1}})$ since all the mixed second differences are nonpositive. Thus $g(x + s) + \Delta_{i_1} g(x) + ... + \Delta_{i_r} g(x + \epsilon_1 + \epsilon_2 + ... + \epsilon_{i_{r-1}}) \leq g(x + s) + \Delta_{i_1} g(x) + ... + \Delta_{i_r} g(x + \epsilon_{i_1} + ... + \epsilon_{i_{r-1}})$.

The inequality holds for any subset of the undetermined coordinates of s^*, so it must hold for the coordinates of s^* which are actually equal to 1. In this case the right-hand side equals $g(x + s^*)$. The left-hand side, however, is indeterminable since we do not know the subset $\{i_1, ..., i_r\}$. However, the left-hand side is clearly bounded below by $g(x + s) + \sum_{i=1}^{k} \min(0, \Delta_i g(x + \epsilon_1 + \epsilon_2 + ... + \epsilon_{i-1}))$ and we do know the set $\{1, ..., k\}$.

If the lower bound for $g(x + s^*)$, $g(x + s) + \sum_{i=1}^{k} \min(0, \Delta_i g(x + \epsilon_1 + \epsilon_2 + ... + \epsilon_{i-1}))$, is greater than $g(x)$, we contradict the optimality of s^* and Step 6 "up" is complete.

If Part 2 does not succeed, we go back to Part 1 to try to prove other coordinates are necessarily 1, using the more powerful linear programming test. We then return again to Part 2. Step 6 "up" has failed when an iteration of Part 1 fails to convert any coordinate of s^* from an undetermined status to a necessarily 1 status.

In practice, at the end of Part 1 the number of undetermined coordinates was $.15n$ or less, where n is the number of variables of g. The lower bound we obtained in Part 2 never failed, so that returning to Part 1 was never necessary.

4.3 Example of an "Up" Reduction of $S(x)$

We consider our inventory model with four items and $m = 2$. Thus the function we want to minimize is

$$g(x) = \sum_{j=0}^{2} (1 - \prod_{i=1}^{4} F_i(x_i + j)) + \sum_{i=1}^{4} \sum_{j=3}^{\infty} (1 - F_i(x_i + j)) + \lambda \sum_{i=1}^{4} C_i x_i$$

and

$$\Delta_i g(x) = -\sum_{j=0}^{2} ((F_i(x_i+j+i) - F_i(x_i+j)) \prod_{k \neq i} F_k(x_k+j) - 1 + F_i(x_i+3) + \lambda C_i$$

The steady-rate Poisson demand rates are .45, .55, .65, and .75, respectively, for the four items so that

$F_1(0) = .63763$	$F_2(1) = .89427$
$F_1(1) = .92456$	$F_2(2) = .98154$
$F_1(2) = .98912$	$F_2(3) = .99753$
$F_1(3) = .99880$	$F_2(4) = .99973$
$F_3(1) = .86138$	$F_4(2) = .95949$
$F_3(2) = .97166$	$F_4(3) = .00271$
$F_3(3) = .99555$	$F_4(4) = .99894$
$F_3(4) = .99944$	$F_4(5) = .99987$

Also $\lambda = .001$, $C_1 = 500$, $C_2 = 70$, $C_3 = 93$, and $C_4 = 25$. We assume that we are at Step 4 of the alogrithm, and we want to show that the point $x = (0, 1, 1, 2)$ is optimal.

Intiially $S(x)$, the set of potentially improving displacements of x, will be $\mathop{X}_{i=1}^{4} M^{+i} \cup \mathop{X}_{i=1}^{4} M^{-i}$, where $M^{+i} = \{0, 1\}$ for $i = 1, 2, 3$, and 4, $M^{-1} = \{0\}$ and $M^{-i} = \{0,-1\}$ for $i =$ 2, 3, and 4.

Here we used Theorem 6. In the "up" phase we want to reduce $S(x)$ to $\mathop{X}_{i=1}^{4} M^{-i}$. To do this we will show that there is no point $x' = x + s$ for $s \in \mathop{X}_{i=1}^{4} M^{+i}$ such that $g(x') < g(x)$.

Step 5 "up"

We apply Theorem 4 beginning with $w = +1 \in M^{+1}$. Because of the nonpositivity of the mixed second difference we know that the vector s that minimizes (4) of Sec. 3 is the "largest" possible, (0, 1, 1, 1). Thus we calculate $\Delta_1 g(x+s) = \Delta_1 g(0, 2, 2, 3) = -.27165 - .06404 - .00967 - 1. + .9988 + .5000 = .15444$, which is greater than zero so the test works and +1 are eliminated from M^{+1}.

We now apply the test for $w = +1 \in M^{+2}$. The vector s which minimizes (4) is $s = (0, 0, 1, 1)$. In this case $\Delta_2 g(0, 1, 2, 3) < 0$, and the test fails. The test also fails for $+1 \in M^{+3} (\Delta_3 g(0, 2, 1, 3) < 0)$ and $+1 \in M^{+4}$. We proceed to Step 6 "up".

Step 6 "up"

To find the "weakest link" among candidates 2, 3, and 4, we calculate the ratios

$$\Delta_2 g(0, 1, 2, 3)/\Delta_2 g(0, 1, 1, 2) = -\ .00081/.00732$$
$$\Delta_3 g(0, 2, 1, 3)/\Delta_3 g(0, 1, 1, 2) = -\ .00194/.00675$$
$$\Delta_4 g(0, 2, 2, 2)/\Delta_4 g(0, 1, 1, 2) = -\ .00197/.00215$$

The variable 2 maximizes this ratio and we hypothesize that the vector s^* which maximizes $g(x+s)$ for $s \in \mathop{X}_{j=1}^{4} M^{+i}$ has $s_2^* = 1$, and that $g(x+s^*) < g(x)$.

We move to Part 1 and try to show by contradiction that $s_2^* = 1$ implies $s_3^* = 1$. The parameters are

$$\begin{aligned}
b_0 &= -\Delta_3 g(0, 2, 1, 2) = -.00051 \\
b_1 &= -\Delta_2 g(0, 1, 1, 2) = -.00732 \\
\lambda_1 &= 1.1\, b_0/b_1 = .077 \\
c_0 &= -(F_3(2) - F_3(1))/F_3(1) = -.12803 \\
c_1 &= -.02458 \\
c_2 &= -.00390 \\
a_0 &= -(F_2(2) - F_2(1))/F_2(2) = -.08891 \\
a_1 &= -.01603 \\
a_2 &= -.00220
\end{aligned}$$

The U_i were not needed, but $U_0 = F_1(0)F_2(2)F_3(1)F_4(3) - F_1(0)F_2(2)F_3(1)F_4(2) = .53517 - .51726 = .01791$.

For $i = 0, 1, 2$, $c_i - c_i\lambda_1 < 0$, so that $\lambda_i^- = 0$ for all i. By the weak duality theorem

$$\lambda_1 b_1 + \sum_{i=0}^{2} U_i \lambda_i^- > \phi > b_0 + \Delta_3 g(x+s^*)$$

Hence $1.1\, b_0 \geq b_0 + \Delta_3 g(x+s^*)$ or $-.000051 \geq \Delta_3 g(x+s^*)$, which contradicts the hypothesis that $s_3^* = 0$ ($\Delta_3 g(x + s^*) \geq 0$). Thus we have shown that $s_2^* = 1$ implies $s_3^* = 1$.

We repeat Part 1 and show that $s_2^* = 1$ implies $s_4^* = 1$. We go to Part 2.

Since all the coordinates of s^* are determined, we simply check to determine whether $g(x + s^*) = g(0, 2, 2, 3) < g(0, 1, 1, 2)$. It is not, so we have eliminated +1 from M^{+2}. We now return to Step 5 "up".

Step 5 "up"

We apply Theorem 4 beginning with $w = +1 \in M^{+3}$. The s that minimizes (4) is (0, 0, 0, 1). Thus we calculate $\Delta_3 g$ (0, 1, 1, 3), which is positive, so the test works. We eliminate +1 from M^{+3}. Also $\Delta_4 g(0, 1, 1, 2) > 0$, so that +1 is eliminated from M^{+4}.

We have thus reduced $\overset{4}{\underset{i=1}{\mathrm{X}}} M^{+i}$ to the zero vector and this concludes the "up" phase.

4.4 Computational Results

In order to test the efficiency of the algorithm just described, the function $g(\cdot)$ was minimized for various numbers of variables. Each variable (item) is described by a unit cost C_i and a mean demand λ_i. These parameters were generated from uniform distributions using random numbers. The value of $k(n)$, the value of the parameter used in the Step 1 initialization procedure, was chosen beforehand and was a "best guess." The value of $\lambda(n)$, the multiplier in the function $g(\cdot)$ when there are n variables, was chosen beforehand to make the problems difficult to solve.

By a difficult problem we mean one for which our algorithm obtains improving vectors using Step 3, which is to say one for which our algorithm finds stationary vectors that are not optimal. A difficult problem also means that Step 6 (both "up" and "down") is employed often, which is to say that many of the elements of the sets M^{+i} and M^{-i} which describe $S(x)$, the set of potentially improving points of x cannot be eliminated by Theorem 4. None of the problems tested was so difficult that it went to Step 7.

	Number of Variables n			
Item	25	50	100	500
Average running time in seconds	5.5	14	27	170
Range	4-9	7-18	15-44	82-340
Average number of times Step 3 yields an improving vector	0.6	1.4	0.4	0.8
Range	0-3	0-3	0-1	0-1
Average number of times Step 6 is needed to eliminate a w	3	10	18	15
Range	0-16	0-27	0-47	0-60

The programming was done by the author in SIMSCRIPT II on the IBM 360-65. Five tests were made for each value of n considered. The value of m in the function $g(.)$ was 9.

5. FURTHER RESEARCH

As stated in the Introduction, the study of Problem (1) has two main topics: identifying functions for which a local optimum is a global optimum, and developing a general solution strategy for finding the optimum in those promising cases.

It would be gratifying if verifiable direct conditions for determining the discrete convexity of a function was developed. For example, it would be interesting if one could make some assertions about a function looking at its matrix of second differences.

Further computational experience with different discretely convex functions would undoubtedly bring insights about the characteristics of an "ideal" general solution strategy.

REFERENCES

1. Brooks, R.B.S., and A. M. Geoffrion, *Finding Everett's Lagrange Multipliers by Linear Programming*, The RAND Corporation, P-3315, February 1966; also published in *Operations Research 14*, pp. 1149-1152, 1966.

2. Geoffrion, A. M., *Elements of Large-Scale Mathematical Programming*, Western Management Science Institute, UCLA Working Paper 144, 1968.

3. Sherbrooke, C. C., *METRIC: A Multiechelon Technique for Recoverable Item Control*, The RAND Corporation, RM-5078-PR, November 1966; also published in *Operations Research 16*, pp. 122-141, 1968.

4. Feeney, G. J., and C. C. Sherbrooke, *The (s-1,s) Inventory Policy Under Compound Poisson Demand: A Theory of Recoverable Item Stockage*, The RAND Corporation, RM-4176-PR, September 1964; also published in *Management Science 12*, pp. 391-411, 1966.

5. Simon, R. M., *Stationary Properties of a Two-Echelon Inventory Model for Low Demand Items*, The RAND Corporation, RM-5928-PR, May 1969.

6. Miller, B. L., *A Multi-Item Inventory Model with a Joint Backorder Criterion*, The RAND Corporation, RM-5901-PR, April 1969.

DISCUSSION

Kay: There is a mathematical reason why a Poisson type of distribution should apply in part failures, but this does not necessarily mean that the simple Poisson distribution applies in any given case.

I have recently found in a related field, concerned with machine tools linked in automatic transfer lines, that the best fit for the available data gave the following distribution for times between failures:

$$f(t) = \sum_{i=1}^{u} \alpha_i \lambda_i e^{-\lambda_i t}$$

with $\sum_{i=1}^{u} \alpha_i = 1$

The proportions α_i are related to classes or types of failures, and in no case did we find a value of $u > 3$.

Using $f(t)$ renders the algebra very tedious, but not necessarily more difficult.

Miller: Your technique is beyond what we have developed at RAND. However we are aware of the problem you mention and have made provisions in METRIC to allow the use of the compound Poisson process. Another procedure provided by METRIC allows the variance-to-mean ratio of the demand rate to differ (invariably greater) from 1.

Seger: You mentioned the existence of a theorem, which appeared in a German technical journal during the 1920's. The theorem provided justification for the use of the Poisson distribution.

1. Were you referring to a theorem given by Palm, i.e., that the number of items undergoing repair has a Poisson distribution, with mean proportional to the repair time mean and independent of the repair time distribution, given that the arrival rate of failed units is exponential?

2. If not Palm's theorem, then which do you refer to?

Miller: Yes. There have been further generalizations of this by Sherbrooke which have appeared in *Management Science*.

Monhemius: Some years ago a paper has been published by Cox and Smith in Biometrika on the so-called Pseudo-Poisson distribution. Here they study the situation in which, under certain conditions, N sources are each creating events in an entirely periodical way (but each source having a different period). They study the summed series of events and show, that already for very low values of N:

a. the distribution for the interval between events can be approximated very well by a negative exponential distribution.

b. the number of events in short periods follows a Poisson distribution.

c. the number of events in longer periods follows a distribution with far lower variance than Poisson (which is intuitively clear).

Do you know this paper?

If yes, can you comment on its relevance?

Miller: No, but it sounds quite relevant.

A Linear Programming Model of an Army Supply System.

G.C. Beare
J.D. Curtis

1. INTRODUCTION

Despite developments in tactics, equipment and organization, the importance of logistics has remained unchanged. Throughout the evolution of warfare the aim of military supply systems has been to ensure that a commander's plan is not limited nor delayed by the lack of essential supplies.

The linear programming model described in this paper was developed as one of a number of computer models designed to investigate various aspect of battle at Corps level. It provides a means of evaluating the capabilities of a supply system in a variety of combat situations and hence of studying the way in which the course of battle might be influenced by problems of re-supply. At present, only the supply of ammunition and fuel are considered as these two commodities constitute the bulk of expendable supplies and their consumption causes the greatest fluctuation in the overall demand on a supply system. However the model can take account of other commodities.

Figure 1 illustrates a simple and hypothetical military system of supply for ammunition and fuel. Generally the structure of a supply system corresponds to the organization of the force it is designed to support. Starting with the Base Depots the chain of supply in this example goes from the Corps Supply Point to three intermediate Divisional Supply Points, then forward to an intermediate Brigade Supply Point in each brigade area, and finally to the consumers, i.e. the individual combat units. Units in the vicinity of the Corps Supply Point and Divisional Supply Points replenish direct from these points. In the divisional area a dump has been included which is not in the normal chain of supply, but which provides a reserve stock that can be drawn on by brigades to supplement normal supply. It must be stressed that the system represented in Figure 1 is purely hypothetical and that while most Army systems follow the general pattern, there are within NATO many detailed variations to this system.

At the beginning of operations each supply point in the system would have an initial stock normally expressed in days or tons. As stock is consumed, units collect from the Brigade Supply Points which in turn are re-supplied from the Divisional Supply Points, and so on up the chain. At all times the main aim is to re-supply the combat units but it is also important to maintain adequate stocks at intermediate supply points.

The amount of available transport and the way in which it is organized are fundamental to any supply system. In Figure 1 a possible transport allocation is shown, with unit transport collecting from the nearest supply points; divisional transport operating between Divisional Supply Points and Brigade Supply Points; and Corps transport moving supplies from the Corps Supply Point to the divisions. The rate of delivery that can be achieved by any transport force depends on the distances involved and approximate distances in km are shown on the diagram.

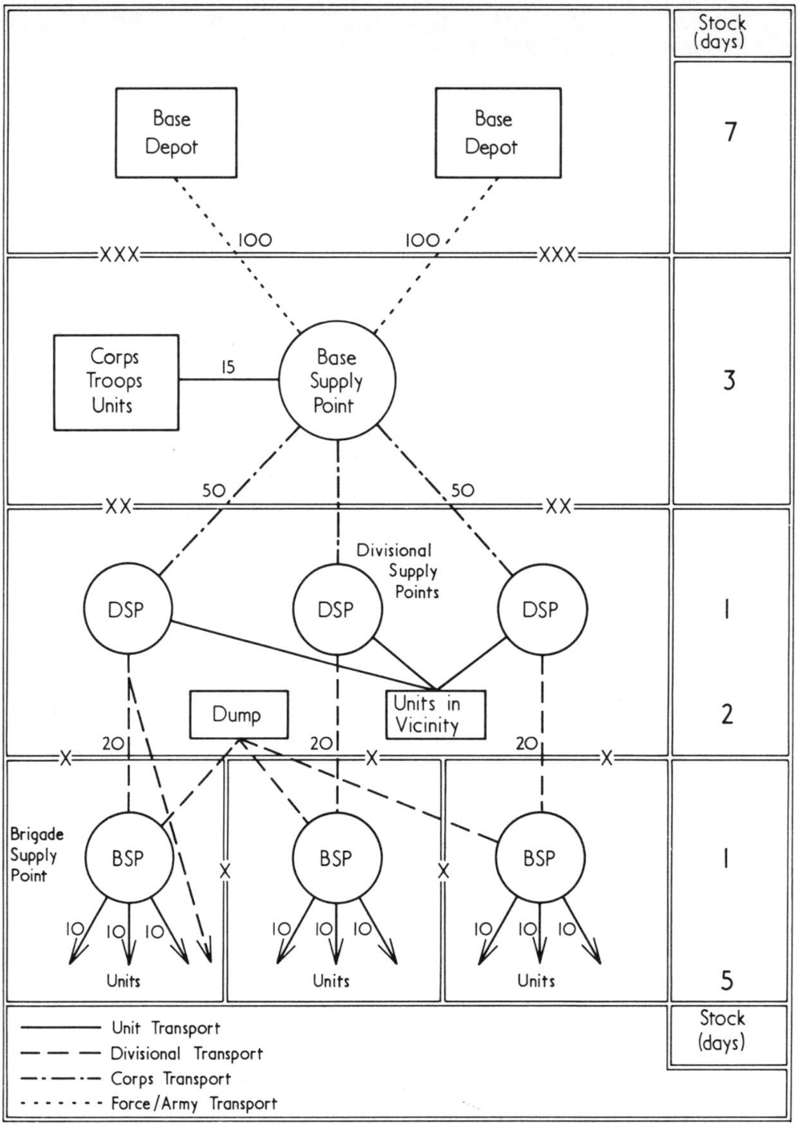

Figure 1 A military supply system

A factor that is important in any supply system and cannot be shown in Figure 1 is expenditure–how much fuel and ammunition are consumed by combat units in a given period of time, and therefore, how much stock must move through the chain of supply.

2. CONCEPTS OF THE MODEL

The features of a supply system that have been represented in the model are:

(a) *Structure* – the number of supply points and their inter-relation.

(b) *Transport* – the total transport available, the method of allocation, and movement rates.

(c) *Stocks* – the amount of ammunition and fuel at each supply point and in units.

(d) *Distances* – the distances between supply points.

(e) *Expenditure* – the expenditure rates of ammunition and fuel in terms of a activity, *e.g.* advance, attack, defence, withdrawal, etc.

Structure. The structure of a supply system was represented in terms of a network in which the supply points correspond to the nodes of the network, and the routes joining two supply points are reduced to a single link. Figure 2 illustrates a hypothetical Corps Supply System in terms of network. To show the different types of supply points, a consumer point is represented as a triangle; an intermediate supply point as a circle; and a source of supply as a square.

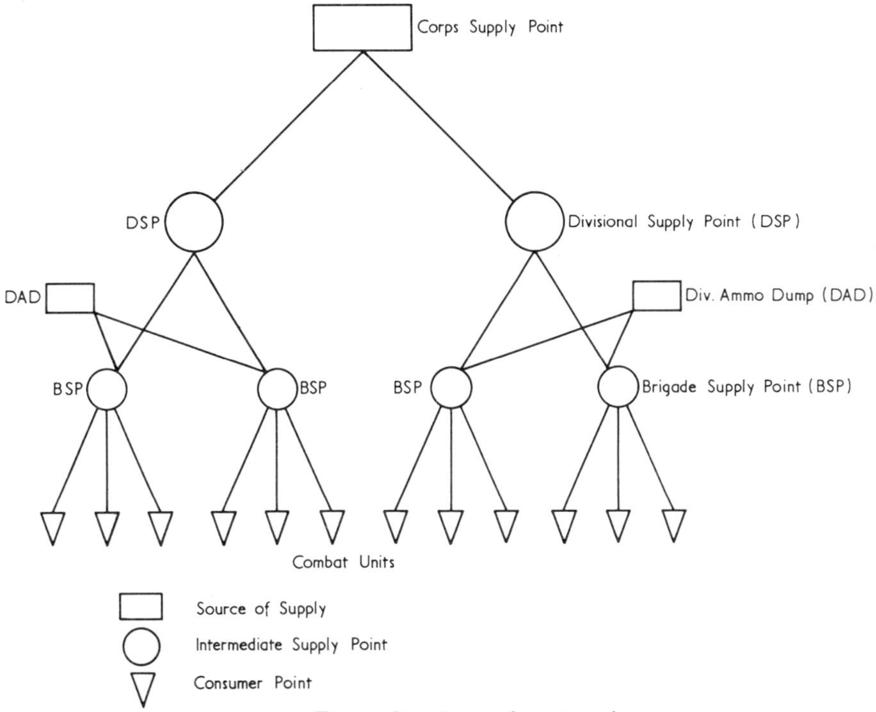

Figure 2 A supply network

Transport. The model can represent a supply system using any form of transport but has been developed primarily to consider road transport. Within a system there is a varying degree of flexibility in the allocation of transport, depending upon the level at which the transport is used. In the network shown in Figure 3 Corps transport can be allocated to the re-supply of either of the divisional supply points; divisional transport is limited to the re-supply of either of the brigade supply points within its own division; and unit transport is restricted to its own unit re-supply tasks. To represent in

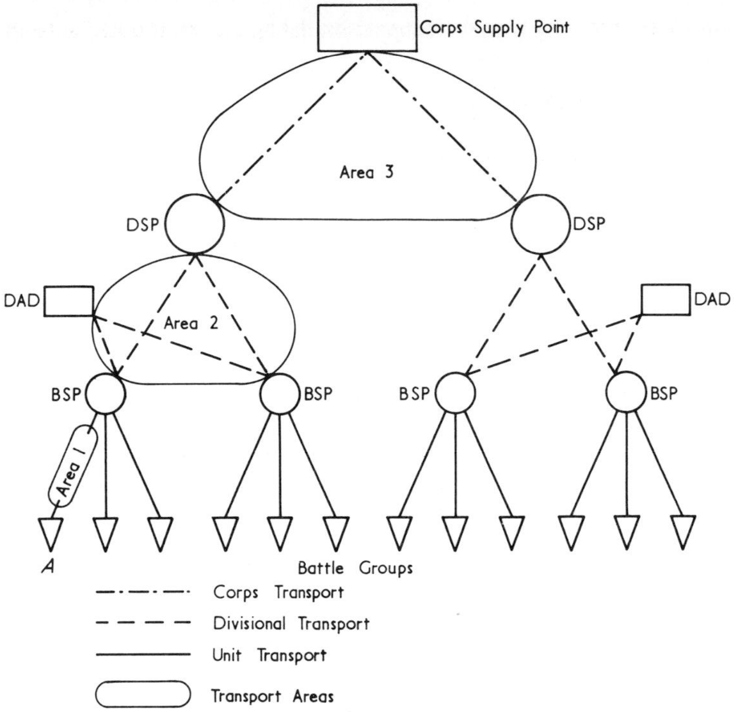

Figure 3 Network showing transport areas

the model the limitations in the reallocation of transport, each link in a system is assigned to a transport area which may contain one or more links. For example, there would be only one link in Transport Area 1 and this area would be allocated the unit transport for Node A. Transport Area 2 would contain four links and would be allocated the divisional transport.

Stocks. Stocks of commodities (in this study limited to ammunition and fuel) are held at all points in the system. It is not necessary for both commodities to be held at every point and, for example, there would be no fuel held at a divisional ammunition dump. Ammunition is not broken down into different types and it is an important assumption in the study that the correct mix of ammunition will be held at all supply points in a system. Each supply point has a planned holding of stock, and the difference between the planned stock and the actual stock of a commodity is referred to as the *deficit*. The model assesses the re-supply of a formation over a specified period of time called the *re-supply period.* Within this time the combat units require a certain amount of re-supply to replace stocks expended. The amounts of stock that they require are called the *demands* and these are set at the beginning of a re-supply period.

3. THE LINEAR PROGRAMMING MODEL

3.1 General

A linear programme model has been used to optimize the flow of stock through the system during a re-supply period as allowed by the available transport. Stock is moved so that deficits are reduced to a minimum and are kept as far as possible to the rear of the system.

3.2 Choice of Objective Function

Initially it appeared that the objective should be to minimize the deficits in stock of each commodity at combat units. However, the ability of a system to meet demands over an extended period of time depends on the maintenance of adequate stock holdings at all points in the system. Thus if the objective function included only the deficits at combat units, these deficits might be met by severely reducing stocks at the supply points immediately behind, and no account would be taken of the ability of the system to meet future demands.

This leads to a second possible objective function—to minimize the deficits throughout the system. However, this would provide no incentive to keep the deficits to the rear of the system and therefore some method of weighting must be introduced that makes a shortage of stock at a combat unit or forward supply point more critical than an equal deficit farther back in the system. In a military supply system, planned stock holdings are generally smaller at forward supply points than at those farther back, and in fact there is a progressive increase in planned stock holdings from front to rear of the system. The necessary weighting can therefore be introduced by expressing the deficit at each point as a fraction of the planned stock holding at that point. In this way, a deficit of say one ton at a forward supply point will be a larger fraction of that planned holding than an equal deficit at the rear of the system, and will therefore have a higher priority for re-supply. These fractional deficits will be referred to as *shortfalls.*

Therefore a third possible definition of the objective function is to minimize the sum of the shortfalls in the system. This seems a satisfactory definition until a situation is considered where, for example, two divisional supply points in the same transport area have different planned stocks. In this case the supply point with the lower planned stock will have priority for re-supply and there will be no delivery to the other point until the first is fully stocked. To overcome this, supply points that should have equal priority for re-supply have been grouped into levels. The aim is then to reduce the size of the largest shortfall within each level to a minimum and the objective function that achieves this is to minimize the sum of the maximum shortfalls in levels.

A similar problem still exists for the shortfalls at a single supply point. If the holdings of fuel and ammunition at a point are not equal, the available transport will be used solely to supply that commodity with the lower planned stock holding. Therefore the sum of maximum shortfalls must include *not* the maximum shortfall in fuel *and* the maximum shortfall in ammunition in a level, but only whichever of the two is the greater.

This objective function is still inadequate for the situation in which the supply points in a level lie in different transport areas.

Figure 4 shows two such points *A* and *B* with different shortfalls at the start of a re-supply period. If the maximum rate of re-supply in Area 1 leaves *A* at the end of the period with a higher shortfall than *B*, the previous objective function provides no incentive for reducing the shortfall at *B* even though transport is available in Area 2.

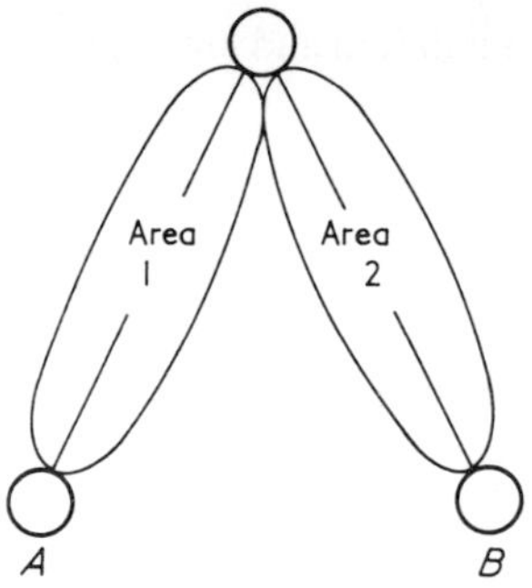

Figure 4 Points in a level in different transport areas

This incentive is provided by including in the objective function a term for the sum of the maximum shortfalls at all nodes in the system. The final form of the objective function is then:

$$\sum_{\text{levels}} S_{\max} + C \sum_{\text{nodes}} S$$

where $\sum_{\text{levels}} S_{\max}$ is the sum of the maximum shortfalls in levels

and $\sum_{\text{nodes}} S$ is the sum of the maximum shortfalls at nodes.

In this objective function the weighting factor C ensures that the sum of the maximum shortfalls at nodes is always the smaller of the two terms and thus prevents it from dominating the objective function. The actual value chosen for C is not important, although it will always be fractional. It should be noted that the shortfalls at dumps are not included in the objective function as no re-supply is required at these points.

4. CONSTRAINTS USED IN THE MODEL

4.1 Transport Constraints

The most important constraints are those governing the use of transport. In the model, transport is described in terms of the available lift, rather than numbers of vehicles, because of the wide variation in capacity between different types of vehicles. Two types of transport are considered, general service vehicles, which can carry either fuel or ammunition, and bowsers which can carry only fuel. For each link in the system a transport coefficient R is calculated, which is the number of tons lift needed to move one ton of a commodity along a link in a re-supply period. The time taken by a vehicle to make a delivery and return to its starting point can be divided into four parts:

(a) The time taken in administration, waiting and loading at the starting point.

(b) The outward journey time.

(c) Waiting and unloading time at destination.

(d) The return journey time.

If it is assumed that a constant dead time d is spent waiting, loading, unloading, etc. for both the outward and return trips, then the round trip time may be expressed as

$$2\ [^{l}/_{v} + d]$$

where ℓ is the link length and v the average speed of movement of the vehicle. The transport coefficient is the round trip time divided by the available time t. The transport coefficient is therefore

$$2\ [^{l}/_{v} + d]\ /t$$

In designing the model a constant delivery rate has been assumed throughout the round trip time, rather than a single delivery halfway through this time. For example, if a 5-ton vehicle was operating along a link for which the round trip time was five hours, then a delivery rate of one ton per hour is assumed, rather than the whole of the five tons arriving after 2½ hours.

There are two transport constraints for each transport area. The first states that the amount of ammunition flowing along the links in a transport area cannot exceed the amount that could be carried by the general service transport in that area. The second limits the total quantity of all commodities carried along these links to the amount that the total available transport could carry. The combination of these two constraints automatically limits the quantity of fuel carried in a transport area to the quantity that can be carried by the bowsers plus the part of the general service transport not carrying ammunition.

4.2 Conservation and Other Constraints

The other constraints in the linear programming model are mainly of the bookkeeping type. For each node in the system there are two conservation constraints; one for fuel and one for ammunition. These ensure that stock is not created anywhere in the system and that it only leaves the system through controlled demand flows. In the conservation constraints for fuel extra terms are introduced that account for the fuel expended by the logistic transport. The remaining constraints determine the shortfalls at each node and from these calculate the terms of the objective function.

All variables in a linear programming model are restricted to be non-negative and the constraints are written in such a way that this restriction prevents stock flowing backwards along a link. Another effect is that the actual stock holding at a node cannot exceed the planned holding at that node.

4.3 Re-Supply Periods

The model assesses the ability of a supply system to meet demands for a prolonged period by breaking up this period into a number of smaller consecutive *re-supply periods*. The lengths of these re-supply periods are chosen to correspond to phases of the battle during which changes in the distances between supply points and in expenditure rates are sufficiently small to allow average values to be used. At the beginning of each re-supply period the model allows commodities to be taken out of the system at the consumer nodes, thus representing stocks that have been expended and for which re-supply is now required. Each run of the model described assesses one of these re-supply periods, and so the performance of a supply system in the course of a battle must be determined by a sequence of runs. Such a sequence of runs might be carried out in one of two ways:

(a) Each re-supply period could be assessed independently, the only connection being that the final conditions for the one run provide the starting conditions for the next. This would mean that the model optimizes the movement of stocks within each re-supply period without reference to events occurring outside that period. This would represent a situation in which a logistic commander was receiving no intelligence, and was unable to make any forecast of the requirements for supply in future re-supply periods. Used in this way the model could not adequately represent, for example, the situation in which one division was going to counter-attack while another was to fight a holding engagement. In this case the first division should have priority for re-supply, as it is known that its expenditure will increase substantially in the near future.

(b) The second way of carrying out a sequence of runs is to set up a multi-time period model, which organizes the movement of stocks within each re-supply period to optimize the performance of the system over the whole course of the battle. This means that forward planning is represented in the model. The extent to which this occurs will depend on the way in which the reallocation of transport has been restricted.

4.4 Changes to the System

In both the single time period and the multi-time period model changes to the system can be made between re-supply periods. For example, an attacker whose lines of communication were extending rapidly might find it necessary to establish new intermediate supply points during the course of the battle. Conversely, a defender who was falling back on his own lines of communication might at some stage need to remove an intermediate supply point from his system of supply. The model allows for both the inclusion of new nodes and the removal of existing nodes at the beginning of a re-supply period. Transport is allocated to a transport area at the beginning of a re-supply period and may be reallocated at the beginning of the next to conform with changing distances, changes to the structure of the system, etc.

4.5 Attrition

Attrition to the system can be taken account of in the model by making simple adjustments to the parameters at the beginning of a re-supply period. For example, damage to routes affects the system by causing delays, and in the model such delays can be incorporated by increasing link lengths. Attrition to stocks can be simulated by introducing an additional expenditure, and attrition to vehicles by reducing the transport allocation to the relevant transport areas.

5. USE OF THE MODEL

The data required for each re-supply period are shown below:

(a) The specification of the network.

(b) Stocks planned to be held at each point (and for the first re-supply period, the stocks actually held).

(c) Expenditures.

(d) Link lengths.

(e) Duration of the re-supply period.

(f) Transport areas and the allocation of transport to them.

(g) Average movement rate and turnround and waiting time.

(h) Fuel consumption for logistic transport.

The model then uses the available transport in the best possible way to move supplies through the system to minimize:

(a) The discrepancy between supply and demand for combat units.

(b) The discrepancy between planned and actual stock levels at all points in the system.

The results produced by the model show for each re-supply period:

(a) The amount of each commodity moved along each link.

(b) The amount of transport used in each area.

(c) The level of stocks at each supply point.

The model has been constructed in a way that allows the parametric variation of the expenditure, the transport available, or both. For example, on reaching a solution that described the performance of a given system with an initial set of expenditures, these could then be gradually increased and the variation in stock levels, commodity flows and transport utilization followed.

The results obtained from the model make it possible to identify the cause of any shortfalls that arise in a system. A general picture of the relationships between stocks, transport and distances can be built up. Conclusions can be reached on the size of the stock-holding at intermediate points, the value of dumped stock can be assessed, and the adequacy of the total transport fleet and its distribution through the system can be investigated.

The model has been used to examine a range of existing and proposed military supply systems, to identify the factors that limit their performance and to test suggested improvements to them. It is proposed that the model should also be used to carry out an assessment of the vulnerability of a Corps re-supply system to enemy attack. In conclusion it must be emphasized that a study of re-supply cannot stand on its own, but must be related to the performance of the units it supports. There would be no value in making costly improvements in the capability of supporting units if this did not significantly increase the effectiveness of combat units.

DISCUSSION

Thwaites: The model assumes an expected consumption rate, should not one think in terms of "confidence limits" on possible rates of consumption particularly of ammunition, with different confidence limits for different weapons in various types of military operation?

Rickers: It was realised, of course, when this model was first constructed, that there were considerable fluctuations in the demand for ammunition and fuel from individual units. It will be seen from fig. 1. that combat units hold approximately 5 days stock themselves. This exists to provide self-sufficiency without resupply, and will act as 'buffer stock' to almost completely damp out short-term fluctuations in expenditure. Long term differences in expenditures between combat units will tend to be minimized by the commander in order to reduce the effects of battle fatigue. This was our justification for using average values of demands over short resupply periods.

Un Systeme de Gestion des Stocks 3e Echelon

R. Sahnoun
A. Péguy

1. GENERALITES

Le Système d'approvisionnement Auto-Contrôlé à Programmation Modulaire est une application particulière de la Méthode AIMS de la BULL GENERAL ELECTRIC.

Il résulte de la collaboration des équipes de travail B.G.E. et SUD-AVIATION, usine de BORDEAUX-MERIGNAC.

Il a été étudié mis en place dans le cadre de cette usine.

Il constitue un ensemble complet de Gestion Scientifique automatisé et à ce titre, il nécessite:

(a) La comptabilisation des stocks par la prise en compte des Entrées et Sorties de Magasins.

(b) La connaissance de données statistiques intéressant l'utilisation du matériel sorti de Magasins.

(c) La connaissance des besoins futurs qu'il évalue successivement:

- à partir des programmes (besoins programmés)

- à partir des sorties prises en compte (besoins aléatoires)

(d) La connaissance des mesures prises pour satisfaire ces besoins, c'est-à-dire, la prise en compte des commandes passées par le Gestionnaire. En fait, ces commandes ont été élaborées et proposées par la SAAPM au Gestionnaire qui les entérine ou les modifie.

Il convient donc de présenter ces données qui constituent l'information avant de décrire le traitement informatique qu'elles subissent et le matériel utilisé pour ce traitement.

1. L'INFORMATION ET SES SUPPORTS

1.1 La comptabilisation des Stocks

Elle nécessite la saisie des Entrées et Sorties des Magasins par le moyen des documents suivants:

Entrées en Magasins

Désignation du document	Fonction
Etat F (1)	Entrée comptable de tout matériel neuf livré au Client Etat
Bon de reversement	Reversement en magasin du matériel provenant d'une sortie excédentaire ou d'une récupération sur une réparation en cours
Avis d'arrivée du matériel	Entrée en magasin d'un matériel appartenant à l'Etat

Sorties de Magasins

Désignation du document	Fonction
Bon de sortie magasin	Sortie de matériel pour utilisation sur un programme de réparation
Bon de mouvement	Sortie de matériel pour tout autre raison (expédition, ferraillage, programme de fabrication)

Ces documents sont présentés en annexe.

1.2 Les données statistiques

Le système s'articule autour d'un code charnière constitué par un Ordre d'Exécution caractérisé par son numéro significatif. Celui-ci est rapproché d'un "Code Matériel" identifiant chaque programme de Réparation ou de Fabrication exécuté dans les ateliers de SUD-AVIATION. Par un traitement approprié, les sorties de magasins affectées aux numéros d'Ordre d'Exécution sont exploitées pour permettre l'établissement de statistiques de base, qui, simultanément intéressent le programme concerné et constituent l'historique de consommation de l'artcile (besoins aléatoires).

1.3 Connaissance des Besoins Futurs

Les programmes identifiés au paragraphe 1.2 sont déterminés par le Gestionnaire. Ils sont transmis par le document donné en annexe. Les besoins aléatoires ne sont pas déterminés par le Gestionnaire, mais établis par le SAAPM à partir des mouvements de sorties.

(1) Pour accélérer sa transmission, l'information est prélevée sur la demande d'émission d'Etat F ou sur le bordereau de réception.

A partir des Programmes connus et des besoins aléatoies qu'il a déterminés, le SAAPM établit les prévisions des Besoins Futurs; il les transforme alors en Propositions de Commande soumises au Gestionnaire pour approbation.

1.4 La prise en compte des commandes

La commande peut être de deux types:

1) Commande interne adressée aux Ateliers de Fabrication-Réparation de l'Entreprise.

2) Commande externe adressée:

   a) A un Fournisseur (commande commerciale)

   b) A un Entrepôt de l'Etat (demande de mise en place)

A leur prise en compte par le destinataire (Fournisseur ou Atelier), les commandes sont confirmées.

En fait, cette procédure constitue un maillon de liaison entre deux chaînes informatiques dont l'une est le SAAPM, l'autre étant le système de Gestion Comptable de l'Entreprise.

Désignation du document	Fonction
– Fiche de fourniture	Document de base déclenchant la commande ou la fabrication d'un matériel
– Fiche de remise en état	Document de base déclenchant la mise en réparation d'un matériel
– Commande commerciale	Document de confirmation signalant la prise en compte de la commande par un Fournisseur
– Fiche de délai	Document de confirmation de la commande signalant la prise en compte de la commande par l'Atelier
– Demande d'approvisionnement	Demande de mise en place de matériel appartenant à l'Etat

Ces documents sont présentés en annexe.

2. MATERIEL INFORMATIQUE UTILISE

– Marque : BULL GENERAL ELECTRIC

– Constitution :

– *Phase 1* (en service depuis le 1.11.69)

  – Ordinateur

    – 1 Unité Centrale type 115 capacité 8 K Octets
    – 3 Unités de disques DS 130
    – 1 Imprimante MB 50

  – Télégestion

    – 2 Télétypes TTY 33

– *Phase 2* (en service en Mai 1970)

  – Ordinateur

    – 1 Unité Centrale type 115 capacité 16 K Octets
    – 4 Unités de disques DS 130
    – 2 Dérouleurs de bandes magnétiques
    – 1 Imprimante MB 50

  – Télégestion

    – 2 Télétypes TTY 33
    – 8 écrans cathodiques 6402 1 A

3. CARACTERISTIQUES DU STOCK GERE

– *Nature du Stock*

  – Pièces de rechanges de matériel aéronautique (ensembles, sous-ensembles, pièces élémentaires de catégorie réparables ou consommables).

– *Importance du stock*

Le système assure simultanément:

  – La gestion scientifique de 40.000 postes intéressant des avions en service actif.
  – La comptabilisation de 60.000 postes peu mouvementés intéressant des avions d'un type ancien et dont un petit nombre reste en service.

– *Identification des Articles et Informations complémentaires*

Il est intéressant de préciser que les articles gérés sont identifiés par leur *numéro de référence* de Constructeur (Part number) et non par un numéro de codification propre à l'Entreprise. Ainsi est écartée cette coûteuse et pénible servitude qui attribue à un article parfaitement connu et internationalement identifié, un numéro matricule qu'il abandonnera à sa sortie de l'Entreprise.

Les informations complémentaires apparaîtront lors de la consultation des fichiers d'accès direct ou de gestion. Il faut citer entre-autres:

- Le numéro de code OTAN, le FSN ou CSD, la catégorie de l'article, sa classe, ses interchangeables, les programmes auxquels il appartient, son prix, son délai d'approvisionnement, son taux de service, son stock en état et celui à réparer, les commandes en cours etc . . .

Il faut mentionner spécialement l'identification du magasin et le numéro du casier qui loge le matériel. L'exploitation de ces dispositions a permis la suppression des fichiers de magasins dont la mise à jour constante constitue une opération fastidieuse et onéreuse.

4. ORGANISATION TECHNIQUE DU SAAPM

De conception totalement originale, le SAAPM établit des propositions de commande par le moyen de deux systèmes indépendants.

Les propositions émises sont comparées et mathématiquement analysées pour déterminer un coefficient de fiabilité caractérisant les divers résultats observés au cours de l'analyse.

L'un des systèmes traite les Estimations à partir des Séries dites *Aléatoires*, l'autre à partir des Séries dites Programmées.

Les organigrammes de ces traitements sont donnés ci-après (Figs. 1, 2). Ils appellent les commentaires suivants:

4.1 Organigramme des Séries Aléatoires

- *Module HOMOBS* (homogénéisation des observations)

  Le module ramène à une période de base fixée, une observation faite sur une période quelconque.

- *Contrôle de l'observation*
  C'est l'un des points de l'analyse précédemment citée. Il examine la stabilité de l'observation par rapport à la série. Le cas échéant, il enregistre une alarme.

- *Module KELEST* (choix de l'Estimation)

  Le SAAPM se réfère aux lois de distribution de GAUSS et POISSON.
  Le module KELEST choisit la loi de distribution qui devra être utilisée pour l'estimation. Pour les cas ambigus, le module se détermine par référence à la précédente estimation.

- *Module EST, POI* (estimation Poissonnienne)

  Le module estime le besoin selon la loi le distribution de POISSON.

- *Module EST, NOR* (Estimation normale ou Gaussienne)

  Le module estime le besoin selon la loi de distribution de Gauss

  NOTA

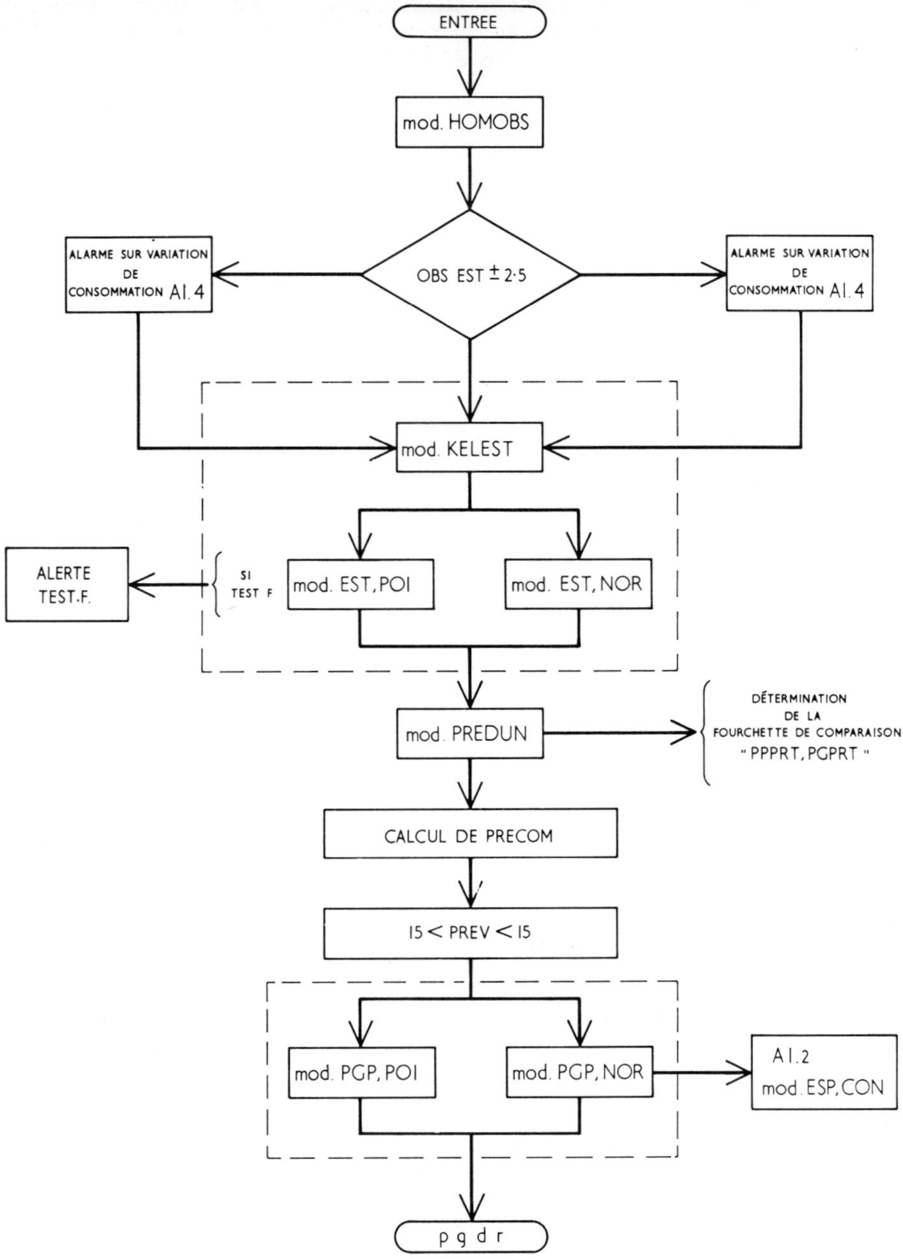

Figure 1 Programme d'estimation, series aléatoires

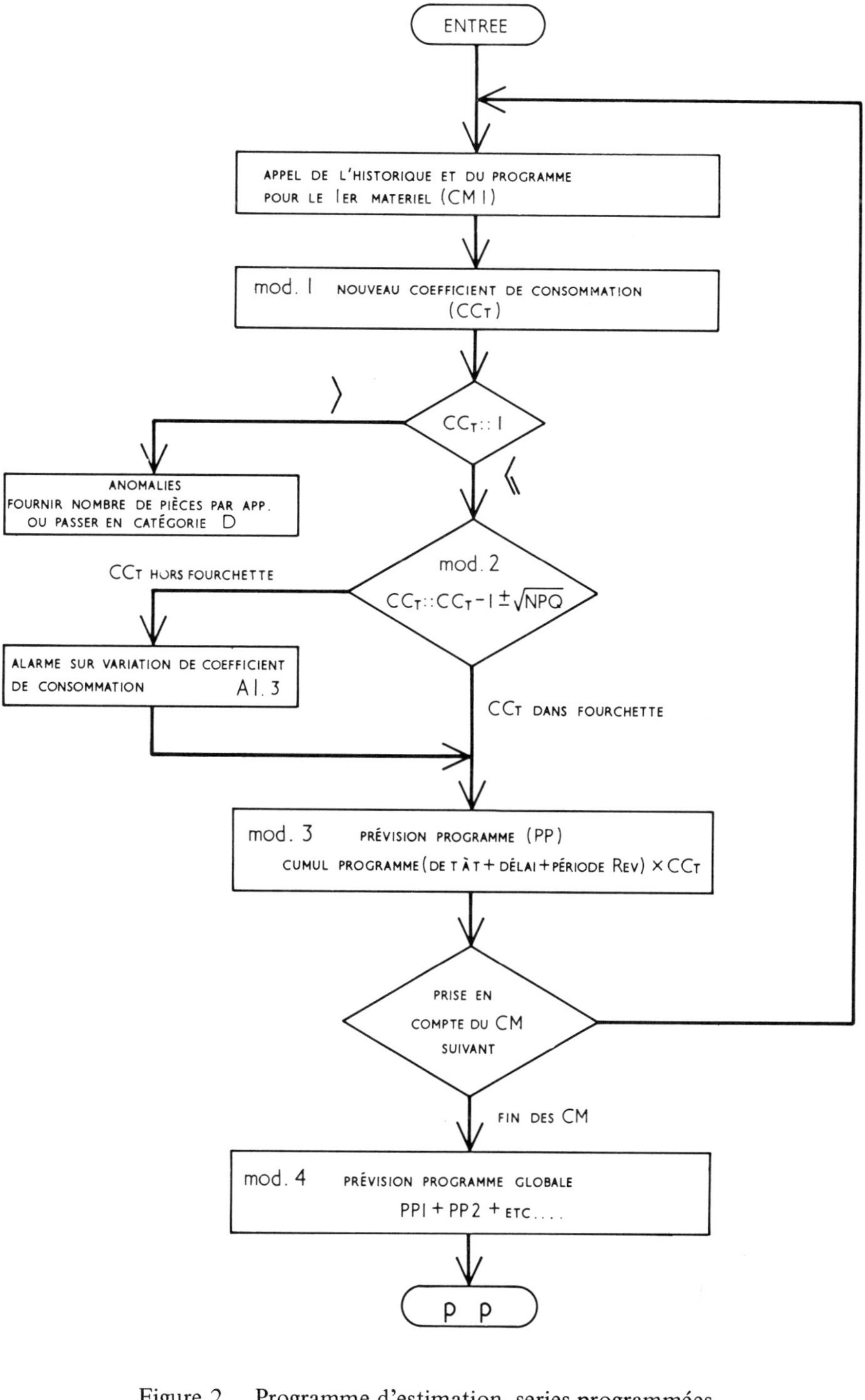

Figure 2 Programme d'estimation, series programmées

Pour ces deux derniers modules, la méthode d'estimation est dérivée du lissage exponentiel et par un test statistique, analyse l'écart entre les estimations et les observations (Modèle A. PEGUY)

– *Module PREDUN* (Prévision décimale unique non saisonnière)

Le module donne la prévision relative à la période considérée. Il l'établit en fonction d'un intervalle de confiance dépendant d'un coefficient de sécurité propre à la loi de distribution utilisée, de la confiance que l'on désire accorder à la prévision, de l'aléa statistique de la prévision dépendant lui-même de l'aléa statistique Delta de la série observée et d'un coefficient fonction de la période de prévision.

– *Calcul de PRECOM* (Prévision de commande)

PRECOM correspond à la prévision de consommation dans un cycle égal à une période de révision.

– *Modules PGP, POI et PGP, NOR*

Ils ont pour but de calculer pour des séries distribuées en loi de POISSON ou en loi de GAUSS, la Plus Grande Demande Raisonnable, c'est-à-dire la quantité d'approvisionnement à assurer pour atteindre un taux de service annuel donné comme objectif.

– *Module ESP, CON (Espoir de Consommation)*

Ce module calcule la prévision de consommation pendant une période couvrant le délai d'approvisionnement. Le besoin non couvert déclenche une Alerte.

4.2 Organigramme des Séries Programmées

La méthode des programmes habituellement utilisée dans les centre de maintenance est reprise et optimisée. Elle détermine pour chaque programme planifié, les prévisions de consommation des articles gérés.

Le SAAPM prend en compte les programmes réalisés, analyse l'Historique de consommation des articles gérés.

– *Module 1* (Calcul du coefficient de consommation)

Il fait suite à l'appel de l'Historique de consommation de l'article considéré pour le programme étudié. Le coefficient de consommation est le rapport du cumul des sorties affectées à ce programme au nombre d'unités d'oeuvre enregistrées pour ce programme dans la même période de temps (8 trimestres maxi).

– *Module 2* (Alerte sur consommation)

Le coefficient de consommation de la dernière période est comparé à celui de la précédente période afin de déclencher une éventuelle alerte.

– *Module 3*

Il exécute le calcul de la prévision de consommation pour le programme considéré

– *Module 4*

Il cumule les prévisions de consommation des divers programmes pour en faire une prévision globale appelée “Prévision Programmée” (PP).

4.3 Organigramme de Comparaison

Le programme compare la “Prévision Programmée” (PP) à la Prévision Totale Aléatoire (PREVT). Il la situe par rapport aux valeurs extrêmes retenues pour l'intervalle de confiance de celle-ci.

La résultat de cette comparaison est combiné avec les alertes signalées par les modules au cours de l'exécution de leur fonction. Cette combinaison permet d'une part de classer dans différents cas les deux prévisions (PP et PGDR) et d'autre part d'en préférer une au dépend de l'autre, laissant toutefois subsister les deux aux fins de comparaison par les soins du Gestionnaire.

L'organigramme de contrôle et le tableau comparatif sont donnés ci-après (Figs. 3, 4).

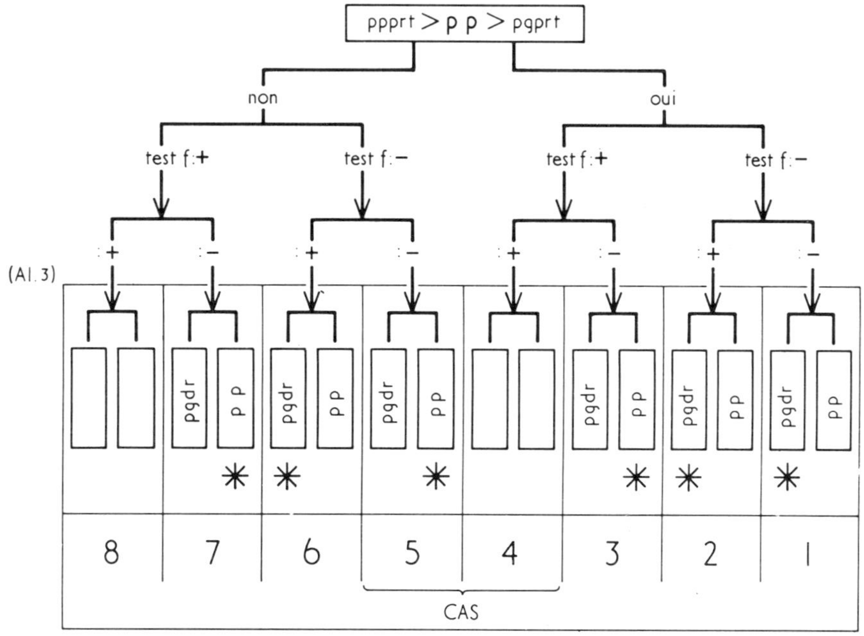

✱ LA PROPOSITION DE COMMANDE À RETENIR EST INDIQUÉE PAR L'ASTÉRISQUE

Figure 3 Programme de comparaison

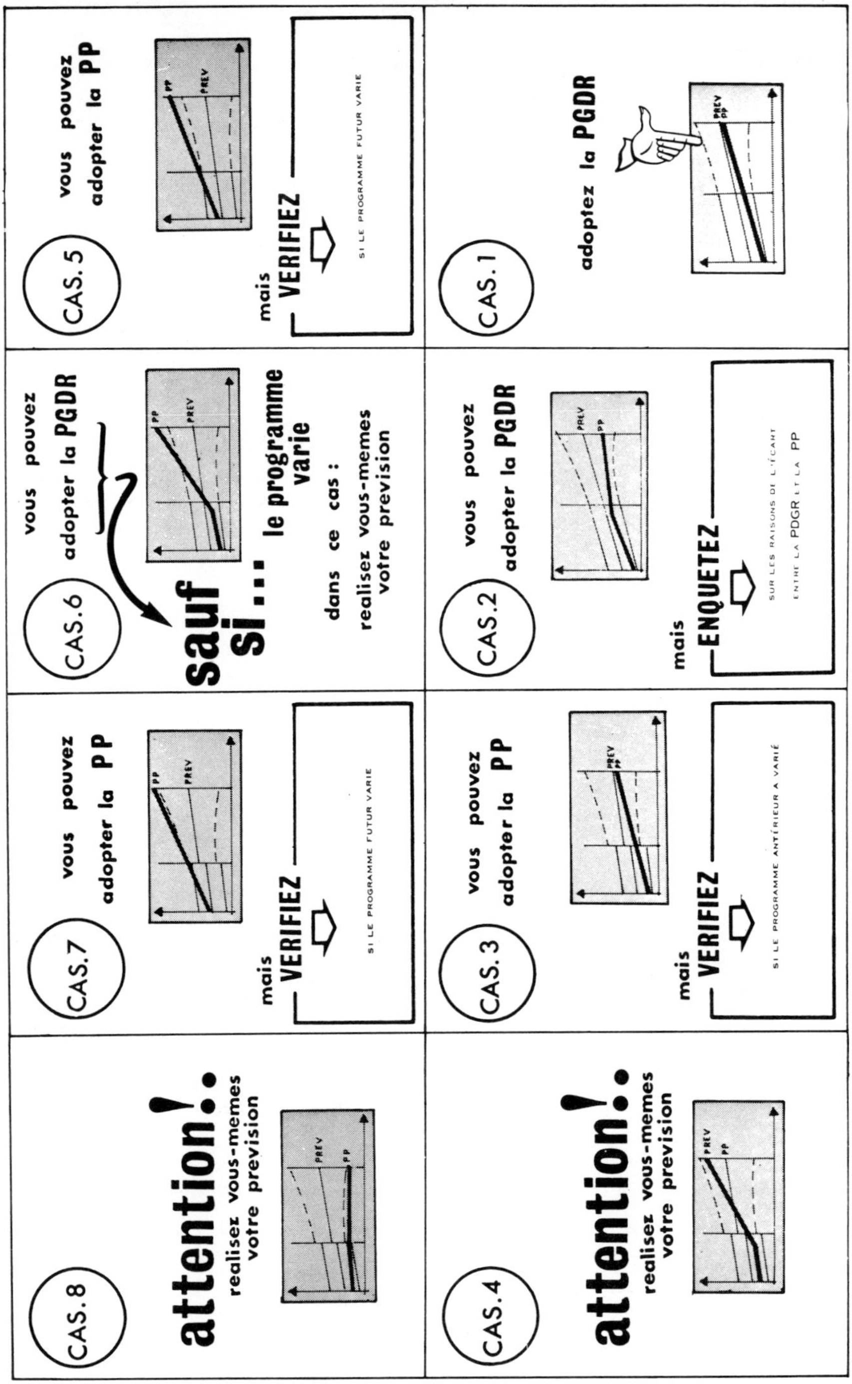
CAS.8
attention!..
realisez vous-memes votre prevision
PREV
PP
CAS.7
vous pouvez adopter la PP
PP
PREV
mais VERIFIEZ
SI LE PROGRAMME FUTUR VARIE
CAS.6
vous pouvez adopter la PGDR
sauf si...
PP
PREV
le programme varie
dans ce cas :
realisez vous-memes votre prevision
CAS.5
vous pouvez adopter la PP
PP
PREV
mais VERIFIEZ
SI LE PROGRAMME FUTUR VARIE
CAS.4
attention!..
realisez vous-memes votre prevision
PREV
PP
CAS.3
vous pouvez adopter la PP
PREV
PP
mais VERIFIEZ
SI LE PROGRAMME ANTÉRIEUR A VARIÉ
CAS.2
vous pouvez adopter la PGDR
PREV
PP
mais ENQUETEZ
SUR LES RAISONS DE L'ÉCART
ENTRE LA PDGR ET LA PP
CAS.1
adoptez la PGDR
PREV
PP

Figure 4

5. LE TRAITEMENT INFORMATIQUE

5.1 Organisation

Il est organisé en travaux journaliers et en travaux trimestriels.
Les organigrammes sont donnés en annexe.

5.2 Description des fichiers

5.2.1 Fichier de Gestion

(a) *Composition du fichier*

L'ensemble des informations intéressant le fichier de gestion fait l'objet du détail ci-après:

(b) *Eléments communs aux consommables et réparables*

		Informations Alphanumériques	*Informations Numériques*
– Code type d'article		1	
– Chaînage		4	
– Part number sans tiret		15	
– Place des tirets dans le part number		4	
– Designation		10	
– 2 Parts number interchangeables		36	
– FSN – CSD		14	
– Code OTAN		5	
– Code catégorie (C.R. ou M)		1	
– Code sans catégorie (P, I, etc...)		1	
–Code avion			2
– 3 validités avion			6
– Code stock maison			1
– Classe (A B C D)			1
– 2 numéros (magasin – casier)			10
– Prix			8
– Code prix			2
– Délai d'approvisionnement			2
	Code matériel (4 K)		
	Coefficient consommation (2 K)		
	Nombre pièces par unité programme (2 K)		220
– 4 fois	Nombre points observés (1 K)		
	8 zônes sorties trimestrielles (40 K)		
	Cumul des 8 zônes sorties (6 K)		
– 24 zônes de sorties banalisées (5 K)			120
– Nombre de points banalisés observés (2 K)			2
– Stock de pièces en état			5
– Reste à livrer en neuf			5
– Limite de stockage			2
– Prêts			5

(c) Eléments propres aux réparables

– Volant théorique	3
– Volant réel	3
– Stock en réparation	3
– Stock hors état	3
– Délai de réparation	2

(d) Paramètre A.I.M.S.

– Code calendrier	1
– B 1 : Coef. de lissage (pour valeur non nulle) (Loi de POISSON)	2
– B 2 : Coef. de lissage (pour valeur nulle) (Loi de POISSON)	2
– SM 1 : Termes récurrents de calcul de coef. de lissage (Loi de POISSON)	4
– SM 2 : Termes récurrents de calcul de coef. de lissage (Loi de POISSON)	4
– VARI 1 : Variance pondérée entre EST et valeur non nulle	6
– VARI 2 : Variance pondérée entre EST et valeur nulle	6
– EST 1 : Estimation relative à la valeur traitée non nulle	6
– EST 2 : : Estimation relative à la valeur traitée nulle	6
– Z : Compteur des valeurs nulles en loi de POISSON	2
– M : Termes récurrents du calcul de coef. de lissage (loi de GAUSS)	4
– N : Termes récurrents de calcul du coef. de lissage (Loi de GAUSS)	5
– O : Termes récurrents du calcul de coef. de lissage (Loi de GAUSS)	6
– W : Termes récurrents de calcul de coef. de lissage (Loi de GAUSS)	7
– EST : Point résultant après lissage de la dernière observation	8
– R : Trend mathématique	7
– DELTA: Aléa statistique entre observation et son résultat lissé	8
– EPSIL : Aléa statistique entre 2 observations successives	8
– B : Coefficient de lissage pour loi de GAUSS	3
– ITEST : Coefficient de confiance de la tendance	2
– ILOI : Indicateur de loi utilisé	1
– Anomalie sur consommation	1
– Compteur d'anomalie sur consommation	1
– Anomalie par test F	1
– Compteur d'anomalie par test F	1
– PGPRT	5
– PPPRT	5
– PGDR	5
– TREND – Tendance significative à employer pour extrapoler (Egal soit à R ou O suivant stabilité de la courbe)	7
– Période de révision (période de commande)	2
– Stock d'alerte	5

– Taux de service		4
– NZG nombre de zéro consécutifs en loi de GAUSS		2

(e) Paramètres séries programmées

– Prévisoin de commande (PP)		5
– 4 alertes sur coef. de consommation (par programme)	—	1
Total des caractéres numériques		546
Total des octets	91	273
Total général	364	
	soit 5 secteurs dont 11 octets disponibles	

(f) *Bilan d'Occupation*

Un cylindre par disque réservé au Soft et à l'indexation, 99 cylindres restent disponibles par disque, soit 19.800 Octets.

Donc possibilité de: $\frac{19.800}{75}$ = 3.960 articles par unité de disque

Soit pour les 40.000 articles, 11 unités (compte tenu des zônes OVERFLOW)

5.2.2 *Fichier d'Accés Direct (Télégestion)*

(a) *Composition du fichier*

L'ensemble des informations visualisées sur écran ou machine télétype, fait l'object du détail ci-après.

Il est toutefois nécessaire de préciser que dans le but de limitation de place nous avons décidé de décomposer en:

1 zône principale ou fichier principal: qui regroupe les éléments communs à l'ensemble des articles.

3 Zônes secondaires: ou sous fichiers qui comprennent les éléments seulement particuliers à certains articles.

(b) La zone principale

(mode d'enregistrement : 8 bits caractère)

		Caractères alphanumérqiues	*Caractères numériques*
– Code type article	Soft	1	
– Zône de chaînage	Soft	4	

– Part number sans tiret	15	
– Place des tirets dans le part number	4	
– Désignation	10	
– Code catégorie (consommable		
(réparable		
(modification	1	
– Code avion		2
– Tests (présence d'interchangeables		
(présence de commandes		
(présence de prêts		
(présence stock société		1
– Classes (A –B –C ou D)		1
– Codes matériels		16
– 2 numéros, magasins + casiers		10
– Prix		8
– Code périodicité de commande		2
– Code prix		2
– Délai d'approvisionnement		2
– Délai de réparation		2
– 4 coéfficients de consommation		8
– Taux de service		4
– Stock des pièces en état		5
– Stock de pièces hors d'état		3
– Stock d'alerte (A1. 2)		5
– Estimation de consommation trimestrielle		5
– Volant réel (réparable)		3
– Test d'alerte signalée		1
Total des caractères numériques		80
Totaux des octets	35	40
Total général	75 Soit 1 secteur	

(c) Les zones secondaires

– Les éléments relatifs:

Aux Interchangeables : (mode d'enregistrement 6 bits/caractère)

		Caractères Alphanumeriques
– Code type article	Soft	1
– Zône de chalînage		1
– Part number sans tiret		15
– Part number interchangeable sans tiret		15
– Part number interchangeable sans tiret		15
Total		50 caractères Soit 1/2 secteur

Aux Commandes : (mode d'enregistrement : 8 bits/caractère)

		Caractères Alphanumériques	*Caractères Numériques*
– Code type article	Soft	1	
– Zône chaînage		4	
– Part number sans tiret		15	
– 4 commandes comprenant chacune			108
– numéro des documents (02. 03. 14. 01 ou 10)			
– quantités			
– dates de livraisons prévues			
– référence des documents			
– codes responsibilité (achat lancement GER)			
– codes informations entrées en "réception"			
Total des caractéres numériques			108
Total des Octets		20	
Total général		74 Octets soit 1 secteur dont 1 octet disponible	

Aux prêts : mode d'enregistrement : 8 bits/caractère)

– Code type article	Soft	1	
– Zone de chaînage		4	
– Part number sans tiret		15	
– 5 zônes de prêt comprenant chaucune :			104
– quantité			
– référence du bon de prêt			
– numéro du magasin emprunteur			
Total des caractères numériques			104
Total des Octets		20	
Total général		72 Octets soit 1 secteur dont 3 octets inutilisés	

(d) *Bilan d'occupation*

Sur les trois disques dont nous disposons en accès direct, nous avons réservé deux disques sur lesquels seront enregistrés les éléments de la zône dite principale, le troisième étant utilisé:

- pour enregistrement des zônes secondaires
- pour les programmes de mises à jour et de consultation

(e) *Disques d'Enregistrement de Zones Principales.*

Deux cylindres par disque étant réservés au Soft et à l'indexation, 98 cylindres restent disponibles par disque le fichier, soit: 19.600 secteurs. Nous disposons de 39.200 secteurs donc d'une possibilité de visualisation de 39.200 articles.

Disques d'Enregistrement de Zones Secondaires et Programmes

En ce qui concerne les zônes secondaires nous avons retenu:

- 3200 références possédant des interchangeables, soit 1600 secteurs
- 3000 références en commande soit 3000 secteurs
- 600 références objet de prêt soit 600 secteurs
- 800 secteurs réservés pour le Soft et l'indexation correspondants

Au total 6.000 secteurs pour les zônes secondaires.

En ce qui concerne les programmes de mise à jour et de consultation 14.000 secteurs restent donc disponibles.

5.3 Exploitation–Elle comprend:

a) La consultation en accès direct à partir des bureaux des gestionnaires dans lesquels sont implantés les télétypes établissant à la demande la "Fiche de position" de l'article demandé.

b) L'exploitation quotidienne qui signale aux gestionnaire les anomalies dans les mouvements de la veille relevées par les divers contrôles informatiques.

c) L'exploitation trimestrielle établissant les propositions de commande soumises aux gestionnaires.

5.3.1 Temps de traitement

- Consultation en accès direct : Heures ouvrables de l'usine
- Traitement journalier : 1 H 30
- Traitement trimestriel : 60 H 00

5.3.2 Les états mécanographiques

Leur nombre est volontaitement réduit pour éviter l'ensevelissement des gestionnaires sous le flot de papier.
Ils comprennent:

Deisgnation de l'Etat	Fréquence d'impression	Observations
– Etat des mouvements refusés et discordance des commandes	Quotidienne	(1)
– Contrôle des noveaux de stock sur mouvement de la veille	Quotidienne	(1) (2)
– Etat d'approvisionnement et contrôle des consommations	Trimestrielle	
– Fiche de position d'article	A la demande du gestionnaire	

(1) Cet état n'est émis que pour signaler les anomalies relevées par les contrôles informatiques. Dans le meilleur des cas, il n'est donc pas émis. L'expérience permet d'estimer que 3% des postes mouvementés font l'objet d'anomalies ou de remarques concernant leur niveau de stock (30 cas par jour environ)

(2) Seules les anomalies sur stocks sont signalées.

5.3.3 Les contrôles significatifs

Il serait maladroit de ne pas signaler les contrôles suivants, véritables clignotants au service des Gestionnaires.
Par ses programmes, le système établit un réseau d'alertes dont les significations sont les suivantes:

Fréquence	Désignation	Signification
Trimestrielle	Alerte test F	Depuis 3 periodes, la consommation est stabilisée au niveau anormal qui est dès maintenant utilisé pour les prévisions futures.
	Alerte 4	Variation anormale du coefficient des consommations aléatoires.
	Alerte 3	Variation anormale du coefficient des consommations programmées.
Quotidienne	Alerte 2	La consommation n'est pas couverte durant le délai d'approvisionnement.
	Alerte 1	Rupture de stock à prévoir avant le rentrée de la prochaine commande.
	Alerte 0	Stock à zéro

5.3.4 Les indices de priorité

Ils sont établis à partir des alertes précédentes. En outre, pour permettre les contrôles, enquêtes et décisions relatives aux propositions de commandes, le SAAPM établit un indice de priorité répartissant ce travail sur trois mois et couvrant la consommation des article sur cette même période.

6. LE DIALOGUE HOMME-MACHINE

Le SAAPM est un outil perfectionné dont le contrôle et l'animation ont été volontairement confiés à son Exploitant, le Gestionnaire de stocks.

En matière d'approvisionnement, comme en management, la décision est difficile à prendre en l'absense des informations élémentaires et de synthèse.

Le SAAPM a confié à l'ordinateur le rôle pour lequel celui-ci a été conçu, la saisie de l'information élémentaire et son traitement, seul moyen d'établissement des données de base préliminaires à toute synthèse. Il lui a également confié la préparation de la synthèse mais a réservé au Gestionnaire, la conclusion de cette synthèse et la décision qu'elle entraîne.

L'affectation de ces domaines réservés, l'un objectif, l'autre subjectif, implique forcément le dialogue entre l'homme et la machine.

C'est la raison pour laquelle notre entreprise est équipée en télégestion. Il faut voir dans cet équipement, non pas le moyen d'interroger les stocks, mais la seule voie qui amène le Gestionnaire au coeur du stock qu'il anime.

La constitution, la gestion ou la mutation d'un stock en cours de correction est une opération qui s'étale sur plusieurs années et dans ses méandres se perdent les efforts et les espoirs des Gestionnaires.

L'animation d'un stock est une fonction directe du dialogue; il faut donc créer celui-ci et lui offrir un moyen d'expression qui pour notre Entreprise est le SAAPM.

7. LE BILAN D'EXPLOITATION

Aux fins de contrôle de l'exploitation et de la gestion des stocks, un bilan annuel d'exploitation est élaboré.

Il permet à la Direction et aux Responsables des Approvisionnements, l'établissement de conclusions sur l'Exercice passé afin de prendre en conséquence les mesures éventuelles nécessaires à l'amélioration de l'Exercice en cours.

Ce bilan présente les volets suivants:

- Valeurs des stocks et des commandes en cours
- Inventaire des stocks mettant en relief les écarts sur chaque poste, les inactifs, etc . . .
- Paramètres de gestion, soit taux de service moyen optimisé, nombre et nature des alertes, etc . . .
- Informations relatives à l'animation des stocks par type d'avion et par catégories d'articles.

ANNEXE

COMPTABILISATION DES STOCKS 1) DOCUMENTS D'ENTREE EN MAGASIN

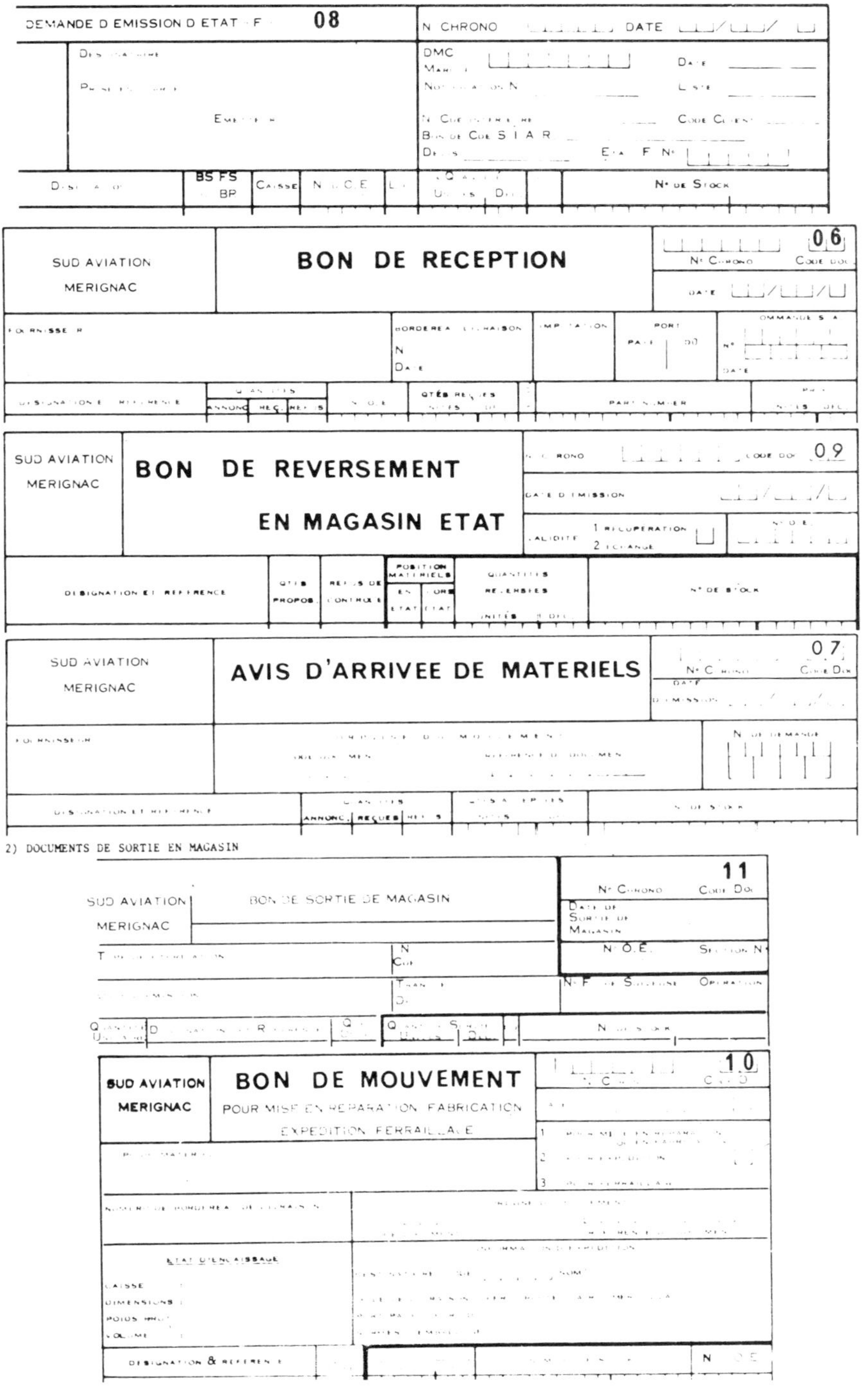

DEMANDE D EMISSION D ETAT F 08
N CHRONO DATE
DMC
BS FS BP
CAISSE
N° DE STOCK

SUD AVIATION MERIGNAC
BON DE RECEPTION
06
N° CHRONO CODE DOC
DATE
FOURNISSEUR
BORDEREAU LIVRAISON
N
DATE
PORT
QTÉS REÇUES
PART NUMBER

SUD AVIATION MERIGNAC
BON DE REVERSEMENT EN MAGASIN ETAT
09
CODE DOC
DATE D EMISSION
1 RECUPERATION
2 ECHANGE
DESIGNATION ET REFERENCE
QTES PROPOS
POSITION MATERIELS
QUANTITES RECENSEES
N° DE STOCK

SUD AVIATION MERIGNAC
AVIS D'ARRIVEE DE MATERIELS
07
N° CHRONO CODE DOC
DATE D EMISSION
FOURNISSEUR
N° DE DEMANDE
DESIGNATION ET REFERENCE
ANNONC. REÇUES
N° DE STOCK

2) DOCUMENTS DE SORTIE EN MAGASIN

SUD AVIATION MERIGNAC
BON DE SORTIE DE MAGASIN
11
N° CHRONO CODE DOC
DATE DE SORTIE DE MAGASIN
N° O.E.
SECTION N°
N° F
OPERATION

SUD AVIATION MERIGNAC
BON DE MOUVEMENT
POUR MISE EN REPARATION FABRICATION EXPEDITION FERRAILLAGE
10
ETAT D'ENCAISSAGE
CAISSE
DIMENSIONS
POIDS BRUT
VOLUME
DESIGNATION & REFERENCE
N

Figure A1

CONNAISSANCE DES BESOINS FUTURS

SUD-AVIATION MERIGNAC	**PREVISIONS DE PROGRAMMES**	Nº CHRONO — CODE DOC. 16 DATE D'ÉMISSION GESTIONNAIRE : VISA — CODE

SI LA PRÉVISION EST RÉALISÉE DURANT

☐ 1. LES 3 OU 4ÈME TRI. DE L'ANNÉE EN COURS

ANNÉE EN COURS		ANNÉE 1				ANNÉE 2				ANNÉE 3	
3°tr.	4°tr.	1°tr.	2°tr.	3°tr.	4°tr.	1°tr.	2°tr.	3°tr.	4°tr.	1°tr.	2°tr.

☐ 2. LES 1 OU 2ÈME TRI. DE L'ANNÉE EN COURS.

ANNÉE EN COURS				ANNÉE 1				ANNÉE 2			
1°tr.	2°tr.	3°tr.	4°tr.	1°tr.	2°tr.	3°tr.	4°tr.	1°tr.	2°tr.	3°tr.	4°tr.

CODE FONCTION C S OU M — CODE MATÉRIELS — PRÉVISIONS PROGRAMMES

PRISE EN COMPTE DES COMMANDES/DOCUMENTS DE BASE

(1) OUVERTURE / OU / MODIFICATIF Nº ...... D'ANNEXE A COMMANDE INTERIEURE — COMMANDE CLIENT — Date — Numéro

NUMERO DE CDE INTERIEURE — FICHE DE REMISE EN ETAT — D.M.C. Etat — Privé

ORDRE D'EXECUTION	Indices IC	IR	CODE CLIENT	MODELE "63" NUMERO	BORDEREAU DE RECEPTION Numéro	Date	CODE MATERIEL	QTE	NUMERO DE STOCK OU P.N.	DESIGNATION	Responsabilité

(2) OUVERTURE / OU / MODIFICATIF Nº ...... D'ANNEXE A COMMANDE INTERIEURE — COMMANDE CLIENT — Date — Numéro

NUMERO DE CDE INTERIEURE — FICHE DE FOURNITURES — D.M.C. Etat — Privé

ORDRE D'EXECUTION	Indices IC	I.	CODE CLIENT	CODE ARTICLE	QUANTITE Unité	Déc.	DESIGNATION	REFERENCE	Responsabilité

SUD-AVIATION MERIGNAC	APPROVISIONNEMENT ÉTAT DEMANDE Nº — DATE / / — CODE DOC. 04
	RÉFÉRENCE DOCUMENT ÉMIS — CODE — NUMÉRO
	DATE DE LIVRAISON PRÉVUE / /

DÉSIGNATIONS & RÉFÉRENCES	QTÉS DEMANDÉES UNITÉS	DÉC		Nº DE STOCK

PRISE EN COMPTE DES COMMANDES/DOCUMENT DE CONFIRMATION

(14) OUVERTURE / OU / MODIFICATIF Nº ...... A ADDITIF Nº ...... A ANNEXE Nº ...... A COMMANDE INTERIEURE — BON DE COMMANDE ETAT — Date — Numéro

NUMERO DE CDE INTERIEURE — FICHE DE DELAIS ET DE LOTISSEMENT

Ce document peut être utilisé communément pour les 3 types d'annexes à commandes intérieures.

ORDRE EXECUTION	DELAIS											
	Nº lot	Echéance	Quantité Unité	Déc	Nº Lot	Echéance	Quantité Unité	Déc.	Nº Lot	Echéance	Quantité Unité	Déc.

Voir en page suivante la commande commerciale

Figure A2

SUD-AVIATION

SOCIÉTÉ NATIONALE DE CONSTRUCTIONS AÉRONAUTIQUES

Société Anonyme au Capital de 316.500.000 F

Régie par la Loi du 24 Juillet 1966 et le Décret du 23 Mars 1967 sur les Sociétés Commerciales

Siège Social : 37, boul. de Montmorency - PARIS XVI°

DIVISION INDUSTRIELLE

G. E. R.

GROUPE ENTRETIEN RÉPARATIONS

AÉRODROME DE BORDEAUX-MÉRIGNAC

B. P. N° 2 — 33-MÉRIGNAC

Tel. : 44.72.88 et 44.74.76
C.C.P. : 3405.98 Bordeaux
N° Entr. 270.33.281.9.003
— R. C. Seine 57-B-9451 —

Télégramme
AÉROSUDÉO-Bordeaux
Télex
Indic. : SUDAÉRO-SUREN
N° Appel : 206.46

IMPORTANT
Référence à rappeler
COMMANDE N°

Date :

Page :

ADRESSE DE LIVRAISON
SUD-AVIATION — USINE SFERMA — 33-MÉRIGNAC

MODE DE LIVRAISON :

CONDITIONS DE RÉCEPTION :

CONDITIONS DE PAIEMENT :

AFFECTATION :

IMPUTATION :

D. A. N°

DEMANDEUR :

FACTURATION :

LE CHEF DES ACHATS :

SERVICE SURVEILLANCE ÉMETTEUR

SOUS COMMANDE : VOIR N° DE Cde

MARCHÉ D'ÉTAT N° :

N° Cde ADMINISTRATION CENTRALE

SERVICE SURVEILLANCE S/COMMANDIER

D. M. C. N° :

USINE D'EXÉCUTION :

POSTE	DÉSIGNATION DES MATÉRIELS	DATE DE LIVRAISON	N. O. E.	QUANTITÉ	N DE STOCK	PRIX HORS TAXE UNITAIRE	PRIX HORS TAXE TOTAL

S.A.G.E.R. 84.000 70 332 bis - Bilion 362E9-69

FOURNISSEUR

Figure A3

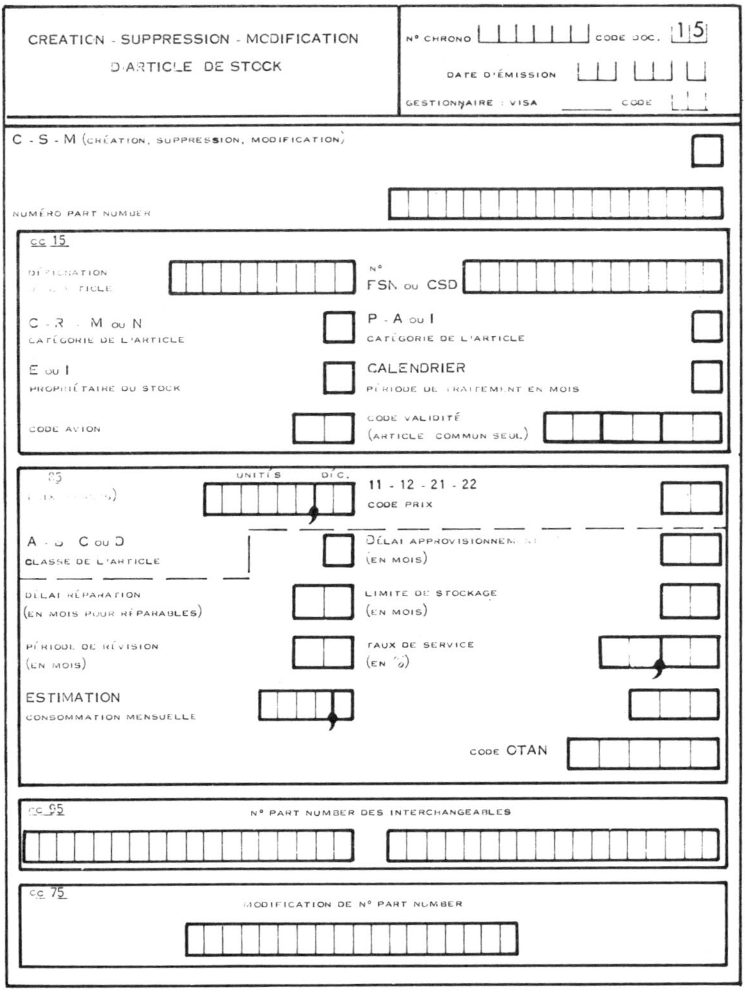
CREATION - SUPPRESSION - MODIFICATION
D'ARTICLE DE STOCK
N° CHRONO
CODE DOC. 1 5
DATE D'ÉMISSION
GESTIONNAIRE : VISA
CODE
C - S - M (CRÉATION, SUPPRESSION, MODIFICATION)
NUMÉRO PART NUMBER
cc 15
N°
FSN ou CSD
C - R - M ou N
CATÉGORIE DE L'ARTICLE
P - A ou I
CATÉGORIE DE L'ARTICLE
E ou I
PROPRIÉTAIRE DU STOCK
CALENDRIER
PÉRIODE DE TRAITEMENT EN MOIS
CODE AVION
CODE VALIDITÉ
(ARTICLE COMMUN SEUL)
UNITÉS
DÉC.
11 - 12 - 21 - 22
CODE PRIX
A - B C ou D
CLASSE DE L'ARTICLE
DÉLAI APPROVISIONNEMENT
(EN MOIS)
DÉLAI RÉPARATION
(EN MOIS POUR RÉPARABLES)
LIMITE DE STOCKAGE
(EN MOIS)
PÉRIODE DE RÉVISION
(EN MOIS)
TAUX DE SERVICE
(EN %)
ESTIMATION
CONSOMMATION MENSUELLE
CODE OTAN
cc 55
N° PART NUMBER DES INTERCHANGEABLES
cc 75
MODIFICATION DE N° PART NUMBER

Figure A4

DISCUSSION

Bureau:

1. 40.000 articles gérés scientifiquement.
   60.000 pas
   % dans le budget annuel?

2. Il est présumé que le système a introduit un niveau de sécurité tenant compte des écarts dans les valeurs des paramètres.

   Est-il tenu compte *à la fois* des écarts de consommations *et* des écarts du temps d'acheminement?

Sahnoun:

1. En réalité il existe 2 types de pièces. Celles appartenant à l'Etat et celles appartenant à Sud-Aviation (100.000 et 40.000).

   Actuellement seules les pièces appartenant à l'Etat sont gérées par le système présenté. Pour permettre la gestion des stocks Sud-Aviation l'augmentation du nombre des unités de disques est nécessaire et a été décidée.

   Pour répondre plus précisément à la question, les 40.000 pièces du stock Etat gérées représentent environ 80% du Budget, les autres constituant essentiellement des stocks morts, ou ceux affectés à des avions en voie de disparition (C 47–Beech etc).

2. Le système peut tenir compte des écarts des temps d'acheminement. De plus, il faut souligner que le système constitue en lui-même un modèle de simulation. Ainsi il est possible de mesurer l'influence d'un paramètre sur le niveau du stock. Cette possibilité est d'ailleurs exploitée pour la Formation des Gestionnaires.

Olivier: Vous utilisez une abaque précise pour finalement faire un approvisionnement périodique.

Plus généralement je crains que des contraintes (stocks de guerre, politique de marchés périodiques) extérieures aient sur l'économie de la gestion une influence beaucoup plus grande que les destructions quelquefois subtiles des modèles mathématiques.

Avez-vous étudié l'intérêt économique de votre modèle en tenant compte des contraintes qui vous étaient imposées par ailleurs?

Sahnoun: Non, le système est mis en place depuis trop peu de temps pour que nous ayons pu mesurer l'influence relative des gains dus à l'optimisation et de ceux dus à de meilleures conditions de gestion. Ainsi l'ingénieur en chef M. Danicaux nous a autorisé à commander les approvisionnements (dans un cadre budgétaire) sans attendre l'autorisation de dépense. Sur certains articles plus de six mois de délais administratifs ont été gagnés et ce gain risque d'être plus important que celui dû aux optimisations. D'autant plus qu'optimiser ne sert parfois à rien. Ainsi nous avons trouvé qu'une pièce de très faible valeur et de très faible consommation ne devait être réapprovisionnée que tous les 2000 ans. Nous avons évidemment adopté une fréquence annuelle et nous fûmes stupéfait car dans la réalité le fournisseur ne livre que par 500 pièces ce qui représente 4000 ans de stocks.

Cependant, les algorithmes d'optimisation sont très performants et fort bon marché puisque 15000 articles peuvent être traités en une heure avec ce modeste équipement. Aussi n'a-t-on aucune raison de ne pas les utiliser même si après avoir procédé à de minutieuses optimisations nous concluons au vu des alarmes en demandant au gestionnaire de réaliser lui-même sa propre prévision.

De Belder: Quels sont les problèmes que vous avez rencontrés au moment de la conversion, dans les magasins, vers le teleprocessing?

Y a-t-il eu des problèmes de recyclage ou de remplacement de personnel inadapté et comment les avez-vous résolus?

Sahnoun:

a. Problèmes de nature secondaire mais d'effets très importants.

- Erreurs dans la rédaction des Documents.
- Incompréhension des relations entre l'information élémentaire et le résultat obtenu dans d'autres secteurs très différents.
- Crainte de la machine.
- Mise en doute de la réalité des résultats jusqu, à l'obtention de la preuve.
- Impression complète que le problème ne peut être pris en compte par l'ordinateur.
- Attentre d'être "au pied du mur" avant de s'intéresser au système.
- D'une manière paradoxale, ce comportement étant manifesté en même temps que l'espoir et la croyance de voir l'ordinateur réaliser des miracles et réduire les difficultés réelles et connues.

b. *Seules méthodes retenues*:

- Traitement quotidien de vraisemblance des documents.
- Période à blanc non publiée. Pendant ce temps, retourner les documents par petites quantités à leurs auteurs pour correction obligatoire.
- Contact constant avec les représentants informatiques. Il faut inclure parmi ceux-ci le responsable *de plus haut niveau hiérarchique possible*.
- Petites réunions de travail avec les magasins et gestionnaires de tous niveaux et des représentants informatiques toujours épaulés par le responsable hiérarchique de plus haut niveau.

En un mot, il faut que *chacun sente le vent souffler* et sache qu'il n'y aura *pas de retour en arrière*.

c. Un déchet de personnel est à prévoir. Il touchera:

- Le personnel très intéressé par l'application, plein de bonne volonté mais mentalement inadaptable (jeunes).
- Le personnel ne voulant pas s'intéresser à l'application mais prêt à la subir sans trop de dommage (vieux).

Optimal Stock Levels for Spare Parts within the Logistic System of a Division

Klaus A. Fent

1. STATEMENT OF THE PROBLEM

The customer, the German Ministry of Defence, is interested in proposals defining techniques which improve the present efficiency of storage and supply of spare parts pertaining to the maintenance system of a division. This should be achieved by a so-called "Optimal Authorized Stock Level Calculation Technique". Moreover, a military requirement concerning system performance is defined which requires that the combat readiness of trucks, tanks and tracked vehicles of the division should not fall below a certain limit.

With this very long term "Optimal Authorized Stock Levels Calculation Technique" the customer had two things in mind:

(i) To determine the authorized stock level quantities that are optimal because they imply a minimum of spare part stocks within the system;

(ii) To find a technique of calculating these quantities which involve minimal material requirements and training of the military personnel at the storage places.

The currently used calculation is as follows:

$$ASL_{cp} = \tfrac{1}{2} \left\{ \frac{D}{N_{cp}} \quad N_{rsp} + ASL_{cp-1} \right\}$$

ASL_{cp} = Authorized Stock Level of the next calculation period

D = Demand expected for the next c.p.
= consumption quantity in the past c.p.

N_{cp} = Number of days of the past c.p.

N_{rsp} = Number of days of the required stocking period

the required stocking period is

30 days for 1st echelon stocks
45 ,, ,, 2nd ,, ,,
60 ,, ,, 3rd ,, ,,

ASL_{cp-1} = Authorized Stock Level of the past calculation period

The calculation must be done manually at each storage place and for those spare-parts that have to be stocked according to logisitical regulations.

The required stocking period is a quantified term for a certain supply range of a given stock, based on general logistic principles. This term involves many factors which are represented by one value for all stocks of a certain echelon, or by three values for all stocks of the whole system. These factors are essentially:

(i) A safety support against fluctuation of demand, for example, in consequence of field maneuvers or general test of the vehicles.

(ii) An emergency reserve to be stocked at peace generally to guarantee a certain independence of supply.

(iii) An estimated value for the different times of requisition and delivery (=stock-bound supply-time data).

(iv) An additional reserve to make up the insufficiencies of the calculation method.

The customer assumed that the general stock level of spare-parts within the logistic system of a division is too high, although it was known that the storage of spare-parts proved to be incapable–again and again–to meet the demand of the maintenance shops in due time, therefore, a study has been investigated by the OR-Gruppen of the IABG at Ottobrunn.

2. MEASUREMENT PROBLEMS–TWO MODELS

While searching for possible ways to carry out this study it was found that two models would have to be designed.

Concerning the organization and operation of maintenance it should be mentioned that storage and supply of spare parts for maintenance is a subsystem of the whole maintenance system: the "Spare Part Stockage and Supply System" (SPSSS). Its aim is to keep maintenance shops supplied with spare parts.

Two significant and measurable criteria for judging this logistic system was given by the 'authorized stock levels' as the optimization parameter and the 'mean value of all stocks' as the input measure. These two measures are applicable to the subsystem. There was also an output measure in that the contractor required that the combat readiness, here termed 'serviceability', of the vehicles should no fall below a certain limit. But there was no defined quantity for the output of the subsystem: a performance measure for holding spare parts available at maintenance shops.

There seemed no chance to design a fast-running analytical model representing the whole maintenance system of a division. This could possibly be done in a reasonable time by a simulation model. But a simulation model appeared to be inadequate for optimization. Hence, an analytical model was designed representing the SPSSS. This model accomplishes the optimization process and finds a new technique of calculation with which it is possible to propose approximate values for optimal authorized stock levels. And a simulation model too was designed which represents the whole maintenance system. It was used to determine the serviceability in respect to alternative values of authorized stock levels.

3. SIMULATION MODEL

Designing the simulation model and making it run turned out to be a very difficult job. First of all, the task to acquire and compile pertinent data caused a lot of trouble and took much time without yielding sufficient results.

To develop computer simulation of the whole maintenance system of a division is very complicated, and the first runs took about 40 minutes each on the CDC 3800. In this simulation model, 700 to 1000 vehicles of a certain type (deployed in up to 44 battalions or units) are taken into consideration, also 50 maintenance units and the stocks of 15 to 25 selected spare parts at 51 stores on three echelons. 240 working days per annum are being considered.

First the simulation model for each day calculates the serviceability of the vehicles in respect to the values of authorized stock levels which are calculated using the formula mentioned. In so doing, the maintenance capacities–on the various echelons and at the various maintenance shops of a certain echelon–are adjusted to obtain the required serviceability.

This happens in accordance with the general regulations for military maintenance, and with regard to the fact that only the number of vehicles of one type and a selected number of spare parts of this one type are taken into account.

The model is capable to determine serviceability depending on variations of authorized stock levels, with all other influence factors being constant except the demand for upkeep, repair and spare parts which are kept variable in accordance with the data obtained from a comprehensive statistical survey of this demand.

4. ANALYTICAL MODEL

As mentioned before the analytical model represents only a subsystem (SPSSS) of the whole maintenance system. In addition to the given applicable measures of 'authorized stock levels' (as the optimization parameter) and 'mean value of all stocks' (as the input measure) the 'quality of stockage' in relation to a certain spare part has been defined as the performance measure of the subsystem.

Spare parts differ in respect to rate of demand per day, r, at a certain storage place and in regard to rate of demand during the calculation period, R, in the whole system. If the quality of stockage is related to a certain storage place, it is defined as the expectation value of the time delay, $\bar{D}$, in satisfying a spare part demand. If the quality of stockage is related to all storage places that directly serve a place of upkeep or repair, it is defined as the expectation value, $\bar{W}$, as the sum of all $\bar{D}$ multiplied by the rate of daily demand r: $W = \Sigma \; \bar{D} \cdot r$. The exact definition is as follows:

$$\bar{W} = \lim_{t_0 \to \infty} \frac{1}{t_0} \; E\left(\sum_{j=1}^{n_2} \; \sum_{k=1}^{n_1(j)} \; \sum_{l=1}^{n_{t_0}(j,k)} D_l(j,k) \right)$$

t_0 = time period
E = expectation value
j = index of the 2nd echelon stocks
k = ,, ,, ,, 1st ,, ,,
l = ,, ,, a specific demand
n_2 = number of 2nd echelon stocks
$n_1(j)$ = number of 1st echelon stocks, supported by stock number j
$n_{t_0}(j,k)$ = demand in the period at a certain stock $S_{1,j,k}$
$D_1(j,k)$ = time delay to satisfy the demand l at stock $S_{1,j,k}$, (see Fig 1).

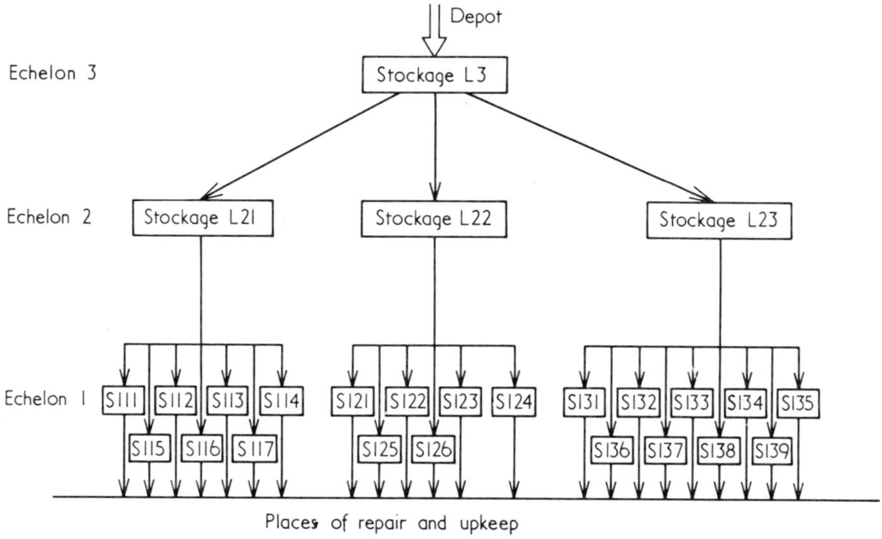

Figure 1 Spare Part Stockage and Supply System (SPSSS)

With this the performance measure of the subsystem is defined. It would also be desirable to have a specific quantity for this value to represent a minimum requirement that guarantees the required serviceability of the whole system. But it is very difficult to be successful in this way. In the first place there is no statistical material available which could serve as the basic value for a correlation ratio between the very global output measure of serviceability and $\overline{W}$ which is defined for a specific spare part only. Consequently, the correlation ratio between all $\overline{W_I}$ (I=index of a specific spare part) and the serviceability had to be considered. And there are quite a few possibilities to generate a certain $\overline{W_I}$ which is composed of different stockage qualities $\overline{D}$ at one stock on the third echelon of four stocks on the second echelon and of 44 stocks on the first echelon.

For this reason, the analytical model at first disregards the requirement of the serviceability (which should not fall below a certain limit) but simplifies the problem by requiring that the quality of stockage is equal to all stocks on a certain echelon.

Considering (for a specific spare part in $0 \leqslant t \leqslant t_0$) the dependence of time delay $\overline{D}$ and mean value of a stock $\overline{S}$ on the authorized stock level *ASL*, we find (see Fig. 2):

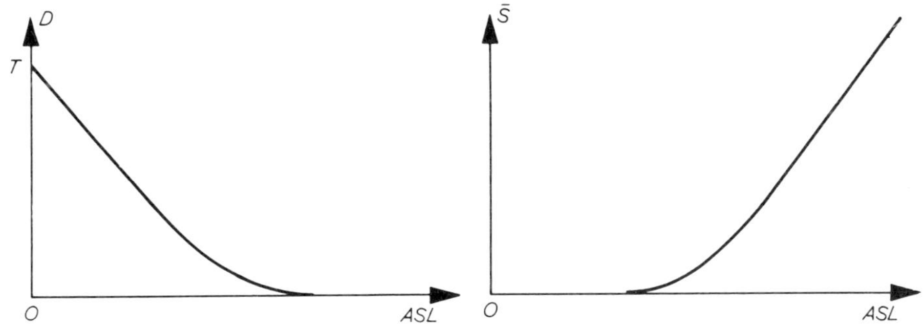

Figure 2 Dependence of time delay and of stock on authorized stock level

If $ASL = 0$, then $D = T$ (T=time of requisition and delivery, if the spare part is in store at the authorized stockage place of the next echelon). As *ASL* increases, $\overline{D}$ decreases to a point closely approaching zero. If $ASL = 0$, then $\overline{S} = 0$. As *ASL* increases, S still remains zero–until *ASL* reaches a certain value–then it increases.

If $\overline{D}$ is equal for all stocks on a certain echelon and if $\overline{D}$ is constant, it is possible to operate with one parameter F_i on each echelon i ($i = 1, 2, 3$), which means a maximal allowable $\overline{D}_i$ of the stocks on echelon i. With a given vector $\vec{F} = (F_1, F_2, F_3)$, the *ASL* at any stock on echelon i can be so determined that the corresponding $\overline{D}$ is still $= F_i$, i.e. $\overline{D}(ASL) = F_i$ and $\overline{D}(ASL-1) > F_i$. For the computation, the demand for spare parts at the store is defined as piecemeal, Poisson-distributed and time-independent.

Hence, the assignment of a certain "*ASL*-stockage" (in the system to a certain vector $\vec{F}$ is unequivocal (although not reversibly unequivocal) and the amount of all *ASL*-stockages of a spare part is identical with the amount of all $\vec{F}$-stockages ($F_1, F_2, F_3 > 0$). And the investigation can now be confined to the $\vec{F}$-stockages.

$\overline{W}$ is defined in $0 < \overline{W} \leqslant \overline{W}_{max}$ where $\overline{W}_{max}$ means that there is no stockage of the spare part concerned in the system. To a certain $\overline{W}_x$ we can assign a certain vector $\vec{F}(\overline{W}_x)$, with the result that the amount of all $\vec{F}_x$-stockages of a spare part with $0 < \overline{W}_x \leqslant \overline{W}_{max}$ is the 'characteristic curve' of this spare part, which is the function $\overline{S}_t(\overline{W})$ for any spare part of a certain R (see Fig. 3).

Now, there is a $\vec{F}_0$ as a $\vec{F}$-stockage which is identical with the optimal stockage to be found. And there also is a $\overline{W}_0 = \overline{W}(\vec{F}_0)$ which is $\overline{W}$ at optimal stockage. We find that $\overline{S}_t$ for small values of $\overline{W}$ decreases very steeply, and this effect is the more noticeable the higher rate of demand R is. This fact is considered during the optimization.

Starting the optimization, first the quality of stockage (of a certain spare part) is fixed close to zero, say about 0.1 days (expectation value of delay D) for all stocks on the first echelon, i.e. $F_1 = 0.1$ days. Then F_2 and F_3 are computed by reducing the stocks of this spare part on the tree echelons.

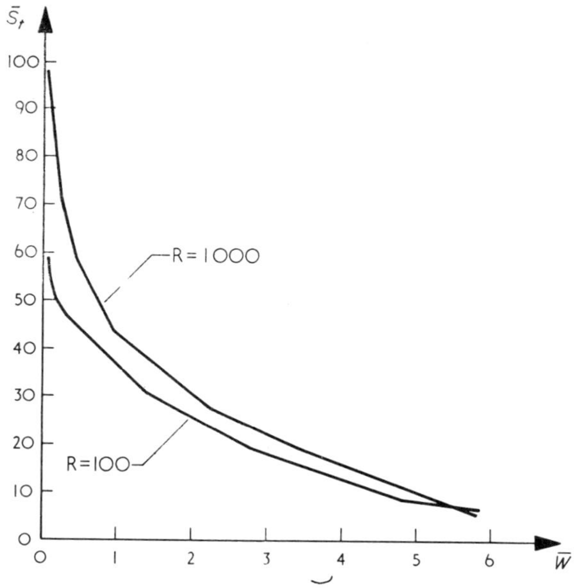

Figure 3 Characteristic curves

For that purpose we start from sufficiently high values of *ASL*. Stocks on the 1st echelon are to be reduced to the point where $\overline{D}_1(ASL) \leqslant F_1$. Then the stocks on the 2nd echelon are handled in the same way as long as $\overline{D}_1$ (ASL) is still $\leqslant F_1$. The corresponding F_2 is noted. This is also the way we get F_3 on echelon 3. So the optimal $\overrightarrow{F_0}$ is determined and we know the corresponding *ASL* on all stockage places where this spare part I is to be stocked, and we also know $\overline{W_I}$ (see FIG. 3: characteristic curves).

If a limit G is given with $\bar{W}_t \leq G$ and $\bar{W}_t = \sum_{I=1}^{N} \bar{W}_I$ (N being the number of characteristic spare parts) then the optimization operates by steps occasionally preferring that (characteristic) spare part or spare part curve at which a step achieves the highest savings, i.e. where $\Delta \bar{S}_t / \Delta \bar{W} \rightarrow \max$.

Then the analytical model pays attention to the fact that the serviceability should not fall below a certain limit, and fixes G so that $\overline{W_t^*} \leq G$ and $\overline{W_t^*} = \sum_{I=1}^{N} \overline{W_I}$, the total sum of the time-delay expectation, if *ASLs* are computed by the calculation method used so far.

5. RESULTS

Some interesting results of the study are shown using the spare parts set of a certain track. The results are valid also for the sets of the other trucks, tanks and tracked vehicles considered. Figure 4 shows that the savings effect increases as R increases. As compared with the method used, the new method achieves savings in $\overline{S_t}$.

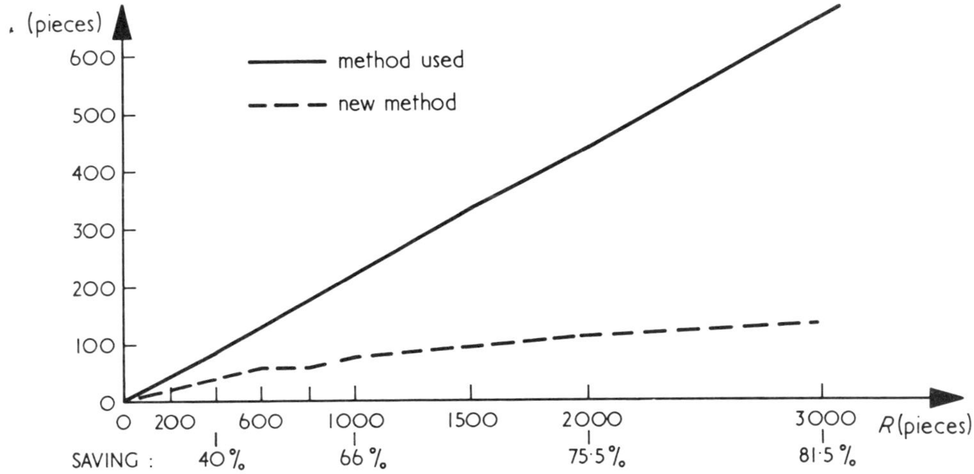

Figure 4

At the same time, the new method achieves a quality of stockage,measured in $\overline{W}/R_t$ that is nearly the same as that achieved by the method used, sometimes even better (see Fig. 5).

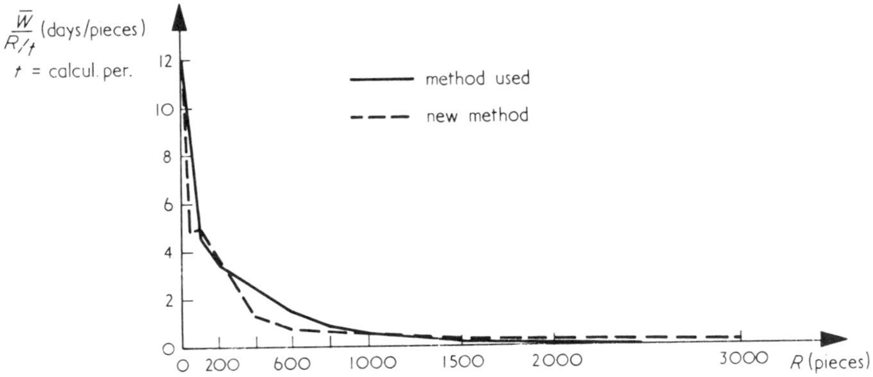

Figure 5 Mean value of time delay as a function of periodic demand

Regarding the various numbers of spare parts classified for the different values of R the savings effect in $\overline{S_t}$ was found to be 40 to 65 percent. Figures 6, 7 and 8 show that the saving is especially effective on the 2nd and 3rd echelon. These curves are the results of the simulation model; so is chart of the serviceability (according to the used and new methods) and the stress on the vehicles of the specific truck type (FIG. 9).

This brief paper had to omit a number of facts. In a résumé of the study it may be said that the savings effect shown is possible because the general requirement (by the method used) of 30, 45 or 60 days for the stocking period is substituted by the

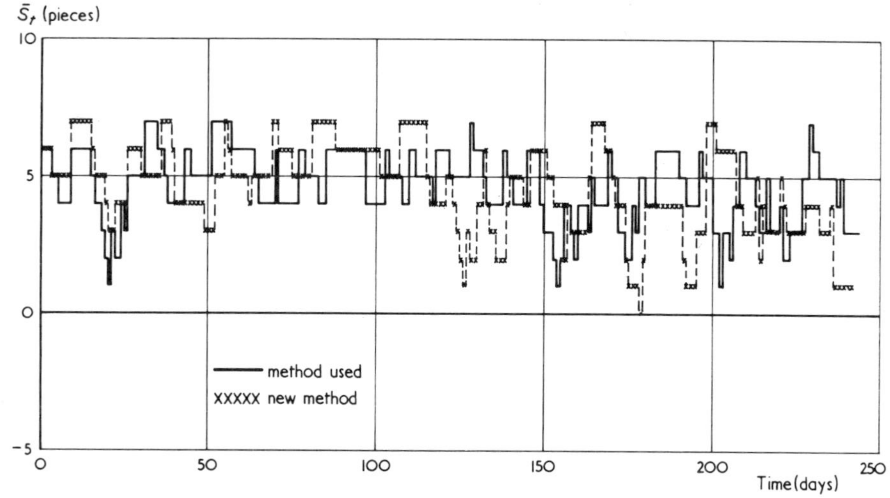

Figure 6 Stockage level chart at the first echelon

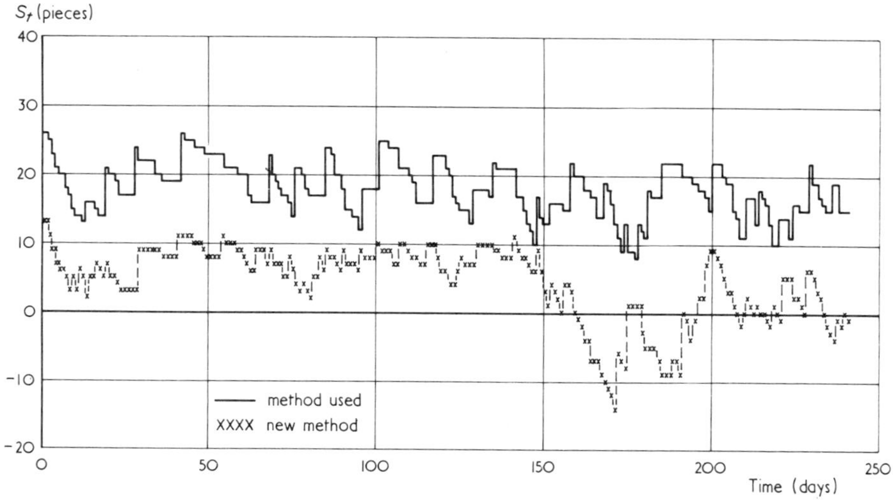

Figure 7 Stockage level chart at the second echelon

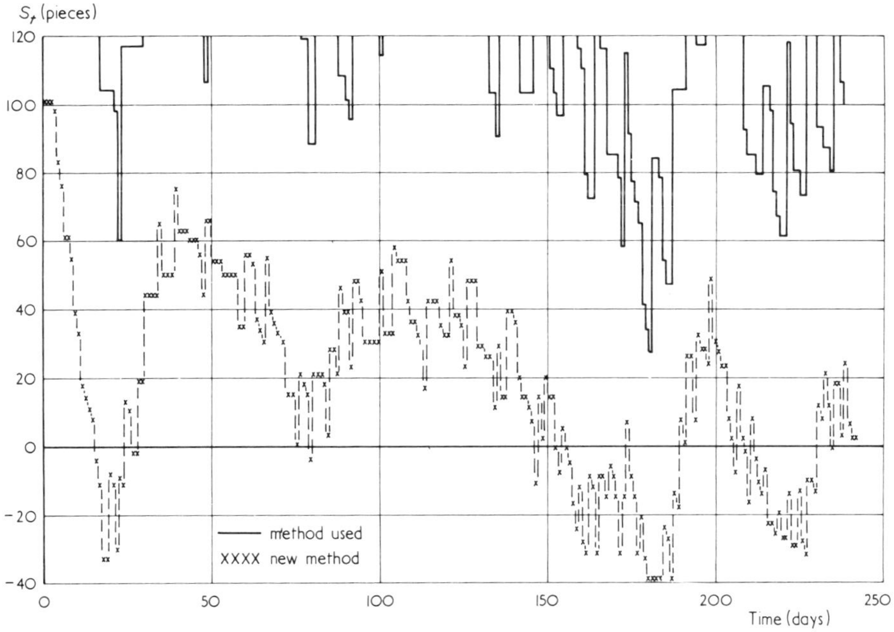

Figure 8 Stockage level chart at the third echelon

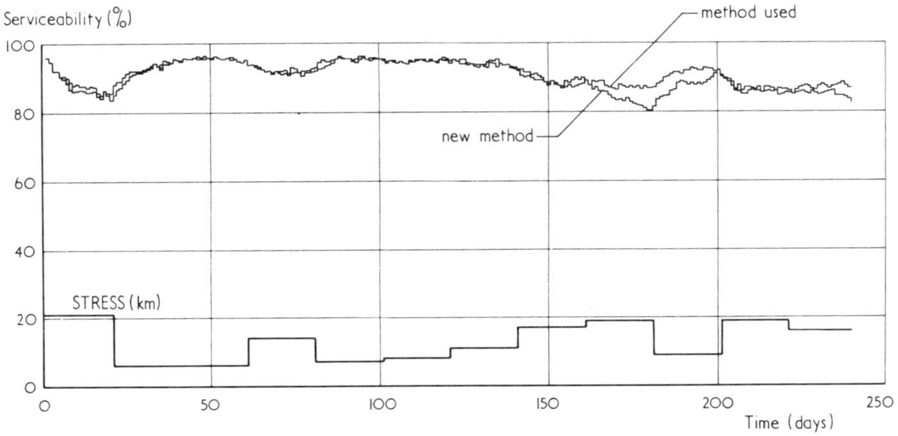

Figure 9 Serviceability and stress chart

consideration of specified requirements for 'stockage quality' of a certain stock in connection with stock-bound supply-time data.

Hence, this analysis achieves the objective of a general requirement for an equality of stockage for all stocks on a certain echelon. Now the whole system (all stocks on three echelons) is taken into consideration and a total requirement for stockage quality is determined that equals the results–measured in the same terms–from the stock level calculation method formerly used.

In consequence of this consideration of the entire system, stocks on the second, and especially on the third, echelon become a kind of transit stocks. So a remarkable reduction of stocks is possible.

This result can be obtained by a more complicated method of authorized-stock-level calculation, and this minimum of stock level can be applied only in peacetime. The dependence on supply of battalions, brigades or divisions is much higher than by the former method and, therefore, the given supply range of safety support is correspondingly lower.

But these essential factors are directly influenced by the stock-bound supply-time data. It allows a specific quantitative steering of the system as an adaption of the performance to various logistic objectives, e.g. the quick change from peace standard to the emergency demand for spare parts is feasible, or the achievement of a certain reliability level for the spare-parts supply.

This advantage to both, the logistic command and the planners, could be one reason for the customer to start investigating the possibility of a centralized system of stock bookkeeping and supply steering that uses electronic data processing equipment.

REFERENCES

Grossberger, H.
Analytisches Modell zur Optimierung der Ersatzteilbevorratung in der Versorgungskette einer Divison, Techn. Bericht 363/3, Operations Research Gruppen der IABG, Ottobrunn bei München, 1969.

Metzenroth, M.
Simulationsmodell zur Untersuchung aufwandsgünstiger Bevorratungshöhen von Ersatzteilen zur Instandsetzung von Kfz des Heeres, Techn. Memorandum 363/4, ORG der IABG, Ottobrunn bei München. 1970.

DISCUSSION

Fogden: Has this system been accepted for use by the German Army?

Fent: Not yet.

The Development and Test of a Mathematical Model for Setting Cost-Effective Stock Levels of Aircraft Repairable Components

A.W. Ayliffe
D.J. Green

1. INTRODUCTION

One problem in inventory control is concerned with two decisions: when to order and how much to order. For those items whose cost is high or whose demand rate is low, or both, the optimal policy is to order a replacement immediately after a demand occurs. This important class of items includes most of the repairable components found on an aircraft. The performance of an inventory under this policy is determined entirely by the number of spare items held as a protection against having unsatisfied demands; if the number of spares is increased the effectiveness of the inventory will increase. However, increasing the number of spares automatically raises the cost of the inventory. The task is, therefore, to relate inventory effectiveness to investment and to present the figures as a management aid.

In the Royal Air Force inventory repairables account for only 20% by quantity but for 80% by cost and they cause about 80% of all aircraft defects. A repairable is defined as an item which can be removed, complete, from an aircraft and serviced either on the station or at a central repair base. This definition therefore includes items such as engines, radios, wheels, instruments, electrical generators, etc. The factors to be considered when deciding how many spares of each item to stock at a particular servicing echelon include:

(a) The demand rate of the item.

(b) The time taken to obtain a serviceable replacement, either by repair or re-supply (the replenishment time).

(c) The cost of the item.

Traditional methods of setting stock levels of repairable components usually consider the demand rate and sometimes the replenishment time (normally as a fixed pre-determined period). If cost is considered at all it is only in very broad bands and the relative advantage to be gained by stocking more of one item rather than another is normally never considered. The mathematical model described in this paper considers all three of the factors listed above and then uses the concept of marginal utility to arrive at a scale of spare repairables which, for a stated number of aircraft at a given location, will meet an effectiveness target at minimum cost. In other words, it is a cost-effective or optimal scale.

2. THE MODEL

Much of the early work in this area of inventory analysis was carried out by the Rand Corporation of America on behalf of the United States Air Force [see References 1, 2 and 3]. The model described in this paper is based on Rand's Base Stockage Model but we have adapted it to the RAF situation and improved on it in a few places. The basic assumption of the model is that demands for repairable components are described by the geometric Poisson distribution.

2.1 Compound Poisson Distribution

The compound Poisson distribution is a generalisation of the simple Poisson distribution. Figure 1 shows a time sample from both distributions. The compound Poisson has batches of demands rather than single demands, but the distribution of time between demands for the simple Poisson is the same as that between batches in the compound Poisson distribution. In both cases the time distribution is exponential.

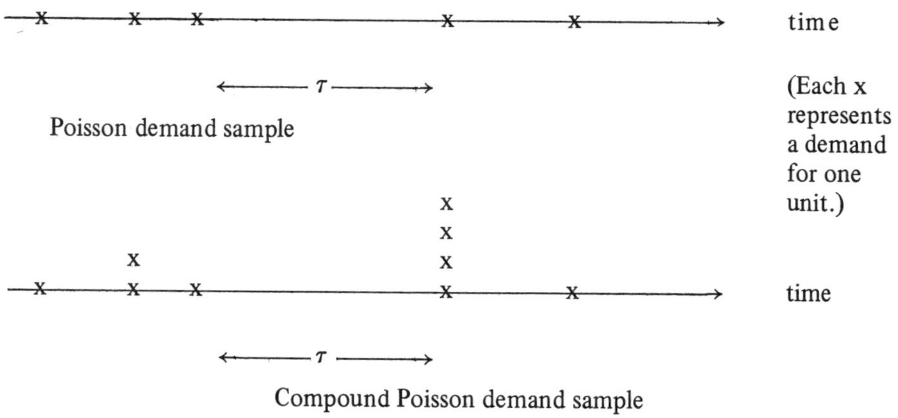

Figure 1 Comparison of Poisson and Compound Poisson demand samples

In both cases the probability that the time between demands is τ is $\lambda e^{-\lambda\tau}$

The compound Poisson distribution can be visualized as a series of customers arriving in accordance with a Poisson distribution, each customer demanding a number of items according to an independent discrete distribution f_j; for example, the compounding distribution f_j may be logarithmic, geometric, or another simple Poisson distribution.

Any compound Poisson distribution has a variance that exceeds or equals its mean; when the variance equals the mean the compound Poisson reduces to the simple Poisson distribution. Another feature of compound Poisson distributions is that the number of demands in a time period does not influence the probabilities of demands in any other non-overlapping time period. The reason for assuming a compound Poisson distribution for demands is that data gathered at stations has indicated that the variance of demand exceeds the mean demand.

2.2 Geometric Poisson Distribution

If a compound Poisson process is required such that the probability distribution of time between demands is independent of the time at which earlier demands occurred, a unique solution exists; the compounding distribution is geometric, and f_j is given by:

$$f_j = (1 - \rho)\,\rho^{j-1} \qquad j = 1, 2, 3 \ldots.$$

where f_j = probability of j demands by one customer.

ρ = constant dependent on the variance to mean ratio of the geometric Poisson distribution.

The mean and the variance to mean ratio of the geometric Poisson distribution can be shown [Ref. 2] to be:

$$\text{Mean} = \frac{\lambda T}{1-\rho} \quad \text{where } \lambda = \text{Customer arrival rate}$$

T = Time period considered.

$$\text{Variance to Mean ratio} = \frac{1+\rho}{1-\rho}$$

The geometric Poisson distribution is often called the "stuttering Poisson"; after each demand occurs there is a probability ρ of a "stutter" or another simultaneous demand, and probability (1-ρ) that a simple Poisson process with customer arrival rate λ is started.

Calculation of the probability of a given number of demands in a time period, given a geometric Poisson distribution, is eased by the use of recursion formulae. Feeney & Sherbrooke [3] derive the following formulae:

$$\theta(x-y/y) = \rho\theta(x-y-1/y) + (1-\rho)\,\theta\,(x-y/y-1)$$

where $y = 1, 2, 3, \ldots.$ and $x \geqslant y$

$$\phi_{y,x} = \rho\,\phi_{y,x-1} + (1-\rho)\frac{\lambda T}{y}\,\phi_{y-1,\,x-1}$$

where $\phi_{0}, = e^{-\lambda T}$, $\phi_{y,0} = 0$ for $y \neq 0$,

and $\phi_{0}, x = 0$ for $x \neq 0$

$$P(x/\lambda T) = \sum_{y=0}^{x} \phi_{y,x}$$

where $\theta(x-y/y)$ = the probability that $(x-y)+y = x$ demands will be made, given y customers.

$\phi_{y,x}$ = the probability that y customers will arrive and make x demands.

$P(x/\lambda T)$ = the probability of exactly x demands in time period T given customer arrival rate is λ.

The formulae are suitable for computer programming.

2.3 Measures of Supply Performance

Knowing the probability of a particular number of demands in a time period, supply performance can be measured. Apart from the parameters of the geometric Poisson distribution, supply performance depends upon the spare stock, s, of a component, and the component's mean replenishment time, T.

The mean replenishment time, T, is defined as the mean elapsed time taken to obtain a serviceable replacement; it is the average of repair times, resupply times, or both; it includes administration, transit, and delay times, and is measured in days.

A queueing theorem due to Palm, which has been extended by Feeney and Sherbrooke [3] states that if demand is compound Poisson, then the number of units in repair, resupply, or both, is also compound Poisson in the steady state condition–for any type of distribution of replenishment times. This implies that the number of units in repair, resupply, or both, depends on the mean replenishment time but is independent of the replenishment time distribution.

Define Ready Rate with scale s, $R(s)$, as:

Ready Rate $R(s)$ ≡ the probability that an item observed at a random point in time has no unsatisfied demands, given that the scale for the item is quantity s.

≡ the probability that s, or less than s, demands have been received for the item during the preceding replenishment time period.

We have, $$R(s) = \sum_{x=0}^{s} P(x/\lambda T)$$

Thus, $$R(s) = R(s-1) + P(s/\lambda T) \qquad \text{for } s \geqslant 1.$$

Defining Fills with scale s, $F(s)$, as:

$F(s)$ = the expected number of demands per time period that can be filled (satisfied) immediately from stock on hand, given that the scale for the item is quantity s.

We have:

$$\text{Number of satisfied demands (fills) per customer} = \sum_{x=1}^{s} P\begin{pmatrix}\text{Customer}\\ \text{makes } x\\ \text{demands}\end{pmatrix} \cdot P\begin{pmatrix}X \text{ items}\\ \text{are in}\\ \text{stock}\end{pmatrix}$$

$$= 1 . \sum_{x=0}^{s-1} P(x/\lambda T) + (1-f_1) \sum_{x=0}^{s-2} P(x/\lambda T) + (1-f_1-f_2) \sum_{x=0}^{s-3} P(x/\lambda T) + \ldots$$

where f_j = probability that a customer makes j demands = $(1-\rho)\,\rho^{j-1}$

Thus, substituting for f_j

Fills per customer = $R(s-1) + \rho R(s-2) + \rho^2 R(s-3) + \ldots\ldots$

Within the component's mean replenishment time, T, the expected number of customers is simply λT. Thus:

$$\begin{aligned} F(s) &= \lambda T \, [\, R(s-1) + \rho R(s-2) + \rho^2 R(s-3) + \ldots . \,] \\ &= \lambda T \, R(s-1) + \rho F(s-1) \end{aligned}$$

Again, the recursion formulae derived for *R(s)* and *F(s)* are suitable for computer programming.

F(s) gives us a measure of a scale's performance–the number of demands we expect to satisfy from stock, in a given time period, with a given scale of spare components. In isolation, *F(s)* is of little use, and for this reason a comparative measure, Fill Rate, is used

$$\text{Fill Rate} = \frac{F(s)}{\text{Expected number of demands in the same time period.}}$$

The Fill Rate, therefore, shows the proportion of demands that will be satisfied if a component has scale s; a fill rate of 0·95 would signify that 95% of demands would be satisfied from stock on hand.

2.4 Marginal Analysis

Turning now to a range of components, rather than one component in isolation, a suitable measure for the overall effectiveness of a scale is the overall fill rate.

$$\text{Overall fill rate} = \frac{\text{Expected fills for the range of components}}{\text{Expected demands in the same period}}$$

Thus, an overall fill rate of 0·95 would signify that the overall scale for the range would satisfy 95% of all demands–not 95% for each component, but 95% of demands received across the range in any time period.

The economist's technique of marginal analysis can be used to maximise effectiveness with respect to cost, thus producing a cost-effective scale for a range of components. The technique compares the value of incrementing the scale of one component, in terms of cost-effectiveness, against that of any other component; the comparative factor used is termed the Component Improvement Factor, *CIF*.

$$CIF = \frac{F(s+1) - F(s)}{C.\ T}$$

where C = component cost in £s
T = component mean replenishment time, measured in days.

Thus the *CIF* is the expected increase in fills per day, per £ spent, when the component's scale is increased by one.

Note: An amended *CIF* could be used for determining the contents of spares kits for deployment (fly-away packs). Instead of considering the cost of each component we could consider its weight or volume.

The marginal analysis technique is easy to apply:

(a) Calculate the *CIF* for each component in the range.

(b) Select the component with the largest *CIF*, say component XX.

(c) Increment XX's scale by one.

(d) Calculate overall fill rate and overall cost for the range.

(e) Update XX's *CIF*.

(f) Repeat (b) to (e), termed optimisation technique, until an overall fill rate of the desired effectiveness, or an overall cost limitation, is reached.

By incrementing the scale of the component with the largest *CIF* we are maximising effectiveness with respect to cost; more satisfied demands per day per £ spent result from incrementing this component's scale than would result from incrementing any of the other component's scales.

The optimisation technique assumes that each component's *CIF* decreases as its scaled quantity increases. This is not always the case. If the mean demand rate of a component is greater than one within the component's replenishment time, the component's *CIF*, when plotted against scaled quantity, generates a unimodal curve of positive skew. The simplest way of overcoming this difficulty is to scale a quantity equivalent to the modal value, termed the initial scale of the component, before the optimisation technique is used.

Having determined each component's initial scale, the optimistaion technique is used until the target overall fill rate is achieved for the range of components. Comparing the scale generated by the optimisation technique with the overall initial scale that was set may show that some components have scales that have remained at their initial settings. If this is the case, the scale generated by the optimisation technique may not be the most cost-effective scale for achieving the target overall fill rate. A suitable method of checking that the most cost-effective scale has been found is as follows:

(a) Store the latest scale generated by the optimisation technique.

(b) Compare each component's initial scale with the scale given by the optimisation technique; form the group of components whose initial scale values have not been increased. For this group of components, calculate Component Degradation Factors (*CDF*s) where

$$CDF = \frac{F(s) - F(s-1)}{C.\ T} \quad \text{(Note: Compare with } CIF \text{ above)}$$

= The expected decrease in fills per day per £ spent when the component's scale is decreased by one.

(c) Select the component which has the lowest *CDF*, say component YY, and decrease its scale by one. This is the component which gives the smallest number of fills per day per £ spent.

(d) Recalculate the overall fill rate and overall cost of the scale; both figures will decrease since YY's scale has been reduced by one.

(e) Use the optimisation technique to regain the target overall fill rate; compare this scale with the stored scale.

(f) Repeat paragraphs (a) to (e), termed "rescan" technique, until one of the following criteria applies:

(i) A component whose scale has been reduced is rescaled immediately.

(ii) The overall cost of the scale is greater than that of the stored scale.

(iii) The scale of each component has increased above its initial setting.

In each case the stored scale is the most cost-effective for the given target fill rate.

The use of the re-scan technique just described is the major difference between our model and Rand's Base Stockage model. Using the Base Stockage Model technique it is not possible to set the stock level of a component between zero and the modal value of demand within the replenishment time. The resulting scale therefore may not be optimal. The re-scan procedure overcomes this problem.

By producing a series of scales for different target overall fill rates, a series of cost-effective points are produced which can be plotted as a management aid. A typical curve is shown in Figure 3.

2.5 Computer Program

The recursion formulae derived and the marginal analysis techniques described have been programmed for a computer. A simple flow diagram is shown in Figure 2. The inputs required for the program fall into two groups; system data and item data.

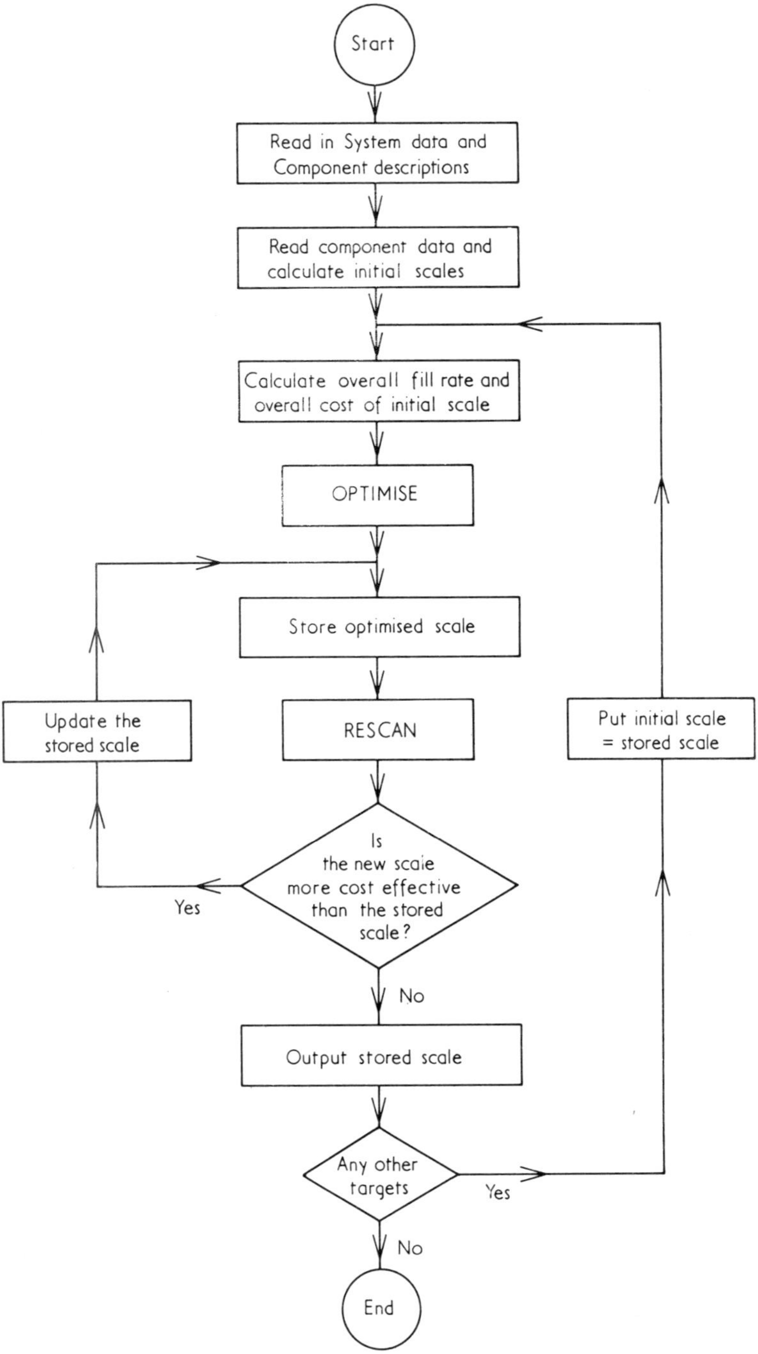

Figure 2 Computer flow diagram

2.5.1 System Data.

(a) Number of items to be considered.
(b) Station and aircraft names.
(c) Forecast flying rate per aircraft per month.
(d) Number of aircraft.
(e) Average number of flying days per month.
(f) Variance to mean ratio of the geometric Poïsson distribution.
(g) Target fill rate(s) to be met (or a ceiling cost(s) which must not be exceeded).

2.5.2 Item Data.

(a) Vocabulary section, reference number and description.
(b) Price.
(c) Mean Time Between Replacements.
(d) Replenishment Time.
(e) Number of aircraft to which fitted.
(f) Number fitted per aircraft.
(g) Minimum scale (if management decides that at least one, or two, etc of an item must be stocked).
(h) Duty cycle (ratio of operating hours to flying hours, normally 1).

The standard output of the program is a list of all the items with their principal parameters (mean demand per day, replenishment time and cost), the minimum scale considered and the recommended stock level for each item. The heading indicates station and aircraft names, target fill rate, forecast fill rate, total cost and the variance to mean ratio used.

The model can perform the following functions:

(a) Produce a single cost-effective scale to meet an effectiveness (fill rate) target.
(b) Produce a single cost-effective scale within a stated cost restraint.
(c) Produce a series of scales from which can be plotted a cost-effectiveness curve as a management aid.
(d) Produce scales as above which are optimal with respect to weight or volume.
(e) In any of the foregoing functions, accept a minimum scale on any or all items as a management input.
(f) Evaluate a scale for a range of items in terms of fill rate.

3. TESTING THE MODEL

The model was tested at two RAF stations simultaneously during 1969. Both stations, which we shall call A and B, operated about 50 single engine jet training aircraft and both had roughly similar tasks in terms of flying hours and numbers of sorties. The servicing arrangements at the two stations were similar except for electrical, instrument and radio components, for which special repair bays have been set up serving their own and several other stations on a regional basis.

The test was split into two phases each of six months duration. During Phase 1 from 1st January to 30 June two airmen recorders at each station collected accurate data on all replacements of the 241 repairable components fitted to the aircraft, completing a card in every case. If a serviceable spare item was not immediately available on the station this was recorded as a 'non-fill' (an unsatisfied demand). A second card was completed for every arising and on this the progress of the removed component through the repair and re-supply processes was noted. Analysis of the first set of cards revealed the Mean Time Between Replacements for each component while analysis of the second set provided the average replenishment time for each component. Some historical data was also gathered from previous records for use when actual data collected during Phase 1 was sparse or non-existent.

At the end of Phase 1 the collected data was analysed and input to the model. Scales were produced for each station for overall target fill rates ranging from 0·5 to 0·99. From these outputs were produced cost-effectiveness curves, one of which is reproduced in Figure 3. The observed effectiveness and the value of the stocks held at

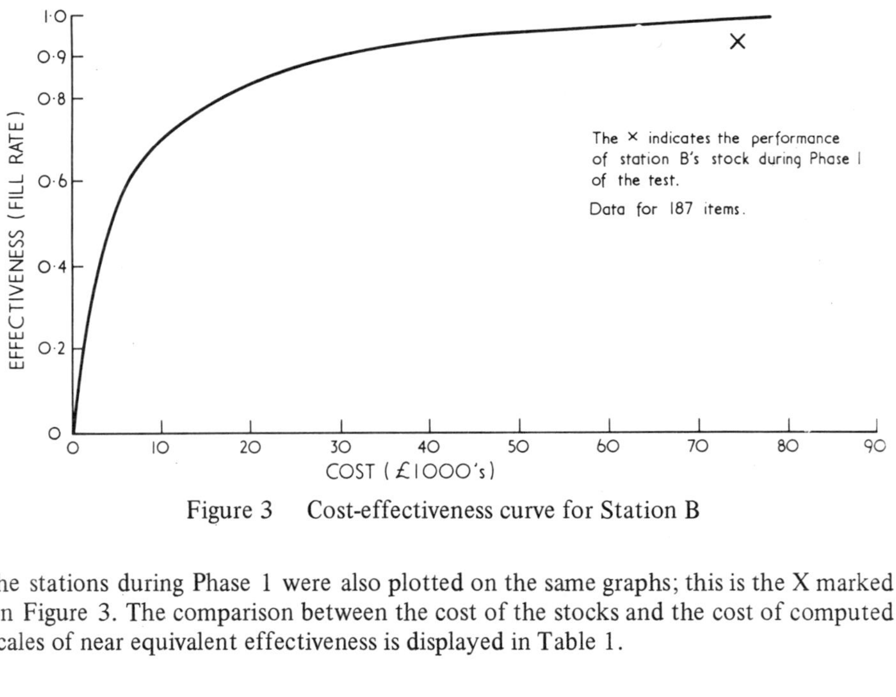

Figure 3 Cost-effectiveness curve for Station B

the stations during Phase 1 were also plotted on the same graphs; this is the X marked on Figure 3. The comparison between the cost of the stocks and the cost of computed scales of near equivalent effectiveness is displayed in Table 1.

	Holdings		*Computed Scales*	
	Cost (£)	Actual Effectiveness %	Cost (£)	Actual Effectiveness %
Station A:	69,120	92.6	41,288	95.1
Station B:	73,519	93.3	45,372	95.1
	£ 142,639		£ 86,660	

Table 1: Cost Comparison – Phase 1

The £55,979 difference in costs between the stations' holdings and the computed scales represents a total saving of 39% and there is a forecast increase in effectiveness at both stations. It should be mentioned here that from the original list of 241 items five have been deleted. The two types of engines were deleted on the grounds that such very high cost items should be hand-managed and three other items, main and emergency batteries and the emergency oxygen bottle, because they were replaced at regular calendar intervals; their replacements could not be described by the geometric Poisson distribution and setting levels would be a simple matter.

During Phase 2 of the test, from 1st July to 31st December, the task was to measure the effectiveness of the computed scales; would they achieve what they forecast? If they performed well it would confirm the model's predictive capability and justify the underlying assumption of geometric Poisson demand. There were several ways of measuring the effectiveness; we could, for instance, have physically stocked the stations to the suggested levels. We rejected this method because it would have involved a tremendous upheaval with items being transported back and forth and also because the resulting comparison would have been between two sets of stocks over two different periods. It was finally decided to write a simulation program which would model the repair/re-supply process, accept actual arising dates as inputs and compare the performance of any three scales in terms of fill rate.

The program was written and tested with the arising dates from Phase 1 as inputs. The results showed that the simulation was a good model of the real-life situation, registering fill rates within 1%–2% of the observed figures. We were therefore confident that the simulation could be used as a yardstick at the end of Phase 2 of the test.

The same data was collected during Phase 2 as in the first phase. At the end of the year the simulation program was run to compare the performance of the stocks held and the computed scales in the light of the Phase 2 arisings.

These were the results:

	Holdings		*Computed Scales*		
	Actual Effectiveness %	Simulation Effectiveness	Forecast Effectiveness %	Simulation Effectiveness %	No. of Arisings
Station A	93.1	93.69	95.1	94.74	856
Station B	94.3	94.01	95.1	91.93	1202
					2058

Table 2: Effectiveness Comparison – Phase 2

The performance of the computed scale at station A was very close to the predicted performance but there was a 3.17% difference at station B. This was investigated further. A close scrutiny of the data used in the production of the scale for station B revealed that for two items, nose wheels and brake units, the replenishment times in

Phase 2 had more than trebled compared with Phase 1. As the lower Phase 1 figures had been used to produce the scale and the higher Phase 2 figures to evaluate it and as both items were fairly high demand rate items, it is easy to see how the 3.17% discrepancy occurred.

Further investigation revealed the reasons for the very large increase in replenishment times. There had been a local shortage of piece-part spares used to repair the wheels and brakes at station B during Phase 2. This did not, however, cause much concern since stocks of wheels and brakes were large. Unserviceable units were therefore left to accumulate until the piece-parts' supply situation eased and this led to the replenishment times being trebled.

However, if the station had been working with the much lower computed level of these items it is probable that more energetic action would have been taken to overcome the piece-part shortages and replenishment times would have been much closer to those recorded during Phase 1. To test this point the simulation was run a second time with the replenishment times reduced to the Phase 1 figures. The effectiveness registered then became 94.01%, much closer to the predicted figure.

3.1 Sensitivity Analysis

An important factor to be considered when deciding how and when to use a model such as this is the model's ability to deal with the unreliable engineering and supply data that is available at the time of initial provisioning of spares for a new aircraft. We therefore carried out a sensitivity analysis varying the principal inputs, Mean Time Between Replacements (MTBR), replenishment time and variance to mean ratio. The results are shown graphically in Figures 4, 5 and 6 for station B. Figure 4 indicates that if every one of the MTBRs used in the model proved in practice to have been 30% optimistic the resulting scale would achieve 92.5% effectiveness rather than the forecast 95%. From this point the effectiveness falls off rapidly as MTBRs are reduced.

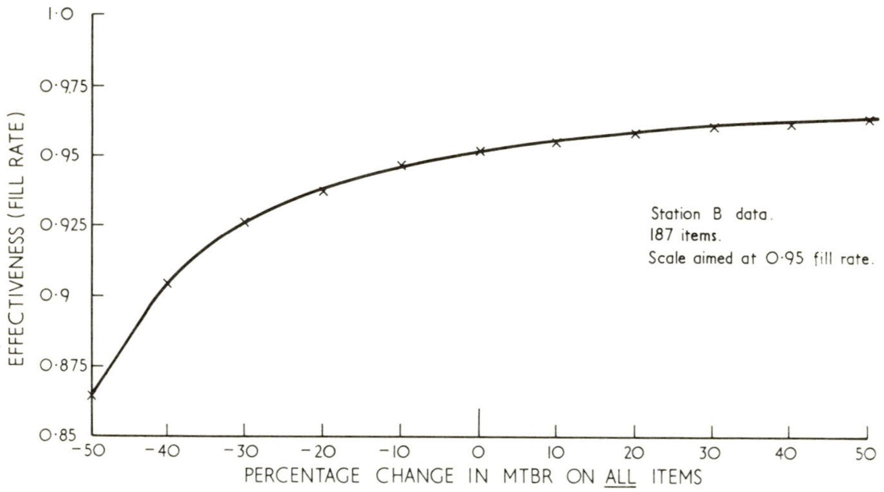

Figure 4 Effect of change in MTBR

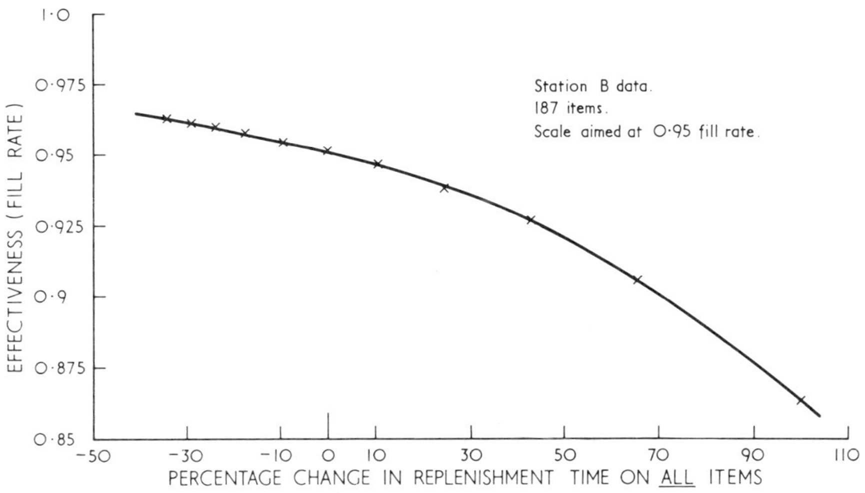

Figure 5 Effect of change in replenishment times

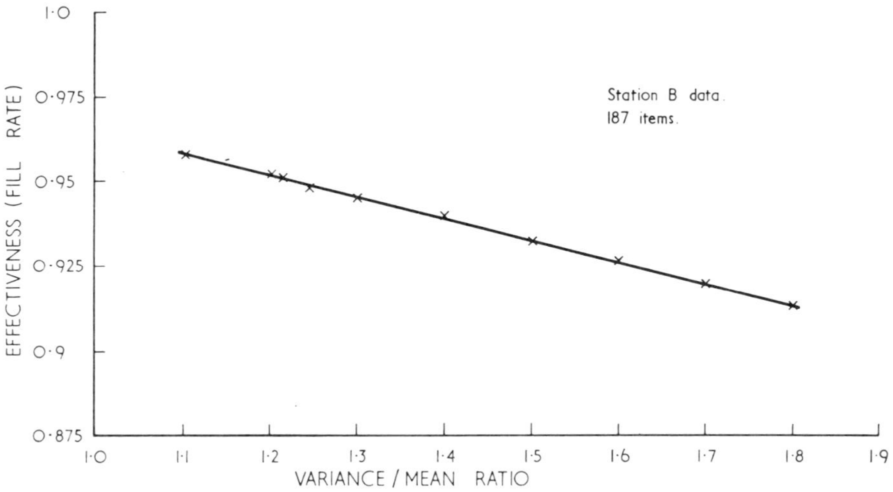

Figure 6 Effect of change in variance/mean ration (datum: $v/m = 1.215$)

Figure 5 shows a similar effect if all replenishment times prove to have been under-estimated, though the degree of degradation is less. The model seems less sensitive to changes in replenishment time than to changes in MTBR.

The variance to mean ratios used in the production of the scales were derived from a preliminary analysis of the data from each station. The figures used were:

Station A – 1.116
Station B – 1.215

Figure 6 shows, using station B data, the change in effectiveness to be expected if, having assumed a value of 1.215 the variance to mean ratio is found to be different in practice. It will be seen that over the range 1·1 to 1·8 the effectiveness varies from 9·957 to 0·912 and we conclude that the model is not very sensitive to changes of this order.

3.2 Effect of Ignoring Cost

A series of runs of the model using data from station B was made to show the effect of ignoring item cost in the optimisation process. For this series all item costs were assumed to be equal. The effect of this is that the resulting scale is effective—it meets the target set—but it is not the cheapest possible scale. The results of that series of runs are shown graphically at Figure 7 together with the cost-effective curve for station B.

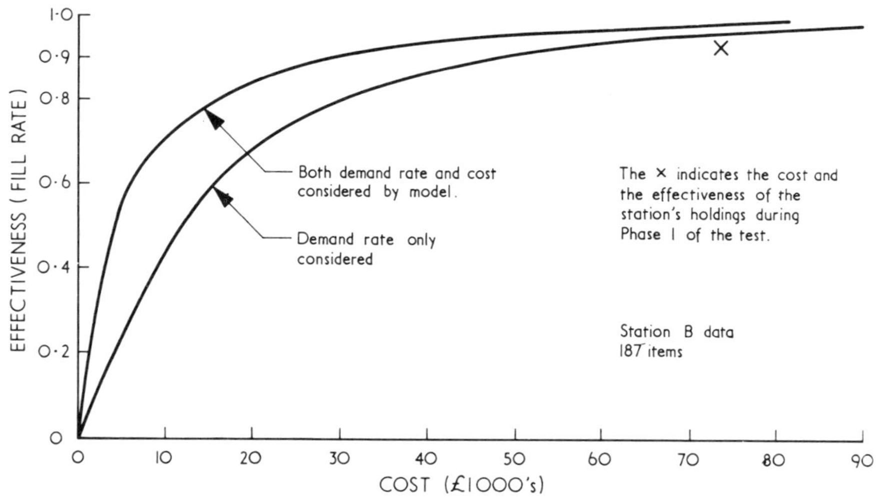

Figure 7 Effect of ignoring cost in the model

They show that optimising against cost produces the greatest percentage saving over the 0·6 to 0·8 effectiveness range (eg 58% at 0·7 effectiveness) but even at the 0·99 point the saving stands at 15·8%.

4. DISCUSSION OF THE RESULTS

The results of the test confirmed the early hopes for the model. It can produce, given accurate input data, scales of repairable components that meet effectiveness targets for substantially lower investments in spares than do the methods in use in the RAF today. The potential savings if the model were used from an early stage in the procurement of all new aircraft could amount to millions of pounds. The question then arises, how best to introduce the model into the procurement process?

The initial provisioning of spares for new RAF aircraft starts about two and a half years before the first aircraft enters service. At this time the reliability data available for the aircraft components is sparse and of very doubtful quality; sometimes even the design is not finalised. Nevertheless, if spares are to be available for use when the aircraft enters service, purchasing must begin. Now implicit in our model is the fact that all data for all items must be available before the level of any item can be set. This situation does not usually occur until late in the initial provisioning stage so to use the model one would have to consider groups of items rather than all items. Inevitably this will lead to sub-optimisation so care must be taken to ensure that the final objective, a total cost-effective buy, is not compromised. This can be achieved by arranging the whole provisioning process into stages.

The stages should be set in the light of the build-up plan for the aircraft and the components' total lead times. Thus stage 1 for the very long lead-time items would take place at, say, two years before in-service date. The number of aircraft to be considered at this first stage should be limited to the number which will be in use at the time: in-service date plus one year plus component lead-time. When stage 1 for the next group of components takes place the stock levels obtained from the use of the model for the first group could be injected as minimum scales and both groups considered together. This process should be repeated until an initial order has been placed for all components.

The second stage of provisioning should take place one year after in-service date. At this time there will be available one year's worth of more reliable data and the model can be used with more confidence. Now a larger number of aircraft will be considered and this should ensure that for any component the total buy suggested by the model is not less than has already been purchased. In this way a cost-effective total buy can be arranged. Depending on the total number of aircraft bought there may be a requirement for a stage 3 or even a stage 4. At each stage the input data to the model would be further refined and better results obtained.

4.1 The Fly-away Pack Problem

It has already been mentioned that the model can accept weight or volume as the optimisation criterion rather than cost. This, of course, has direct relevance to the problem of setting stock levels in kits of spares to support short-term deployments of small numbers of aircraft. The logistic support of such deployments is normally more concerned with the air-life capability required than with cost. The use of the model, optimising against volume, can ensure that the spares kit is the smallest possible to meet the staged effectiveness target.

4.2 Conclusion

We feel that not least of the advantages to be gained by the use of the model is the discipline it imposes on logistic staff. It demands consideration of all relevant factors and encourages a system approach to provisioning. In these times of spiralling prices the potential savings offered by the model make it a valuable tool for those armed services operating on low budgets who want the best value for their money.

REFERENCES

1. Feeney, G.J., and Sherbrooke, C.C., A System Approach to Base Stockage of Recoverable Items. Rand Corporation Memorandum RM-4720-PR December 1965.

2. Sherbrooke, C.C., Discrete Compound Poisson Processes and Tables of the Geometric Poisson Distribution. Rand Corporation Memorandum RM-4831-PR July 1966.

3. Feeney, G.J., and Sherbrooke, C.C., The (*s*-1, *s*) Inventory Policy Under Compound Poisson Demand. Rand Corporation Memorandum RM-4176-PR September 1964.

ACKNOWLEDGEMENTS

The advice and assistance given during the development and test of the model by Mr J. K. Wood and Mr D. J. Eggleton of the staff of the Deputy Chief Scientist (RAF), Ministry of Defence, are gratefully acknowledged.

DISCUSSION

Kay: The model for cost effective stock-levels suggested by the authors, might be acceptable in commercial practice when goods are stocked for sale and it is usually impossible to estimate the cost of lost sales. The best one can do therefore, is to minimise the cost of stock holding subject to an acceptable Fill Rate.

In the case of stocking aviation spares, whether military or civil, all the items in the range are, by the authors' definition, of equal operational importance, but of a wide technical variety. Thus one must assume that the replenishment time T also varies widely. Non-availability of any one item when demanded will ground the aircraft for an average time equal to T, if $T \leqslant 1/\lambda$, that is replenishment time is smaller than average interval between demands, and same calculable value, less than T, for the case $T > 1/\lambda$.

The average cost of a unit stock-out of the ith item in the range is, therefore, also calculable as a function of aircraft type or mission and average replenishment time for that item, say $C(T_i)$.

The true cost of any given stocking policy is the sum of the cost of stock holding and cost of stock-out.

Stock holding costs depend in the main on the long term average of stock held, which in the case discussed by the authors is given by

$$s = \text{Average stock held} = S - \lambda T_i/(1-\rho)$$

The average policy cost for the ith item is thus

$$C_i = g(S_i) + D_i C(T_i)$$

where D_i is the average deficiency of the ith item.

From the authors' definition of Fill Rate, $F(s)$, – Fills with scale s, – is less than expected demand per time period. Thus:

$$D_i = \frac{\lambda T_i}{1-\rho} - F(s)$$

The most cost-effective policy is given by

$$\text{Average Policy cost for all items} = \text{Min} \sum_{i=1}^{N} [g(S_i) + D_i \ C(T_i)]$$

where N is the number ot items in the range.

This cost will be a minimum when each element,

$g(\overline{S_i}) + D_i \ C(T_i)$, under the summation is a minimum.

Clearly $g(\overline{S_i})$ increaaes and D_i decreases with s.

In certain, admittedly rare, cases the form of functions, $g(S)$, $F(s)$ and $C(T_i)$ may be such that the minimum can be obtained by analysis.

In general, however, each items minimum average cost can be calculated by a stepwise procedure, similar to that suggested by the authors.

May we suggest, that if cost-effectiveness is the accepted criterion, and all relevant costs are calculable, Fill Rate is a function of cost, and the optimal Fill Rate is given by the least cost solution. The Overall Fill Rate is, therefore, not a good indicator of performance.

Ayliffe: We agree that this method would achieve a completely cost-effective result but, unfortunately, we cannot cost a stock-out in the military environment. On one occasion it might mean no penalty (because the aircraft was not due to fly) and on another occasion it could mean a lost war. We have suggested in our report on our field test that more research be carried out into the relationship between supply effectiveness and aircraft availability and in this respect we are eagerly awaiting the RAND Memoranda on this subject that was mentioned by Mr. Miller.

Kay: One must accept that shortage costs can not be estimated for military aircraft. Nevertheless, the overall fill rate does not seem a proper criterion, because, by definition, each component controlled by this system is of equal importance, i.e. shortage causes grounding of an aircraft.

On the other hand the re-supply time of a shortage varies from component to component. Hence "time grounded" is a measure of shortage, that ought to be brought into the calculation, perhaps via individual fill rates.

Ayliffe: Mr Kay is probably right but we can foresee difficulties here. For instance, we could not use the mean replenishment time that we use in the main calculation because a back-order means that priority repair and re-supply procedures would be invoked. A knowledge of depot stocks then becomes necessary and we must then move to a multi-echelon model such as METRIC. Our model is designed for one station (or one echelon) and overall fill rate is about as far as we can go with any degree of accuracy.

Incidentally, a stock-out does not necessarily mean that an aircraft is grounded. The requirement may have arisen on an aircraft that was already in scheduled maintenance or on one not due to fly for some time.

Monhemius:

1. In section 3 you say that you studied the 241 repairable components during half a year. This way, you determined the Mean Time Between Replacements for each component. To get some idea of the accuracy of these calculations my question is: is:

   – in what range did the demand per component lie?

   – have you found any components with demand zero?
   If yes, what has been your estimate?

2. In section 3.1, you give the variance-to-mean ratios for each station.
   Could you justify why you *do* pool the data for this calculation for all components in one station and why you do *not* pool them for the two stations?

3. In section 2.5.1 you include in the system data the "average number of flying days per month".
   Do you mean averaged over all months from a period of six months?

4. Do you feel that it would pay to use information about total flying hours for each individual month as one of the data to forecast demand during replenishment time (if these times are not too long) (and if this total number of hours differs from month to month)?
   I ask this question because we found in some cases that both theoretically *and* empirically it can be shown that correlation *from month to month* between flying hours and demand exists but is not very strong.

5. For how many components has your re-scan procedure been active? (i.e. decreased initial scale).

Ayliffe:

1. The range of demands was between zero and about 200, the latter figure giving an MTBR of 25 hours. During Phase 1, the first six months of the field test, about 35% of the complete range of components showed zero demand rates. For these components we supplemented our Phase 1 data by two years historical data. If we still had no demand data after this exercise, we excluded components from the range considered by the model–in other words we gave components which showed

no demand history a zero stock level. I must point out that all 241 components were considered by the simulation conducted at the end of Phase 2. In two or three cases, components to which we had allotted zero scales were the subject of demand arisings. These demands were counted as non-fills against our proposed scale. During Phase 2 of the test another 35% of the range of 241 components showed no demand arisings. The majority of these components were those for which no demands had been observed during Phase 1.

2. The model is designed to give a stock level for a single station. Using two stations, therefore, gave us two different tests of the model. Bearing this in mind, we could not combine data from the two stations. The variance-to-mean ratios that we used–1.116 at station A and 1.215 at station B–were derived by fitting regression lines to data collected at each station.

3. You will recall that our test was carried out on a training aircraft. The stations that were observed were working five days per week. We had to make allowance for this in the model to determine the correct mean demand rates. In the field test we equated the expected number of flying days with the expected number of working days in the Phase 2 period–this gave us our forecast monthly figure.

4. The question implies that the model should be run every month. This period is too short. We agree that a correlation from month to month between flying hours and demands is not strong. The use of the six months period improves the correlation considerably, and it leads to the steady state demand rates which are required by the model. Once the steady state demand rates have been determined the model can be run as frequently as desired–it could, therefore, be used to show the number of additional spares that would be required to cope with an increased flying rate.

5. This is a difficult question to answer. The Rescan procedure could operate on any component with a demand rate greater than one demand within the replenishment time. We had about thirty of these items in the range. How many components it does operate on is largely determined by the target fill rate that is set. The Rescan procedure might not operate at all when target fill rates are high because each component might require a high fill rate to achieve the high target overall fill rate. We may be able to provide you with more detail after carrying out some supplementary computer runs–perhaps the questioner would contact us in this respect.

Peguy: Sur des simulations réalisées sur 200 articles (Mr. Dassault) nous avons testé 2 modèles de prévisions

1. en tenant compte du nombre d'Heures de vol et d'atterrissages;

2. d'une manière purement aléatoire.

Les résultats ont été parfaitement farfelus dans la premiere simulation et nettement plus satisfaisants dans le second cas.

Geraerds: Would Mr. Peguy be issuing the results of the studies he had cited?

Peguy: I have no intention of doing so, but could answer any questions posed to me directly.

O'Connor: Your model, and the other inventory models we have described this week, all assume that rejection rates are constant. Of course, this is not true for a new system, until it is typically about 5 years old. Therefore such inventory models do not optimise stocks over the life cycle of the system, but only at one point in time or over the steady state period. Do you consider that Monte Carlo simulation models would not provide a better tool for scaling Hi-value repairable components?

Ayliffe: We don't see simulation as a means of scaling Hi-value repairables since the advantages of optimisation would be lost. We feel that the five years you quote are not the result of the original items settling down to a constant failure rate but rather the end of a process of modification designed to improve the original failure rates. After five years most possibilities of improvement are exhausted. The over-riding consideration is that one has to order spares before the aircraft has flown. One can really only consider the steady state; the answers may be wrong–but they will be wrong more cheaply than were previous models. As we suggested, staging the procurement process helps to overcome the effects of poor data (and the burn-in period) as does the transition from estimated to actual failure rates in the sequential use of the model. Mr Baglow has pointed out that there is a lack of information regarding manufacturer's initial estimates of failure rates and subsequent actual performance. Work in this area, to determine relationships, would greatly assist models such as ours.

Baglow: Very small fleets may display very small utilisation rates. A steady state observed rate may not be obtained before the aircraft leaves active service.

Buckland: A general point arises from a study of the papers presented at this Conference, and, indeed, from much of the literature associated with the subject-matter field of the Conference. Far too many writers and speakers say, for example "... the demand (or arising rate) follows (or is assumed to follow) a Poisson distribution". As there are at least three statistical models for which this distribution form is valid it makes for immediate difficulties of interpretation of the author's intentions in relation to the subsequent developments and consequences contained in the paper. It is somewhat disconcerting to find the above-mentioned loosely drafted statement appear to indicate the inevitable result of demand being generated by a Poisson process and later in the paper for it to emerge as the statistical description of data over a suitably long block of time within which the demand sequence is anything but that of a Poisson process. There is even a case in the literature where the demand was truly binomial in form and the author used the Poisson distribution as a convenient approximation but in his initial drafting managed to convey the impression that he was dealing with a Poisson process.

In short, there still seems insufficient appreciation that with a Poisson distribution of Demand incidents need not arise from a stream of events where the intervals of time are negative exponentially distributed–a Poisson process. This comment is really a plea for a much closer regard to the specification of the essentially statistical features of a logistic support system.

Conclusions of the Conference

A. Tyler Port

Ladies and Gentlemen,

I am faced with an impossible task. I am expected to summarize in 15 minutes that which consumed 23 hours and 15 minutes of weighty dissertation and discussion. I would be prone to assume that someone badly overestimated my capacity to fulfil my assignment were it not for the fact that I know that our Chairman and host was fully aware that I am not a mathematician or a scientist or even a common ordinary manager. From the outset I have suspected that he never really was counting on me to crown your four hardworking and illuminating days with a scintillating tour de force to solve all your future problems.

Simply stated, I am an adminstrator temporarily turned diplomat and occupying the rather unique position of being an international civil servant. I work for all of the NATO countries. Your bosses viewed collectively, are my bosses; and it is in that context that I think I am really expected to address you today.

There is no question but that we have come a long way from the days when we sought to solve our management problems by keeping things in such a frenzy of reorganizational turmoil that we could count on moving on to our next posts before the birds all come home to roost and our thinking had to be recalculated. The papers we have listened to during the past four days have ranged from efforts to "exaggerate the vertical" to examples of how "small boys climb trees", to borrow two expressions from Admiral Ferier's memorable opening remarks. Viewed from the perspective of the irrepressible doodler for whom Prof. Conrath proposed to design a nonsense input button, the first thing that strikes me about this Conference is a large number of participants.

While we must make allowance for the touristic attractions of this pleasant country, the Grand Duchy of Luxembourg, the large attendance seems to me to be a sign that there is an expanding interest in logistic support problems in both civil and military circles.

I have also been struck by the extremely wide variety of presentations. Such variety obviously points to the magnitude of the problem and the difficulty of grasping it as a whole.

Some of the papers have been presented in a scientific, or at least, mathematical form; others, on the contrary, are more pragmatic, and some come between these two categories.

This seems to indicate that logistic support has out-grown the empirical age and is gradually developing into an applied science, although it is still difficult to define its scope.

At all events, I think I can see several trends worth mentioning in this impressive mass of ideas, suggestions and projections at different stages.

The *first trend* concerns the evaluation and comparison of various possible patterns in the general organization of logistic support. There can be no doubt that progress is being made, for, instead of continuing, more or less by force of habit, to accept a certain type of so-called conventional pattern, we are studying and trying out models designed to produce optimum solutions within, of course, the framework of the assumption inputs. I believe this to have been one of the important features of the Conference. It is encouraging to see that good work is being done in this direction.

The *second trend* which emerges from some of the papers is the interdependence of development, production and logistic support. It is becoming increasingly obvious that an equipment item or a weapon system cannot be designed exclusively with an eye to operational performance and production costs. The idea is gaining ground that operational performance is inherently bound up with easy maintenance and readily obtainable spares. It is also agreed that the cost of production is only one of the elements entering into the cost of equipment and that the cost of support is an equally important factor.

From this, it follows that one lesson is quite clear: logistic support must be taken into account at its true value at the equipment development stage. The logistician must no longer be regarded as the faithful servant who can be called in to help at the last minute. He must participate in the design and development of systems.

A *third point* I have noted is the growing tendency to feed more numerous and more detailed controlling data into support operations with a view to increasing their efficiency and reducing their cost.

This is linked with a *fourth trend*: the fast-spreading influence of electronic data processing in logistics. The problem is not just one of operational control and timely resupply. The aim is also to assemble maintenance data with a view to continually improving preventive maintenance methods, either from the standpoint of operational security or from the standpoint of cost. In addition, this powerful tool must be used for an a priori comparison of patterns and for an a priori attempt to identify the optimum pattern.

My tentative summary is obviously incomplete. Some of the experts in the audience must also have been disappointed with the shortage of new material as far as certain special aspects of the general subject of the Conference are concerned.

I think it is in the nature of things that this should be so. The comparatively recent development of aviation and electronics, in particular, has confronted the logistician with problems of a new dimension. These sciences have also provided him with new means of solving such problems: for example, communications are virtually instantaneous, urgent transport movements take only a few hours or, at the most, a few days, and data processing methods are becoming more and more refined. The adaptation of methods to resources and requirements is a constant challenge. The logistics edifice is built brick by brick and this Conference will have contributed one such brick. We must press on with the search for the optimum solution to the various problems confronting us. This leads me to the second series of observations.

It struck me that the most interest and discussion was aroused among us when speakers sought to relate experiences they had gained in coping with real life situations. Those dealing strictly in theory found the task of selecting assumptions frequently left them out on limbs which the pragmatists among us were all too ready to saw off. While I have heard some references to the lack of new material uncovered here, each of us has benefitted greatly from just learning about what the others are up to. I have even gathered the impression that this Conference has proved a useful source of information for military services which have not known what is going on in the sister services of the same military establishment. The same thing goes for the insight some MODs have gained of what their lower echelons are about. All of which is very helpful, for the real name of the game is communication, and the reinvention of the wheel, to use one of Mr. Geraerds' very apt expressions, is to be avoided.

This leads me to suggest that what has happened here during the past four days should not go down in history as a one of a kind thing, a rare and fascinating bringing together of articulate logistical rapporteurs, a unique happening. Although similar Conferences were held in 1964 and 1967, this 1970 Conference quite obviously revealed a growing need of the member nations of the Alliance to exchange views on matters which vitally concern the capacity of their respective military establishments to field and operate their armed forces. The matter should not end there.

We are indeed indebted to the NATO Advisory Panel on Operational Research for sponsoring this meeting. However, while we owe this debt to the science side of NATO's house, and while we all recognize that logistics is rapidly becoming a science, what we are in fact dealing with is more directly concerned with the realm of management and the evolution of techniques than science as defined in the classic sense. Taking my cue on the one hand from the welter of methods and procedures discussed here, and on the other from the oneness of purpose underlying each speaker's exposé, I would venture the thought that some means should be found for converting this symposium into a continuing dialogue. A logical place for such a dialogue to begin might be under the auspices of the Conference of National Armaments Directors.

For example, the NATO Group of National Directors of Codification, which operates under the CNAD, has been working for a number of years at perfecting a method of identifying and segregating individual items of supply. Just as our efforts to establish a uniform system for numbering pieces and parts has proven the value of cooperating in a fundamental mechanical aspect of logistics, I can foresee the advantage of extending efforts of this type into a coordinated search for better methods of provisioning, screening, procuring, maintaining, transporting and communicating about things material. I emphasize the word "coordinated" to allay the fears that immediately spring to mind when a "NATO type" seems to suggest an assault on the classic sobriquet that "logistics is a national responsibility". In this connection I would refer you to the fact that the equipping of NATO forces is likewise a "national responsibility". However, this has not deterred the member nations from charging the Conference of National Armaments Directors with the task of finding ways to increase cooperation among the NATO partners in research, development and production of military equipment. Today there are more than a hundred groups, committees, subcommittees and panels of various sizes busily engaged in exchanging information on current and future concepts and designs throughout the entire spectrum of weaponry and associated subjects. Each inquiry is directed toward expanding the nations' knowledge of what is going on in one or more weapons areas or technical fields and toward the discovery of a common ground for working together on a new development or production effort.

Where these latter efforts are crowned with success a joint logistics arrangement invariably follows involving pooling arrangements for the purchase of spare parts and certain aspects of maintenance. What happens in specific cases involving established NATO projects like HAWK, for example, leads to an additional thought I wish to leave with you. Strictly speaking, what I have in mind is not part of the conclusions of the Conference, but may be regarded as a kind of sequel.

On the whole, in the military field and especially in logistics, the pooling of NATO resources is intended to achieve one common goal. Viewed either from the standpoint of cost effectiveness or military potential, the problem is how to provide the best overall service for the total pooled resources of the community.

Various NATO members are having to face the problem of logistic support for weapons systems common to several countries or even to all the countries of the Alliance. I am speaking here of weapons systems which became common to more than one country other than by virtue of their first having become a NATO project. In such cases many practical difficulties have to be overcome.

The first hurdle is the coordination of the maintenance concepts and organizational pattern to be adopted by participating countries. It is true, of course, that member countries have different military organizations. They have reached different levels of technical and industrial development. It is clear, however, that if each country takes a unilateral decision on maintenance concepts, taking only its own criteria into consideration, it would really be a fluke if total costs corresponded to the optimum overall solution.

There is surely no need to enlarge on the many arguments in favour of the coordination of maintenance concepts: standardisation of documentation, standardisation of test tools and equipment, coordination in respect of supply requirements, and an increase in lateral support capability. In addition to these purely material factors, we should not forget the advantages of standardising personnel training, nor the benefits of mutual aid in dealing with technical personnel questions.

All this considerably increases the problem of determining the optimum pattern. It is no longer sufficient to look for the maintenance concept best suited to one country and one given army; the task now is to introduce new parameters and to coordinate and harmonise countries which have different military organizations and have reached different levels of industrial development.

If the first obstacle, consisting in the coordination of maintenance concepts and patterns, is successfully cleared, we have to tackle the next one, which consists in the coordination of supplies.

I should be repeating myself if I were to stress the advantages of such coordination.

I should likewise be repeating myself if I were to say that, here again, the problem has taken on a new dimension, for the solution must suit not one, but several countries.

Since I have mentioned the problem of the cost-effectiveness of weapon systems, common to several countries of the Alliance, I should round off my remarks by drawing attention to a matter which is only distantly connected with logistics, the subject of the Conference, but belongs rather to the sphere of politics and economics.

The process of centralising what can be centralised should be conducive to the reduction of costs. Personnel training comes to mind, as well as the letting of overall contracts for maintenance operations. Think of joint procurement of supply items. Think, too, of a pool of so-called safety spares which can represent as much as 80% of storage investments within a national support framework.

However, let there be no illusions. These common place ideas are not easy to put into practice. Apart from the cost-effectiveness parameter on which the line of argument is based, there are two other operative criteria: national habits and a country's concern to safeguard the legitimate interests of its industry. You will perhaps accuse me of speaking undiplomatically and certainly of speaking bluntly, when I put it this way: but it is as if, in some cases, countries preferred to pay more as long as their own industry got the order.

The solution to this problem seems to lie in centralisation of contracts let to the lowest bidder, their allocation being decided however, not on a day-to-day basis, but when a larger number can be handled together, covering a larger number of programmes. Nevertheless, the fact remains that this would be a compromise between two criteria and that a strong community spirit would have to exist before the system could be applied.

I must apologise if I have allowed myself to raise issues which do not, strictly speaking, come within the framework of the Conference, but it really is difficult to speak of cost-effectiveness in logistics without making some reference to the larger issue of orders and contracts.

In closing I wish to express on behalf of the Secretary General, His Excellency Mr. Manlio Brosio and the Assistant Secretary General for Scientific Affairs Prof. Randers, my warmest congratulations to General Bureau and to the members of his staff for the magnificent arrangements for this Conference and for its detailed planning and day to day operation. I want to express appreciation which NATO feels to the Grand Duchy of Luxembourg for the use of this wonderful building and to Minister Schaus for his warm welcome and for his bountiful hospitality. I would not conclude without also expressing my deep admiration for the tremendous work performed by the authors and speakers and participants at this Conference. It has been a pleasure for me to attend and I hope that we all may convene in less than three years time. Another brick has indeed been added to the logistics structure on which we are all working and has opened the door to new developments.

Thank you.

Closing Address

Allocution de Cloture

General-Major R. Bureau

Mesdames, Messieurs

Il m'appartient, avant de vous libérer de vos contraintes, d'adresser quelques mots de remerciements. Mais que les impatients se rassurent, je serai très bref.

Je commencerai par m'excuser pour quelques points.

Le classement des présentations sous différentes rubriques n'a pas été chose facile et prête certainement à remarque.

Le cadence de ces journées a été quelque peu rapide, les interruptions trop courtes: c'est que j'avais moi-même mes propres contraintes et que j'ai enregistré un écart entre les prévisions et la réalité quant au nombre de présentations.

Le drapeau du Canada est logistiquement "obsolete". L'horlege de la salle est particulariste, car elle s'obstine à ne pas s'aligner sur l'heure de Greenwich, etc.

Quoi qu'il en soit de tout ce qui aurait pu être, je remercie le *Gouvernement luxembourgeois* de son accueil qui se traduit notamment par la mise à notre disposition de ces installations à titre gracieux et par la magnifique réception de mardi soir.

De nombreux noms devraient être cités: le Commissaire Général *Ferrier* pour son brillant discours d'introduction, ses conseils et ses interventions toujours sages.–Je dois ici présenter les excuses du Commissaire Général Ferrier qui a dû nous quitter hier soir et a exprimé l'espoir de vous retrouver en 1972 ou 1973.

Je citerai aussi Monsieur *Tyler Port* qui a succombé à mes charmes en acceptant imprudemment la charge ingrate de tirer les conclusions de la Conférence.

Monsieur *Lestel*, président du "Panel de la Recherche Opérationnelle de la Division des Affaires Scientifiques de l'OTAN" qui m'a beaucoup aidé dans la sélection des présentations.

Monsieur *Knight*, qui avec sa modestie bien connue, a conseillé, a suggéré les dispositions d'organisation.

Monsieur le Capitaine *Stoltz*, Directeur administratif qui m'a ouvert les portes du pays-hôte.

Monsieur *Manning* de NAMSA qui me seconde dans le jeu radiophonique des "questions et réponses" et, hélas pour lui, dans la publication des Proceedings.

Les présidents de session, les orateurs, les interprêtes, les points de contact et aussi les participants, car bien entendu, il n'y aurait pas de conférence sans participants.

Je remercie aussi le personnel luxembourgeois du Ministère d'Etat en charge de ces installations et l'opératuer du Service de l'Education Nationale.

Je remercie enfin mon personnel NAMSA et en particulier Mlle. Rüssler et Monsieur Deroover pour leur activité inlassable.

Un mot aussi pour ceux que j'ai oubliés, car pour moi une chose est sûre, j'ai certainement oublié quelqu'un.

Je n'aurai malheureusement pas la possibilité de vous saluer chacun individuellement. Je vous souhaite donc et terminant un agréable retour dans vos pays respectifs et je forme le voeu de vous rencontrer à nouveau.

J'espère que cette Conférence, prolongement de celle d'Athènes, aura été pour vous, comme elle l'a été pour moi, un enrichissement, un réconfort pour nos propres difficultés et une aide pour les résoudre.

Mesdames et Messieurs, la Conférence est close.

List of Participants

NATO

ADLINGTON, T., Lt. Colonel (UK), Defense Support Division, NATO, 1110-Brussels.
ALDINGER, Gerhard, Colonel (G), NAMSA–NSC, Commander, NATO Supply Center, Capellen, Luxembourg.
AMICK, Roy W., Colonel (US), Chief, Plan and Policy Branch Logistics Division, SHAPE– Casteau, B 7010, Belgium.
ASH, John (UK), NAMSA–NSC, Chief Electronics Programme Division, NATO Supply Center, Capellen, Luxembourg.

BALLOW, John C., Major (US), CANDE Division (CSB-M) SHAPE–Casteau, B 7010, Belgium.
BROECKER, Klaus (G), NAMSA, Marketing-Branch, Procurement Division, 17, rue Auguste Lumière, Luxembourg.
BROWNING, Philip Y., Colonel (US) Chairman of the Board of Directors, NAMSO, 17 rue Auguste Lumière, Luxembourg.
BUREAU, Robert, Général-Major (B), Directeur Général de la NAMSA, 17, rue Auguste Lumière, Luxembourg.

CALCAGNO, Luigi, Général (I) Directeur Logistique Adjoint, Bureau de Gestion OTAN-HAWK, 20, rue Galliéni, 92- Rueil Malmaison, France.
CALCAGNO, Riccardo P. (I), NAMSA, Chef de la Branche des Effectifs, Division de l'Administration et de l'Organisation, 17, rue Auguste Lumière, Luxembourg.
CAUCHIE, Marc-Antoine, Colonel (F), NAMSA, Directeur de l'Administration et de l'Organisation, 17, rue Auguste Lumière, Luxembourg.
COLLU, Mario, Major General (I), Defence Support Division, NATO, 1110-Brussels.
COMBES, José, Ingénieur en Chef (F), Chef de la Maintenance, Bureau de Gestion OTAN-HAWK, 20, rue Galliéni, 92- Rueil-Malmaison, France

DE BELDER, Jean-Marie, (B), NAMSA–NSC, Depot Operations Division, NATO Supply Center, Capellen, Luxembourg.
DRAGOTTI, Giovanni, Colonel (I), NAMSA–NSC, Brokerage Programme Manager, NATO Supply Center, Capellen, Luxembourg.

FALCH, Johannes H. (N), Telecommunication Engineer, CANDE Division (CSB-MS) SHAPE– Casteau, B 7010, Belgium.

GOUILLON, Maurice, Colonel (F), NAMSA–NSC, Electronics Programme Division, Chief Maintenance Section, NATO Supply Center, Capellen, Luxembourg.
GUNEY, Ihsan, (T), NAMSA, Logistics Directorate, 17, rue Auguste Lumière, Luxembourg.

HINTZ, Torben, (D), NAMSA–NSC, NADGE Project Office, NATO Supply Center, Capellen, Luxembourg.

HUBEAUX, Robert, (B), NAMSA, Chef de la Branche Organisation, Division de l'Administration et de l'Organisation, 17, rue Auguste Lumière, Luxembourg.
HUXLEY, John, Group Captain (UK), NAMSA, Chief Maintenance Branch, Logistics Directorate, 17, rue Auguste Lumière, Luxembourg.

JOORIS, Daniel, (B), NAMSA–NSC, Chief of Personnel Division, NATO Supply Center, Capellen, Luxembourg.

KLOSE, Hans W. (G), Directeur Logistique et Directeur Général Adjoint HAWK, Bureau de Gestion OTAN–HAWK, 2o, rue Galliéni, 92-Rueil-Malmaison, France.
KNIGHT, M.A.G. (UK), Scientific Affairs Division, NATO, 1110-Brussels, Belgium.

LOHMANN, K.J. (G), Assistant Scientific Adviser, HQ AFCENT, Brunssum, Netherlands.

MANNING, Roger (UK), NAMSA, Chief NADGE Project Office, Logistics Directorate, 17, rue Auguste Lumière, Luxembourg.

NOUGAROU, André (F), NAMSA, Chef de la Division des Achats et des Ventes, 17, rue Auguste Lumière, Luxembourg.

PLUMIER, Jules (B), NAMSA, Logistics Directorate, 17, rue Auguste Lumière, Luxembourg.
TYLER-PORT, A., (US) Assistant Secretary General for Defence Support, NATO, 1110–NATO–Brussels, Belgium.

RAHBEK JENSEN, Jorgen, Chief Operations Research, 405 Rheindahlen, Mönchen-Gladbach, HQ 2nd Allied Tactical Air Force, Germany.

SEGERSON, Harold (US), Logistics Support, NADGE Management Office, 3, Avenue Palmerston, 1040-Brussels, Belgium.
SHEPHERD, John (UK), NAMSA, Logistics Directorate, 17, rue Auguste Lumière, Luxembourg.
SINGER, Melvin (US), NAMSA, Logistics Directorate, 17, rue Auguste Lumière, Luxembourg.
SKOGLI, Leif, (N), NAMSA–NSC, Electronics Programme Division, Chief Supply Section, NATO Supply Center, Capellen, Luxembourg.
STEINBACH, Guillaume, Colonel (NL), NAMSA–NSC, Chief Procurement Division, NATO Supply Center, Capellen, Luxembourg.
STONE, John (UK), Defence Support Division, NATO, 1110–NATO–Brussels, Belgium.
SYMONS, P.F.J., Major-General (B), Assistant Director, Logistics Division, IMS/LOG. DIV., NATO, 1110–NATO–Brussels, Belgium.

VISINE, François (F), NAMSA, Counseiller Juridique, 17, rue Auguste Lumière, Luxembourg.

WEHNER, Curt, Colonel (G), Staff Officer Plans, IMS/LOG.DIV., NATO, 1110–NATO–Brussels, Belgium.

WHITWORTH, Daniel C. (US), NAMSA, Director, Logistics Directorate, 17, rue Auguste Lumière, Luxembourg.
WIIG, Johan (N), NAMSA–NSC, Sidewinder/Bullpup Programme Manager, NATO Supply Center, Capellen, Luxembourg.
WOLFF, Helmut, Colonel (G), NAMSA, Chief Supply Branch, Logistics Directorate, 17, rue Auguste Lumière, Luxembourg.

YUKSEL, Sureyya, Brig-General (T) Turkish Army, DACOS-Logistics, SHAPE–Casteau, B 7010, Belgium.

INTERPRETERS

LAURIN, (Mrs) Geneviève, NATO International Secretariat, Interpretation Section, 1110–NATO–Brussels, Belgium.
MISSON, Alain, 113, rue Dodonée, Bruxelles 18, Belgique.
SPROUL-BOLTON, R.A., Commander C.E.O.A., 11, rue du Général Pershing, 78 - Versailles, France.
YELIN, Alec, 60, rue du Général de Gaulle, B-1310–La Hulpe, Belgique.

BELGIUM

BEECKAMN, Honoré, Capitaine de Corvette, Etat-Major de la Force Navale, Bureau de Gestion, Caserne Prince Baudouin, Place Dailly, 1030–Bruxelles.
BIEUVELET, Jean, Licencié Sciences Commerciales et Consulaires, IBM, Tour du Midi, 29ème étage, 1060-Bruxelles.
BLANCHART, Jean-Claude, Lieutenant d'Aviation Ir., Ecole Royale Militaire 30, avenue de la Renaissance, Bruxelles 4.
BRIOT, Maurice, Lieutenant Colonel Direction de l'Artillerie, Etat-Major de la Force Terrestre, Place Dailly, 1030 Bruxelles.
BRUNARD, Yves, Major d'Aviation IFM, Centre de Recherche pour la Défense, Direction de R.O., 1, Avenue du 2 Lanciers, 1040 Bruxelles.
BUYS, Prosper, Capitaine de Corvette (T), Etat-Major Force Navale Belge, Service Technique, Quartier Prince Baudouin, Place Dailly, 1030 Bruxelles.
DELHAYE, Jean, Général-Major d'Aviation, Etat-Major de la Force Aérienne, Caserne Prince Baudouin, Place Dailly–1030-Bruxelles.
DE VIRON, Christian, Lt. Colonel Etat-Major Force Terrestre, Caserne Prince Baudouin, 1030-Bruxelles.
DOBBELAERE, André, Colonel d'Aviation, Q.G.–Base Force Aérienne, Brusselse Poortstraat 17, Gent.
DOHET, Louis, Chercheur Opérationnel, Centre de Recherches pour la Défense, ERO, Avenue du 2ème Rgt. des Lanciers No. 1, 1040-Bruxelles.
VAN CALSTER, Marcel, Major BEM, Ecole de Geurre, Avenue de Cortenbergh 115, 1040 Bruxelles.
GALWIC, Robert, Colonel d'Aviation, Ministère de la Défense Nationale, Etat-Major de la Force Aérienne, Caserne Prince Baudouin, Place Dailly, 1030 Bruxelles.
GELUYCKENS, Daniel, Capitaine de Vaisseau, Groupement Logistique de la Force Navale Oosteroever–Voorhaven, 8400 Oostende, Belgique.

GRANDSARD, André, Major Ir., Etat-Major Général, Section Logistique (JS4) Place Dailly, 1030 Bruxelles.
HENNICO, Leopold L., Chef Dept. R.D. au CRD, Centre de Recherche pour la Défense, 1 Avenue du 2e Regt. de Lanciers, 1040 Bruxelles.
JONCKHEERE, Raymond, Lieutenant-Colonel BEM, Ecole de Guerre, Avenue de Cortenbergh, 1040-Bruxelles.
LEBACQ, Jacques, Major d'Aviation BEM, Ecole de Guerre, Avenue de Cortenbergh, 115, 1040-Bruxelles.
LEMAIRE, Henri, Capitaine de Vaisseau BEM, Etat-Major Force Navale, Section Logistique, Caserne Prince Baudoin, Place Dailly, 1030 Bruxelles.
LEONARD, Alexandre, Major EM/Force Terrestre (Direction de l'Ordonnance), Caserne Prince Baudouin, Place Dailly, 1030-Bruxelles.
LOBET, Guy, Commandant IFM, Q.G. Base FT, Caserne Lempereur, 1980 Tervuren.
MAGDELYNS, Clément, Colonel BEM, Etat-Major de la Force Terrestre, Place Dailly, 1030-Bruxelles.
DENIS, A., Lt. Colonel Etat-Major de la Force Aérienne, Place Dailly, 1030-Bruxelles.
PONCIN, Jean, Colonel Ir., Bureau d'Etude et d'Organisation (JSBEO) Forces Armées Belges, Place Dailly, 1030-Bruxelles.
SALLE, E. H. G., Major Base Force Terrestre, Caserne Lt. Colonel Lempereur, 1980-Tervuren.
SCHINGTIENNE, Norbert, Lt. Colonel Etat-Major de la Force Terrestre, GS/GDL, Caserne Prince Baudouin, Place Dailly, 1030-Bruxelles.
TEMMERMAN, Jacques, Chef de la Section Logistique, Etat-Major Force Terrestre, Place Dailly, 1030-Bruxelles.
TOUSSAINT, Albert, General-Major Président de la Commision d'étude et de réforme de la Logistique (CERL), Place Dailly, Forces Armées Belges, 1030-Bruxelles.

CANADA

BAGLOW, Robert L., Director of Logistics Analysis, Defence Research Analysis Est. National Defence Headquarters, Ottawa 4, Ontario.
COLLINS, David A., Commodore, Director General Supply, Canadian Armed Forces, Ottawa 4, Ontario.
CONRATH, David W., Associate Professor of Management Sciences, Department of Management Sciences, University of Waterloo, Waterloo, Ontario.
STEPHENS, Commodore, Robert St. G., Canadian Forces Headquarters, Department National Defence, Ottawa 4, Ontario.
WATTS, Rex Virgo, Engineer, Department of National Defence, Cartier Square, Ottawa, Ontario.
WILLIS, Robert C. Captain (N) DEVIL Program, Canadian Forces Headquarters, 880 Wellington Street, Ottawa 2, Ontario.

DENMARK

BIRKELUND, Hakon, Lt. Colonel, Air Material Command RDAF, 3510–FLY-Vaerloese.
KLEIN, Preben, Capt. RDN Material Command, 1433 Holmen.
EICHEN, Preben H., Major Danish Army Staff, Kastellet, 2100-København.
JAENSSOn, Skjold A., Divisional Engineer, Headquarters RDAF, Henriksholmsallé, 1950-Vedbaek.

JENSEN, Ole Bjørn Mathiasen, Major Danish Defence Command, Kastellet, 2100 -København.
LARSEN, Ernst Thor, Lt. Colonel Danish Delegation to NATO, NATO, 1110-Bruxelles.
RODHOLM, Immanuel B., Captain Naval Staff, Genersgade 20, København.

FRANCE

BARBIEUX, Jacques, Ingénieur, SNIAS (Nord Aviation), 12, rue Béranger, 92-Chatillon s/Bagneux.
BAUGUIN, Gérard, Ingénieur Civil des Mines, SEMA, 9, rue Georges Pitard, Paris XVe.
BERAULT, Guy, Chef de Groupe de Servives, THOMSON-CSF, 1, rue des Mathurins, 92-Bagneux.
BLEHAUT, Philippe, Ingénieur, THOMSON-CSF, Division DSE, 1, rue des Mathurins, 92-Bagneux.
BRESSON, Maurice, Ingénieur en Chef, Société A.E.R.O., 3, Avenue de l'Opéra, Paris Ie.
CRUON, Roger, Ingénieur en Chef, Centre Interarmées de Recherche Opérationnelle, 24, Avenue Prieur de la Côte d'Or, 94-Arcueil.
DARRICAUX, Guy, Ingénieur en Chef de l'Armement, Direction Technique des Constructions Aéronautiques, Service de la Production Aéronautique, 4, Avenue de la Porte d'Issy, Paris XVe.
d'ANGELY, Gaston, Ingénieur, Société ALCATEL, 22, Avenue Edouard Herriot, 92-Le Plessis-Robinson.
DOMPNIER, Paul, Colonel Direction Centrale du Matériel de l'Armée de l'Air, 26, Boulevard Victor, Paris XVe.
DOUGE, Denis, Ingénieur, S. A. Engins MATRA, 78 Velizy.
DUPUIS, Jacques E., Ingénieur, THOMSON C.S.F.-D.S.E. 1, rue des Mathurins, 92-Bagneux.
FERRIER, Jacques, Commissaire Général de la Marine, Directeur du Commissariat de la Marine, 2 Rue Royale, Paris
FLORIT, Modeste, Ingénieur Principal des Etudes et Techniques d'Armement Direction Technique des Constructions Aéronautiques, Service de la Production Aéronautique, 4, Avenue de la Porte d'Issy, Paris XVe.
GOUBLIN, Roger, Directeur, THOMSON-CSF (DSE), 1, rue des Mathurins, 92-Bagneux.
GROSJEAN, Robert, Capitaine, Direction de la Gendarmerie et de la Justice Militaire, 19, Boulevard Latour-Maubourg, Paris VIIe.
BUSCOT, Monsieur, Ingénieur, Laboratoire Central de Télécommunications, 18-20, rue Grange-Dame-Rose, 78-Vélizy-Villacoublay.
KOCH, Andric Marie, Directeur des Projets, Société Européenne de Téléguidage (SETEL) 89 rue de la Boétie, Paris VIIIe.
LABIN, Emile, Président-Directeur-Général, Société d'Etudes et Conseils AERO, 3, Avenue de l'Opéra, Paris Ie.
LAPLAUD, Jean-Pierre, Ingénieur Principal, Département Approvisionnements, Direction du Matériel Air France, B.P. 124 Orley-Aérogare.
MARCOVICI, Claude, Ingénieur, Société Anonyme des Engins MATRA, B. P. 1, 78-Velizy.
MENARD, Jacques, Colonel Etat-Major de l'Armée de Terre, 4e Bureau, 231 Boulevard Saint Germain, Paris VIIe.

OLIVIER, Philippe, Ingénieur en Chef, Département Approvisionnements, Direction du Matériel Air France, B. P. 124, Orly-Aérogare.
PECHMAJOU, René, Ingénieur Général, Directeur du SCA, B. P. 126, Versailles.
PEGUY, Alain G. C., Directeur Scientifique, Société Centramite de Mécanographie, Tour Nobel, 92-Puteaux.
POIRIER, Colonel, Lucien, Centre de Prospective et d'Evaluations, Ministère d'Etat chargé de la Défense Nationale, 10, rue St. Dominique, Paris VIIe.
QUENTIN, Pierre, Chef de Bataillon, Centre de Prospective et d'Evaluations, Ministère d'Etat chargé de la Défense Nationale, 10, rue St. Dominique, Paris VIIe.
ROUGERIE, Armand, Chef d'Escadron, Centre Interarmées de Recherche Opérationnelle, 24, Avenue Prieur de la Côte d'Or, 94-Arcueil.
SAHNOUN, Roger, Ingénieur, Sud-Aviation, B.P. No. 2–33-Mérignac.
SALANIE, Jean, Chef d'escadron, Section de Recherche Opérationnelle de l'Armée de Terre, Ecole Militaire, 1 Place Joffre, Paris VIIe.
TABET, Daniel, Ingénieur, Laboratoire Central de Télécommunications, 18-20, rue Grange-Dame-Rose, 78-Vélizy.

GERMAN FEDERAL REPUBLIC

ARMONAT, Dr. Richard, Dipl.-Ing., Industrieanlagen-Betriebsgesellschaft mbH 8012 Ottobrunn, Robert-Koch-Strasse.
BAX, Jürgen, Dornier AG-München, 8031 Gilching, Starnberger Weg 62.
BREITWIESER, Gert, Ing., Materialinformationszentrum der Marine GmbH, 2940 Wilhelmshaven, Goekerstrasse 50.
BUELTE, Friedrich, Diplom-Ingénieur, Technisches Buero Flugzeuge in der Industrieanlagen-Betriebsgesellschaft mbH, 8012 Ottobrunn, Zaunkoenigstrasse 18.
BUSCH, Hugo W., Gruppenleiter, IABG, Operations Research Gruppe, 8 München 33, Schiessfach 505.
COMMERÇON, Elke, Dipl-Kfm., Operations Research Gruppe der I.A.B.G., 8 München 33, Schliessfach 505.
FENT, Klaus A.F., Dipl. rer. oec., Industrieanlagen-Betriebsgesellschaft mbH, 8012 Ottobrunn; Mailing address: Heilwigstrasse 48, 8 München 82.
FISCHER, Dietrich, Dipl.-Ing., Operations Research Gruppe der I.A.B.G., 8012 Ottobrunn, Robert-Koch-Strasse.
GRIEBELER, Erich, Dipl.-Volkswirt, Zentrale OR-Stelle, 55-Trier, Trevererstrasse 1.
GUSTKE, Hans Harald, Oberst, Bundesministerium der Verteidigung, Fü L IV 4, Bonn.
HENLE, Karl F., Diplom-Mathematiker, Bundesministerium der Verteidigung, Referat Fü S-VIII-5, 53-Bonn, Postfach 161.
SCHROTER, E., Ing. Dornier AG, Werke München, 8031 Gilching, Starnberger Weg 62.
HORMANN, Fritz, Dipl. Ing., Materialinformationszentrum der Marine GmbH, 2940 Wilhelmshaven, Goekerstrasse 50.
HORNUNG, Walter, Oberstleutnant a.D., Operations Research Gruppe der I.A.B.G. 8 München 33, Schliessfach 505.
JAHNIG, Dieter, Technical Adviser, Messerschmitt-Boelkow-Blohm GmbH, Unternehmensbereich Flugzeuge, 8 Muenchen 86, Arabellastrasse 1.
KALLHARDT, Emil, Dr., Department Manager, Messerschmitt-Boelkow-Blohm GmbH, Unternehmensbereich Flugzeuge, 8 Muenchen 86, Arabellastrasse 1.

KLEINSTEUBER, Werner, Dr., rer. nat. habil., AEG-Telefunken, Bereich Forschung und Entwicklung, 79 Ulm-Donau, Postfach 830.

KRUMM, X. A., Lt. Colonel Bundesministerium der Verteidigung, Abt. Streitkräfte, Referat SV 3, Bad-Godesberg.

LAUER, Alois, Dipl.-Kaufmann, Zentrale OR-Stelle, 55-Trier, Trevererstrasse 1.

METZENROTH, Michael, Dipl.-Physiker, Operations Research Gruppe der I.A.B.G. 8 München 33, Schliessfach 505.

MUENZENMAIER, Kurt, Manager, MTU Motoren-und Turbinen-Union GmbH, 8 München 50, Postfach 500640.

NIEMANN, Werner, Engineer, Standard-Elektrik Lorenz AG, 7000 Stuttgart-Zuffenhausen, Hellmuth-Hirtz Str. 42.

PEINEMANN, H., Messerschmitt-Boelkow-Blohm GmbH, Unternehmensbereich Flugzeuge, 8 Muenchen 23, Rheinstrasse 28.

PEKING, Karl, Techn. Proj. Manag, Engineer, Standard Elektrok Lorenz AG, 7000 Stuttgart, Hellmuth Hirtztrasse 42.

REICHERT, Emil, Supply Manager, FEG Flug-Elektronik-GmbH, 8 Muenchen 81, Arabellastrasse 4.

SCHMIDT, Peter, Dipl. Wirtsch. Ing., Dornier Systems GmbH, Planungsberatung, 799 Friedrichshafen, Postfach 648.

SCHON, Johannes, Colonel Industrieanlagen-Betriebsgesellschaft, Operations Research Gruppe, 8012, Ottobrunn; Mailing Address: 811 Penzing, Nr. 89.

SEIDLER, Ferdinand, OTL a.D., Ing., St. Elektrik Lorenz, 53-Bonn, Am Hofgarten 10.

SPIESS, Wolfgang, Dipl. Ing. Instiut für Raumfahrttechnik der TU Berlin, 1 Berlin 10, Franklinstrasse 29.

STACH, Guenter, Program Manager, Messerschmitt-Boelkow-Blohm GmbH, Unternehmensbereich Flugzeuge, 8 Muenchen 86, Arabellastrasse 1.

UTSCH, Dieter, Ing., Dornier Systems GmbH, Planungsberatung, 799-Friederichshafen.

VERBAERE, Jacques, Ingénieur Techn., Standard Elektrok Lorenz AG (WSISATT) 7000-Stuttgart-Zuffenhausen, Hellmuth-Hirth-Strasse 42.

ITALY

FRANCAVILLA, Andrea, Docteur Ingénieur, Società Finanziaria Meccanica FINMECCANICA, Viale Maresciallo Pilsudski 91-00197 ROMA;

GRILLINI, Bruno, Cdr. SELENIA-Industrie Elettroniche Associate S.p.A. Via Tiburtina Km 12,40, Roma;

MALERBA, Sabino, Capitaine Stato Maggiore Esercito, Ufficio Sta. Me. R.O., Roma;

MARINI, Giulio, Major-General Stato Maggiore Difesa, URI-BOD-NAMSO, Via Luigi Rizzo 28, Roma;

MARCIANO, Donato, Colonel Ispettorato Log. dell' Escercito, Roma;

MASSOBRIO, Mario, Colonel Consiglio Tecnico Scientifico della Difesa, Via Luigi Rizzo 28, Roma;

MATACOTTA, Mario, Major-General Ministero delle Difesa, Ufficio Centrale Allestimenti Militari, Via Napoli 2, Roma;

OLIVI, Giovanni, Lt. Colonel Ispettorato Logistico Marina Militare, Roma;

PARENTI, Alfredo, Capitaine SME-Ufficio Statistica Meccanografia e Ricerca Operativa, Via XX Settembre, Roma;

PIU, (Madame) Giovanna, Systems Analyst, VITROSELENIA, Via Tiburtina Km 12,4, Roma;

RUMALDINI, Giuseppe, Prof. A. Aurora, 21, Viale Indipendanza, Perugio,
RUSSO, Giuseppe, Colonel Stato Maggiore Difesa, URI-BOD-NAMSA, Via Luigi Rizzo 28, Roma;
SARCINA, Ruggiero, Capitaine Ispettorato Logistico dell' Esercito, Roma;

NETHERLANDS

ANNEVELINK, Jan, Major, Ministry of Defence (Quartermaster General) v.d. Burchlaan 32, The Hague;
ARENTSEN, E. J. E., Lt. Colonel Dienst van de Kwartiermeester-Generaal, Afdeling Techniek, van der Burchlaan 31, The Hague;
BRACKEL, L. F. H. M., Captain Royal Netherlands Airforce, Directorate of Material Air, le van den Boschstraat 8, The Hague,
COX, A. L., General Ministerie van Defensie, Materieelraad, Wassenarseweg 6, The Hague;
DIETERS, Colonel H., Royal Netherlands Air Force, Directorate of Material Air, le van den Boschstraat 8, The Hague;
GERAERDS, W. M. J., Lt. Colonel Royal Netherlands Air Force, Directorate of Material Air, le van den Boschstraat 8, The Hague,
TEN BRINKS, M. E., Major Office of the Controller for the Army (DAD), van de Burchlaan 31, The Hague;
HESLINGA, Brigadier General Reimer J. W., DEP Quartermaster General, van der Burchlaan 31, The Hague.
MONHEMIUS, Wim, Professor of Operations Research, Technological University Insulindelaan 2, Eindhoven;
HEUTINK, Willem, Major Dienst Kwartiermeester-Generaal (DKMG), van der Burchlaan 31, The Hague;
OORTHUIJS, J. H., Commodore Royal Netherlands Air Force, Directorate of Material Air, le van den Boschstraat 8, The Hague;
POLMAN, L. R. R., Major Royal Netherlands Air Force, Directorate of Material Air, le van den Boschstraat 8, The Hague;
VERBERNE, Harry J. H., Major Office of the Controller for the Army, van der Burchlaan 31, The Hague;
VOLTEN, Hendrik, Major Legerraadstaf, van der Burchlaan 31, The Hague;

NORWAY

ANDRESEN, Alf, Conseiller Adjoint, Délégation de la Norvège auprès de l'OTAN, 1110-OTAN-Bruxelles;
HØGBERG, Kjell T., Major Royal Norwegian Air Force, Air Materiel Command, B. P. 10; 2007 Kjeller;
SAETHER, Wayne, Major A/S Kongsberg Vapenfabrikk, 3600 Kongsberg;
SOLSTRAND, Ragnvald, Research Engineer, Norwegian Defence Research Establishment, P. O. Box 25, 2007 Kjeller;
THU, J. P., Lt. Colonel Norwegian Army Materiel Command, Oslo MIL, Oslo 1;

TURKEY

BILGE, Ahmet, Colonel SHAPE Turkish Military Representative, SHAPE, Casteau, Belgique.
ONAL, Kemal, Colonel Turkish Ground Forces Command, Logistics Department, Supply and Maintenance Section (K.K.K.);

UNITED KINGDOM

AYLIFFE, Arthur W., Flight Lieutenant CSDE, RAF Swanton Morley, Dereham, Norfolk;
BAKER, Brian E., Head of Logistics Group, Deputy Chief Scientist, (RAF) Staff; Ministry of Defence, Main Building, Whitehall, London S.W.1;
BAKER, John, Colonel Ministry of Defence (Army) (Q (AE) 4), Whitehall, London S.W.1;
BENYON-TINKER, Graham, Lieutenant Royal Navy, Director Weapons Equipment Surface, Admiralty Surface Weapons Establishment, Portsdown, Cosham, Hampshire,
BRITTEN, Robert W. T., Brigadier Q(AE), Ministry of Defence (Army), Old Admiralty Building, Whitehall, London S.W.1;
BUCKLAND, William R., Dr. The Economist Intelligence Unit, Spencer House, 27 St. James's Place, London S.W.1;
CURTIS, John, Scientific Officer, Defence Operational Analysis Establishment, Parvis Road, West Byfleet, Surrey;
EGGLETON, David John, Ministry of Defence, DSC (RAF)Staff, Main Building, Whitehall, London S.W.1;
FOGDEN, William P., Wing Commander Ministry of Defence (E50(RAF)), Old War Office Building, Whitehall, London S.W.1;
GRAHAM, William, P.S.O. Reliability, HG D.E.M.E. (A.), First Avenue House, High Holborn, London W.C.1;
GREEN, David J., Flight Lieutenant, CSDE, RAF, Swanton Morley, Dereham, Norfolk;
HANNAFORD, Robert, Wing Commander Supply Control Centre, RAF Hendon, The Hyde, London N.W.9;
HOUGH, John, Operational Research Group, Rolls-Royce Ltd., The Spot, Derby.
IRVING,John N.B., Wing Commander Royal Air Force, Ministry of Defence (Air) Old Admiralty Building, London S.W.1;
KAY, Emile, Senior Research Fellow, Materials Handling Research Unit, Cranfield Institute of Technology, Cranfield, Bedford;
LAMB, Stanley A., Wing Commander Supply Control Centre, RAF Hendon, The Hyde, London N.W.9;
LAMBERT, John D., Squadron Leader CSDE, RAF Swanton Morley, Dereham, Norfolk;
LANDAU, Clive S., Central Studies Establishment, P.O. Box 63, Lyneham 2602 ACT, Australia;
MORRIS, Leonard J., Head of Division, Director General of Supplies and Transport (Naval), Ministry of Defence, RN Store Depot, Copenacre, Hawthorn, Wiltshire;
MORRIS, Roy, Group Captain Ministry of Defence, Main Building, Whitehall, London S.W.1;

NORTON, Frank, Directorate of Naval Operational Studies, Ministry of Defence, Main Building, Whitehall, London S.W.1.;
O'CONNOR, Patrick D. T., Squadron Leader Ministry of Defence (Air), Whitehall, London S.W.1;
PAGE, Malcolm B., Lieutenant Colonel Q(AE)1, Ministry of Defence (Army) Old Admiralty Building, Whitehall, London S.W.1;
RAWSON, Kenneth, Directorate of Naval Administrative Planning, Ministry of Defence, Whitehall, London S.W.1;
RICKERS, Maurice, Principal Scientific Officer, Defence Operational Analysis Est., Parvis Road, West Byfleet, Surrey;
SILVER, William E., Principal Scientific Officer, Directorate of Naval Operational Studies, Ministry of Defence, Main Building, Whitehall, London S.W.1;
THWAITES, Keith W., Head of Management and Logistics Section, Directorate of Naval Operational Studies, Ministry of Defence, Main Building, Whitehall, London S.W.1;
TYLER, Philip G., Group Captain Ministry of Defence (DDE 14(RAF)), Old War Office Building, Whitehall, London S.W.1.
WILLIAMS, Douglas W., Group Captain (Rtd), Hawker Siddeley Aviation Limited, Kingston upon Thames, Surrey.
YOUNG, Hugh W., Commander Royal Navy, Director Weapons Equipment Surface (Naval), Admiralty Surface Weapons Establishment, Portsdown, Cosham, Hampshire;

UNITED STATES

ALBERTS, Henri C., Senior Scientist, The Vertex Corporation, 10400 Connecticut Avenue, Kensington, Maryland;
BARNARD, Robert A., Director of Product Support, Lockheed-California Company, P.O. Box 551 (Dept.61-01), Burbank California 91503;
BLUMBERG, Donald F., President, Decision Sciences Corporation, Benjamin Fox Pavilion, Jenkintown, Pennsylvania 19046;
BUCK, Pierpoint B., Doctor (Ph.D.) Institute for Defense Analyses, 400 Army-Navy Drive, Arlington, Virginia 22202;
DURBIN, Eugene P., Director of Logistics Studies, Rand Corporation, 1700 S Main Street, Santa Monica, California 90402;
HIRSCH, Warren M., Professor of Mathematics, New-York University, Courant Institute of Mathematical Sciences, 251 Mercer Street, New-York, N.Y. 10012;
McGLAMERY, John W. Jr., Manager-Research; Materiel Services, Eastern Airlines Inc., Miami International Airport, Miami, Florida 33148;
MILLER, Bruce, Consultant, The RAND Corporation, c/o C.O.R.E., 54, de Croylaan, Heverlee, Belgium;
ROSENMAN, Bernard B., Chief, AMC Inventory Research Office, Army Logistics Management Center, Frankford Arsenal, Philadelphia, Pennsylvania 19137;
SEGER, Mr., Lockheed-California Company, P.O. Box 551, Burbank, California 91503;
SIMON, Richard, National Institute of Health, 9000 Rockville Pike, Bethesda, Maryland, 20014.
SUTTON, S. Scott, Ph.D., Center for Naval Analyses, 1401 Wilson Blvd., Arlington, Virginia 22209;

TERRY, Frederick H., Technical Director, US Army Logistics Doctrine, Systems and Readiness Agency, New Cumberland Army Depot, P.O. Box 2947, Harrisburg, Pennsylvania 17105;

TRUELOVE, Alan, Dr., Mathematica, Inco.,

WILLIAMS, Henry J., Manager, European Systems Department, Planning Research Corporation, 7 Stuttgart 80, Patch Barracks, West Germany.

WINNER, John K., Senior Associate, Planning Research Corporation, 7 Stuttgart 80, Patch Barracks, West Germany.